White Paper 3
System Level ESD
Part I: Common Misconceptions and Recommended Basic Approaches

Industry Council on ESD Target Levels

December 2010

Revision 1.0

Abstract

Over the last twenty years, an increasing misconception between system level designers (OEMs) and semiconductor component (IC) providers has become very apparent relating to three specific ESD issues:

- ESD test specification requirements of system vs. IC providers;
- Understanding of ESD failures in terms of physical failure and system upset and what causes these failures in terms of system level and IC level constraints;
- Lack of acknowledged responsibility between system designers and IC providers regarding proper system level ESD design.

In White Paper 1 from the Industry Council on ESD Target Levels, which presented a paradigm shift in the realistic and safe IC level ESD requirements, we introduced the importance of separately addressing the system specific and IC specific ESD issues. In White Paper 3 we present the first comprehensive analysis of system ESD understanding including ESD related system failures, and design for system robustness. The main purpose of the present document is to close the existing communication gap between the OEMs and IC providers by involving the expertise from OEMs and system design experts. This will be accomplished by what we describe in this document as "System-Efficient ESD Design" (SEED) which promotes a common IC / OEM understanding of the correct system level ESD needs. White paper 3 will be constructed of two parts. A key finding of Part I of the white paper is the development of a framework for sharing IC / system level circuit information so that best practice ESD protection and controls can be co-developed and properly shared.

Later, in Part II of White Paper 3, the Industry Council will use the information in Part I to establish recommendations for IC and system level manufacturers regarding proper protection, proper controls and best practice ESD tests, which can properly assess ESD and related EMI performance of system level tests. The purpose of White Paper 3, Part II will be to better define the ESD relationship between IC manufacturers and system level OEMs and their respective responsibilities.

About the Industry Council on ESD Target Levels

The Council was initiated in 2006 after several major U.S., European, and Asian semiconductor companies joined to determine and recommend ESD target levels. The Council now consists of representatives from active full member companies and numerous associate members from various support companies. The total membership represents IC suppliers, contract manufacturers (CMs), electronic system manufacturers, OEMs, ESD tester manufacturers, ESD consultants and ESD IP companies. *In terms of product shipped, the member IC manufacturing companies represent 8 of the top 10 companies, and 12 of the top 20 companies, and over 70% of the total volume of product shipped by the top 20 companies, as reported in the EE Times issue of August 6, 2007.*

Core Contributing Members	Advisory Board*
Robert Ashton, ON Semiconductor	Marcus Dombrowski, Volkswagen
Jon Barth, Barth Electronics	Johannes Edenhofer, Continental
Patrice Besse, Freescale	Vsevolod Ivanov, Auscom
Jonathan Brodsky, Texas Instruments	Frederic Lefon, Valeo
Brett Carn, Intel Corporation	Christian Lippert, Audi
Tim Cheung, RIM (* - Advisory board also)	Wolfgang Pfaff, Bosch
Stephanie Dournelle, ST Microelectronics	Wolfgang Wilkening, Bosch
Charvaka Duvvury, Texas Instruments (Chairman) c-duvvury@ti.com	Rick Wong, Cisco
David Eppes, AMD	
Yasuhiro Fukuda, OKI Engineering	**Associate Members (July 2010)**
Reinhold Gärtner, Infineon Technologies	Michael Chaine, Micron Technology
Robert Gauthier, IBM	Freon Chen, VIA Technologies
Ron Gibson, Stephen Halperin & Associates	Ted Dangelmayer, Dangelmayer Associates
Horst Gieser, FhG IZM	Kelvin Hsueh
Harald Gossner, Infineon Technologies (Chairman) Harald.Gossner@Infineon.com	Natarajan Mahadeva Iyer, Global Foundries
	Larry Johnson, LSI Corporation
LeoG Henry, ESD/TLP Consulting	Marty Johnson, National Semiconductor
Masamitsu Honda, Impulse Physics Laboratory	ChangSu Kim, Samsung
Michael Hopkins, EM Test USA	Yongsup Kim, Samsung
Hiroyasu Ishizuka, Renesas Technology	Marcus Koh, Everfeed
Satoshi Isofuku, Tokyo Electronics Trading	Soon-Jae Kwon, Samsung
David Klein, IDT	Brian Langley , ORACLE
John Kinnear, IBM	Roger Peirce, Simco
Timothy Maloney, Intel Corporation	Donna Robinson-Hahn, Fairchild Semiconductor
Tom Meuse, Thermo Fisher Scientific	Jeff Salisbury, Flextronics
James Miller, Freescale Semiconductor	Jay Skolnik, Skolnik-Tech
Guido Notermans, ST-Ericsson	Jeremy Smallwood, Electrostatic Solutions Inc.
Nate Peachey; RFMD	Jeremy Smith, Silicon Labs
Wolfgang Reinprecht, AustriaMicroSystems	Arnold Steinman, Electronics Workshop
Alan Righter, Analog Devices	David E. Swenson, Affinity Static Control Consulting
Theo Smedes, NXP Semiconductors	Nobuyuki Wakai, Toshiba
Teruo Suzuki, Fujitsu	Michael Wu, TSMC
Pasi Tamminen, Nokia (* - Advisory board also)	
Benjamin Van Camp, SOFICS	
Terry Welsher, Dangelmayer Associates	

Acknowledgments:

The Industry Council would like to thank all the authors, reviewers, and specialists who shared a great deal of their expertise, time, and dedication to complete this document. We would especially like to thank our OEM advisors from Audi, Auscom, Bosch, Continental, Cisco, Nokia, RIM, Valeo and Volkswagen who gave us valuable direction, not only addressing important topics for Part I, but also pointing us towards further issues that need to be addressed in more detail in Part II.

Editors:
Robert Ashton, ON Semiconductor
Brett Carn, Intel Corporation

Authors:
Robert Ashton, ON Semiconductor
Jon Barth, Barth Electronics
Patrice Besse, Freescale Semiconductor
Jonathan Brodsky, Texas Instruments
Stephanie Dournelle, ST Microelectronics
Charvaka Duvvury, Texas Instruments
Reinhold Gaertner, Infineon Technologies
Robert Gauthier, IBM
Harald Gossner, Infineon Technologies
Leo G. Henry, ESD/TLP Consulting
Mike Hopkins, EM Test USA
John Kinnear, IBM
David Klein, IDT
Tom Meuse, Thermo Fischer Scientific
James W. Miller, Freescale Semiconductor
Tim Maloney, Intel Corporation
Guido Notermans, ST-Ericsson
Nate Peachey, RFMD
Wolfgang Reinprecht, Austria Microsystems
Alan Righter, Analog Devices
Theo Smedes, NXP Semiconductors
Pasi Tamminen, Nokia
Benjamin Van Camp, Sofics
Terry Welsher, Dangelmayer Associates

Mission Statement

The Industry Council on ESD Target Levels was founded on its original mission to review the ESD robustness requirements of modern IC products to allow safe handling and mounting in an ESD protected area. While accommodating both the capability of the manufacturing sites and the constraints posed by downscaled process technologies on practical protection designs, the Council provides a consolidated recommendation for future ESD target levels. The Council Members and Associates promote these recommended targets for adoption as company goals. Being an independent institution, the Council presents the results and supportive data to all interested standardization bodies.

In response to the growing prevalence of system level ESD issues, the Council has now expanded its mission to directly address one of the most critical underlying problems: insufficient communication and coordination between system designers (OEMs) and their IC providers. A key goal is to demonstrate and widely communicate that future success in building ESD robust systems will depend on adopting a consolidated approach to system level ESD design. To ensure a broad range of perspectives the Council has expanded its roster of Members and Associates to include OEMs as well as experts in system level ESD design and test.

Preface

While IC level ESD design and the necessary protection levels are well understood, system ESD protection strategy and design efficiency have only been dealt with in an ad hoc manner. This is most obvious when we realize that a consolidated approach to system level ESD design between system manufacturers and chip suppliers has been rare. This White Paper discusses these issues in the open for the first time, and offers new and relevant insight for the development of efficient system level ESD design. This effort has been divided into two parts. In Part I, this document will identify and eliminate the misconceptions common in the understanding of system level ESD. This will be followed later by Part II where we will explore realistic system ESD protection requirements and strategies. We would also like to note that in Part I we address direct stress effects on external and internal pins of an IC while the more intricate effects of inter-chip pin coupling will be carefully considered in Part II. This document is intended to be useful for both chip suppliers and OEMs/ODMs. As a final note, we would like to clarify that this document addresses system level ESD issues only, but not Electrical Overstress (EOS) unless they manifest from a system failure.

Disclaimers

The Industry Council on ESD Target Levels is not affiliated with any standardization body and is not a working group sponsored by JEDEC, ESDA, JEITA, IEC, or AEC.

This document was compiled by recognized ESD experts from numerous semiconductor supplier companies, contract manufacturers and OEMs. The data represents information collected for the specific analysis presented here; no specific components or systems are identified.

The Industry Council, while providing this information, does not assume any liability or obligations for parties who do not follow proper ESD control measures.

Table of Contents

Glossary of Terms

AEC	Automotive Electronics Council
ANSI	American National Standards Institute
ASIP™	application specific integrated passive ™
bigFET	bipolar insulated gate field effect transistor
BiCMOS	bipolar complementary metal-oxide semiconductor
CAN	controller area network
CBE	charged board event
CCE	charged cable event
CDE	cable discharge event
CDM	charged-device model
CLK	clock
CM	contract manufacturer
CMOS	complementary metal-oxide semiconductor
DC	direct current
DUT	device under test
DSP	digital signal processor
ECU	electronic control unit
EM	electromagnetic
EMC	electromagnetic compatibility
EMI	electromagnetic interference
EOS	electrical overstress
EPA	ESD protected area
ESD	electrostatic discharge
eSATA	external serial advanced technology attachment
ESDA	Electrostatic Discharge Association; ESD Association
EUT	equipment under test
FM	frequency modulation
ggNMOS	grounded gate N-channel metal-oxide semiconductor
GND	negative voltage supply
GRP	ground reference plane
HBM	human body model
HCP	horizontal coupling plane
HDMI	high definition multimedia interface
HMM	human metal model
HSS (HSSL)	high speed serial link
IC	integrated circuit
ID	identification
IDDQ	component quiescent supply current
IO	input/output
IP	intellectual property
IEC	International Electrotechnical Commission
ISO	International Organization of Standards
JEDEC	Joint Electronic Devices Engineering Council
JEITA	Japan Electronics and Information Technology Industries Association
LC	inductor/capacitor network
LIN	local interconnect network

LU	latch-up
MM	machine model
OEM	original equipment manufacturer
ODM	original design manufacturer
PCB	printed circuit board
PHY	physical layer
PICC	proximity IC cards
PUT	pin under test
RC	resistor capacitor network
RF	radio frequency
RLC	resistor inductor capacitor network
RP	residual pulse
RPS	residual pulse stress
RX	receiver
SAW	surface acoustic wave
SCR	silicon controlled rectifier
SMD	surface mount device
SOA	safe operating area
SPICE	simulation program with integrated circuit emphasis
TLP	transmission line pulse
TLU	transient latch-up
TVP	transient voltage pulse
TVS	transient voltage suppression
TX	transmitter
USB	universal serial bus
VBR	breakdown voltage
VCP	vertical coupling plane
VDD	positive voltage supply
VFTLP	very fast transmission line pulse
UTP	unshielded twisted pair

Crosstalk: Any phenomenon by which a signal transmitted on one circuit or channel of a transmission system creates an undesired effect in another circuit or channel. This phenomenon is usually caused by undesired capacitive, inductive, or conductive coupling from one circuit, part of a circuit, or channel, to another.

ESD Design Window: The ESD protection design space for meeting a specific ESD target level while maintaining the required IO performance parameters (such as leakage, capacitance, noise, etc.) at each subsequent advanced technology node.

External Pin (interface pin): An external pin is one which at the board/card level is exposed to potential ESD threats from the outside world.

Hard Failure: Failure of a system due to physical damage to a system component which can only be repaired by the physical repair or replacement of the damaged component.

IEC-Robustness: The capability of a product to withstand the required IEC ESD-specification tests and still be fully functional.

IEC ESD event: An ESD stress as defined in IEC 61000-4-2.

Internal Pin (non-interface pin): An internal pin is one which is exposed to ESD threats typically only during IC manufacturing.

It2: The current point where a transistor enters its second breakdown region under ESD pulse conditions and it is irreversibly damaged.

Residual Pulse: The resulting voltage/current (after system level ESD protection devices) seen by an IC component from an IEC stress waveform.

SEED: System-Efficient ESD Design - Co-design methodology of on-board and on-chip ESD protection to achieve system –level ESD robustness.

System level ESD Robustness: The capability of a product to withstand the required IEC ESD-specification tests and still be fully functional.

Soft Failure: Failure of a system not due to physical damage in which the system can be returned to a functional state without the repair or replacement of a component. Return to a functional state may or may not require operator intervention. Operator intervention may include rebooting or power cycling. Soft Failures can involve software issues and software fixes but in the context of this document they are primarily due ESD events injecting unwanted signals into the system which place the system into a state in which it does not function as intended.

Executive Summary

Our intention in this document is to work with the OEMs, and with their participation and feedback, eliminate misconceptions about system level ESD while jointly addressing the design of robust ESD systems. **Our aim is to bring suppliers and customers together for a common purpose towards the development of ESD robust systems.**

There is a growing awareness in the electronics OEM community that system level ESD robustness is an important requirement for reliable products. System level ESD testing is today applied to a wider range of products than ever before. Designing ESD robust systems can be very challenging, especially for systems which integrate advanced technology integrated circuit (IC) components. For most system designers, ESD protection strategy and design efficiency are only dealt with in an ad hoc manner. Many of the most severe system level ESD design problems can be traced to misconceptions between system designers (OEMs) and their IC providers. Adopting a consolidated approach to system level ESD design, which addresses these misconceptions, will be key to future success in building ESD robust systems.

This White Paper serves three important purposes. First it provides an overview of system level ESD test and design challenges in the industry today. Second, it identifies and characterizes the primary misconceptions mentioned above. Third, it introduces a new co-design approach called "System-Efficient ESD Design" (SEED) that promotes a common OEM/IC provider understanding of the correct system level ESD needs.

This white paper is the first part of a two part document. Part I will primarily address hard failures characterized by physical damage to a system. "Soft failures", in which the system's operation is upset but without physical damage, is also critical and predominant in many cases. The same soft failures can also refer to system upsets involving recoverable damage to system malfunction. However, these issues are out of the scope of the current document and will be dealt with in detail in Part II of this white paper. Although EOS failures can result from a system failure our focus here is not intended to cover other types of EOS failures that can come from mishandling, etc.

Background and Purpose (Chapter 1)

There is a critical need in the IC industry to directly address the growing division in the understanding of system level ESD between system/board designers and their IC providers. The true nature of system ESD reliability, especially in light of the rapid advances in the IC industry, requires a comprehensive examination. There are three aspects to this study.

> 1. Understanding of the nature of system failures which can be either "hard" or "soft." Hard failures are typically related to physical damage which is not recoverable, while soft failures describe a system upset or malfunction including recoverable damage.
>
> 2. Clarification of misconceptions that often lead to an inefficient approach to system level ESD design. For example, the commonly held belief that IC level ESD specifications (such as the HBM) can ensure robust system ESD design.
>
> 3. Definition of the whole system in the context of which portions of the IC components on a PCB are involved in the protection strategy. For instance, identifying the external (interface) pins that would be in the critical path of an ESD event and require careful

design strategy, versus internal (non-interface) pins which are not as affected and may not require special attention during system design. However, while differentiating internal versus external pins, it must be noted that issues associated with inter-chip pins are also important. Part I will address only direct stress issues while the indirect effects coming in from coupling will be dealt with in more detail in Part II of the white paper. OEM concerns about failures of products in manufacturing and in the field have often led system manufacturers to take their own initiatives, whether effective or not, or to make various demands from the IC suppliers, whether justified or not in each and every case.

Test Methods and Their Field of Application (Chapter 2)

The existing system level ESD test methods and their field of application are discussed in great detail. First, it is noted that IEC 61000-4 is a set of EMC test standards which includes the system level ESD test method, IEC 61000-4-2. It specifies calibration waveforms, procedures and stress points for executing ESD tests on systems. The standard clearly excludes several locations and situations. The standard also explicitly encourages committees, manufacturers and users to derive standards from IEC 61000-4-2 for specific applications. We discuss the IEC 61000-4-2 ESD procedure and present several examples of application specific interpretations. We also discuss examples where people have derived practices to stress locations or situations that were excluded in the IEC 61000-4-2. The most extreme example is the application of system level ESD stress directly to ICs. Several approaches for this application are discussed.

Proven System Level Fails (Chapter 3)

Here we address proven system level failures from actual case studies, examples of both field returns and failures generated during qualification testing. Field failures generated during a system operation are not easy to resolve as to whether they come from ESD or EOS types of events. If a failure is detected, a thorough root cause analysis would be necessary to establish the cause. The examples given try to ascertain whether the failures are related to the device HBM or CDM robustness as well as the type of external protection device implemented and their effectiveness in protecting against an IEC ESD event. By classifying the failure types and establishing the failure statistics a better insight into the system failure phenomena will be obtained. In this document we highlight how system problems are typically solved.

OEM System Level ESD Needs and Expectations (Chapter 4)

Next, we discuss the needs and expectations that OEMs have from their IC suppliers such that the OEM can design products that will not be physically damaged or have their operation upset by ESD stress. Three hypothetical design paths for ESD robust systems are outlined:

1. Design with ESD robust ICs in which ESD is not a concern.

2. Design with a combination of ESD robust and non-robust products, but with clear guidelines, procedures and tools available (i.e., the SEED approach).

3. Design with a combination of ESD robust and non-robust ICs, but without clear ESD guidelines, procedures and tools available to the system designer.

As desirable as path 1 may be, it is typically unrealistic. Instead we describe the information and tools needed to move from path 3 (which often describes ESD system design today) to path 2, which is a realistic goal for the future.

Lack of Correlation between HBM/CDM and IEC 61000-4-2 (Chapter 5)

There is a common assumption in the system design community that the IC level HBM has relevance to ESD performance at the system level. This assumption persists because of a lack of understanding of the differences between the models and a lack of actual data to make valid comparisons. Figure 1 contains data where some IC and system level information is available for a rough comparison. Details of the tests which were done to generate this data are not available as is often the case. However, this limited data serves to suggest that correlation is not likely (though not disproven by such a small data set). In this section, this lack of correlation and why this is expected is discussed. The relationship between IC ESD models (HBM and CDM) and the IEC 61000-4-2 is further explored. By comparing required waveform characteristics, equivalent circuit models and practical realization issues, it is demonstrated that these IC models cannot be expected to correlate to system level ESD. Actual test comparisons that have been reported in the literature are reviewed. While emphasis is placed on common HBM and CDM tests, comparisons to other emerging models are discussed. These include cable discharge events (CDE), human metal model (HMM), transient-induced latch-up, extended pulse length transmission line pulse (TLP) and charge-board events (CBE).

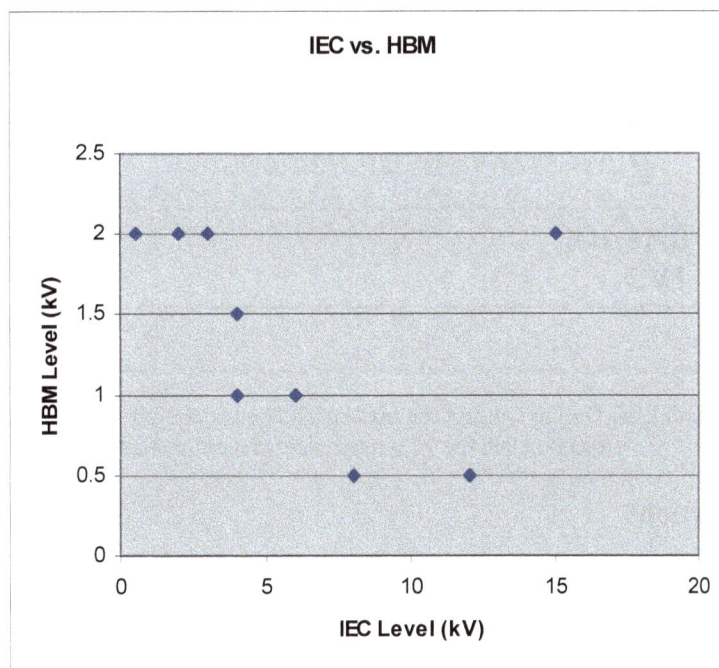

Figure 1: Comparison of IC ESD and System ESD

Relationship between IC Protection Design and System Robustness (Chapter 6)

Finally, we discuss the relation between IC protection design and system design robustness to avoid physical damage. Requirements and constraints of IC level ESD protection design are presented. Starting by highlighting misconceptions in equating IC ESD robustness, like HBM according to JEDEC with system level ESD robustness, a detailed discussion of various system level ESD protection concepts (both on-chip and on-board) is performed. On-board and on-chip protection circuits interact and can even compete. This requires a careful evaluation of the relevant parameters. Using analysis methods like transmission line pulsing, which reflect essential characteristics of IC pins and board protection elements, a systematic design of system level ESD protection can be developed. Essentially the comprehensive on-chip/ on-board protection co-design methodology, referred to as System-Efficient ESD Design (SEED), enables an optimum protection design from building blocks that provide a clear advantage to today's trial and error approach. The generic SEED concept is illustrated in Figure 2. The benefits are discussed for some examples like USB, CAN bus and antenna interfaces.

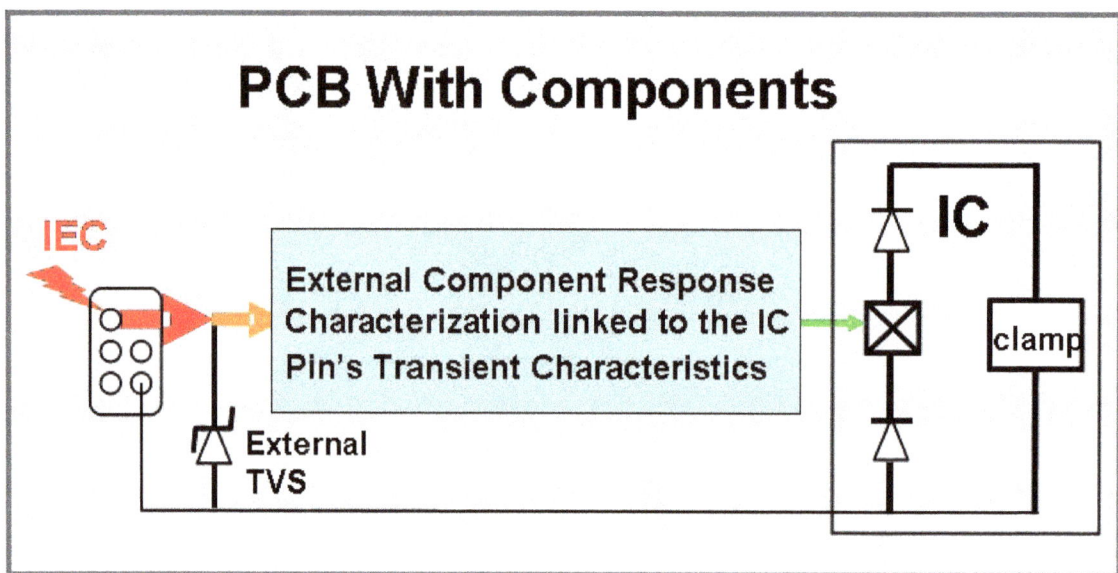

Figure 2: System-Efficient ESD Design concept requires careful consideration of interaction between the PCB protection and the IC pin transient characteristics.

Cost of System Protection

Assessing the cost of design, it becomes clear that a co-design approach is superior to design concepts relying on excessive ESD robustness requirements at the IC level. We show a path that illustrates how IC suppliers and system manufacturers can cooperate in the future, with both parties benefiting from this approach to achieving required system robustness without overly specified IC level ESD targets and performance restrictions.

Frequently Asked Questions

Q1: Why is the Industry Council addressing Non-Correlation issues between Device Level and System Level testing?
Answer: Some OEMs have been under the impression that higher levels of system robustness can be achieved by designing and measuring greater than necessary IC ESD levels. Our focus is to show that the system ESD measurement is relevant only when the IC is placed on the PCB and that stress data obtained at the IC level does not often correlate to system ESD capability when running the current IEC 61000-4-2 test procedure.

Q2: Is there a correlation between device failure thresholds and real world system level failures?
Answer: There is rarely correlation between device (IC level) failure thresholds and real world system level failure in the field. Device failure thresholds are based on a simulated ESD voltage and current directly injected into the device (IC) with the device in a powered down condition. Real world system level failures in the field occur in many different conditions, most of which are powered. In addition, there is no clear definition of soft failure robustness for ICs, and many real world errors are soft failures. First one needs to establish reliable methods for soft failure evaluation of ICs before one can attempt to compare IC level and real world failures.

Q3: Why wouldn't you expect to see correlation between IC level and system level testing?
Answer: Since the tests are done in different environments (unpowered versus powered or stand-alone versus on board) along with the different stress current wave shapes for the two tests, it is not surprising that they would be uncorrelated. In some instances external IC pins with higher IC level ESD robustness may result in less of a load on the on board ESD protection components. However, there are many examples where an improved HBM level for an IC resulted in lower system level ESD as discussed in Chapter 5 and Chapter 6. The approach of relying on IC level ESD for system level ESD protection is not only impractical and unpredictable, it also detracts from the need for an efficient system ESD design in which the on board and on chip protection work together.

Q4: Why are IC manufacturers now being asked to perform System Level testing at the IC level?
Answer: On the surface this seems like a very logical thing to do. If ICs and all other components can survive system level stress, it would appear that there should be no problems when it comes time to test the system for ESD robustness. There is also the desire to reduce component count for economic reasons. Unfortunately this approach may not be the most practical or economic approach and it also may not work. See Chapter 4.

Q5: Will there be a need for an IC ESD Target Level, to confirm System Level performance?
Answer: No. System level performance is a combination of on-chip ESD protection, on-board protection components and system mechanics design. The detailed properties of the IC's ESD protection such as turn on voltage, resistance, and maximum withstand current are much more important than the IC's HBM and CDM level measured in voltage. Automotive manufacturers do require some levels of IC ESD robustness for some bus transceivers but the specific test conditions are specified to be very close to use conditions.

Q6: Can devices really be designed to withstand real world system level events?
Answer: It is certainly possible to design ICs that can withstand system level ESD stress, but it is a complex and often unwise path. It is hard to know the exact details of the stress that will reach

an IC on a board due to circuit board parasitics, making the design difficult and prone to overdesign. Additionally, IC protection for IEC stress consumes considerable area and is likely not to be the most economical path.

Q7: Do all pins on a device need to be tested using system level events?
Answer: <u>External pins</u> (e.g. USB data lines, Vbus line, ID and other control lines; codec and battery pins, etc) need to be tested if the IC is not to be protected with on board components. But if the pin is to be protected by on board components, TLP characterization of the pin is more useful. <u>Some internal ESD sensitive pins</u> (e.g. control pins, reset pins, and high speed data lines, etc.) can be inductively coupled during a discharge to the case and/or to an adjacent trace of an external pin undergoing system testing. These pins need to be identified and may need to be tested using system level events.

Q8: JEDEC publication JEP155 recommended lowering IC ESD levels; will this have an impact on the overall system reliability?
Answer: Many systems, outside of the automotive industry, have been shipped with the IC ESD levels as recommended by JEP155 without a reduction in system level ESD robustness or an increase in other reliability issues (Data in the automotive industry is lacking because that industry has been very strict about maintaining the higher IC ESD levels. Although in the automotive industry ECUs containing devices with lower component ESD robustness are out in the field without known problems). First of all, most pins on an IC are signals that are internal to the system and will not be directly exposed to system ESD stress. Even for externally connected IC pins the relation between IC level ESD and system level ESD is not straight forward. Traditional levels of IC ESD robustness such as HBM 2000 V and CDM 500 V are not enough to protect against the much more severe system level ESD tests. In a system the IC's on chip ESD protection must work in harmony with the system level ESD protection. Increased levels of HBM and CDM can in fact lower system level ESD performance if the IC's ESD circuits begin to conduct at lower voltages than the board level ESD protection (see Section 6.7.1). This is an area that needs further work as stated in Clause 11 of JEP155.

Q9: If system level ESD testing at the IC level does not guarantee system level ESD performance, aren't higher target levels of IC HBM ESD better than nothing?
Answer: This would only give a false sense of security and could result in extensive cost of analysis, customer delays and a circuit performance impact. (Remember, higher HBM ICs may be harder to protect!). System ESD protection depends on the pin application and therefore requires a different strategy. System level ESD is clearly important, but targeting excessive IC level requirements could pull resources away from addressing and designing better system level ESD.

Q10: I often hear that the IEC 61000-4-2 pulse is a superposition of a CDM and a HBM pulse. Can IEC 61000-4-2 ESD testing replace CDM and HBM testing?
Answer: No. Looking at the two peaks in an IEC 61000-4-2 pulse the time duration is indeed comparable to a CDM and HBM pulse. However the required levels and discharge nature are completely different. This is because HBM and CDM is intended for IC level testing while IEC 61000-4-2 is intended for system level testing.

Q11: If CDM methodology and levels are modified would there be more fallout for EOS at the IC or System Level?

Answer: CDM and EOS event fails are completely different in total energy and time duration. Effective CDM protection does not guarantee EOS protection. EOS protection must be provided at the system level. There is no correlation between IC CDM failures and system EOS failures. Please refer to Annex D.1 and Annex D.1.3 for details in JEDEC publication JEP157. The fallout rate due to EOS would not change as a result of modifying CDM methodology and levels.

Q12: Since ICs are now designed for lower IC ESD levels, why would this not be reflected by a sudden change in the overall health of a systems for ESD capability?

Answer: The overall health of a system is dependent on a comprehensive approach to the protection methodology that includes a number of factors including on board protection components, optimized board signal routing, component packaging and, as a last line of defense, the IC level protection.

Q13: If all the recommended approaches listed in this document are followed would we then guarantee that a system will never fail for ESD?

Answer: One cannot guarantee that a system will never fail from ESD because there are many different discharge conditions and levels in the field. ESD sensitivity from one system (OEM product) to another system is always different because of differences in various product designs. Note; the recommended approaches in this document are intended to help produce more robust systems in a more efficient manner.

Q14: How will you reach all the different system designers for their inputs?

Answer: The current document has been reviewed by a number of OEM representatives and they are in agreement with the conclusions of the document. We expect that the publication of this document will result in further input from the system design community. This input will be especially important as the Industry Council works on Part II of this white paper.

Q15: If the system designers who are not involved in this document do not agree that it is a shared responsibility then what is next?

Answer: The system designers need to be educated in terms of system ESD versus IC protection design. Education with regard to these issues is a major focus of the Industry Council and we are convinced that as the benefits of the shared ESD responsibilities become evident more system designers will become convinced of their shared responsibility.

Q16: If an IC with the new lower ESD levels starts showing high levels of system failures how will the industry address this?

Answer: First, an investigation comparing ICs from provider A and provider B should look at the details of the IC level ESD designs, not just the IC failure levels in volts. Second, the OEM should share the system level ESD test results with the IC providers. For example, if IC provider A fails and IC provider B (2nd source) passes. IC provider A needs to investigate why their IC fails. Next, the OEM should review their ESD protection design for further improvement for both IC suppliers. This type of dialogue is important in the future.

Q17: What is the purpose of IEC 61000-4-2?

Answer: The purpose of the IEC 61000-4-2 test is to determine the immunity of systems to ESD events during operation. The document states that it relates to equipment, systems, subsystems and peripherals, without further defining them. Its scope and description clearly indicate the purpose: to test electrical and electronic equipment that may be subjected to ESD from operators directly to the system under test or from indirect discharges from personnel to adjacent objects. See Section 2.0.1 of Chapter 2.

Q18: Which level does my system need to pass according to IEC 61000-4-2?
Answer: Like the international HBM and CDM standards, the IEC 61000-4-2 spec does not prescribe pass/fail levels. It describes the method and procedure on how to perform the tests. Related documents, such as IEC 61000-6-1, state that 4 kV contact discharge and 8 kV air discharge are suitable requirements. See Sections 2.0.2 and 2.0.3 of Chapter 2.

Q19: I cannot apply IEC 61000-4-2 as intended. What can I do?
Answer: Actually the IEC 61000-4-2 states that it serves as a basis to derive suitable standards for situations that are not covered by the document. For such cases, user committees should develop suitable procedures. Chapter 2 discusses several such standards.

Q20: What is HMM?
Answer: HMM stands for Human Metal Model. It is a method to assess the robustness of external IC pins against a system level ESD pulse. See Section 2.2.2 of Chapter 2 for details.

Q21: Is the Human Metal Model related to Machine Model? Should a device be required to have high MM levels to pass the HMM and this help the IEC performance?
Answer No. HMM has nothing to do with MM. Besides, MM is not a relevant IC level test and should not be used for any type of assessment.

Q22: After System Level ESD stress my application needs to be re-set. Is this a fail?
Answer: This is really application specific. Several system levels standards, including the IEC 61000-4-2 give several failure criteria. Which one is applicable depends on the application. E.g. for consumer electronics a manual re-set might be acceptable, but for safety-related applications this is strictly forbidden. See Section 2.0.2 of Chapter 2.

Q23: IEC 61000-4-2 refers to system upset. How do I evaluate this on my module or IC?
Answer: For a module this may be possible, if the module can be operated outside of the full system. Possible methods for modules are discussed in Section 2.0.4 of Chapter 2. For ICs the options are more limited. Latch-up tests such as JESD78B help to show immunity from upset but a standardized test method for transient latchup would provide a useful tool. For application to ICs typically permanent damage is the only failure criterion. See Section 2.2.2 of Chapter 2. Tests such as HMM could be performed on an operating IC but the required test setup would likely be very complicated.

Q24: Do all system level ESD standards use the same waveform?
Answer: In short: No. However, most use the waveform as defined in the IEC 61000-4-2, which is determined by a 330 Ω, 150 pF RC network. An example of a standard that uses a different waveform is ISO 10605, This standard uses the same type of ESD gun but the RC network is modified for some of the tests, using a 2 kΩ resistor and/or 330 pF capacitor instead of the values used in IEC 61000-4-2. See Section 2.1.1 of Chapter 2.

Q25: Is SEED considered to reproduce real, physical behavior of board and IC?
Answer: SEED is a concept to limit damaging current pulses reaching the internal IC pin. So in this sense it represents what the physical effect would be on the IC pin coming from an IEC stress at the external port of the PCB. What it represents for the board depends on how well the scenario is represented during the SEED analysis.

Q26: How can system/board designers get the required information about the IC IO behavior?
Answer: First, both the OEM and the IC supplier must define the 'external pins'. Following this, the IC supplier provides the TLP curve of the pin under interest with either bias applied or without bias which would depend on the pin application in the overall system board. The measured TLP at the pin will not only represent the pin's internal ESD clamp behavior but it will also include the IO design behavior to the transient pulse analysis.

Q27: What is the required degree of accuracy of the simulation models?
Answer: The simulation models can only be as accurate as the measured waveforms at the external clamp under IEC pulses along with the variations, and the internal IC clamp under the TLP conditions. Experience will teach us what level of accuracy is needed. Even if early attempts at simulation do not have the level of accuracy we may desire the simulations will still provide insight into the ESD properties of a design.

Q28: How can snapback devices be handled?
Answer: At first glance it would appear that snapback devices that are part of the on chip ESD protection would present a challenge in the design of an on board protection strategy. If, however, the on board clamp circuit maintains the voltage and current in the IC below the voltage and current failure levels of the snapback device there should be no problem.

Q29: Are different models needed for powered and unpowered conditions?
Answer: This depends on the pin application where the OEM would define whether the external pin would be facing powered conditions. It is generally good idea to provide TLP for both powered and unpowered conditions

Q30: Is it enough to simulate only an idealized IEC waveform or do we have to consider a wider range of discharge waveforms like CDE pulses etc?
Answer: This again depends on the application of the pin in the system board. For example, if it is an Ethernet pin a CDE pulse characterization may be important. It is generally a good idea to use waveforms that represent the expected system level stresses as best as possible.

Q31: What do you mean by hard and soft failures?
Answer: A hard failure is one in which a component of the system is physically damaged and the component must be repaired or replaced to return the system to a functioning state. A soft failure is one in which the system can be returned to a functional state without a physical repair to the system. A soft failure may or may not require operator intervention. Operator interventions may include the shutting down of a program on a computer or the power cycling of a system

Q32: HBM testing seems to be measured differently in different documents or discussions. This is confusing. Why is there a difference and what is it?

Answer: When the first ESD event measurements were made many years ago, the HBM term was used for any electrostatic discharge from a human. Both the discharge from the finger and the discharge from a metal object held in the hand were identified as HBM. Both these tests retained the HBM name for many years with widespread confusion. IC testers (i.e. two pin and multi-pin HBM testers) used the HBM name and hand held system level testers (ESD guns) sometimes also used the HBM name, even though these were two very different applications.

Just a few years ago the ESDA finally chose to differentiate them. The HBM (Human Body Model) definition remains for the current discharge through two pins of an IC provided by 2 pin or multi-pin "HBM" test system.

HMM (Human Metal Model) is a new term which defines the direct IC two pin device testing with the IEC 61000-4-2 waveform. Its lower spark resistance from a metal object held in the hand creates a higher current discharge than HBM testing at the same voltage. This is the threat provided by both ESD guns and the recently included IEC-50 ohm source HMM test systems.

Although the HBM term was originally used for both tests; you can help minimize this confusion by remembering these simple but important differences, and refraining from referring to either the HMM device test or the IEC 61000-4-2 system test as HBM!

Chapter 1: Purpose and Introduction

Charvaka Duvvury, Texas Instruments
Harald Gossner, Infineon Technologies
James W. Miller, Freescale Semiconductor
Pasi Tamminen, Nokia

1.0   Motivation and Purpose

There is a critical need in the IC industry to directly address the growing division in the understanding of system level ESD between system/board designers and their IC providers. The true nature of system ESD reliability, especially in light of the rapid advances in the IC industry, requires a comprehensive examination. There are three aspects to this study.

1. Understanding the nature of system failures which can be either "hard" or "soft." Hard failures are typically related to physical damage while soft failures describe a system upset. Soft failures on other hand refer to system upsets involving recoverable damage to system malfunction.
2. Clarification of misconceptions that often lead to an inefficient approach to system level ESD design. One such misconception is the commonly held belief that IC ESD specifications such as the HBM test can ensure robust system ESD design.
3. Definition of the whole system in the context of which portions of the IC components on a PCB are involved in the protection strategy. For instance, identifying the **external (interface)** pins that would be in the critical path of an ESD event and require careful design strategy, **internal (non-interface)** pins which are not affected and do not require special attention during system design and internal pins which are susceptible to stress via coupling.

The main purpose of this white paper is to address these issues from a variety of perspectives; including IC manufacturers, system board designers and OEMs/ODMs. The target audience spans the range from IC manufactures to board designers to OEMs/ODMs because the solution of system level ESD issues requires the effort and communication from all stages of system development.

As a special note, this document focuses on ESD protection designs, and is not intended to address the full scope of Electro-magnetic Interference immunity designs.

1.1   Scope

This white paper is the first part of a two part document. Part I will primarily address hard failures characterized by physical damage to a system (failure category d as classified by IEC 61000-4-2). Soft failures, in which the system's operation is upset but without physical damage, is also critical and predominant in many cases. We also note that some indirect coupling that might occur between interacting chips on a board are much more complex to describe but these will be considered in more detail in Part II. While preventing hard failures requires a carefully optimized approach, soft failures require an even deeper understanding of their nature before they can be comprehensively addressed. For example, whereas hard failures involve direct current stress leading to damage, many soft failures are associated with the complex nature of EMC. Other types of failures can come from pure EOS events involving a shorted battery, inductively coupled surge from power supplies, or even lightning strikes. These specific issues are not in the scope of this document.

Although some soft failure issues and methods for designing systems against them will be covered in Part I, the issues associated with the numerous types of soft failures will be dealt with in more detail in Part II of this white paper.

1.2   Background

The ubiquity of electrical systems in modern life has led to their increasing deployment in hostile environments. It is well known that system components such as integrated circuits are susceptible to damage or data upset from externally generated electrical overstress, including electrostatic discharge events. Consider for example some common system level EOS/ESD events that may occur:

- A microprocessor IC on a laptop computer damaged when plugging in a charged USB cable

- A system built by cabling several large frames together fails during installation caused by damage on a device pin at the connecting point

- A communications IC on an operating cell phone suffers data state upset when a charged person touches the keypad

- A microcontroller IC on a washing machine is damaged due to inductive voltage spikes as the motor switches on

- A power management IC on an operating mobile phone suffering power loss that only happens in the winter when a user removes it from the holster

Even from the short list above, it is clear that system level EOS/ESD events can be caused by a wide range of application-dependent external stress sources. EOS/ESD test methods have been developed in an attempt to reproduce many of the most common system level EOS/ESD events (IEC 61000-4-2, ISO 10605, etc.). But one must note that while the ESD test methods have been designed to generate repeatable and reproducible results, they cannot address the full range of real world ESD events. Even with each of these stress tests there are still known issues with real world fidelity, test fixturing, etc. that need to be addressed. Therefore to obtain a better perspective we need to address the following issues.

1.   What is the meaning of system level ESD robustness?

2.   Do the system level tests adequately mimic the environment that a component on a board would encounter in the field?

3.   What is the correlation between system requirements and returns from the field?

4. How are proper control methods implemented for the system?

5. Is the IEC 61000-4-2 stress method, which does not directly stress input and output pins, sufficient or do we also need to address the stress at the ports as done e.g. in ISO 10605?

6. How do we prove that the IC ESD design can protect a board?

7. What are the concerns for system level ESD vs. IC level ESD?

8. When is it better to place protection at the IC level? When is it better to place it at the board level?

9. What are the considerations for design speed versus cost effectiveness for the different design scenarios?

10. Under what conditions does crosstalk between signal lines which are directly stressed and other lines, either within a wiring harness or on a circuit board, become important?

11. Why are system level requirements increasing? What is the impact on IC ESD design capability?

12. How do we arrive at a safe and practical strategy that is useful for the electrical/electronics industry?

1.3 Problem

As cited below, concerns about failures of products in manufacturing and in the field have often led system manufacturers to take their own initiatives, whether effective or not:

- Implement increasingly more ESD robust system/board level design practices, perhaps partly because of more demanding RF functions. That is, RF signal integrity requirements are becoming more demanding, forcing OEMs to specify EMC functionality very strictly

- Require increased ESD tolerance from the suppliers of the ICs placed on the board (even if they only apply for the external pins one still needs to define and understand how these pins are categorized)

- Perform EOS/ESD testing using system level test methods

- Implement improved static charge controls in the manufacturing environment

These issues and strategies consequently have led to the following trends in system level EOS/ESD:

- Increased system level EOS/ESD performance targets which often leads to the same target for the component

- Continued requests for fewer on-board components: due to cost constraints driving the reduction of discrete ESD protection elements along with the drive to reduce parasitics in high speed lines

- Increasing competition as a result of expectations to provide more integration and higher performance in components

- Increasing expectations from board designers that IC components must self-protect once in a system

- Increasing difficulty for IC component designers to understand and target a customer's system level EOS/ESD requirements and then translate these requirements into actual pin by pin stress on the IC component

- Increased misunderstanding between system/board designers and IC providers

Based on the trends listed above, it is clear that improving the overall understanding of system ESD protection and establishing more productive dialogue between the IC supplier and the System Board Manufacturer has become crucial.

1.4   OEM Requirements

This "productive dialogue" of communication between the IC supplier and the System Board Manufacturer forms the first step towards improvement in the system level ESD design practice. The most important topic to address in this white paper is: "What do OEMs need?"

The obvious answer would be OEMs want solutions that work. The next question is then: "What are possible approaches to get working solutions?" This is where clear communication of expectations versus performance of the solutions becomes critical. These issues naturally become more confusing when multiple suppliers are involved in the same product. We realize the challenge for OEMs is always to supply solutions at a lower cost to their own customers. The question also arises on what the OEMs expect from their IC suppliers to assist in the design of ESD robust systems. There is inevitably a tangle of interests that need to be filled. To properly address these issues and arrive at a clear solution, a more efficient methodology is needed. One of the main objectives of this white paper is to explore a new approach beneficial to both suppliers and system board manufacturers.

1.5   System-Efficient ESD Design (SEED)

We introduce the concept of System-Efficient ESD Design or SEED. For the design of an efficient system many clarifications are first needed.

1. At best the IC ESD levels provide insufficient information for any system ESD design since IC level tests do not reflect what the pin experiences during the IEC ESD event.

2. An understanding of the stress event seen by the internal pin is paramount to design the system protection.

3. Thus, System-Efficient ESD Design (SEED) can only be achieved after a thorough understanding is obtained about an IC's pin interactions in the system along with the transient behavior of the pins during ESD stress.

4. For such an approach to be applied, efficient characterization methods like TLP data (ANSI/ESD STM5.5.1-2008 Electrostatic Discharge Sensitivity Testing – Transmission Line Pulse (TLP) – Component Level) would be needed to analyze the IC pin and system interaction.

By demonstrating such a new concept it becomes apparent that:

- High levels of HBM protection are not necessary for system ESD design nor do they guarantee ESD-safe system design. Indeed, IO performance requirements may limit the ESD HBM levels that may be obtained.

- While the HBM ESD levels generally accepted for IC handling need to be reduced to a more realistic value, robust system ESD design targets can still be obtained as long as the interactions between the ESD stress and the full system design, including integrated circuits, are understood and addressed in a systematic manner.

1.6   System Definition (Internal Pins versus External Pins)

As mentioned earlier, an optimum system design would first involve defining which pins of an IC may be affected during the design for system protection. Consider first a simplistic representation of a part of a PCB as shown in Figure 3. More importantly, the pins attached to the signal buses connecting several PCBs would be critical since they would be exposed during repair. Obviously, pins attached to external connectors like a USB port are also critical for system ESD. It is worth noting here that although the inter-chip pin connecting one IC to the other IC may not see any coupling during unpowered conditions, there is always the possibility that during a powered condition there would be some coupling that may have to be taken into consideration.

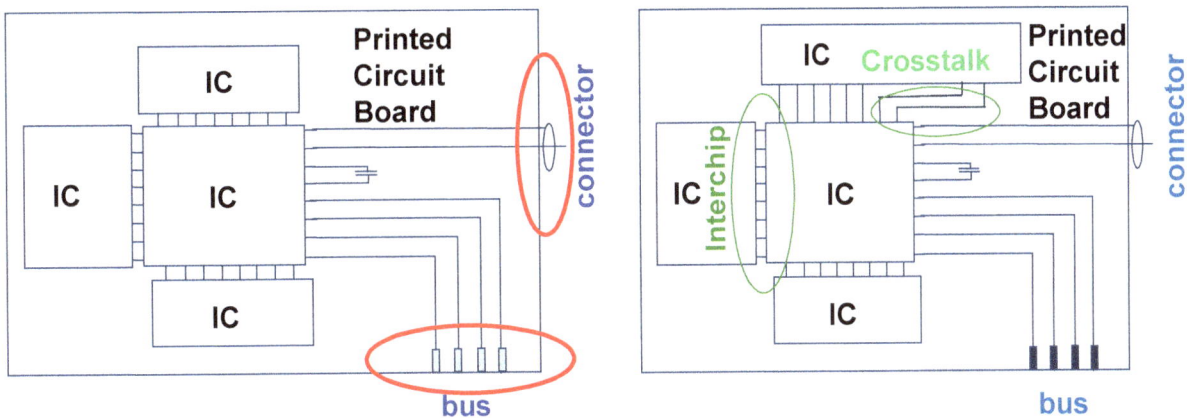

Figure 3: Classes of pins for specific system level ESD considerations including external and inter-chip coupling.

As a second example, consider a more detailed case as illustrated in Figure 4. Here the system design approach must consider the pins and ports where the ESD zaps are indicated. Even with these considerations there is also the uncertainty about the internal pins and if they do see any remnant energy pulses. Can they also get by with the minimum ESD levels that are required for IC handling? At best the IC ESD levels provide insufficient information for any system ESD design since IC level tests do not reflect what the pin experiences during the IEC ESD event. A more detailed analysis is still pending for Part II of the white paper.

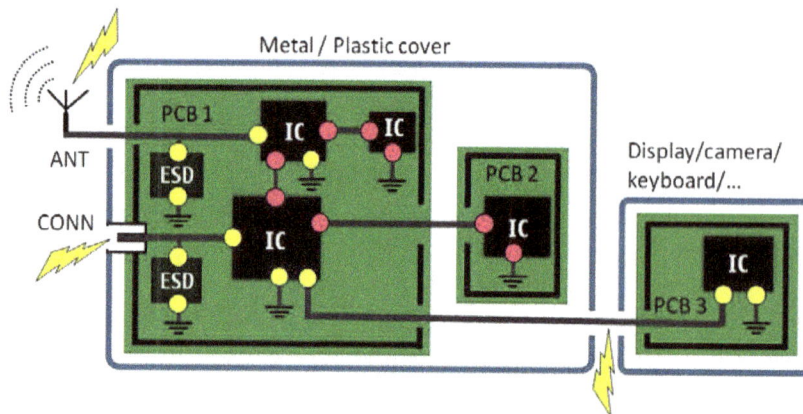

Figure 4: A wired system design with designation for external ESD threat.

In summary, the dialogue between the system builder and the IC supplier can only be improved through a thorough understanding of each type of system to know which types of pins would be susceptible to ESD events and therefore require close attention to system level design protection.

1.7   General Approach and Outline

In order to present the details of this white paper, we have formulated a strategic flow that will give the reader a solid understanding of the current state of knowledge of system level ESD robustness and define test steps that must be taken to develop and implement a highly successful design practice. We start by defining the current practices used for testing system protection performance (Chapter 2). This will be followed by a review of known system test failures (Chapter 3). Once this background is established we then present what the OEMs need and should expect from their IC suppliers (Chapter 4). This will then bring us to the review of common misconceptions that high levels of IC ESD performance will improve system robustness to ESD when tested to the known system level tests (Chapter 5). Dispelling this misconception will be the first step toward better insight into OEM requirements which will yield systems with good ESD immunity without overdesign of integrated circuit external pins. Based on this we will describe the methods of system ESD design that will be most compatible for customer needs and for the suppliers to be able to deliver these requirements (Chapter 6). Finally, we will summarize our findings (Chapter 7). Part II of this white paper will flesh out our proposed design strategy and focus on soft failures.

This document is only Part I. As mentioned above, a sequel to this white paper, Part II, will be documented at a later time. As a preview, the second part will deal with more details on EMC related system design. It will construct a framework of types of soft failures and root causes, followed by a reference methodology for the IC system design. The eventual goal will define the required design targets for system level ESD protection as a whole.

Chapter 2: Test Methods and Their Field of Application

Theo Smedes, NXP Semiconductors
Jon Barth, Barth Electronics
Patrice Besse, Freescale Semiconductor
Mike Hopkins, EM Test USA
Guido Notermans, ST-Ericsson
Nate Peachey, RFMD

2.0 The Basic System Level Test: IEC 61000-4-2

The basic system level ESD test method is described in [1]. The primary purpose of the IEC 61000-4-2 test is to determine the immunity of systems to external ESD events outside the system during operation. The document states that it relates to equipment, systems, subsystems and peripherals, without further defining them. Its scope and description clearly indicate the purpose: to test electrical and electronic equipment that may be subjected to ESD from operators directly or from indirect discharges from personnel to adjacent objects [2]. The scope further describes that the document is a basic reference method and that product committees, users and manufacturers are responsible for appropriate use and severity levels. The document explicitly recommends those groups to consider adopting the method where appropriate. This chapter will discuss several such initiatives, which resulted in formal procedures or standards. Also many companies, vendors and OEMs, have adopted their own internal qualification procedures, often inspired by the published standards and common practices.

The first ESD test method was identified as Human Body Model. It was originally generated in two different spark gap conditions, both of which were called HBM. One ESD discharge came from the bare finger of a charged human, while the other discharge was from a metal rod held in the hand of a charged human. The discharge current was measured with a sensor in the center of a large metal ground plane. The two discharge characteristics had significantly different electrical characteristics. The discharge between the finger and a metal is reflected in the IC level ESD test method Human Body Model. The discharge between two metal electrodes became the IEC 61000-4-2 test standard. It is a more severe current discharge than the HBM Test and is presently generated from a hand-held unit sometimes identified as an ESD gun. To avoid reproducibility issues a contact discharge method was added to the air discharge method. Note that this type of discharge does not reproduce the characteristic spark associated with ESD discharges. Concerning the current shape measurement for ESD generator calibration, contact discharge mode is recommended. No clear relationship between contact and air failure voltages is expected: *It is not intended to imply that the test severity is equivalent between tests methods*" [1].

The original IEC specifications were chosen based on measurements made with 500 MHz – 1 GHz oscilloscopes and a current sensor of unknown time domain / frequency domain response. Although the methods and reproducibility have been improved over time, this does not imply that the method covers all ESD events that may happen in practice. Real ESD events may, for example, have much shorter rise times, especially at lower voltages.

Measured waveforms that meet the specifications are shown in Figure 5. The only IEC specifications for system level current are peak current, rise time and the ratio of current at 30 ns and 60 ns to the peak current. A historic concern in the IEC 61000-4-2 test simulator is that there are no specifications for electromagnetic radiation emitted from the gun. The latest revision of IEC 61000-4-2 contains a considerable Annex which describes the radiated phenomena and provides test engineers with guidance for recognizing and dealing with radiated effects. In order to perform contact mode testing, an internal switch is used in the ESD simulator. This switch is usually a relay designed to provide clean switching operation and a good current waveform with a smooth, fast rise time. Unless controlled, this rise time would be on the order of 600 ps or less. IEC 61000-4-2 calls for a rise time of 0.8 ns +/- 25% which means the rising current must be slowed down to be compliant, and this is typically done by controlling the parasitic inductance and capacitance to free space near the tip of the simulator. This fast rising current in conjunction with parasitic and real components produces a radiated field, the characteristics of which are highly dependent on a simulators' physical design.

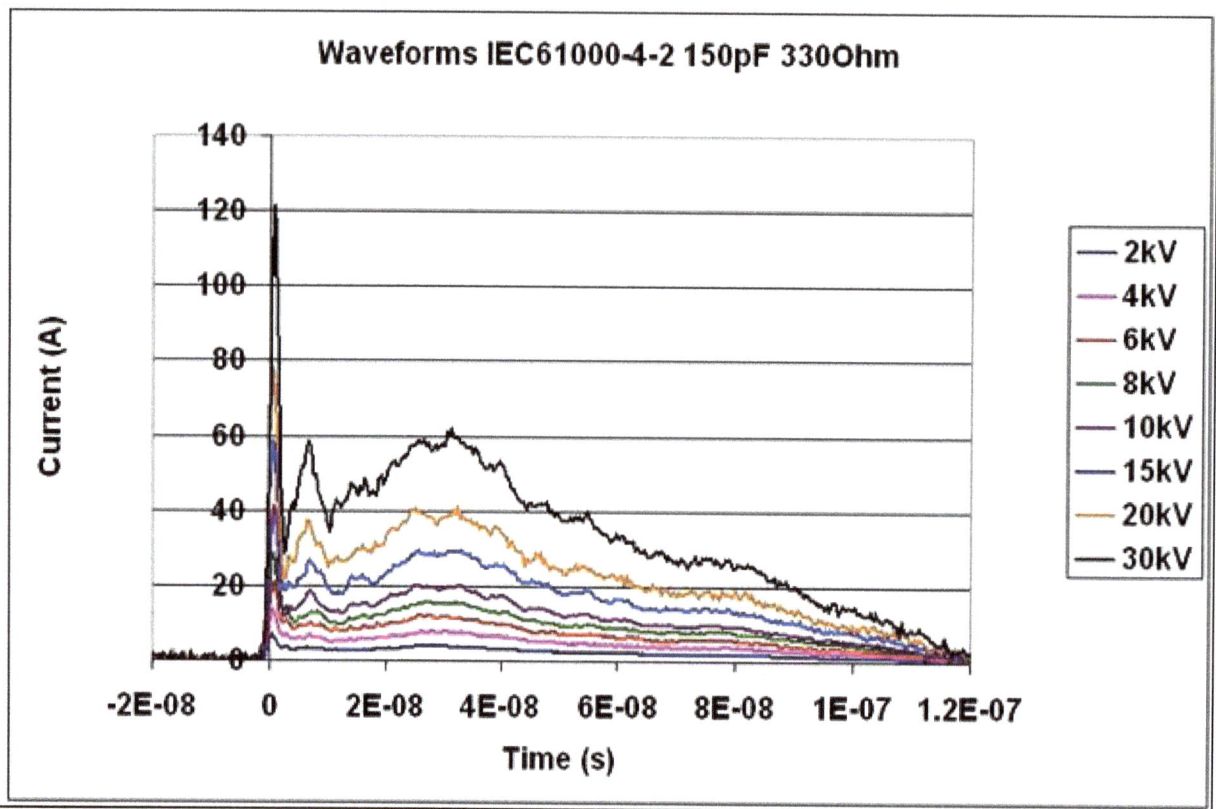

Figure 5: Measured waveforms of contact discharge from an IEC 61000-4-2 ESD gun on the prescribed calibration target

2.0.1   Rationale and Procedure

As mentioned above, the method targets direct and indirect ESD events between a person and a piece of equipment. The waveform described in the standard consists of 2 distinct regions; a sharp, short first current spike, followed by a slower, longer and smaller discharge current. The first peak supposedly represents the discharge through the tool that the person is using, while the slower part represents the discharge of the body through the length of the arm. From the same reasoning the stress locations are also defined. Direct discharges are applied to metal locations accessible to persons during normal use of the equipment, but NOT to maintenance and service points (such as battery contacts) and the contacts of connectors with a metallic shell. For the latter the reasoning is that in any real life situation, the discharge will be to the grounded shell. Most of the above cases are stressed with contact discharge. Only insulated covers and connector pins with a plastic shell are stressed with air discharge. This is summarized in Table 1. Indirect discharges are always done by contact discharge to a coupling plane. Bleeder resistors (470 kΩ) are used to prevent charge built-up for multiple discharges.

Table 1: Stress location and mode for direct discharge

Case	Connector Shell	Cover Material	Air discharge to:	Contact discharge to:
1	Metallic	None	-	Shell
2	Metallic	Insulated	Cover	Shell when accessible
3	Metallic	Metallic	-	Shell and cover
4	Insulated	None	a)	-
5	Insulated	Insulated	Cover	-
6	Insulated	Metallic	-	Cover
Note: In case a cover is applied to provide (ESD) shielding to the connector pins, the cover or the equipment near to the connector to which the cover is applied should be labeled with an ESD warning.				
a) If the product (family) standard requires testing to individual pins of an insulated connector, air discharges shall apply.				

2.0.2 Failure Criteria

The recommended classification in [1] is as follows:

a) Normal performance within limits specified by the manufacturer
b) Temporary loss of function or degradation of performance which ceases after the disturbance ceases. Equipment under test recovers its normal performance without operator intervention
c) Temporary loss of function or degradation of performance. Recovery requires operator intervention
d) Loss of function or degradation of performance which is not recoverable, owing to damage to hardware or software, or loss of data

It is clear that several of those categories do not relate to physical damage, but rather to system upsets. High-speed energy that leaks (conducted or radiated) into a system can cause upsets in circuit operation with false or error information attached to digital signals. Thus, the test identifies the effectiveness of system shielding to prevent or minimize the amount of high-speed currents which get inside a system to develop errors on signal lines. Especially in the case of safety-related systems, those types of fails are much more relevant than the failure criterion normally associated with device level ESD testing.

2.0.3   Typical Requirements

IEC 61000-6-1 [3] and IEC 61000-6-2 [4] prescribe general requirements for products in a residential and industrial environment, respectively. With respect to ESD there is no difference between the two variants. The tests must be carried out in a well-defined and reproducible manner, and they must be carried out in the most susceptible operating mode expected during normal use. The standards prescribe 3 performance criteria:

a) **Performance criterion A**: The apparatus shall continue to operate as intended during and after the test. No degradation of performance or loss of function is allowed below a performance level specified by the manufacturer, when the apparatus is used as intended.

b) **Performance criterion B**: The apparatus shall continue to operate as intended after the test. No degradation of performance or loss of function is allowed below a performance level specified by the manufacturer, when the apparatus is used as intended. The performance level may be replaced by a permissible loss of performance. During the test, degradation of performance is however allowed. No change of actual operating state or stored data is allowed.

c) **Performance criterion C**: Temporary loss of function is allowed, provided the function is self-recoverable or can be restored by the operation of the controls.

These standards prescribe that equipment meets performance criterion B for 4 kV contact and 8 kV air discharge according to IEC 61000-4-2. Note that hard failure is not mentioned in these standards.

2.0.4   Exceptions

This section describes system level ESD situations which are often encountered in practice, but do not fit the standard as defined in [1].

I.   Connector Pins

Although [1] clearly states not to stress connector pins with a direct contact discharge, it is common practice in many companies to do so. Therefore semiconductor suppliers receive requests to deliver components and ICs to let the system survive this kind of stress. Very often it is not well specified how exactly to arrange the connections. If not specified by the requester, an appropriate way to get repeatable results is to implement a configuration as described in [6]. In some cases, such as with (mini-) USB connectors, it is not possible to connect the ESD gun to the connector pin. In such cases it has been suggested to remove the connector and discharge directly to the signal wire or to insert a conducting wire in the connector and discharge to this wire. Obviously the connection to and the properties of the wire may influence the results.

II.   Devices and Components

It is obvious that the system level ESD standards are NOT intended to be used for testing single components or ICs. This even holds for the dedicated components, the so-called ESD diodes, which are added on PCBs to let the system meet the requirements. Nevertheless the semiconductor industry has been confronted with request to prove that ICs are 'IEC-compliant'. Approaches to accommodate such requests are addressed in Section 2.2.2.

III. Printed Circuit Boards

Most system level standards do not specify tests for PCBs. Practically speaking; they are often treated as modules or sub-assemblies, if ESD tests are performed.

IV. Sub-assemblies (modules)

This has always been a gray area in the European EMC Directive [7]. As originally written, the directive only applies to "apparatus", or finished complete products put on the market in the EU. The latest version modifies this by stating that certain components or sub-assemblies should also fall under the directive in certain circumstances. Specifically, it states that a component or sub-assembly intended for incorporation into an apparatus (finished product) by the end user which could either generate or be susceptible to EMC, ***does*** fall under the directive.

That means that items such as video cards, hard drives, or sound cards that a consumer could purchase, take home and install in his or her computer must be CE marked, which in turn means they must be tested for EMC, including ESD.

The basic standard, IEC 61000-4-2, does not give any guidance for testing such products and is open to interpretation. However, if one cannot perform ESD tests as specified in IEC 61000-4-2, the use of other methods is allowed as long as those methods and the reasoning behind using them are clearly documented.

Testing sub-assemblies has two basic problems – what points are tested, and how is the sub-assembly powered or should it be powered at all.

Test points:
From the standpoint of the manufacturer, making sure the product will survive normal handling by the consumer is probably the biggest concern. The consumer is likely not familiar with ESD control procedures, and probably doesn't have a wrist strap or any other means of ESD mitigation. As a result, any part of the sub-assembly is likely to be involved in a discharge in the un-powered state. A cautious manufacturer will probably perform tests to any point on the sub-assembly likely to be handled by untrained personnel. Test levels selected are likely to be in line with those established by product or generic standards for the final product.

Powering the sub-assembly:
In order to test to IEC standards, it is necessary to have the unit powered and operating in a normal manner. This presents some problems and raises a number of questions: If it is installed in an operating system for testing, test points may not be accessible; if installed on extender cards or via cables, is this a valid test since the proximity of the sub-assembly to other parts of the system, shielding and housing may affect the test results. Testing the sub-assembly on a jig will raise the same questions.

Equipment manufacturers are using the three basic methods mentioned in the above paragraph; depending on experience with the testing and the ability to access a sub-assembly installed in a system.

<u>Testing an installed sub-assembly:</u>
If testing can be done on an installed sub-assembly, it is likely the most realistic test possible. However, it only works if test points are still accessible after assembly. In a product like a desk top computer testing a video board or sound card for ESD may still be possible when the side covers of the main unit are removed. One must also make sure it's only the sub-assembly being tested and failures aren't due to radiated effects on other assemblies in the main unit. Some manufacturers use a "golden" unit, which has well understood ESD immunity characteristics so that a failure or upset in the sub-assembly being tested can be distinguished from any other possible failure or upset.

<u>Testing using extenders or cables:</u>
If test areas are not accessible when the sub-assembly to be tested is installed, other methods must be used to both power and exercise the sub-assembly. Extender cards and/or cables can be used to bring the sub-assembly outside the main unit and allow access to test points, but this introduces potential problem that the tester needs to be aware of:

-Cables and extenders add inductance in all lines to and from the sub-assembly, which for purposes of ESD add significant impedances between the sub-assembly and its mainframe.

-The additional impedance reduces any ESD currents that may flow from the affected sub-assembly into the mainframe and increases the susceptibility of the sub-assembly to the radiated effects from the ESD test.

<u>Testing in a jig:</u>
Problems similar to those found when testing with cables or extenders also exist when testing with a jig. Once the sub-assembly is removed from the main housing, the effects of an ESD event can be modified considerably for the reasons noted above.

<u>Summary:</u>
Several methods are commonly used for testing sub-assemblies for the effects of ESD. Although from the manufacturers' point of view handling is a big issue, IEC is **only** concerned with the effects of ESD during operation. Since testing sub-assemblies to IEC standards requires the unit to be operational, it is often necessary to test with unit covers removed which reduces shielding, or with the sub-assembly at the end of a cable or in a jig. In this case exposure to radiated fields and added inductances may significantly alter the ESD susceptibility characteristics of a product.

2.1   System Level Test Methods Based on IEC 61000-4-2

2.1.1   ISO 10605:2008 Road Vehicles [8]

The recent automotive ESD standard, ISO 10605 'Road Vehicles - Test methods for electrical disturbances from electrostatic discharge' [8] includes several test methods detailed in the IEC standard 61000-4-2:2008 with direct and indirect test discharges. It also includes the similar functional performance status (4 classes comparable to classes a-c of [1], while c is split into 2 distinct classes, neither of which allow permanent damage) after tests as well as the calibration test methods. However, several differences between both ESD standards remain. One particular application is the discharge on a coupling structure that provokes coupling to the cable harness. This test is intended to simulate the effect of an indirect ESD to a cable inside cable harness in a car and is detailed in Annex F of [8]

I. Differences Compared to IEC 61000-4-2

The automotive ESD standard ISO 10605 presents test methods for the electronic modules integrated in the vehicle. ESD tests are performed in two conditions: un-powered condition and powered condition, in which the battery is used.

As a consequence, this automotive ESD standard specifies connecting the ESD generator ground to the coupling plane, which acts as the battery ground or the chassis of vehicle. The IEC standard 61000-4-2 in contrast specifies connecting the ESD generator to the reference ground plane. The coupling plane is connected to the ground plane with 2 x 470 kΩ resistors (used to prevent charge built-up for multiple discharges). This is illustrated in Figures 6 and 7.

The automotive ESD standard uses both contact and air discharge modes, while the contact discharge mode is the preferred test method in the IEC standard 61000-4-2.

The ISO 10605:2008 does not specify an upper level of stress voltage. However, the ESD generator characteristics shall be in the range from 2 kV up to 15 kV for contact discharge mode and from 2 kV up to 25 kV for air discharge mode. This can confuse vehicle manufacturers, which sometimes require a 15 kV contact and/or 25 kV air discharge. The IEC standard 61000-4-2 details the preferred range of voltage levels from 2 kV up to 8 kV for contact discharge mode and up to 15 kV for air discharge mode.

A significant difference is that the automotive ESD standard has multiple RC discharge networks, whereas the IEC standard uses a single RC discharge network. The specified capacitance network for [8] depends on the location of the electronic modules in the vehicle. Obviously severity of testing depends on the RC network. Table 2 summarizes the test parameters detailed in both standards.

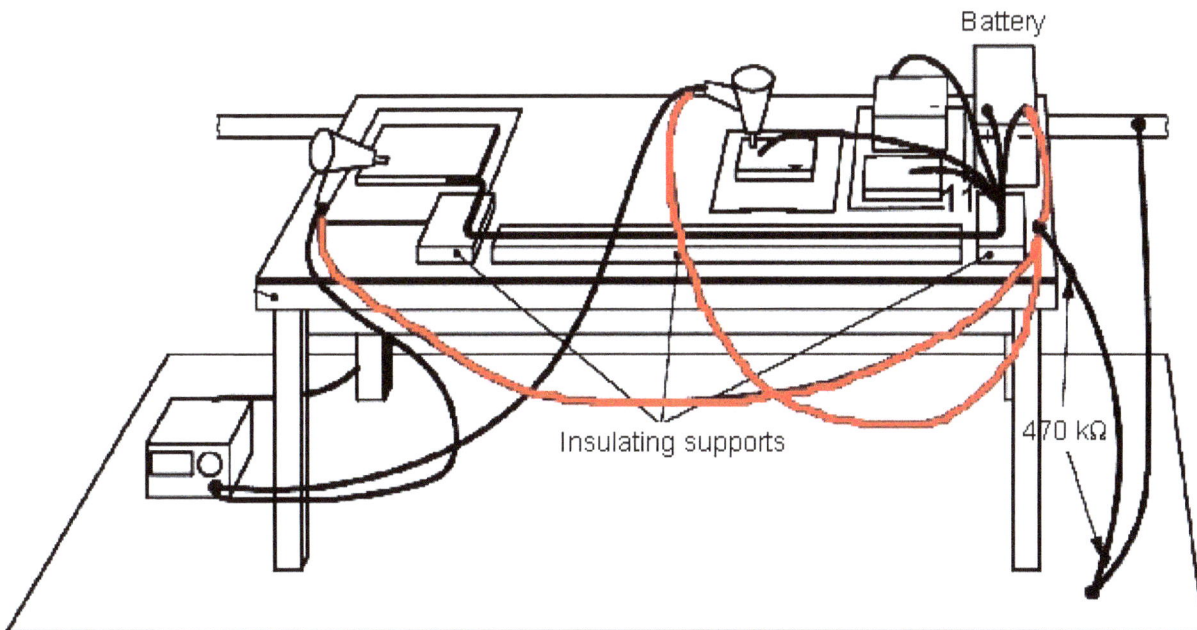

Figure 6: ESD test bench for powered condition from ISO 10605. The ground connections of the ESD gun are highlighted in red.

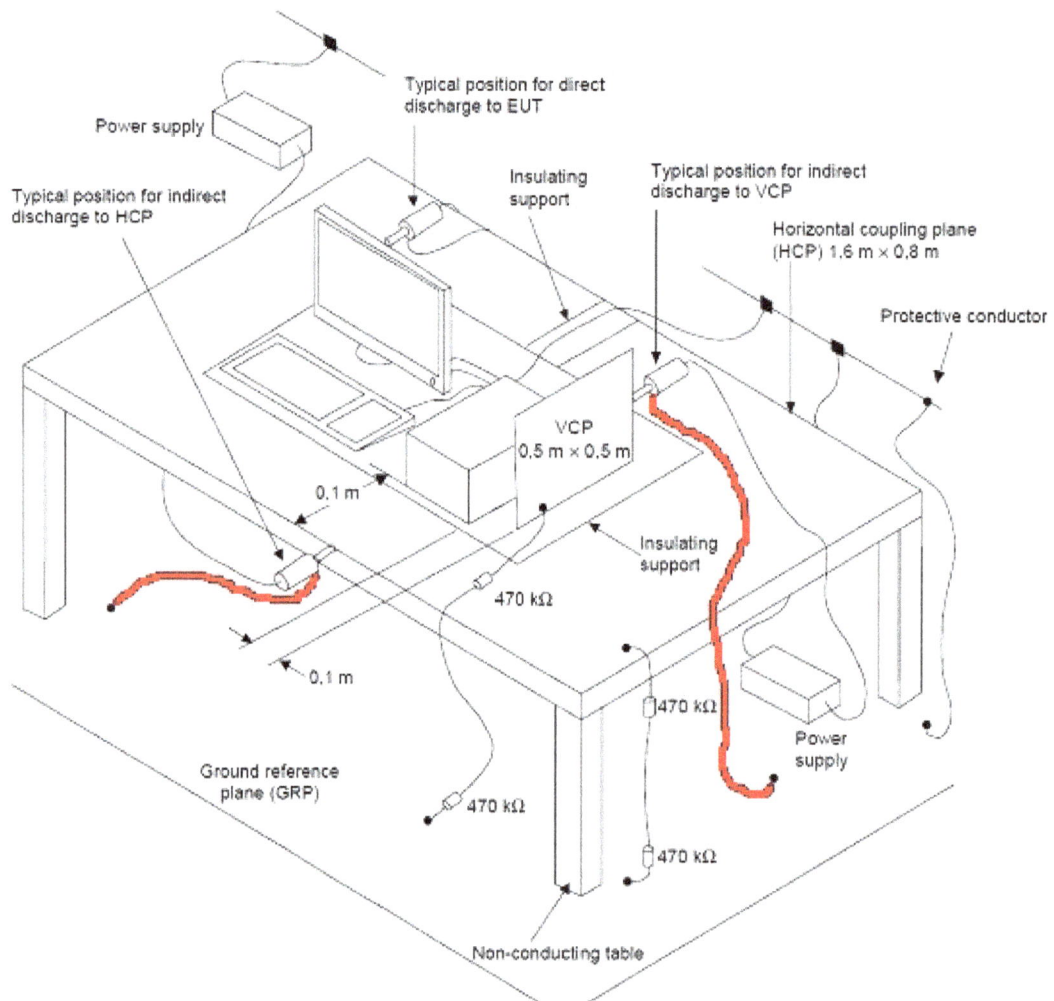

Figure 7: ESD test bench for powered condition from IEC 61000-4-2. The ground connections of the ESD gun are highlighted in red.

Table 2: Test parameters of ISO 10605 [8] and IEC 61000-4-2 [1]

Standards	ISO 10605		IEC 61000-4-2	
Parameter	Contact	Air	Contact	Air
Output Voltage	2-15 kV	2-25 kV	2-8 kV	2-15 kV
Interval Time	Minimum 1 s		Minimum 1 s	
Polarity at each stress voltage level	Positive and negative		Positive and negative	
Network Capacitance	150 pF/330 pF		150 pF	
Network Resistance	330 Ω/2000 Ω		330 Ω	
Number of Discharge pulses	Minimum 3		Minimum 10	
ESD Generator Ground reference	Battery ground		Earth	
Test conditions	Unpowered/Powered with battery		Powered	

II. Consequences

Using multiple RC networks with both un-powered and powered conditions implies a large quantity of samples for testing and it dramatically increases the test time. It has a significant impact on designers/producers for the costs of testing the electronic modules.

The severity of test depends on the RC module network used to create the direct discharge from any point to the ground of the ESD generator. Both peak current and total energy of the discharge can vary considerably compared to the standard IEC waveform. Since the network mainly affects the second peak of the pulse it is expected that test results for different networks relate to each other for hard failures, because of the Wunsch-Bell relation. For the system-upset type of failures this relation is less clear. The standard leaves room for interpretation as to which network needs to be used in which situation.

The automotive ESD standard has a low impact on the test equipment suppliers. They are able to propose an ESD generator with the entire RC discharge network.

2.1.2 DO-160 for Avionics [9]

DO-160, Environmental Conditions and Test Procedures for Airborne Equipment, is the basis for virtually all environmental testing done on non-military avionics equipment. For military avionics, a large part of DO-160 is being incorporated into a forthcoming revision of Mil STD. 461. DO-160 contains test procedures for a number of environmental conditions and Electrostatic Discharge (ESD).

The purpose of the ESD test is to determine the ability of avionics to withstand an air discharge electrostatic event. Because of the very low humidity experienced in high altitude aircraft, air discharge tests are done to 15 kV to any surface of a box that can be accessed by a person. At each location, 10 pulses of each polarity are done and the results recorded.

Although IEC 61000-4-2 includes testing to 15 kV air-discharge, most products don't need to be tested beyond 8 kV for compliance purposes. For avionics, however, 15 kV is the ONLY test level. The discharge network is the same as that used for IEC testing and the specified air discharge tip is identical to that described in IEC 61000-4-2, so in reality the differences are the test voltage (15 kV air discharge only) and the evaluation of the test results.

The compliance requirements appear vague on the surface:
"Following application of the pulses, DETERMINE COMPLIANCE WITH APPLICABLE EQUIPMENT PERFORMANCE STANDARDS, unless specified otherwise."

But the result is actually very similar to IEC failure criteria because the result of a 15 kV test might be acceptable for one box, but not for another. For example, an upset or re-set of critical flight controls during an ESD event is not acceptable; however, the need to re-set the entertainment system due to a 15 kV ESD event is perfectly acceptable. Hard failure of any box at 15 kV is unacceptable; as is any failure that poses a safety hazard. What is allowed is determined by the nature of the box being tested under DO-160, but specified by a product standard in the IEC world.

It should be noted that aircraft and equipment intended for use in aircraft are specifically excluded from the EMC Directive [7].

2.1.3 ISO/IEC FCD 10373-6 for PICC (Proximity IC Cards)

This standard [5] is an example of an interpretation of the generic standard [1] for a specific group of products. The standard is about credit card-like products that contain an IC, which may or may not have external contacts. Examples are banking-cards containing a chip with external contacts or contactless identification cards. The method prescribes the use of a normal ESD simulator as specified in [1].

The method prescribes that a standard sized card is divided in 4 by 5 equal test zones. Other sized cards (e.g. e-passports) are to be divided with a 1 cm x 1 cm grid. Direct air discharges are to be applied successively to each test zone, while the card is positioned on an insulating support on a horizontal coupling plane. If the card includes contacts, the contacts should face up and the zone which includes contacts should not be exposed to discharges. After the test the PICC should operate as intended. The standard does not mention required stress levels.

2.2 Device Level Tests Based on IEC 61000-4-2

2.2.1 Rationale

System manufacturers use the IEC 61000-4-2 discharge waveform to determine the failure level of pins on connector/cable ports. Although the IEC 61000-4-2 is intended only for systems, system manufacturers want assurance that the devices they implement will indeed pass this specification once in the completed system. Consequently, many of them have begun requesting IEC 61000-4-2 test results from devices they design into their systems.

I. Problems Associated with Applying the Stress

Requests for system level tests on devices are typically made for circuitry that is directly connected to external ports or connections. Thus, for devices or components that will be tested using the IEC 61000-4-2, only those pins coming to the exterior of the system are normally tested. These tests are typically performed using contact discharge although air discharge is sometimes used, usually at customer request. Air discharge is not recommended.

Air discharge test results usually add no additional information about the performance of the device in the final system. First, the air discharge test is not as reproducible as the contact discharge test. Second, air discharge testing of the completed system often highlights issues related to the overall shielding or grounding of the electronic system. Since these issues must be addressed at the system level, any air discharge results for individual devices are, in general, not applicable to the final system. Similar arguments also hold for contact mode discharges applied to parts that have a direct conduction path to ground.

Contact mode eliminates some of the problems with air discharge. An ESD pulse is a complex phenomenon that is dependent on environmental parameters. The contact method reduces the number of parameters (mainly the speed and angle of approach, air quality and geometry), ensuring more stable rise-time and peak current. Moreover, with the contact method, the ESD pulse can be reliably delivered to the aimed pin without random sparks to neighboring pins.

II. Problems Associated with Testing for Soft Failures

Product standards that specify ESD testing to IEC 61000-4-2 require that *"loss of function or degradation of performance"* be determined. When testing a device alone it is difficult to determine the operational state of a device, since often the complete system is needed to bring the device to a normal operational state. However, if testing a powered device one can measure voltage/current and determine if something has changed. This is useful as an indication that damage may have occurred, but *"degradation of performance"* may have occurred at a much lower stress level. Therefore, one must be careful ***not*** to assume that a device that passes a stress test, powered or not, will operate properly when tested in a system where *"loss of function and degradation of performance"* is the criteria for pass or fail.

2.2.2 Existing Device Level Test Methods

I. Zwickau

"Zwickau" ESD tests have been developed by the University Of Applied Sciences Of Zwickau in Germany in collaboration with an industrial consortium. The automotive ESD test for bus interfaces, sometimes referred to as the "Zwickau ESD Tests" are specific EMC/ESD tests applied to automotive applications, more specifically on transceivers such as LIN, CAN or Flexray.

A. Applications (LIN, CAN)

LIN and CAN are both communication systems used for vehicles. The LIN (Local Interconnect Network) is a single-wire serial communications system whereas the CAN (Controller Area Network) works in a differential mode at a higher speed. German car makers have described the OEM requirements in [10]. ESD requirements are defined in the EMC parts of the LIN / CAN conformance tests. Two test configurations exist for the CAN: only transceiver and transceiver with CM choke. For the LIN, three test configurations are described: test with transceiver only (no external devices), test with a bus capacitor (220 pF) and test with bus capacitor and indirect ESD coupling (derived for transceiver level from the ECU-test in [8], Annex F). For all cases the failure criterion is a physical (hard) failure (class D). More details are described in [11].

The Zwickau test set-up is similar to the HMM test set-up, which is detailed in Section III, Figure 8. Typical test parameters are as follows: The discharge level is from 1 kV to V_{ESD_DAMAGE}. The discharge voltage step is 1 kV until 15 kV is reached, then the discharge levels are 20 kV, 25 kV and 30 kV. Three positive polarity discharges are applied with 5sec delay, and before each stress a bleed-off resistor is used to discharge the tested Pad. The same sequence is used for the negative zaps. Tests are performed unpowered with a required minimum of three samples. The minimum accepted level is -/+6 kV.

B. Limitations

This test method can be used when results at the board level can replicate the results in the application. It is mainly driven by the European automotive industry. It is limited to transceivers such as LIN, CAN or Flexray. There is no shared document describing the "Zwickau" tests outside the LIN / CAN consortium. It could be replaced by the HMM tests described in Section III, as it is based on similar test procedures. Tests are performed in unpowered conditions; hence it cannot guarantee safe behavior of the system if an ESD zap occurs during normal operations.

II. IEC TS 62228 [6]

This Technical Specification is in fact a formalization of the Zwickau method discussed above. Formally it is restricted to CAN transceivers only. The procedure prescribes unpowered contact discharge, R and C according to [1], while the PCB is connected directly to the ground of the ESD simulator.

The stress levels are prescribed as: discharge voltage levels ranging from 1 kV to the fail level, with a maximum of 30 kV. 1 kV steps are required up to 15 kV, with 5 kV steps at higher levels. A required pass level is not mentioned, but the fail level must be reported. Failure is determined by measurements after the stress that indicates (physical) damage.

III. Human Metal Model (HMM)

Introduction

Workgroup 5.6 of the ESD Association has developed a standard practice for applying IEC 61000-4-2 stress [1] to ICs. This test is called Human Metal Model [11]. The HMM method is based on the Zwickau LIN test [12]. The test methodology is called the Human Metal Model to distinguish it from the well known IC-level ESD test according to the Human Body Model. Note that, unfortunately, some standards (e.g. [8]), refer to an IC-level gun test as an HBM test, although the gun test and IC-level HBM test have nothing in common.

HMM testing is a field under development, in which no single standard is accepted universally. When testing ICs by means of HMM, the test results need to be interpreted with considerable caution, since the IEC 61000-4-2 test was not designed for IC level testing. Test results at the IC level typically do not correlate with system level tests. Indeed, improvement of the IC level performance may even decrease total system performance [13].

When testing a single IC on a special test board, the grounding will typically be different from the grounding in the real application. Many systems have a worst case mode of operation if the IC has a fairly low-ohmic (e.g. capacitive) path to ground. The HMM test differs from the IEC 61000-4-2 test in that the coupling planes associated with the IEC 61000-4-2 have been eliminated. Thus none of the capacitive coupling between the ground plane and the IC under test is present. Redefining the HMM test configuration to be a test where the test PCB is hard grounded to the ground plane of the test setup and to the ESD gun ground to eliminate these capacitances is warranted since the configuration of the final system cannot be predicted. If the DUT is hard grounded to the external ground, the complete IEC 61000-4-2 setup (with ground plane, table and horizontal coupling plane) is not necessary. Instead a simplified configuration (see Figure 8) can be used.

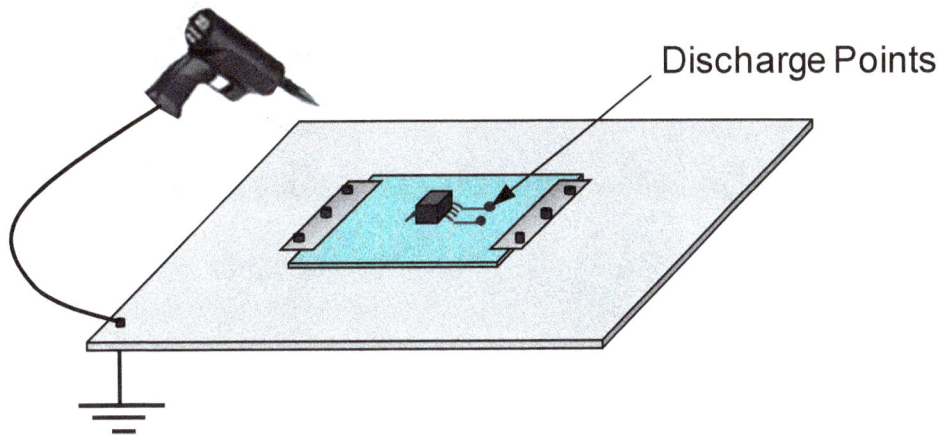

Figure 8: Simplified HMM test configuration

Like the IEC 61000-4-2 discharge, an HMM discharge has two peaks (see Figure 5). The first one is very fast and the second one is slower. The second peak stems from the discharge of the main RC network of the gun. The first peak is generated from the discharge of the tip-to-ground capacitance. Note that (parasitic) tip-to-ground capacitance of the gun tip depends on the orientation of the gun, the thickness of the test board (i.e. distance of the gun tip to the ground plane) etc. Therefore, the rise time and the peak of the initial current spike depend on poorly controlled parameters. The second peak, in contrast, hardly depends on parasitic components and is relatively well defined.

The parasitic tip-to-DUT capacitance produces ringing. The IEC 61000-4-2 waveform is defined to be measured with a 2 GHz oscilloscope or better to give a more accurate view of the ringing in the waveform.

The HMM Standard Practice allows three different configurations for testing. First, the gun tip can be touched to the DUT that is mounted on the test board as shown in Figure 8. The concern is that the electromagnetic field radiating from the gun tip may influence the test results. The second setup is where a hole is provided such that the gun tip can touch the discharge point through a hole in the test plate as shown in Figure 9. The test plate is then placed vertically to allow the gun to touch the discharge point from behind the ground plate. Thus the test plate serves as an EM shield for the DUT. The third setup involves replacing the gun with a 50 Ω pulser. The pulser is connected to the test point on the DUT through a 50 Ω cable. This eliminates the need for a test plate and removes any variability due to excess EM radiated fields from the gun.

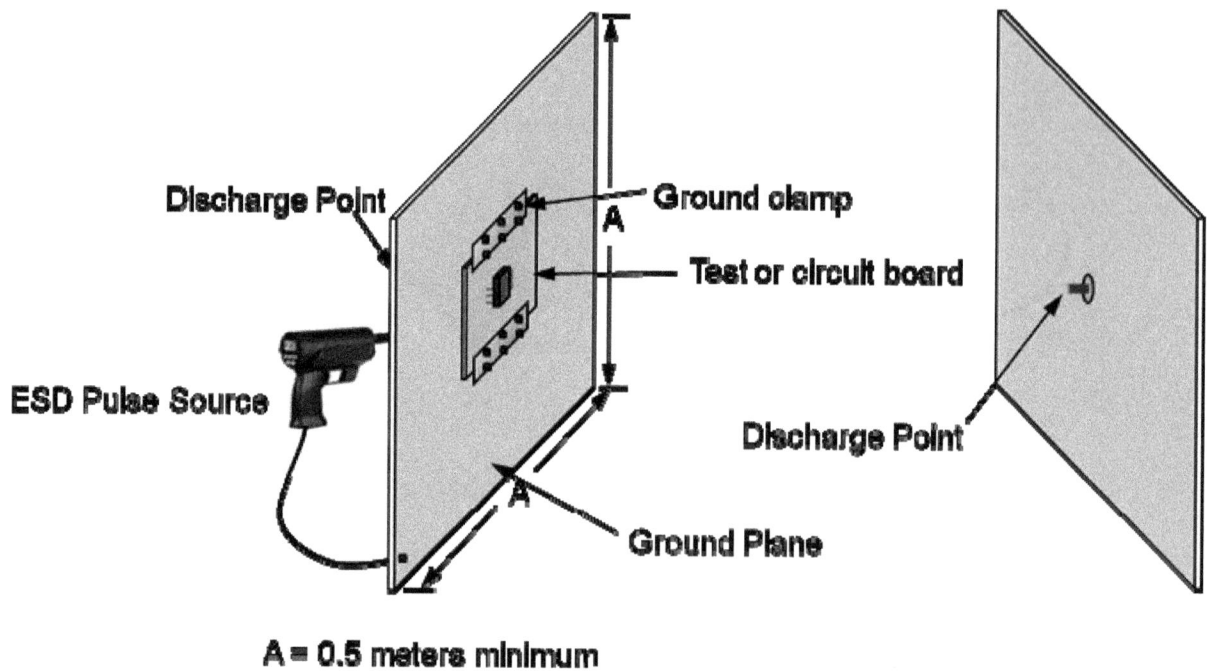

Discharge Point

Ground clamp

A

Test or circuit board

ESD Pulse Source

Discharge Point

Ground Plane

A

A = 0.5 meters minimum

Figure 9: Vertical test plate HMM test configuration

HMM Test Parameters and its Development

As more external components are connected to electronics systems for external data or control, the connecting cables can become ESD charged. When a charged cable is connected to an external data port, the ESD into the external pins of the IC can result in damage. A test was needed to measure the system immunity to these cable discharge events. Since the IEC 61000-4-2 simulators are widely used for system level testing of the effects of ESD, it is a natural extension to use these simulators for testing individual pins on connectors. Although IEC 61000-4-2 specifically excludes the testing of connector pins for compliance purposes, many system manufacturers believe it is necessary to inject ESD events directly into connector pins to determine their level of immunity.

The HMM test injects the test pulse directly into connector pins to determine at what test pulse amplitude damage occurs to the external pins of the IC. The test pulse amplitude which causes external pin damage is an important parameter in HMM device testing. HMM calls for both powered and unpowered testing.

A note on nomenclature: There has been some confusion in the use of the term HBM in the electronics industry. For integrated circuit ESD testing HBM has been used to describe an ESD stress from a human finger. This is the familiar JEDEC/ESDA HBM standard now described by ANSI/ESDA/JEDEC JS-001-2010. HBM has also been used to describe an ESD event to a system in which a person holding a metal tool touches an electrical system, the IEC 61000-4-2 system level ESD test. ESDA Workgroup 5.6 adopted the name Human Metal Model to describe the use of the IEC 61000-4-2 waveform when stressing ICs to distinguish it from the traditional device level HBM test. The IEC 61000-4-2 system level discharge simulator became the HMM tester as well. While these two tests use the same current waveform, they are applied to different samples, which can cause different electrical sensitivities.

4. Challenges in Achieving Repeatable Failure Levels

The IEC 61000-4-2 standard defines the current waveform when it is discharged into a 2 Ω target. However, the fast-changing gun current causes electric and magnetic fields around the injection point which are not defined. These fields are strongly dependent on the design of the gun and the gun tip. These and other factors cause considerable challenges to obtaining a reproducible HMM failure level.

Challenges between Gun Waveform Variations

An extensive characterization study [14] of nine brands of commercially available system level discharge generators (guns) established that all guns show strong high-frequency components in the radiated electric and magnetic fields. These high EM fields stem from unshielded currents in the generators. Depending on the sensitivity of the device-under-test for these fields, these unintended disturbances may severely impact the observed failure level of the DUT. For 'slow' CMOS circuits which react to frequencies below 1 GHz, a factor of 2 variability in observed ESD failure level may still occur depending on the type of gun used. For fast CMOS circuits which are able to react to 50 ps pulses, the observed variability may even amount to a factor of 5.

Challenges between 50 ohm HMM Pulse Sources and Guns

It has been found that IEC 61000-4-2 generators (guns) generate waveforms with generally poor repeatability and emit electromagnetic radiation around the tip. Alternatively, IEC pulses can be applied through a 50 Ω transmission line system which is adapted from a VFTLP pulser [15]. Since all components in the current delivery path can be made to maintain the 50 Ω impedance, the pulse quality is significantly improved. Furthermore, the increased distance between the DUT and the relays which generate the EM radiation virtually eliminates the EM received at the DUT. The transmission line can be operated at 50 Ω or 330 Ω. The 50 Ω transmission line source is identified at the equivalent IEC voltage to produce the same current waveform threat as the 330 Ω hand held ESD gun. When using the transmission line with 330 Ω impedance, there will be large reflections that need be de-convoluted from the signal.

The relays used in IEC guns inevitably generate small displacement currents prior to the main discharge, due to the increase in capacitance when the relay contacts approach. This very small current, which is not specified in the IEC 61000-4-2 spec, can cause considerable charging of an isolated device-under-test in a regular IEC test. Such a pre-pulse may disrupt the protection of the device against system level discharges, in particular if the protection is slew rate triggered (which is well-known from HBM testing of ICs).

In other cases the pre-pulse may lead to a delay in breakdown of a high-voltage p-n junction, which in turn may lead to a reduction of the safe-operating-area. It has been shown that the pre-pulse voltage varies strongly between different IEC generators, which are another source of irreproducible IEC test results.

The low source resistance of the 50 ohm system eliminates most of the pre-pulse threat in device testing. Using a 100 Ω transmission line (50 Ω both on the high voltage pin and ground) it is possible to control the biasing of the DUT independently, which separates the effect of the biasing and results in more reproducible IEC test results.

In summary, using a transmission line to deliver the IEC pulse, potentially improves the reproducibility of the IEC test results. The ESDA HMM Standard Practice allows use of either an IEC 61000-4-2 compliant ESD gun or a 50 Ω source capable of supplying an IEC 61000-4-2 current waveform. However, the equivalence of the two pulse sources during device testing has not yet been verified.

Challenges between Testing Devices on a Test Board and in a System

There is always danger in expecting such tests as the HMM test to provide information about how a particular IC will perform once it is in the completed system. One type of testing where this is particularly problematic is air discharge testing. The IEC 61000-4-2 prescribes both contact discharge and air discharge testing to be done on systems. The HMM is only a contact discharge test. Occasionally customers will request air discharge test results from devices or components. However, the air discharge test is an attempt to find unprotected paths for ESD energy to get inside the completed system. When there are air discharge failures, typical solutions include improvement of the shielding within the system or improving the ground paths of the system. These can only be addressed at the system level, and any air discharge results of a particular component have no relevance once the component is placed on a board in the system.

Challenges from Radiation Conversion to I or V on PC Boards

Electromagnetic radiation directed at a conductor will induce currents into the conductor. This is the same result as was found with an antenna converting RF radiation into electrical signals. Radiation near unshielded leads or PCB traces connected to devices will produce currents and voltages on those conductors. Unspecified amounts of radiation with uncontrolled amplitude and time variations can create similar currents in conductors connected to devices leads. The uncontrolled amount and type of radiation can create unknown effects on devices being HMM tested. An additional concern is that the radiation from the gun reaches the test area before the current test pulse arrives. In the real event the current threat and radiation from the discharge begin at the same time. High speed voltages of a few volts can be high enough to turn the external pins of the IC on before the main current pulse arrives. This unusual turn-on condition, which would not occur in a system because of shielding, can cause it to operate in a manner different from what the ESD protection is built to protect against.

The high speed radiation begins to be emitted as soon as the high speed currents begin to pass through the complex shaped conductors used to prevent their passage. The amount of effect that excessive radiation has on HMM device testing remains to be determined. ESDA Working Group 5.6 is presently working on details of the HMM device test to help identify these effects. Round robin device testing will be made on typical devices which can be subjected to the HMM threat. Some answers to these questions are expected in 2010.

2.3   System Level ESD Tests under Development

2.3.1   Cable Discharge Event (CDE)

Low voltage ESD generated during hot plugging can produce Cable Discharge Events which are randomly spaced electrical pulses leading to data, or soft failures. The electrical signal produced by connecting charged cables to a system connector can also damage external pins of ICs, so it is included in this section. Experiments have determined that the amount of voltage which can build up on cables during flexing can be hundreds of volts. Physically long cables however can produce long discharge pulses. This alone can be sufficient energy to damage external pins of ICs which are identified as hard failures.

The discharge which forms between the system and cable metal connector pins forms a low resistance spark. ESD protection clamps typically have low resistance I-V characteristics. When the charged cable impedance is greater than the spark and protection clamp resistance, the discharge pulse will circulate back and forth between the system and an open ended cable. The "ring down" or damped current waveform will dissipate most of its energy charge in the silicon clamp and spark resistances. The mismatch between the source and load for CDEs increases the ringing and adds to the possibility of IC damage. There is a distinct difference between the effects in unshielded vs. shielded cables. In the latter case the location of the discharge – shield or conductor- plays an essential role [16].

The ESDA is preparing a standardized test method to determine failure levels from CDE threats at different amplitudes. Experiments by the members of this working group have identified many different discharge waveforms. Some example waveforms are shown in Figure 10. The test pulse rise time must be identified by a high speed measurement chain. Because the discharge occurs between metal electrodes in cable and system connector pins, the test pulse speed can be very fast at the typical charging voltages found in this threat. Determining the test simulation waveforms will require more experiments with sufficient bandwidth sensors and oscilloscopes to capture the real world event. These highly variable CDE waveforms include multiple waveforms. This standard has been in discussion for over one year, with concerns focusing on providing a test specification which can be used in commercial testers. Because of limitations on participant's time to develop and define a reasonable and effective test method, this information may not be available for many months.

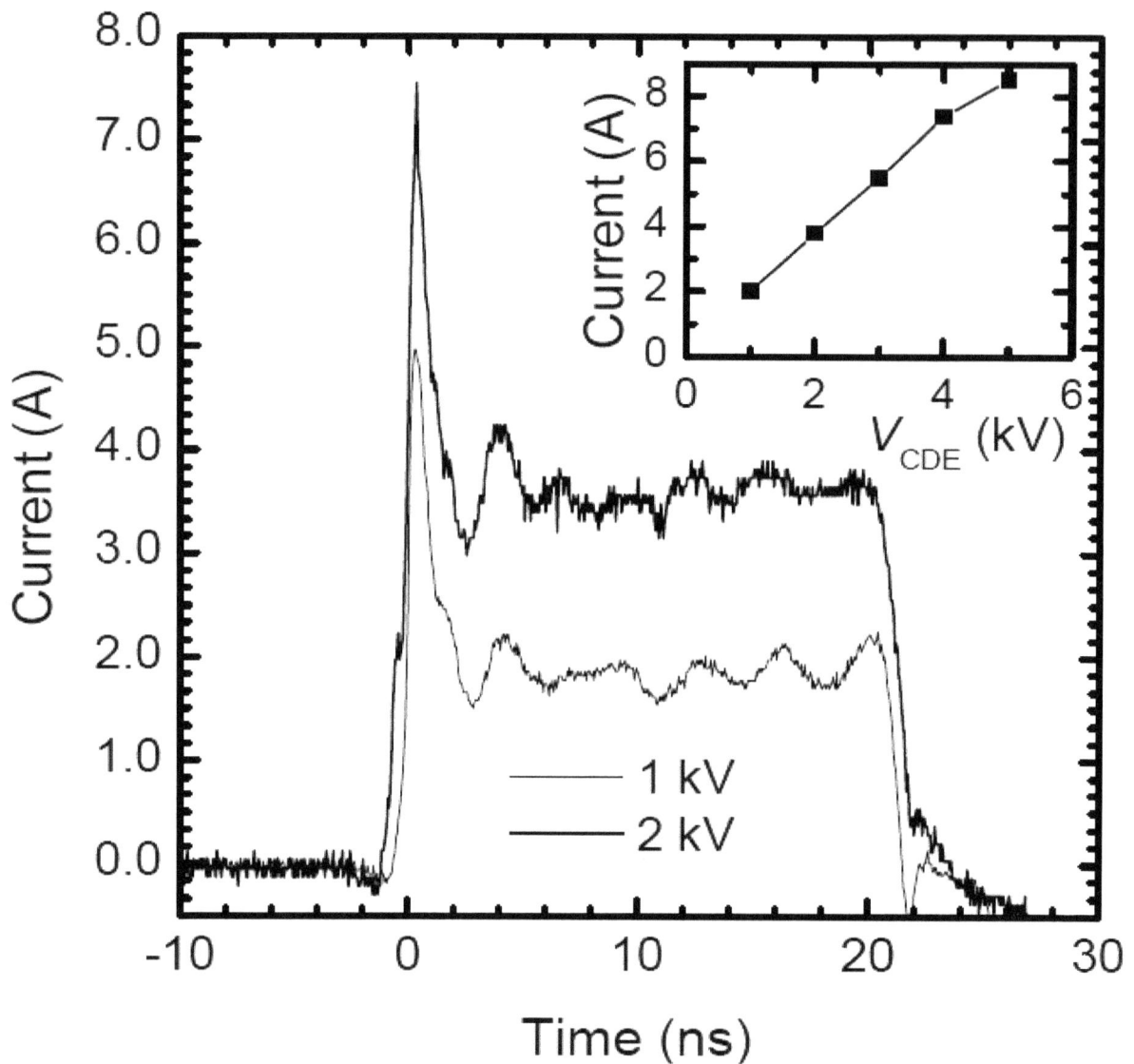

Figure 10: Measured CDE currents for a USB cable with V_{CDE} = 1 kV and 2 kV, respectively. Inset: Current amplitude of the rectangular part of the waveform as function of the pre-charge voltage V_{CDE}.

2.4 Discussion and Conclusions

2.4.1 Failure Mechanisms

The four levels of system response during IEC 61000-4-2 ESD testing, with only one of the levels related to physical damage, indicates that upset rather than hard failure may be the likely outcome. This is also related to the original specification of air discharge only, where the most likely effect would be a disturbance in the ground system; either via a direct arc or through electromagnetic coupling. Note that air discharges were supposed to be done to non-conducting surfaces or metallic (connector) shields. Such a disturbance of the ground potential can easily result in unwanted behavior of the system. However since the disturbance is already at the ground connection not much energy will be able to flow into internal devices, therefore actual damage is less likely.

When contact discharges are applied to any other location than ground-related points, the energy will be conducted to the ground via a, best-case intended, internal path. This path must be robust enough to sink the current. If this is not the case, fatal damage is likely. If the path is robust enough, the risk of a system upset is still present if the voltage excursions put the system in an unwanted state. This will depend heavily on the complete system design.

In all situations that deviate from the generic description in the IEC 61000-4-2, the dominant failure mechanism depends very much on the test method and type of application. Several example cases are discussed in Chapter 3.

2.4.2 Conclusions

The goal of the IEC 61000-4-2 is to assess, for final applications, the immunity to electrostatic discharges of locations which people have access to in normal use. The second goal is to serve as a basis to derive standards for situations where the IEC 61000-4-2 is not applicable. Several examples have been discussed, the most deviating case being the HMM, where system level pulses are applied to individual ICs. Whereas originally the focus was on 'system upset' as a failure mechanism, the changes to contact discharge and the use of system level ESD on modules, PCBs and ICs have increased the significance of physical damage as a failure mode. Requirements are typically set by users or committees and are inherently very application dependent

References

[1]   IEC 61000-4-2, 'Electromagnetic compatibility (EMC) – Part 4-2: Testing and measurement techniques – Electrostatic discharge immunity test', Ed. 2.0, 2008.

[2]   M. Honda, 'Evaluation of System EMI immunity using indirect ESD testing, Proceedings EOS/ESD Symposium, pp. 185-189, 1988.

[3]   IEC 61000-6-1, 'Electromagnetic compatibility (EMC) – Part 6-1: Generic standards – Immunity for residential, commercial and light-industrial environments', 2nd edition, 2005.

[4]   IEC 61000-6-2, 'Electromagnetic compatibility (EMC) – Part 6-1: Generic standards – Immunity for industrial environments', 2nd edition, 2005

[5]   ISO/IEC FCD 10373-6, 'Identification cards -- Test methods -- Part 6: Proximity cards', 2001.

[6]   IEC TS 62228, 'Integrated circuits – EMC evaluation of CAN transceivers', 1st Edition, 2007.

[7]   EMC Directive, 2004/108/EC, 2004.

[8]   ISO Standard 10605, "Road vehicles -- Test methods for electrical disturbances from electrostatic discharge", International Organization for Standardization, 2008.

[9]   DO-160 rev. F, Environmental Conditions and Test Procedures for Airborne Equipment, RTCA, Inc., June 2007

[10] Hardware Requirements for LIN, CAN and FlexRay Interfaces in Automotive Applications, Dec. 2009.

[11] ANSI/ESD SP5.6-2009, 'Electrostatic Discharge Sensitivity Testing - Human Metal Model (HMM) - Component Level', 2009.

[12] LIN Conformance Test Specification LIN EMC Test Specification", Version 1.0; August 1, 2004

[13] Industry Council on ESD Target Levels, 'Aspects of system-level ESD', 2008.

[14] K. Wang, D. Pommerenke, R. Chundru, T. van Doren, F. Pio Centola, and J.S. Huang, 'Characterization of Human Metal ESD reference discharge event and correlation of generator parameters to failure levels - part II:

Correlation of generator parameters to failure levels', IEEE Trans. Electromagnetic Compatibility, Vol. 46, 4, 2004.

[15] E. Grund, K. Muhonen, N. Peachy, 'Delivering IEC 61000-4-2 current pulses through transmission lines at 100 and 330 ohm system impedances', Proceedings EOS/ESD Symposium, pp. 132-141, 2009.

[16] T.J. Maloney, 'Primary and Induced Currents from Cable Discharges', Proceedings EMC Symposium, pp. 686-691, 2010.

Chapter 3: Proven System Level Fails

Reinhold Gaertner, Infineon Technologies
Tim Maloney, Intel Corporation
John Kinnear, IBM
Leo G. Henry, ESD/TLP Consulting
Wolfgang Reinprecht, Austria Microsystems

3.0   Introduction

There are many discussions going on in the industry between suppliers and customers about ESD problems which occur in the final system ESD test or in the field at the end-user. Why did the failure happen? Is it due to a HBM-weak device or maybe pre-damage during production and transport of the device or PCB? The root cause for field failures generated during system use (handling or operation) is not easily proven and the confirmation whether it is coming from an ESD or an EOS event is very difficult to establish.

In White Paper 1 [1], regarding HBM, the Industry Council collected extensive data on field failures with a root cause of ESD or EOS and demonstrated little relation to the HBM qualification value of the devices. This evaluation was based on approximately 21 billion devices. A similar study was done for CDM [2]. In both studies the number of field failures has been compared with the number of sold devices and correlated to the HBM/CDM qualification voltage. System level failures are due to various root causes, such as ESD generated noise (to be reproduced by an ESD gun stress), Cable Discharge Events (CDE) during installation of systems or exchange of boards, Charged Board Events (CBE) during installation or exchange of boards, or simply EOS events, that can be generated by spikes on the power supplies, wrong polarity of power and so on. For most of these root causes a test method does not exist, therefore there are no qualification target values.

On the other hand, problems in the field are not only due to damage to devices/systems but also to so-called soft failures like system lockups, where the system is not damaged but its functionality is interrupted temporarily. These problems happen more often than damage but most of them are not reported back to the board and IC manufacturer. Most of these problems are resolved by resetting the device by rebooting or repowering and may often be blamed incorrectly as software bugs.

In the following sections, case studies of actual system problems in the field or during system qualification tests have been collected. The root cause for the failures was evaluated as well as the type of failure. A very interesting point was the question of how system problems are typically solved.

3.1   How to Prove a System Level ESD Fail?

Verifying that a failure coming back from the field is a system level ESD fail can be difficult. Normally when a failure occurs in the field, the end user will send the failing equipment back to the manufacturer for analysis, especially when the failure mechanism occurs often, when the equipment is large or has safety problems or data integrity failures. The system manufacturer then tries to reproduce the failure in the lab using one of the methods described later in this document. The reaction of the equipment under test (EUT) to ESD or ESD-like pulses can be categorized in four classes:

A.   EUT continues to function normally
B.   EUT has an upset condition but recovers automatically
C.   EUT has an upset condition and needs manual interference to recover
D.   EUT is damaged

If case D happens with the same failure mode, the device can be analyzed and the results of the physical failure analysis can be compared with failures found in the field returns. If it is the same, a system level ESD failure can be confirmed.

However, it is not always that straight forward, especially taking into account that soft failures (case B and C) are often difficult to analyze. When the system (or subsystem) comes back from the field, the system manufacturer checks the system and the result is often: "No Trouble Found". These soft failures are also difficult to reproduce by ESD system stress, since the occurence often depends on the complete test setup. In such cases it is difficult to clearly identify a system level ESD event as the root cause for the field problem. This must be kept in mind when drawing the right conclusions from the case studies analyzed later on.

3.2   System Level Fails – Case Studies

The Industry Council has collected 58 system level case studies. Some studies have completed a deeper evaluation where the failure was duplicated with existing or new engineering test methods in the lab. Others have been reported but never resolved as they only occurred once or twice and did not result in physical failures.

Figure 11 shows the type of failure, for example whether physical damage (blue bars) or a soft failure (red bars) were reported from the field. As can be seen in Figure 11, more physical damage was reported than soft failures, although system manufacturers report that usually soft failures are the dominant failure mode. The reason for this could be that soft failures, while they happen more often, are in general not reported back to the manufacturer. Typically these soft failures are resolved during system development by the system design engineer prior to product launch. This is consistent with the results that EMC engineers from OEMs and EMC test houses obtained while doing qualification tests on systems to IEC 61000-4-2. If soft failures occur in the field (at the end user), they are often not reported back to the system manufacturer. This may be another reason why we tend to see more returns for physical damage than soft failures.

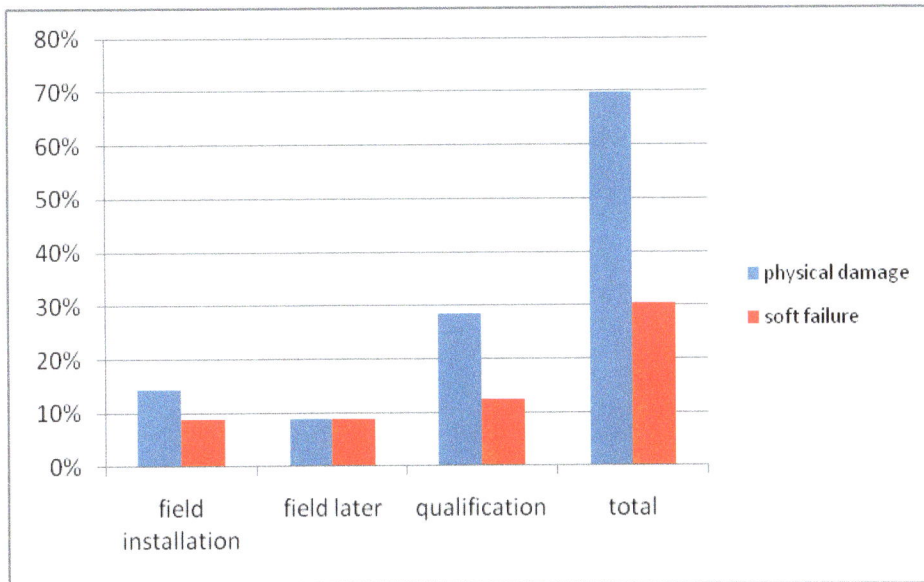

Figure 11: Type of failure with failure location (percentage of cases with given information)

When system qualification tests are performed according to IEC 61000-4-2, discharges are done only to the housing of external connectors as the standard states and not to the product or the pins of the connector. By following the requirements of the standard, hard failures resulting in device damage are very, very rare if they appear at all. However, many system manufacturers not only want to discharge to the housing but to the stress connector pins. Failures during this type of stress are definitely reported back to the IC manufacturer, another reason why we have more hard failures than soft failures in the graph.

Also, the automotive industry does stress the connector pins and therefore has reported more hard failures during qualification than in the non-automotive industry. Nevertheless they experience more soft failure issues but they have difficulty quantifying (and reporting) it. In addition, Figure 11 highlights in which situations failures occurred. Most physical damage was reported during qualification, while in the field, damage occurred mainly during installation. For example, when different parts of a system are connected to each other or when any upgrades are installed. Out in the field, after the system has run a while, the main failure mechanism is a soft failure.

Figure 12 shows the details of the failures; whether the damage or the soft failure had its origin in a CBE, which can happen when two sub-assemblies of a system are mounted together, or in a CDE, which can happen when a system is mounted in the field or when new components are attached. Other root causes can be any Electrical Overstresses (EOS), that can have various causes or true system level ESD events (mainly reproduced by an ESD gun test). Figure 12 shows that soft failures happen mainly during system ESD events with a low chance of occurrence during CDE and CBE and not at all during an EOS event (too much energy). Physical damage is mainly due to EOS and system ESD events even though CBE and CDE may contribute.

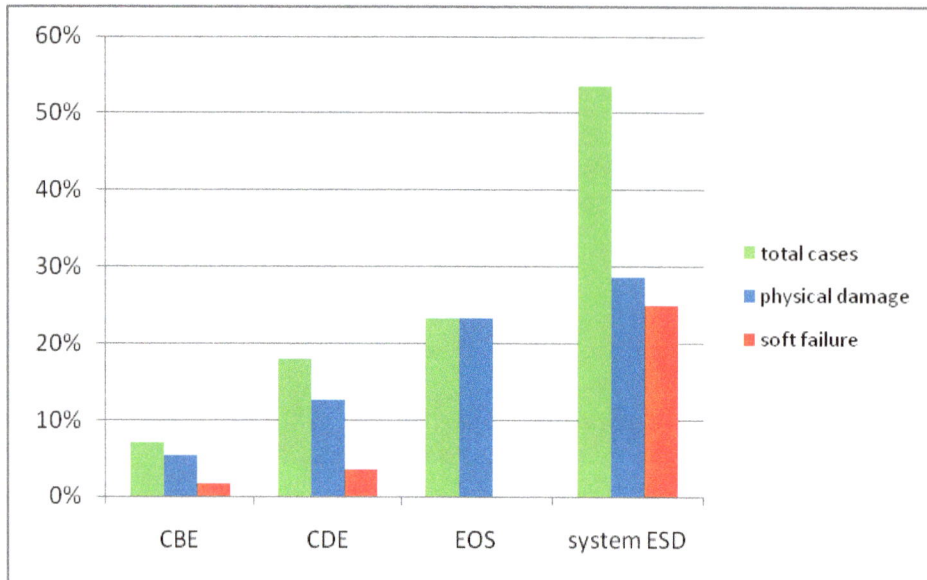

Figure 12: Type of failure – details; physical damage and soft failure included (percentage of cases with given information)

In Figures 13 and 14, the previously shown failures are depicted in relation to the affected pin. Physical damage appears at the external pin or power pin with few exceptions (Figure 14). Only one internal pin was damaged during a CBE event and one internal pin was damaged by an ESD event in the field, where the pin was rather robust with respect to HBM (see below).

For soft failures, more problems are seen on internal pins; but since only two cases have been reported with information about the affected pin, statistics might be too low to draw a conclusion.

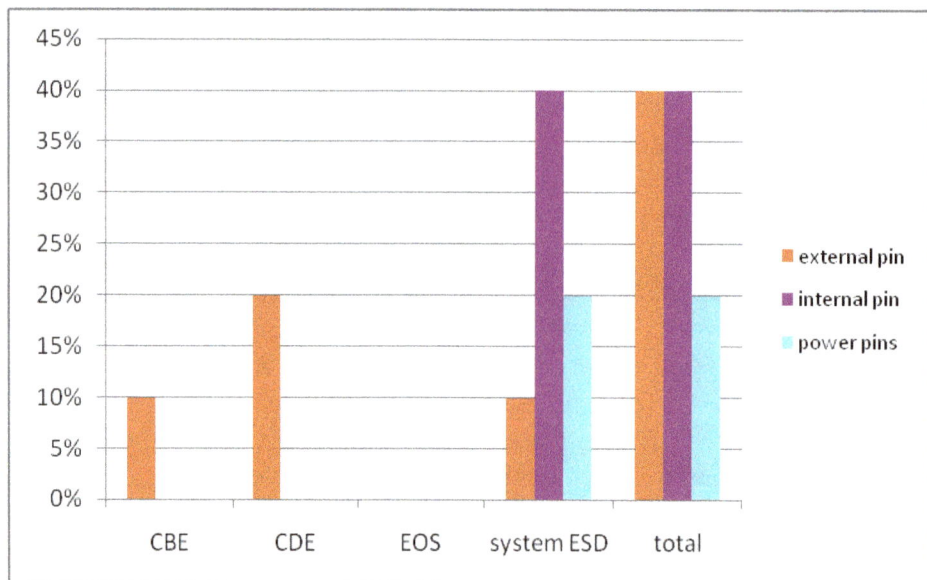

Figure 13: Root cause for soft failures; affected pin included (percentage of cases with given information)

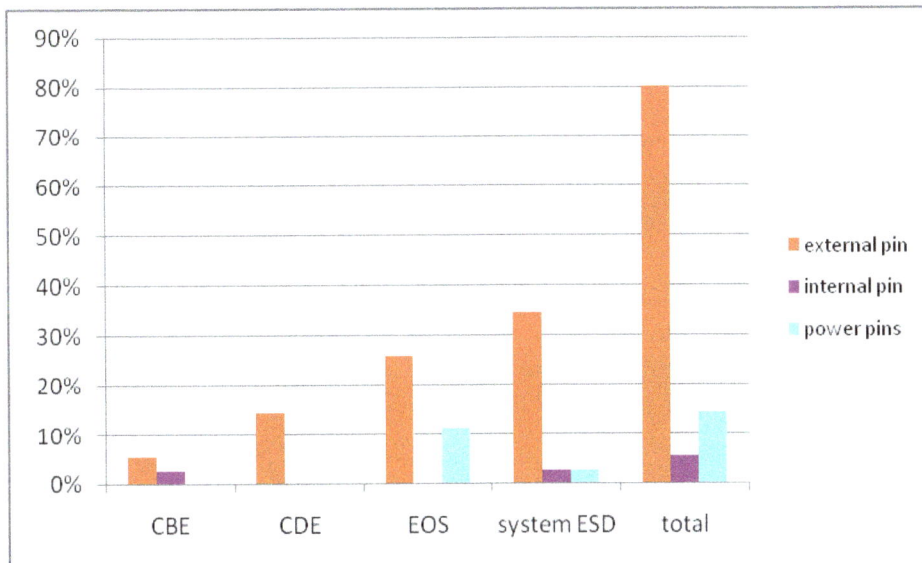

Figure 14: Root cause for physical damage; affected pin included (percentage of cases with given information)

Figure 15 shows how the reported case studies have been solved. Most of the problems have been solved by improving the process in the field, by improving the board layout or the board protection. Only four cases reported that the problem was solved by improving on-chip ESD-protection. All these pins had an HBM robustness of 2 kV before the failures occurred, but since these were pins connected directly to external connectors, this was not enough as one might expect. In one case, improving the HBM robustness on the IC level from 1500 V to 2 kV even reduced the system level robustness from 4 kV to 3 kV. It is also important to note that when failures occur a quick solution is often extremely important, making chip redesign a last resort.

The fourth category of solving the problem, through a software change, was only reported for two case studies, but is a very helpful tool for soft failures. It is especially used in the early test phase of a product.

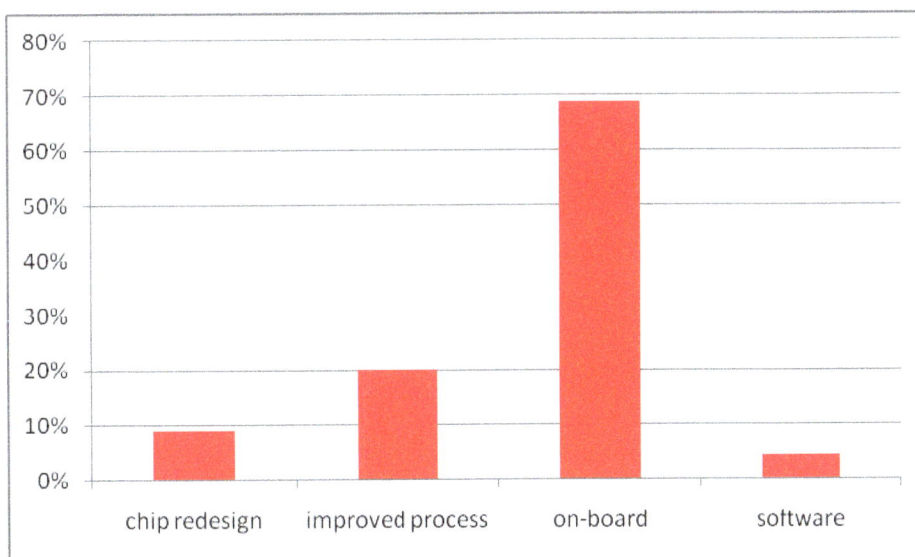

Figure 15: Solution of problem (percentage of cases with given information)

3.3 Detailed Case Study

System Level Damage – Cable

A large server during installation was experiencing failures when the machine was powered on. Parts were replaced and upon receipt back at the factory, the failure was duplicated at the system level. The parts were operational upon leaving the factory.

The installation consisted of several large frames that had to be cabled together to form a system. Investigation indicated that the part that failed and the device on the part that failed had a direct connection to the cable pin. If fact, the cable pin was in the top corner of a multi-connection cable. At this point, ESD was not suspected as a problem as the HBM sensitivity was in excess of 5,000 V.

Further failure investigation showed that all of the failures occurred during the cold time of the year. The data centers where these machines were installed had a very low relative humidity level at the time of the installation. While duplicating the installation, it was found that the outside of the cable could retain a significant charge and thus induce a large charge on the cables inside.

To verify that the cable could cause the problem, a direct discharge was applied to the connector pin at the frame. The failure was duplicated in the experiment. The conclusion was that the cable could charge to a level such that if the corner pin made first contact, there would be enough energy to cause the part to fail.

The fix for this problem was quite simple. These cables had a ground shield that would typically be connected after the cable was connected to the frames. The sequence was changed so the shield was connected first and then the cable was plugged into the system. In this case, the discharge from the cable occurred on the shield and the signal pins did not experience a damaging discharge.

After this change of sequence, there were no additional installation failures due to ESD discharges from the cables.

System Level Upset – Cables

Large mainframe servers have the ability to be upgraded in customer locations. In fact, some of the upgrades doubled the size and capacity of the original machine. One of the features of this upgrade was the ability to complete the mechanical upgrade and verify the upgrade while the customer continued running applications on the installed machine. Only when the machine upgrade was confirmed would the customer have to interrupt his machine to merge the two systems together into one large system.

This was accomplished by a process known as logical fencing in this machine. While the upgrade was being installed and tested, the existing machine would ignore all signals from the upgrade, effectively "fencing" them off. As long as those signals met the expectations, the machines upgrades occurred without a problem.

It was found in a few systems that during the upgrade, the machine that was running had an upset condition that caused it to go off line. The machine had to be restarted and lost the jobs that were currently running on the system. In addition, it took 45 minutes to 1 hour for everything to go back on-line. There was no damage and no parts had to be replaced; just a restart of the machine.

It was found that the errors occurred during a particular operation where cables were connected from one side of the machine to the other. These were relatively short cables that routed from one system board to another. The cables were Teflon™ coated, a charge was measured on them but the energy level was quite low.

Measuring the discharge showed that the energy level would not be sufficient to damage the parts as the part had a sensitivity level in excess of 5,000 V Human Body Model. However, there was enough charge to send a signal on the line when connected. It was thought that the logical fencing would be enough for the signal to be ignored but the noise that was generated got past the logical fencing and caused the system to be upset.

The solution was quite easy. The Teflon™ coating on the cables was changed to be in the static dissipative range. Since the installation team would be wearing wrist straps, the exterior of the cables would discharge by handling. After this change was made in the cables, upgrades were made without any machine upset.

3.4   Summary

- Failures that result in physical damage are reported more often to IC suppliers than soft failures. Typically, soft failures are resolved after a reset (reboot or repower) and are considered an annoyance and are therefore not analyzed in detail.
- Physical damage failures were found during qualification tests or initial installation while soft fails were found during qualification and post installation.
- Most of the reported fails could be reproduced by an ESD gun test, but CDE and CBE are also systematic problems.
- Most of the problems have been solved on the system side by improving board layout/protection, system level packaging, software or the handling process.
- On-chip improvement was only reported for system pins and typically these failures only occurred during qualification testing or initial system installation.
- Pins affected by system fails are mainly external pins.
- Fails of internal pins are mainly soft failures. Only two occurrences of physical damage of internal pins are reported in the survey. One of them involved a pin with more than 2 kV HBM robustness.

3.5   Conclusion

Different types of system failures due to different types of root causes can be found during ESD system testing as well as in the field. Due to the variety of root causes, there are also a variety of solutions or protection strategies. These problems can normally not be resolved by increasing the device level ESD robustness but rather by improving the handling process, changing the housing of the system (system level package) or the on-board layout. Other ways to solve the problem are to improve the signal to noise ratios, the ability to reject random signals in the device, design recovery methods in the system operating system or machine controlled software (microcode).

Another way of solving the problem in the future may be to develop a systematic and efficient co-design method where the IC manufacturer and the system manufacturer would cooperate at an early stage. This method will be explained later in this document.

References

[1]   Industry Council on ESD Target Levels. "White Paper 1: A Case for Lowering Component Level HBM/MM ESD Specifications and Requirements," August 2007, at www.esda.org;

[2]   Industry Council on ESD Target Levels. "White Paper 2: A Case for Lowering Component Level CDM ESD Specifications and Requirements," Revision 2, April 2010, at www.esda.org.

Chapter 4: OEM System Level ESD Needs and Expectations

Robert Ashton, ON Semiconductor
David Klein, IDT
Leo G. Henry, ESD/TLP Consulting
Stephanie Dournelle, ST Microelectronics

4.0   Introduction

Component suppliers must comply with an OEM's needs in terms of functionality, performance, form factor and reliability. The OEM's needs may be supplied by formal specifications or from the component manufacturer's knowledge of the component's end use. In addition, OEMs have another set of implicit requirements for system level ESD: OEMs want ESD solutions that work, minimally affect the functional performance of the system and cost as little as possible to implement. These explicit and implicit requirements are often in contention. A successful solution requires that the OEMs not only understand each component's characteristics, but also how different components work together in the system. This last requirement, particularly in the context of the overall system, is the focus of this chapter. First we describe three hypothetical paths for ESD design. This is followed by a descriptive list of OEM system level ESD needs and a discussion of the realities of current ESD solutions, their physical limitations, and how they measure up against those requirements described in the list.

> Note: The majority of this chapter is written assuming a simple OEM to component supplier relationship which would be expected to exist for a consumer product such as a mobile phone or a laptop computer manufacturer. This has been done to more easily contrast different design approaches without going into the details of an individual industry. Section 4.4 addresses the concern that may exist in other industries such as the automotive industry.

4.1   Paths to ESD Robust Systems

It is worthwhile to repeat the first implicit requirement from Section 4.0; OEMs want system level ESD solutions that work. The ESD needs must, however, fit within the general needs of the system. Due to construction of the system chassis or the physical ports into the system, different parts of the system may be exposed to different levels of stress during a system level test as described in the IEC 61000-4-2 specification. Even in a system where all the individual ICs have robust HBM and CDM levels, the system can fail system level ESD stress.

There are three general paths that can define the way to a robust ESD solution:

1. The components, including integrated circuits, chosen for the system are all inherently robust to system level ESD and the OEM does not need to think about system level ESD at all. (A component is considered to be robust to system level ESD if it is able to survive stress with an IEC 61000-4-2 current waveform to a specified level without physical

damage. Since component robustness to ESD stress is generally only tested for physical failure this path can only be expected to work for hard failures.)

2. Not all components, including integrated circuits, chosen for the system are inherently robust to system level ESD, but component suppliers provide clear rules and procedures for using a set of system level ESD robust and non system level ESD robust components that will produce an ESD robust system.

3. Not all components chosen for the system are inherently robust to system level ESD, and the OEM has to find a solution on their own to design an ESD robust system.

At first glance, path 1 looks quite attractive for an OEMs' system design. However, as discussed below, cost and performance reasons might prevent this approach, which inevitably results in significant overdesign of the ESD measures. Additionally, even a "robust" IC may suffer from soft errors when it is integrated into a full system. Most OEMs would probably find path 2 an acceptable alternative. It is certainly more desirable than path 3. Unfortunately most OEMs would probably consider their current situation somewhere between paths 2 and 3. Board designers are faced with the trial and error approach of path 3, while some experienced designers have developed tools and experience that bring them closer to path 2. Today there are many system level ESD robust components that promise to make systems robust to ESD, but there is certainly no guarantee of first pass success. The development of design tools to use these components and the proper characterization of components will make path 2 a reality. This is what we refer to as System-Efficient ESD Design, SEED, introduced in Chapter 1 and further detailed in Chapter 6.

While path 1 may seem desirable, economic and technical constraints will move the industry to the SEED approach because of its distinct advantages to path 3. This will be illustrated in the following two examples.

First, a high level example - a mobile phone OEM and a IC supplier that provides a hypothetical single chip mobile phone solution. The single chip includes all of the electrical functionality of the phone and is robust to system level ESD stress. The OEM only needs to supply an attractive case, keyboard, display, microphone, speaker, antenna, battery and charging connector. Still a bit of work, but seemingly not a hard design challenge. This may not be the best path for a variety of reasons. The single IC may have grown too large to fit easily into the phone's desired form factor. The single IC may also be too expensive because high cost, state of the art, silicon area may not be used efficiently. There are many reasons for this.

- Filter ICs built on chip use large areas of silicon
  - Off chip filtering may be more economical
- Driver circuits for USB and speakers may not be economical in advanced technologies.
  - Separate driver chips may be more economical
- System level ESD structures are not ideal in state of the art integrated circuits
  - It may be hard to produce low C, highly robust protection in advanced silicon technologies with high doping levels
  - System level ESD structures are large and expensive in advanced silicon technologies
  - Dedicated ESD protection structures placed on the board may be more effective and lower cost than including the protection within a state of the art IC.

These points suggest that multi component solutions may be more economical than a single IC design. Path 1 also assumes that it is possible to provide a single IC that can guarantee a passing ESD testing result. Most system level ESD failures are in fact due to system upset rather than physical damage. Even with a very robust system on a chip solution, the OEM must still design a system's case, connectors and circuit boards to defend against system upset and damage during an ESD event. Even in a single chip solution the SEED approach becomes valuable. The value of the SEED approach increases when multiple components are used.

The next example will deal specifically with high speed integrated circuit IOs. As discussed in Chapter 1 the first priority in system level ESD protection is for IC pins which connect directly with system level IOs. This includes high speed serial interfaces such as HDMI, eSATA and USB 2.0/3.0. Protecting these IOs is a particular challenge. Capacitance budgets for protection components on high speed lines are very low and often very little impedance can be tolerated between the physical cable port on the chassis and the transceiver without degrading the port's performance. Simply adding a stand-alone ESD clamp between the physical port and the transceiver IC might appear to be an obvious solution as shown on the left side of Figure 16 below. However, care must be used in the selection of the component. In addition to selecting a protection component with capacitance low enough that it will not distort signals, the turn on voltage of the protection component is critical. If the turn on voltage of the protection is not lower than the turn on of the transceiver, the protection will not be able to perform its function. This is illustrated on the right side of the Figure 16.

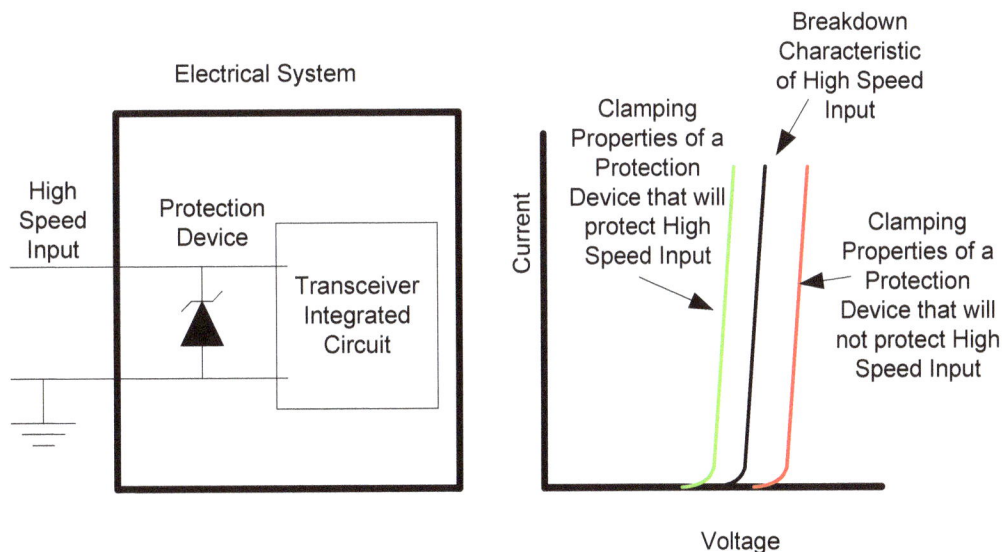

Figure 16: Example of high speed interface system level ESD protection clamping requirements.

This high speed interface example illustrates the three paths to ESD design. In path 1 the transceiver would need to be system level ESD robust, but as we have seen this may not be possible or economically the best solution. In path 3 a protection element can be added, but without knowledge of both the protection component's capabilities and the transceiver's properties there is no guarantee that the combination will produce an ESD robust system. Only path 2, the SEED approach, will guarantee a successful design, since both the properties of the protection and the properties of the transceiver are understood and can be shown to work effectively together.

4.2   An OEM's ESD Needs and Expectations

The challenge for OEMs is to drive down costs and speed up design time while providing current and relevant solutions to their customers. As discussed in the previous section, unless the OEM is happy using a single source vendor and that vendor's components are designed to work in the configuration needed by the OEM, there is no guaranteed single solution. Often, the OEM and its component vendors have to work in an environment where not all the details of the design are available, either because the design is still a work in progress or due to confidentiality concerns. However, there are a common set of requirements that OEMs tend to have that are described below:

Requirement	Defense of requirement [why it is necessary]
Components well characterized for their ESD performance	Without knowledge of how a component will perform during an ESD stress it is impossible to predict the component's or the system's behavior during an ESD stress
Good app notes with respect to ESD.	These instructions tell the OEM's engineers how the component's creators envisioned it being protected in a system. This gives color to the standard data sheet table of values such as those discussed in the next section and can define component specific improvements that can be incorporated into the system design.
Software for simulating ESD current flow and voltage levels in a system during an ESD event.	This software will provide the ability to highlight trouble spots and implement fixes during the design phase.
ICs integrated with error detection indicators for reset, latch-up and other soft failures.	Different products use different system designs. The failure modes and levels are different. More robust error detection indicators must be used to allow the processor to refresh the failure functions and locations by software and/or hardware. Software which detects and notifies of upsets during ESD testing can be very helpful.

In the characterization of ICs for ESD it is desirable that OEMs refer to international standards when requesting certain IC properties. For example TLP validation, which is able to provide more detailed information of the IC behavior during stress, is not yet commonly in use [1]. More detailed standards are needed before OEMs can request and suppliers provide information that fully enables "SEED" operation mode.

An OEM cannot, however, always use a single source supplier. A major challenge is when an IC from supplier A will pass the system level ESD test while supplier B's IC will fail. The OEM will often require an improved IC ESD level from supplier B in an effort not to disrupt the product launch schedule. Care must still be used here, since an improved ESD level for supplier B's product may not solve the problem if the root cause was really supplier B's IC ESD protection triggering at a lower voltage than an off chip protection product.

4.3   ESD Characteristics of ICs and Systems

If it is accepted that system designs will typically be a combination of system level ESD robust and non ESD robust products, what can an OEM reasonably expect from their IC suppliers? The answer depends on how the ICs behave during an ESD event. This may be a more complex situation than it first appears. Today, many OEMs may ask for the IEC 61000-4-2 level of the IC. Moving beyond the fact that IEC 61000-4-2 does not describe how to test ICs, a single survival voltage is a very incomplete description of the system level ESD properties of an IC. Survival of an IC to a particular level of IEC 61000-4-2 stress is usually interpreted as lack of physical damage. This ignores the fact that most system level ESD failures involve a system upset. Even if the system level ESD failure criteria is restricted to physical failure, an IEC 61000-4-2 level sheds very little light on how the product will perform in conjunction with other ICs. For example a system level ESD protection component may be able to survive a large IEC 61000-4-2 stress but may not have a low enough turn-on voltage or on-resistance to protect sensitive circuits. The list below outlines some of the component properties that may be needed to design ESD robust systems.

- Properties for system upset
  - Active components
    - Susceptibility to ground voltage disturbance (possibly on a per ground pin group level)
    - Susceptibility to supply voltage disturbance (also possibly on a per supply pin group level)
    - Susceptibility of IOs to state change due to transients
    - Susceptibility of internal circuits to EM fields
    - Integrated circuits with built in self recovery and resets for lock-up
    - Susceptibility to electric and magnetic field coupling into the IC's leadframe / heatsink etc.
  - Passive (including protection) components
    - Capacitance
    - Inductance
    - Resistance
    - Spark Gap
    - Turn-on voltage and on-resistance
      - TLP I-V curves provide basic knowledge
      - TLP I-V curves with multiple pulse lengths provide additional information
      - Ideally a high current Spice model
- Properties for physical damage to system
  - HMM (IEC 61000-4-2) passing level for ESD stressed pins
    - Full TLP characterization may make this unimportant or it may help define the passing level
  - Turn on voltage and on resistance for ESD stressed pins
    - TLP I-V curves provide basic knowledge
    - TLP I-V curves with multiple pulse lengths provide additional information
    - Ideally a high current SPICE model is also available

The above component properties appear to be enough to give an OEM a good start at predicting system level ESD behavior. However, there is another issue; what is the actual system level ESD stress, and how does it propagate through a system? At this point OEMs, and the electronics industry as a whole, need to take some responsibility. Interfaces need to be developed for communicating the ESD withstand capabilities of ICs to OEMs. How a system level ESD pulse propagates through a system cannot be a IC supplier's responsibility, although use guidelines may be given, especially for suppliers of protection components. The OEM should have the best understanding of how signals travel within a system. 3D ESD simulations are now available but only the largest companies are able to use them due to complexity and cost.

A particular concern is the issue of EM fields causing upset. It is only recently that measurement techniques have been developed to determine the susceptibility of ICs to EM fields [2]. Standardized measurement procedures would be helpful in evaluating susceptibility to EM fields. At present, the level of EM fields that a product or component should be immune to is not well understood. IEC 61000-4-2 compliant ESD guns emit considerable EM fields but the intensity of those fields is not controlled by the IEC 61000-4-2 standard and the intensity of the EM fields emitted by the guns varies considerably from manufacturer to manufacturer. This issue must be addressed by the industry by updating existing test standards and/or creating new test standards.

4.4   Industry Specific Concerns

Specific industries have more complex business models than has been assumed in this chapter. For example in the automobile industry the automotive OEM does not typically deal directly with suppliers of integrated circuits. The OEM typically obtains electrical subassemblies such as radios or engine control modules from a subsystem manufacturer, often referred to as a Tier 1 supplier. It is the Tier 1 supplier that designs the subsystem to the OEM's specifications and obtains electrical components from integrated circuit manufacturers. A main difference compared to OEMs of stand alone devices is that the subsystems are spread all over the vehicle and are connected via a cable harness with around 1000 single wires with a total length of 2 km. Therefore one emphasis is indirect ESD pulsing. Another difference in the automotive business is that automobile assembly and repair is not normally done in an ESD controlled environment. A main concern of the automotive OEMs is the risk of field fails (damage while assembling/repairing, malfunction during customer-use) and the cost impact for the modules. Subassemblies such as engine control modules must be robust to ESD and cannot rely on the automobile to provide physical and electrical protection. Subassemblies and subsystems must therefore be robust to ESD. For this reason, as discussed in Chapter 2, the automotive industry has developed a separate set of ESD test methods which include the stressing of subassemblies and subsystems in both the powered and unpowered states.

This does not mean, however, that the approaches outlined in this chapter and the White Paper do not apply in the automotive industry. The SEED approach has simply moved from an interface and design strategy involving OEMs and electronic IC suppliers to an interface and design strategy that involves Tier 1 suppliers and electronic IC suppliers.

4.5   Summary of Realizable Needs and Expectations

The needs and expectations of OEMs in terms of the ESD properties of ICs is a complex subject. ESD solutions should be designed within the complex constraints of technical needs and economic necessity. OEMs need detailed information on the properties of ICs when they are stressed with ESD in order to determine how they will perform within a system. OEMs also need to have tools that allow them to predict the system's behavior when exposed to an ESD stress. It is not realistic in most cases to rely totally on IC suppliers to provide a full ESD solution in their ICs. The SEED approach will provide a disciplined and effective tool for designing systems which are robust to hard failures from ESD. The improved understanding of ESD events within systems that SEED encourages will also be important to the understanding and elimination of soft failures as well.

References

[1] ANSI/ESD STM5.5.1-2008 Electrostatic Discharge Sensitivity Testing – Transmission Line Pulse (TLP) – Component Level

[2] Impact of ESD Generator Parameters on Failure Level in Fast CMOS System, Aug 18-22, 2003 IEEE EMC Symposium

Chapter 5: Lack of Correlation between HBM/CDM and IEC 61000-4-2

Terry Welsher, Dangelmayer Associates
Alan Righter, Analog Devices
Robert Gauthier, IBM
Leo G. Henry, ESD/TLP Consulting
Tim Maloney, Intel Corporation

5.0 Introduction

System level designers have legitimate concerns about the vulnerability of their systems to ESD malfunction or damage. Design organizations are always searching for the most efficient design techniques for avoiding these problems. In particular, they often try to identify a design platform that can be used across many products, thus minimizing design time and cost. Over the last two decades of EMC design, there has been a significant movement towards pushing design decisions further upstream and to embed re-usable platforms in the design. It was natural that this trend resulted in system designers looking for solutions at the chip level. The idea that at least some portion of ESD "immunity" could be "built-in" at the device level was an attractive option to be pursued. In many cases this meant that the responsibility for finding these solutions was transferred, at least in part, from the EMC engineer or designer to supply chain or component engineering. This transfer was also enabled by the fact that the discussion of ESD effects often blurs the distinction between reversible system level malfunction and actual irreversible hardware (device) damage. In any case, the first impulse then was to target IC ESD ratings as the first step towards "better system level performance", even though there was no evidence for making this connection.

In this chapter we discuss the current limited data on correlation and why using IC ESD ratings cannot be expected to work. We also discuss the relationships among other ESD and similar stress tests and where correlation might or might not be expected. Many references of comparison studies and studies of other "system"-like stress models are included.

5.1 Correlation between ESD Models

It is important to understand what is meant by "correlation". One approach is to look for statistical correlation. To address the statistical definition of correlation one needs to perform testing using the two models in question over a range of products to establish whether there is any useful mathematical relationship between the two models. As of the writing of this document, little data was available to firmly establish the question of correlation on empirical grounds alone. However, the data available can be used to get a sense of whether the position made here - that the models will not correlate – is likely to be true.

In Figure 17, IEC 61000-4-2 system rating [1] and HBM device rating (withstand threshold) (JEDEC A114) [2] are plotted for the cases where they have been reported on the same system. These data points were taken from the larger data collection discussed in Chapter 3. This scatter plot of these nine data points shows no indication of correlation between the IEC and HBM data sets. More data is of course needed to firmly make this point. We can say at this point that the likelihood of good *statistical* correlation between the IEC ratings and the HBM thresholds of a device is small.

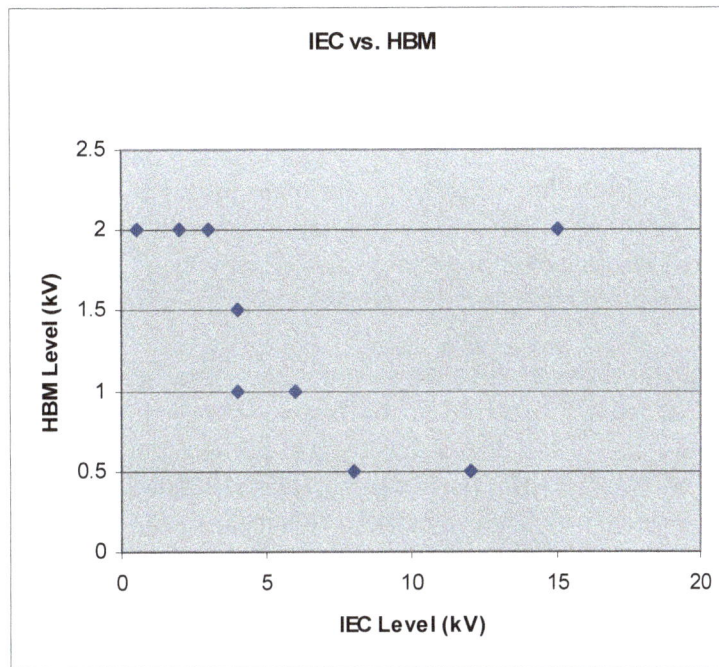

Figure 17: IEC 61000-4-2 System Ratings vs. HBM Device Withstand Thresholds for systems where both have been reported (In some cases only passing levels were available. However, when lower passing levels are reported, it usually means that the failing level was only a few hundred volts higher. This is sufficient for this rough analysis).

These results suggest that the standard HBM rating of a device is not likely to be a useful predictor of system level robustness. This may seem to be a trivial point. However, much of the reluctance to adopt the 1000 volt HBM target proposed in White Paper 1 has been based on the assumption that lowering these targets will result, statistically, in the erosion of system level ESD performance. There are of course other dimensions of correlation that could be explored. There are two device models (HBM and CDM) and there are two main categories of failure or malfunction at the system level (irreversible hardware damage and reversible upset or data transmission errors). It is clear that, in attempting to correlate HBM or CDM to the second category of system failure, one is trying to correlate two entirely different mechanisms. This alone suggests that any "correlation" should be very weak. A slightly stronger case can be made for trying to relate the IC level HBM and CDM thresholds to the relative "immunity" of a system to hardware damage *of that particular device*. However as we shall see even this correlation is likely to be minimal. The following experience is illustrative: the first version of a device passed 1.5 kV HBM, and passed 4 kV in an automotive customer's system. A single pin of the device was redesigned and the part then passed 2 kV HBM, but the part in the system only passed 3 kV IEC and failed 4 kV IEC, and failed on the fixed pin. This pin was protected at the system level by a varistor that had a higher trigger voltage than the IC protection, and the IC protection improvement meant more IEC current could be carried by it and it failed. Experiences such as these demonstrate how counter productive it can be to depend on device-level HBM for system level improvement.

Most attempts to use device level information for system level purposes are based on the HBM. This is due to its similar scope and name and the fact that the HBM thresholds are available for almost all integrated circuits. Even though CDM is responsible for virtually all production-related ESD failures, CDM threshold data is still not universally available. Thus most of the discussion about correlation is centered on the relationship or lack thereof between HBM and the IEC method.

5.2   Differences among Device Level Tests (HBM/CDM) and IEC 61000-4-2

The idea that device level ESD thresholds could provide information about a system level test such as IEC 61000-4-2 probably comes from incomplete understanding and knowledge of how either or both of the methods are actually defined and implemented. In this section we explore the apparent similarities and the fundamental differences.

The impression that the HBM device level test should be similar to the ESD gun based test may be widespread simply because the descriptions of the two stresses in the standards themselves are similar. The original intention of the device level HBM test was to simulate handling (touching) of a device by a charged person while the scope of IEC 61000-4-2 refers to the "electrical and electronic equipment subjected to static electricity discharges, from operators directly, and to adjacent objects" and that it represents "… a [charged] human body holding a metallic object such as a key or tool". The high level schematic descriptions of each model as they are commonly shown are very similar as shown in Figure 18. These representations leave the impression that the two models might only differ in the choices of the RC-network chosen (device HBM: R=1500 Ω, C=100 pF; IEC 61000-4-2: R=330 Ω, C=150 pF) and thus one might conclude that the results of the two tests would be similar. However, a more careful look at the details of the methods shows that this is not the case.

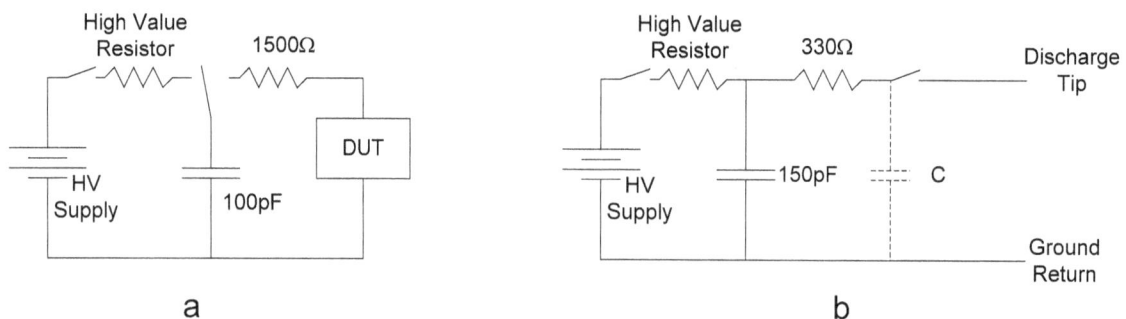

Figure 18: Common simple circuit representations of the a) device level and b) system level methods

5.2.1   Differences in Test Procedure and Threshold Definition

Standardized HBM testing of devices specifies stressing each pin of a device in several configurations where other groups of pins are grounded. This is summarized in Table 3, which is the pin combination table from ANSI ESDA/JEDEC JS-001-2010 [3]. For a typical device there may be hundreds or thousands of different pin / grounded group configurations stressed, each would in general have its own failure level. The device level HBM ESD threshold that is assigned is the minimum of all these failure levels. In practice, devices often are step-stressed and testing is terminated when the first failure is obtained. Thus, to make a relevant comparison between the device level HBM and the IEC method, one would have to select a pin combination which closely resembles the stressing of the device at the system level. Even if this connection can be made, there are still distinct differences between the stresses as described in the next section. A summary of the differences between the system level and HBM methods is given in Table 4 (originally published in White Paper 1 [4A])

Table 3: HBM Pin Combinations for Integrated Circuits (ANSI ESDA/JEDEC JS-001-2010)

Pin Combination Set	Connect Individually to Terminal A	Connect to Terminal B (Ground)	Floating Pins (unconnected) (Must include no-connect pins)
1	All pins one at a time, except the pin(s) connected to Terminal B	First supply pin group	All pins except PUT* and first supply pin group
2	All pins one at a time, except the pin(s) connected to Terminal B	Second supply pin group	All pins except PUT and second supply pin group
N	All pins one at a time, except the pin(s) connected to Terminal B	Nth supply pin group	All pins except PUT and Nth supply pin group
N+1	Each Non-supply pin one at a time	All other Non-supply pins collectively except PUT	All supply pins
* PUT = Pin under test.			

Some other distinctions are important to mention here. The dotted-line capacitance in Figure 18b represents the unspecified capacitance of the ESD gun to its environment. It leads to the initial peak in the IEC standard waveform. This capacitance is not important in the device-level HBM model as the relay position leaves this capacitance uncharged before the stress. The initial current pulse has been shown to produce CDM-like (gate oxide) failures [5, 6, 7]. However, there is no indication that the device-level CDM would be a good predictor of behavior in the system level test. The rise time of this initial peak can have a dramatic effect on the propensity of a system to experience soft errors during the IEC stress. However, since the device level CDM is done on an unpowered isolated device (as is the HBM), correlation to system soft error susceptibility is not expected.

Table 4: Comparison of IC Level HBM (ANSI ESDA/JEDEC JS-001-2010) and System Level ESD (e.g., IEC 61000-4-2, ISO 10605)

	IC level ESD test	System level ESD Test
Stressed pin group	Multitude of pin combinations	Few special pins
Supply	Non-powered	Powered & non-powered
Test methodology for 'HBM'	Standardized	Application specific using various discharge models
Test set-up	Commercial tester & sockets	Application specific board
Typical qualification goal	1 ...2 kV HBM	8 ...15 kV
Corresponding peak current	0.65 ... 1.3 A	> 20 A
Failure signature	Destructive	Functional or destructive

5.2.2 Differences in Pulse Characteristics (Hardware Failure View)

One way to investigate the lack of correlation in failure mechanisms between System Level ESD and the other ESD models is to consider differences in the electrical signature of the model stimulus, namely the rise time, peak current, energy and power parameters of the different models.

The total power and energy of stresses in the total event times of ESD levels was initially described by Wunsch and Bell [8] and extended to the very short (adiabatic) and very long (equilibrium) timeframes by Tasca [9]. The resulting plot of failure power density versus log (time) and the different ESD event representations are shown in Figure 19. Figure 19 represents the entire time spectrum of EOS. ESD is a subclass of EOS events which results from triboelectric charging and is generally characterized by short pulses. However, some ESD events have a very high power density, and can result in EOS-like events even for short duration time scales. Examples of these ESD events (Charged Cable Events, Human Metal Model and Charged Board Events) are discussed in Section 5.3.

• Wunsch-Bell 1D *Electro-Thermal Failure* Model for Square Pulse.

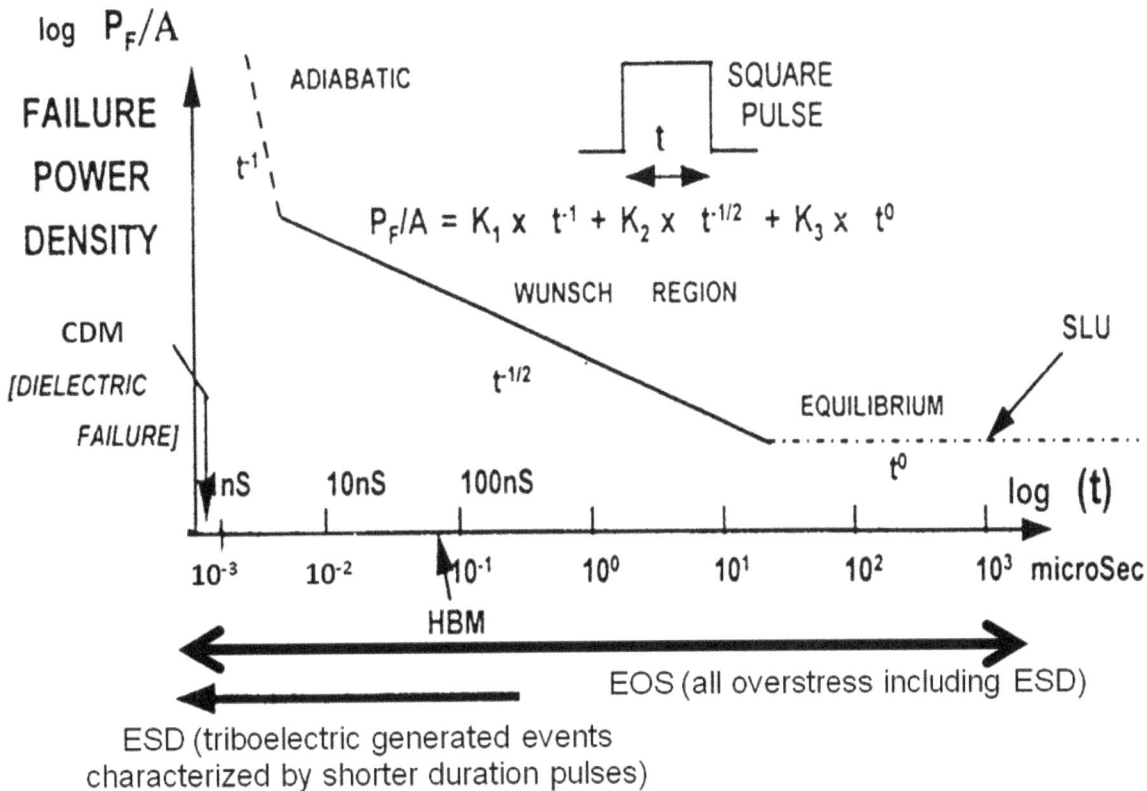

Figure 19: Wunsch-Bell Power Density / Time plot for ESD / EOS level pulses.

The Industry Council on ESD Target Levels authored White Paper 2 on CDM which gives relevant information considering a 1 kV event from each model [4B]. Comparison values for HBM, CDM and IEC pulses from White Paper 2 for a 1 kV ESD event from each model are shown below in Table 5. Total peak current values for HBM and IEC ESD events of this magnitude are 0.67 and 3 A, based on the fixed circuit models for these methods. However, for CDM, which is dependent on the total device size / capacitance, the peak current can range between approximately 1 and 25 A. It should be noted however that the values in this table represent the event through the circuit model itself, and the actual event parasitics below are dependent on the stressed system circuit under test.

Table 5: Comparison of some typical network values and electrical quantities for a 1 kV stress

	C (pF)	R (Ω)	τ (ns)	Q (nC)	E (μJ)	Pavg (kW)	Ip (A)
HBM	100	1500	150	100	50	0.33	0.67
CDM	1-100		1-2	1-100	0.5-50	0.25-25	1-25
System	150	330	50	150	75	1.5	3

One major difference among the models which is not included in the Wunsch-Bell plot is the initial rise time of the measured pulse. The standard HBM circuit model has a specified rise time between 2 and 10 ns, while the CDM rise time is typically in the range 50 – 500 ps, depending again on the effective device size / capacitance. However, the IEC initial pulse has a rise time of between 0.6 and 1 ns while the secondary larger total energy pulse has a rise time between 10 and 20 ns. These differences are critical to the effectiveness of on-chip protection structures. Most ESD on-chip circuits are dependent on the rise time of the initial pulse for turn on response, in addition to the total power / peak current handling capability. Thus they can be expected to perform differently for different model stresses. Thus when one attempts to use the HBM (or CDM) threshold, even if one has identified the external pin in the system, one is making the incorrect assumption that the protection structures will operate the same over a wide range of rise times. It may be necessary to define a set of waveforms that an IC may be exposed to in a system during an ESD event in order to fully evaluate how the IC will behave and design countermeasures at the board level to prevent failure.

The concept of simulation fidelity has been described by Pierce [10]. Particular ESD events which occur in the field can only be well-simulated if the particular ESD model is precisely described. Thus, when conducting failure-mode analysis and investigating root causes of failure, it is often necessary to adjust the model parameters to produce the exact physical and electrical signatures.

Average power during the ESD pulse through the circuit is difficult to calculate without knowing the effective resistance and peak internal voltage of the circuit ESD path. But a few general comparisons can be made for the IC case from assumptions of ESD path resistance and breakdown voltage.

The HBM circuit model consists of a 100 pF capacitor and a 1500 ohm series resistor. The HBM current results from the resistive divider between the 1500 ohm resistor and the path resistance of the device under test. Path resistances within a device under test may be on the order of 1-10 ohms and device breakdown voltages are on the order of 10V, sometimes more, sometimes less. A peak circuit voltage in this case for a 2 kV HBM ESD event could be in the 10-12V range. For CDM, the resistance varies depending on the circuit area and connectivity to supply and ground. Also, due to the ns-time scale of the CDM event, the dv/dt of the CDM voltage is much higher than HBM, well under a nanosecond, which is very close to the turn-on time of the ESD protection, and the fast ramp results in a peak voltage that is higher for the short CDM time compared to HBM. An approximate in-circuit voltage value / path resistance could be 20 V and 5 Ω for a 500 V CDM discharge. An IEC pulse could take a similar path (in the unpowered state) to that of an HBM pulse, as its stressed and grounded points (IO and system ground respectively) may be similar to that for the IC itself. Assuming this is true, the path resistance would be similar, but the peak voltage would be much greater. The resistive divider from the IEC circuit results in a higher peak voltage (by 4.7X) compared to the HBM pulse at an equivalent voltage (the peak current is greater as well). So peak voltages for the IEC case for the same protection would be much higher (40-50 V) for a 2 kV pulse compared to a 2 kV HBM pulse. This means that a typical IEC pulse, if applied directly to a device, will produce stress far in excess of the level for which the protection circuitry is designed.

However, actual system level testing is even more complicated. To discuss the energy of an applied pulse, distinction must be made between air and direct discharge. In the latter case, a high portion of the energy is injected directly into the *system*. If the pulse is applied to the outer casing, the energy arriving at the board can be greatly diminished by proper shielding and current diversion techniques, and the energy delivered to the IC will be lower by an unknown amount. For air discharge, the fraction of the energy coming in to the system directly is significantly lower, again by an uncontrolled amount. This is counteracted somewhat by the fact that the specification for air discharge immunity is typically set higher (15 kV vs. 8 kV for direct discharge). However because much of the energy is radiated, there is more chance of stray fields entering the system and reaching board components through other paths. Thus designing robust system inputs, even if focusing only on hardware failure, requires a system approach as is described in Chapter 6.

5.2.3 Differences in Pulse Characteristics (System Upset View)

It is clear that the current waveforms associated with the IEC requirement and the IC specifications differ significantly. The amount of energy in an IEC pulse is much higher than in an IC level HBM or CDM pulse, as is the peak current. In addition, the frequency spectrum is very different. It includes high frequencies comparable to CDM, lower frequencies comparable to HBM and everything in between. But even if the exact applied IEC waveform could be well defined, the waveform reaching the IC in a particular system is usually unknown. Furthermore, the system level test is much more complex than a pin-to-pin case; due to EMC radiation and other coupling effects, both capacitive and inductive, such as "crosstalk". This means that multiple IC pins can receive spurious stresses more or less simultaneously during an IEC test. This can complicate understanding the system and IC responses greatly.

Frequency Domain and the IC. The frequencies of the applied pulse tend to be lower when reaching the IC because of interaction with the parasitic impedances of the system. In general, the high frequency first peak will be lowered and stretched in time. Likewise, the more energetic second pulse tends to extend in time. These trends are not universally true however, and dependent on board and application.

Switching states in an IC can create sudden redistribution of the system level energy, causing secondary high frequency pulses to arise. One such example is the discharging of a stabilizing capacitance in response to the triggering of a snapback-based clamp. In this case, the fast rise time of the superimposed pulse will not cause any additional issue for the ESD protection, as it happens after triggering. Therefore, the main issue concerning the pulse frequency modulation is the unknown <u>time duration of the pulse</u> arriving at the IC. For high speed pins which can be touched directly that have much less parasitic impedance between the connector and the IC, the high frequency part of the IEC pulse can reach the IC. A closer look at the details of the two peak IEC waveform is discussed in the next section.

5.2.4 IEC Two Peak Waveform as Applied to Chassis Metal

The IEC 61000-4-2 system ESD test method describes a current waveform into a standard target load. This waveform has been described by mathematical expressions and circuit models in professional literature. K. Wang et al [11] have developed a mathematical reference discharge waveform given by

$$i(t) = \frac{i_1}{k_1} \cdot \frac{\left(\frac{t}{\tau_1}\right)^n}{1+\left(\frac{t}{\tau_1}\right)^n} \cdot \exp\left(\frac{-t}{\tau_2}\right) + \frac{i_2}{k_2} \cdot \frac{\left(\frac{t}{\tau_3}\right)^n}{1+\left(\frac{t}{\tau_3}\right)^n} \cdot \exp\left(\frac{-t}{\tau_4}\right)$$

with the following constants

$$k_1 = \exp\left(-\frac{\tau_1}{\tau_2}\left(\frac{n\,\tau_2}{\tau_1}\right)^{1/n}\right)$$

$$k2 = \exp\left(-\frac{\tau_3}{\tau_4}\left(\frac{n\,\tau_4}{\tau_3}\right)^{1/n}\right)$$

and the following parameter values for a 5 kV pulse

$i_1 = 21.9$ A	$\tau 1 = 1.3$ ns	$\tau 3 = 6$ ns	$n = 3$
$i_2 = 10.1$ A	$\tau 2 = 1.7$ ns	$\tau 4 = 58$ ns	

References 11-13 also present comparable circuit models for simulating the IEC pulse. Cannigia and Maradei [13] present a circuit comparing results to field simulations and having a distributed model of the strap as in Figure 20 below. The construction of models such as these shows the level of detail that is left out of the superficial description of the models in the standards.

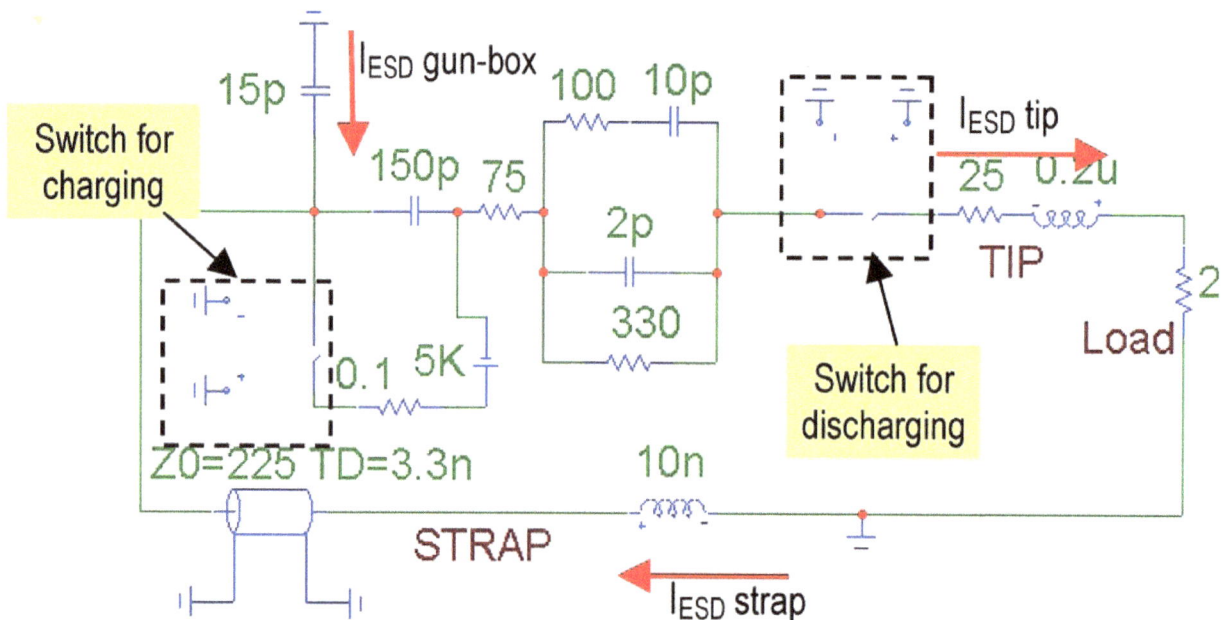

Figure 20: SPICE model of the IEC generator, from Reference 11

5.3 Relationships among Different Possible Realizations of System-Type Stresses

In order to address concerns about the failures of devices in higher levels of assembly and in systems, considerable effort has been applied to create stress models (tests) which simulate specific situations. The intent of these approaches is to simulate real-world environments that the existing models do not. In this section we briefly review the models and methods that are of current interest and where possible indicate how they may relate to each other.

5.3.1 Cable Discharge Event (CDE)

A cable discharge event often originates from a cable being charged triboelectrically. The discharge occurs when the cable is connected to a system and the charge stored by the cable is discharged through a connector pin into the system. Early literature on cable discharge events are included in references [14-16]. Mitigation techniques include ensuring the ground part of a connector is connected into the system first to avoid discharging of the cable onto boards/components inside the system that are electrically connected back to pins on the connector of the system. Other mitigation techniques can involve board-level solutions or chip/integrated circuit solutions. Some details of attempts to standardize a CDE test were introduced in Chapter 2. The rise/fall and pulse width of CDE pulses are very different than IEC 61000-4-2 or Human Metal Model (see Section 5.3.3) type pulses. The CDE pulse width is dependent upon the cable length. The rise/fall times of the CDE pulse are usually highly dependent on the parasitics of the cable and connectors. There has been no correlation to date shown between CDE system level results that can occur from cable plugging and the IEC 61000-4-2 system level results. In addition, there has been no correlation found between CDE results with powered systems and unpowered ESD HBM, MM or CDM levels for ICs. However, there has been some effort to correlate CDE to long pulse TLP (Section 5.3.4). Some insights into CDE and the resultant stresses for shielded and unshielded cables are provided in a recent study [17]. For example, the strong pulse that can occur on the shield of a plugged-in cable, when charged objects are neutralized, is found to produce a fast bipolar induced pulse on the interior lines of the cable. This will have much reduced magnitude but can still be hazardous.

5.3.2 Transient Latch-Up

Transient latch-up can occur when a transient signal is applied to the power supply and/or signal line pins on an IC. The transient stimulus distinguishes this approach from the standard JEDEC JESD78 [18], in which the signal pads see static (DC-like) signals for the current injection test and power supply voltage ramp testing on the supply pins. While the focus of both of these tests is to stimulate CMOS latch-up, almost all reported latch-up problems in the field are likely to be caused by the transient events. Transient latch-up trigger currents/voltages are a function of the pulse characteristics such as pulse width and rise/fall times [19]. Figure 21 shows the typical cross sections of test structures utilized for characterizing external latchup using static or transient triggering sources where both injectors of holes (positive mode external latchup) and injectors of electrons (negative mode external latchup) structures are shown [20].

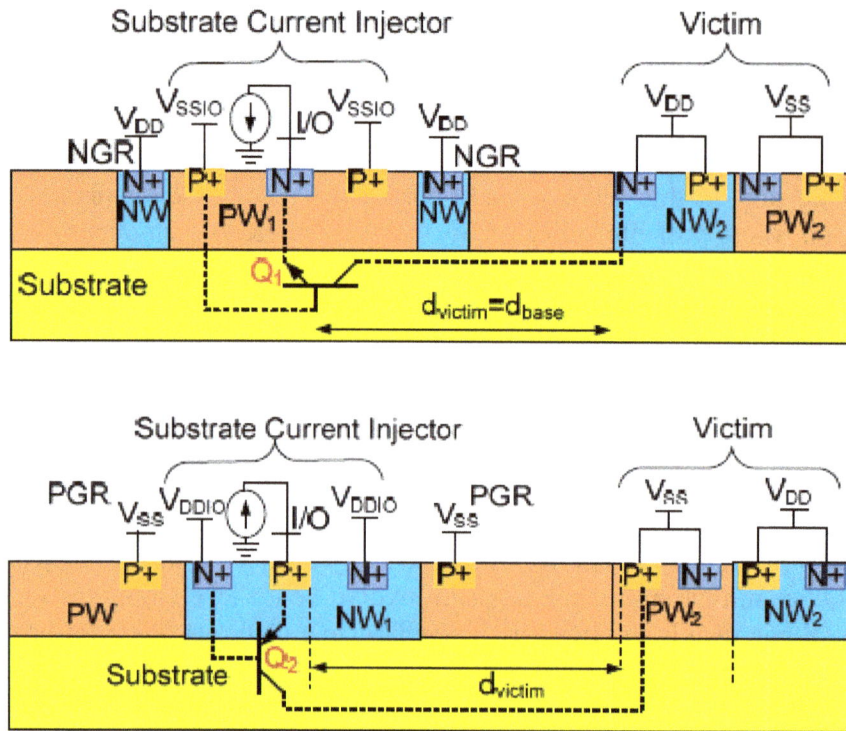

Figure 21: External latchup test structure cross sections

Figure 22 shows the results from negative external latchup testing acquired in a 180 nm bulk CMOS technology node using p-type starting wafers. The data shows the strong trigger current dependence on the applied pulse width for 1us and below pulse widths [19].

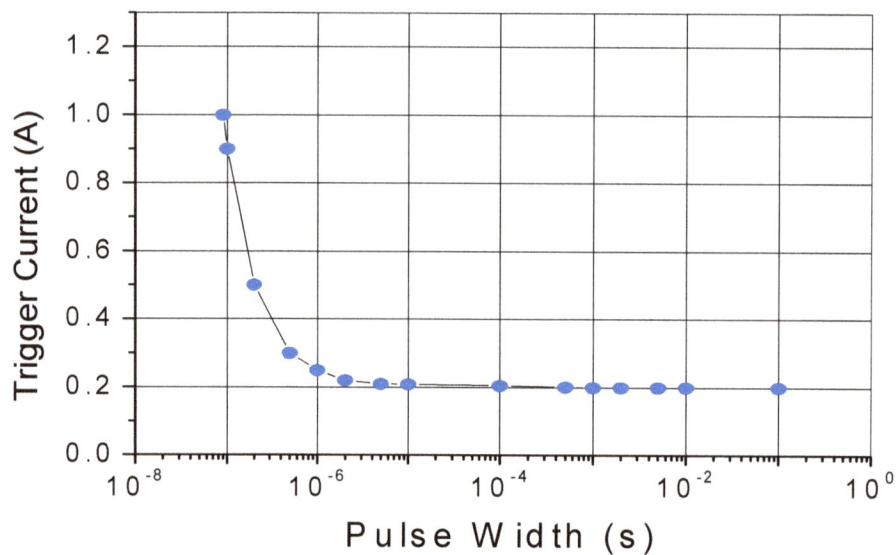

Figure 22: Transient latchup signal pad (negative mode) trigger current vs. pulse width (180 nm bulk CMOS technology node) [19]

More recent work is shown in Figure 23 where positive and negative current injection testing was completed for a 130 nm bulk CMOS technology node [20]. The similar study as shown in [19] was completed for negative mode injection results for varying pulse widths with similar results. In addition, as shown in Figure 23 the work also included positive mode injection results for varying pulse widths, showing very little rise in positive I_{trig} down to pulse lengths of about 20 ns. As seen in Figure 21, the base length of the vertical Q_2 for positive mode, is usually considerably shorter than the lateral base length Q_1 for negative mode, thus explaining the fast response of positive mode. The aforementioned bipolar induced CDE pulse [17] can thus be hazardous unless there is a good p+ collector to raise positive I_{trig}.

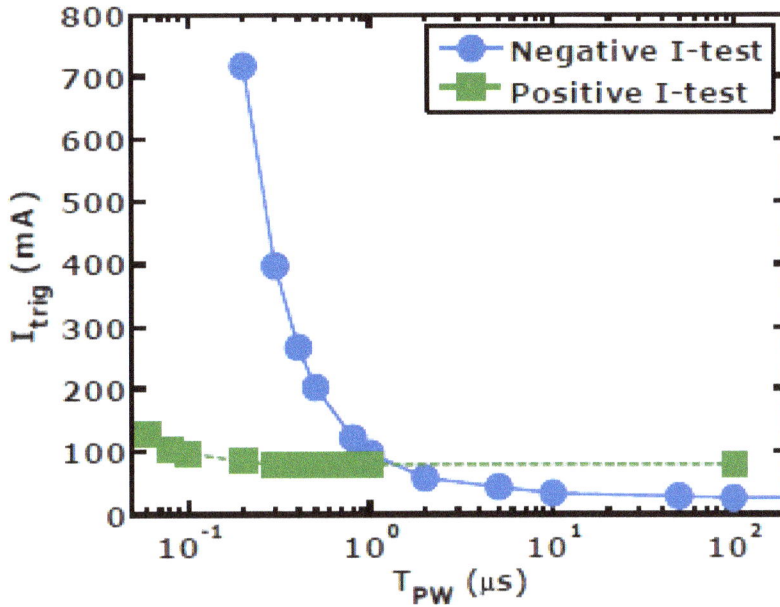

Figure 23: Transient latchup signal pad (negative and positive mode) trigger current vs. pulse width (130 nm bulk CMOS technology node)

Figure 24 shows data acquired using different stimulus sources performing powered latchup testing [19]. The different trigger pulses used have different pulse widths and rise/fall times and give different trigger current values when these pulses are used as the stimulus on the I/O signal pad latchup test (negative mode).

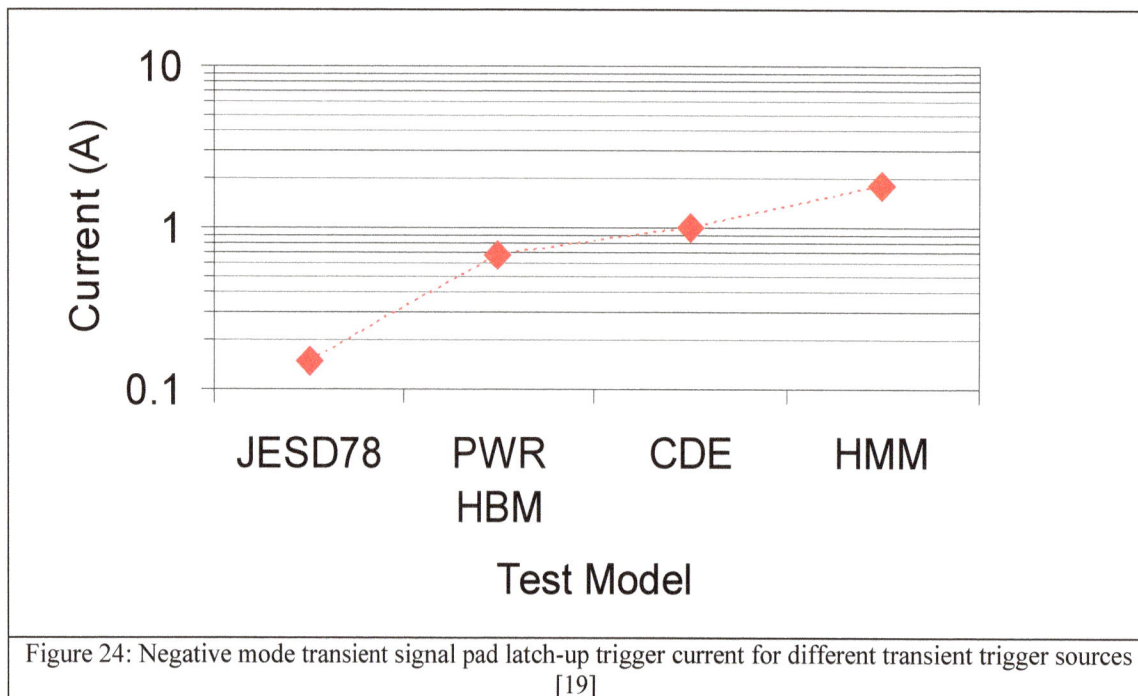

Figure 24: Negative mode transient signal pad latch-up trigger current for different transient trigger sources [19]

One major conclusion, based on a review of published prior work, is that there has been no evidence found showing linkage between reduced ESD targets for HBM, MM or CDM and reduced latchup results [19, 20].

5.3.3 Human Metal Model

We have argued that the standard HBM and CDM thresholds were not good predictors of system level performance. Many system designers and manufacturers have come to accept this fact and have begun looking for a "better" device level test. As was discussed in Chapter 2, many OEMs have begun to require that IC suppliers apply the IEC-defined stress directly to the system external pins of a component, even though direct ESD stress of external system pins is explicitly excluded in the IEC 61000-4-2 standard method. The IEC 61000-4-2 testing of system ports is carefully defined. If the port has a metal shell or cover, contact discharge is to be done to this metal shell, but not directly to the pins. If the port connector has an insulating body, air discharge is to be done to that area of the port. It should also be noted that the ESD and EMI transients from application of this pulse may have an impact anywhere within a system, affecting numerous ICs, and does not permit evaluation of any one particular IC in a system. Additionally, the IEC 61000-4-2 test document identifies several modes of failure, including soft failures, in addition to hard failures. Therefore, it is expected that it will be very difficult to predict the particular pass / fail system level voltage of the IC itself from the system level application of this test.

Round robin testing of this method has begun and is expected to be completed before the end of 2010. If the results indicate the test can be repeatable the standard practice may be elevated to a Standard Test Method. However, this will not change the fact that this procedure is unlikely to be a good predictor of how the IC will react to system level tests according to IEC 61000-4-2.

5.3.4   Extended Pulse Length TLP Testing

There has been investigation into assessing the correlation between the system level ESD pulses and an extended length pulse generated from a TLP tester. TLP pulse lengths in the 400-500 ns range have been used to generate and detect damage and determine if correlation may exist between the voltage of the TLP pulse and the voltage of the IEC pulse. A 2006 study [21] described 500 ns – 1000 ns extended pulse TLP testing to simulate cable discharge events from UTP standard cables on 0.25 um technology ESD structures. The 500 ns TLP compared well with the 56 m cable discharge pulse time of 475 ns in the paper. A 2008 study [22] evaluated the correlation between cable discharge events and a 490 ns TLP pulse on an Ethernet PHY transceiver. Correlation of failure to peak voltage between CDE and the long TLP pulse was achieved.

5.3.5   Charged Board Events (CBE)

Another source of confusion between device and higher-level assemblies relates to failures which occur at the circuit board level and may be caused by electrical overstress (EOS). Indeed the physical failure analysis often suggests this.

It has long been known that ICs and other ESD-sensitive components remain at risk when they are mounted onto printed-circuit boards and other assemblies. However, most ESD testing and characterization of these ICs has been done on stand-alone parts. Further, IC failure analysis data, which is based on knowledge of failure signatures seen in standard HBM and CDM tests, has caused many to conclude that ESD failures are relatively rare when compared to other electrical failures commonly classified as electrical overstress. Recent data and experience [23, 24] now suggest that many failures previously classified as EOS may instead be the result of ESD failures due to Charged Board Events. A charged board stores much more energy than a device (IC) because its capacitance is many times larger. In fact, the charge (energy) transferred in the event can be so large that it can cause EOS-like failures to the ICs on the board. While it is likely that many of these failures occur during handling in manufacturing, some do occur during installation, maintenance and repair. This can be confused with system failures since the failure only becomes obvious when the board is installed and the system is powered up.

5.4   Review of Other Published Case Studies and Investigations

A review of studies done through 2007 was given in White Paper 1. In this section we survey some of the work since then.

In one study [24] the researchers used a field-induced charged-board event stress method to get information on the IC on the board. The waveform had a ringing shape similar to the MM waveform. Stress levels of 500V to 3 kV were investigated. The study did not attempt to find a connection with the IEC stress. Such a comparison was made in [25]. The authors compared stresses from HBM (JEDEC A114) IC level (from 1 to 8 kV) with IEC system level stresses (from 2 to 8 kV) and found no test result correlation between the models.

In [26], an IEC pulse generator and a TLP test system were both used to deliver a pulse thru a coaxial cable to the DUT, mounted on a test fixture board. The IEC generator used a 330 Ω series resistance and the TLP used both 50 Ω and 100 Ω series resistances. The TLP produced a more repeatable pulse shape than the gun. The waveform had the same general shape; that is, the gun produced a "camel hump" shape but the TLP had a flat plateau shape instead of a hump. The TLP method applied low voltages (up to 450V with I = 1.63 A for the measured pulse current) compared to the IEC Gun (1200V with I = 2.0 A for the measured pulse current). The TLP method showed increasing pulse current at stress levels close to failure. The IEC method did not detect these changes.

In [27], off-chip surface mount protection devices were stressed based on the IEC ESD gun (contact discharge) principle, but the Human Metal Model method, which uses the IEC waveform, was used to get the stress data. These SMDs are used to protect antenna switching pins, which are connected to the outside world. The three different type SMDs passed 8KV using the contact discharge mode.

In [28], the authors used TLP (with parameters assumed to be similar to IEC system level requirements) to stress the high voltage pins of Lateral DeMOS based protection devices. The pulsed voltage ranged up to 50-70 V, with corresponding pulsed current up to 12-15 amps. The authors were testing for transient latch-up and hot plug-in type failures. They reported on the mechanisms, but no comparisons were made to the actual IEC system level procedure.

Finally, in [29], the authors actually define a System to Component Correlation Factor and use thermal failure as the correlatable mechanism. They used the HBM and HMM methods, where the ratio of the two failure voltage levels ranged from 11% to 39% depending on the technology. They further identified the range of similar ratios for transient voltage overshoot as 11-14%, and the range for thermal failure as 29-39%. This study showed potential correlation between HMM and HBM but does not provide any evidence of correlation of either model with the IEC method.

5.5   Conclusion

ESD ratings obtained at the device level using the standard HBM and CDM tests have no useful relation to the impact the device may have in system level stress testing. The differences in the stress testing procedures and the electrical characteristics of the different pulse waveforms [30] make any correlation between the methods difficult and unlikely. Early correlation data confirms this hypothesis. More realistic models and test methods such as cable discharge events, charged board events and others have been or are being developed which are designed to simulate more realistic system level environmental stresses. These new methods should be regarded as complements to the IEC 61000-4-2 standard since the IEC method does not address many of the expected stresses.

References

[1] IEC 61000-4-2, 'Electromagnetic compatibility (EMC) – Part 4-2: Testing and measurement techniques – Electrostatic discharge immunity test', Ed. 2.0, 2008.

[2] JEDEC Standard – "Electrostatic Discharge (ESD) Sensitivity Testing Human Body Model" – JESD22-A114 (1995)

[3] ESDA/JEDEC Joint Standard for Electrostatic Discharge Sensitivity Testing – Human Body Model – Component Level, ANSI ESDA/JEDEC JS-001-2010.

[4] Industry Council on ESD Target Levels. A) "White Paper 1: A Case for Lowering Component Level HBM/MM ESD Specifications and Requirements," August 2007, at www.esda.org; B) "White Paper 2: A Case for Lowering Component Level CDM ESD Specifications and Requirements," Revision 2, April 2010, at www.esda.org.

[5] T. Smedes and N. Guitard, 'Harmful Voltage Overshoots due to Turn-on Behaviour of ESD Protections during fast Transients', Proceedings EOS/ESD Symposium, pp. 357-365, 2007.

[6] W. Stadler et al., 'From the ESD robustness of Products to the System ESD Robustness', Proceedings EOS/ESD Symposium, pp. 67-74, 2004.

[7] T. Smedes et al., "Relations between system level ESD and (vf-) TLP", Proceedings EOS/ESD Symposium, pp. 136-143, 2006.

[8] D.C. Wunsch and R.R. Bell, 'Determination of threshold failure levels of semiconductor diodes and transistors due to pulse voltages', IEEE T. Nucl. Sci., NS-15, pp 244, 1968.

[9] D. M. Tasca, "Pulse Power Models for Semiconductors," IEEE Trans. Nuc. Sci., Vol. NS-19, December 1972.

[10] D. Pierce, "ESD Failure Mechanisms", 1997 ESD Tutorial, 1997 EOS/ESD Symposium.

[11] K. Wang, D. Pommerenke, R. Chundru, T. Van Doren, J. L. Drewniak, A. Shashindranath, "Numerical Modeling of Electrostatic Discharge Generators", IEEE Trans. Electromagnetic Compatibility, 45, pp. 258-271 (2003).

[12] F. Zhou, J. Huang, Y. Gao, S. Liu, X. Wang, L. Wang, "Analysis of HBM-ESD Current Rise Time and its Deciding Factors", 4th Asia-Pacific Conference on Environmental Electromagnetics (CEEM 2006), pp. 529-533.

[13] S. Caniggia, F. Maradei, "Circuit and Numerical Modeling of Electrostatic Discharge Generators", IEEE Trans. Industry Appl., 42, No. 6, pp. 1350-1357 (2006).

[14] Static Discharge Between LAN Cabling and Data Terminal Equipment, Category 6 Consortium, TIA December 2002

[15] Cable Discharge Event in the Local Area Network Environment, White Paper, Intel Order Number 249812-001, July 2001

[16] R. Brooks, "A Simple Model For a "Cable Discharge Event"" , IEEE802.3 Cable Discharge Ad-hoc Committee, March 2001

[17] T.J. Maloney, "Primary and Induced Currents from Cable Discharges", 2010 IEEE Electromagnetic Compatibility Symposium, Ft. Lauderdale, FL, July 2010, pp. 686-691.

[18] JEDEC Standard, "IC Latch-Up Test, JESD78B, December 2008

[19] B. Reynolds, M. Muhammad, R. Gauthier and J. Coutinho. "A Test Method to Determine Cable Discharge Event Sensitivity at the Module Level, presented at International ESD Workshop, IEW May 2007.

[20] F. Farbiz and E. Rosenbaum, "Understanding Transient Latchup Hazards and the Impact of Guard Rings", 2010 IRPS, paper 4D.2 (May 2010).

[21] Ming-Dou Ker; Tai-Xiang Lai, "Dependence of Layout Parameters on CDE (Cable Discharge Event) Robustness of CMOS Devices in a 0.25-μm Salicided CMOS Process", International Reliability Physics Symposium Proceedings, 2006, Page(s):633 – 634.

[22] Y. Lin et al, "The Challenges of On-Chip Protection for System level Cable Discharge Events (CDE)," 2008 EOS/ESD Symposium, pp. 125-131.

[23] A. Olney, B. Gifford, J. Guravage, A. Righter, "Real-World Printed Circuit Board Failures," EOS/ESD Symposium Proceedings, EOS-25, pp. 34-43, 2003.

[24] T. Reinvuo, T. Tarvainen, T. Viheriäkoski, EOS/ESD Symposium Proceedings, EOS-29, pp. 318-322, 2007

[25] D. Robinson-Hahn, J. Chapin, B. Lyons, Fairchild Semiconductor. "Evaluating IC components utilizing IEC 61000-4-2", presented at International ESD Workshop, IEW May, 2008, Port D'Albret, South France.

[26] Evan Grund, K. Muhonen, N. Peachey. "Delivering the IEC 61000-4-2 Current Pulse Through Transmission Lines at 100 ohm and 330 ohm System Impedances", presented at .International ESD Workshop, IEW May, 2008, Port D'Albret, South France.

[27] N. Peachey, S. Muthukrishnan, "Protection of Mobile Phone Antenna Ports against System Level ESD stresses", presented at International ESD Workshop, May, 2009, So Lake Tahoe, CA, USA

[28] V.A. Vashchenko, "System Level and Hot Plug-in Protection of HV pins. , National Semiconductor Corp., presented at International ESD Workshop, May, 2009, So Lake Tahoe, CA, USA

[29] S. Thijs, M. Scholz, D. Linten, C.Russ, W. Stadler, S Sawada, G. Groeseneken. "System to Component Level correlation Factor. Presented at IEW 2010, Tutzing, Germany.

[30] P. Besse, "ESD/EMI in an automotive environment", presented at IEW, 2010, Tutzing, Germany.

Chapter 6: Relationship between IC Protection Design and System Robustness

Jonathan Brodsky, Texas Instruments
Charvaka Duvvury, Texas Instruments
Harald Gossner, Infineon Technologies
David Klein, IDT
Guido Notermans, ST-Ericsson
Theo Smedes, NXP Semiconductors
Pasi Tamminen, Nokia
Benjamin Van Camp, Sofics

6.0   Introduction

There is a perennial misconception that an IC's ESD protection is intrinsically and critically related to the system level ESD protection when the IC is placed in a PCB application. In Clause 8, this non-correlation of results between IC and system level stresses has been presented. However, when the issue is related to the **external pins**, an IC pin's ESD protection, designed for handling during production, assembly, and test, will influence the effectiveness of the system protection design as indicated in Figure 25. Here, an important distinction is necessary:. if the external pin protection involves an on-chip design strategy, then the design has to directly meet the IEC waveform stress standard. Moreover, designing for excessive HBM and CDM ESD levels does not guarantee system protection against an IEC stress. If on the other hand the external pin's protection is provided by an external clamp then a thorough understanding of the interaction between the IC ESD clamp operation and the external clamp efficiency is required. In this latter case it would be important to note again that the IC's protection is designed only to meet the ESD specifications in a safe, ESD controlled environment. However, to achieve system protection for the external pins to the outside world, rigorous analysis of the interaction between the IC pin's ESD and the external clamp design is necessary.

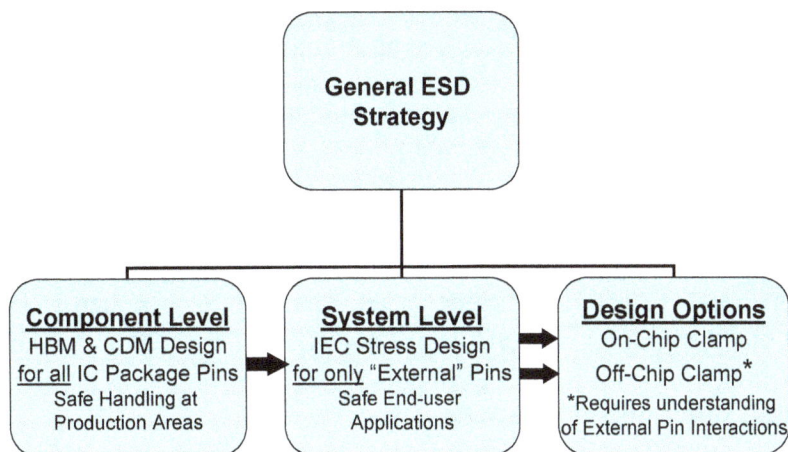

Figure 25: Component vs. System ESD

After a brief background of IC ESD protection methods, the main focus of this chapter will shift to the system protection design techniques from the IC's point of view. Important concepts will include:

1) The relevance of the IC application to the system ESD design

2) PCB protection methods

3) Interaction between the IC pin clamp and the on-board protection

4) Characterization of the IEC pulse

5) Various types of external pins and how the parasitic and components on a board affect design

6) Pros and cons of an on-chip IEC protection strategy

7) The effect of the parasitic components on the IEC pulse stress

Finally, an approach for integrated and compatible design will be offered based on all of these aspects. This will effectively involve a co-design effort between the IC ESD designer and the system board designer.

6.1 IC ESD and Latch-Up Protection Methodologies and Irrelevance for System Robustness

The ESD protection design for an IC package is known to be critical for safe production and handling. It is commonly understood and accepted that this protection design is expected to meet or exceed the required ESD specifications when these ICs are handled in an ESD-safe area also known as ESD Protected Area (EPA). The control method techniques for EPA have progressed sufficiently over the past 30 years such that ICs now require only the minimum specified protection levels [1]. Typical IC level ESD protection requirements, addressing both the HBM and CDM, commonly specify 2 kV and 500 V, respectively. To meet the HBM level of 2 kV the IC pin must survive an equivalent current transient pulse of 1.3 A in magnitude with a rise time between 2 ns to 10 ns, and a decay time of 150 ns. The energy under this pulse roughly corresponds to a 100 ns wide square wave pulse with a magnitude of 1 A. The CDM level in contrast refers to a current pulse with a rise time of ~200 ps and a pulse width of 1 ns, but the current magnitude can vary widely due to large variations in the IC package sizes [2].

ICs also require immunity to latch-up as tested by the JEDEC latch-up test standard JESD78B. While the device is placed in standby mode the IOs are pulsed with +/- current 100 mA injection with a compliance voltage of 1.5×VDD. The test is typically performed at both nominal and high temperature. The device is considered to have failed if the IDDQ current (after the stress is removed) exceeds 140% of its pre-stress value. Therefore the IC protection devices designed at the pins must pass both the ESD requirements and latch-up requirements. Latch-up can have significant impact on the system level ESD design, as will be described in the later sections.

6.1.1 Constraints From Silicon Advances

The IC ESD protection design technique involves protection clamp design to handle the corresponding range of current pulses for both HBM and CDM tests. These clamps essentially are triggered by either type of pulse discharging the current; while preventing any voltage buildup that might cause damage to the input gate oxide or to the output buffer transistors. However, during the last few years, as silicon technologies have rapidly advanced, compatibility between the IO protection circuit and the increasing demands for IO speed have become an extremely challenging aspect. For data rates above 20 Gb/Sec, HBM and CDM levels cannot be met in a practical situation. The impact on CDM robustness is more serious for the following reason; IC package size has a direct correlation to the discharge current at a given CDM specification voltage. Many of the high speed serial link buffers (HSS) are now commonly designed in high pin count IC packages, producing relatively large peak discharge currents which make CDM protection design difficult. This effect, combined with rapidly shrinking silicon technologies that result in lower oxide breakdown voltages and sensitive transistor junctions, leads to a drastic reduction in the ESD Design Window. For example, large microprocessor chips (used for internet application) at the 45 nm node cannot tolerate CDM peak discharge current levels higher than 4-5 A. In these specific cases, achieving a 500 V CDM passing level becomes impossible to reach, and this situation obviously gets worse as the technologies are scaled further. Detailed and documented work done by the Industry Council on ESD Target Levels addressed these issues and concluded that due to improvements in manufacturing ESD controls and awareness, HBM levels can be dropped to 1 kV and CDM levels to 250 V while still keeping ICs safe during production and assembly [1,2]. The new levels are steadily receiving more support and OEMs are generally starting to accept the new IC ESD specifications.

Silicon technology scaling does not have such a directly negative impact on latch-up immunity. However, the trend towards bulk technologies with high resistance substrates, to achieve better RF performance and cost-effectiveness, can increase latch-up sensitivity. The design for latch-up loses some margin as the devices become more susceptible to accidental latch-up effects, especially when tested with system level IEC pulse tests.

6.1.2 HBM/CDM Events Compared to System Level IEC ESD Stress Pulses

Historically, there has been confusion about the difference between the HBM stress and the system level IEC ESD stress pulse. The models are different: HBM refers to a human discharging through the skin to a pin of the IC, whereas system level IEC ESD stress refers to a human discharging through a metallic tool to the system. As a result, the two models are represented very differently. For HBM, a 100 pF capacitor is charged to 1 kV and discharged through a 1500 Ω resistor. For system level IEC ESD stress, a 150 pF capacitor is charged to 8 kV and discharged through a 330 Ω resistor. When compared to the HBM event, the system level peak current is much higher – 0.7 A/kV vs. 3.75 A/kV, although both event time domains are approximately the same. With the system level requirement of 8 kV vs. 1 kV for HBM, the corresponding IEC peak current level translates to greater than 30 times larger in magnitude. The energy under the pulse for the system level IEC test requirement is thus more than one order of magnitude larger than the required HBM level for safe handling. With this difference in energy, any generated physical failures and their failure modes would be distinctly different. For the system level IEC stress it is not uncommon to see failures resembling an EOS failure mode. The event rise-times are also different: 2-10 ns for HBM compared to 0.6-1.0 ns for the system level IEC stress. But the similarity comes through for the CDM event where the rise time is typically in the 0.2-0.3 ns range. The actual system level IEC pulse has a shape with a CDM-like initial pulse (~30 A at 8 kV) followed by an HBM-like tail pulse (~15 A at 8 kV). Depending on the design, physical damage can be found which mimics EOS damage or HBM/CDM like failure modes. However, the physical damage in total only represents a minority of the relevant system failures due to system level ESD stress. 'Soft' or resettable failures are more prevalent system level failures . Soft failures do not correlate to any failure mode known from IC-level ESD tests.

6.1.3 Marginal Impact of HBM/CDM Robustness on System Level IEC Stress Robustness

HBM stress testing is intended to gauge the ability of an IC pin to protect against handling in an ESD protected area. The most relevant pin combination is stressing to GND. However when the same IC is now placed on a system board, the internal pins do not directly encounter the HBM event. Even if some *residual noise voltage* appears, it has no relation to the HBM event and thus the reduction in IC ESD levels would play no role for these types of pins. Another distinct difference between HBM and the system level IEC stress testing is that HBM is an unpowered test, while the system level IEC test is usually performed with power applied to the system. As a result, the discharge paths are different. Therefore, predicting the survival of an internal pin during system level ESD events has no correlation to its measured IC HBM performance. Some comments on the CDM reduction are also in order. When an IC IO pin is designed for CDM performance, care is taken to suppress transient voltage overshoots from the very fast rise times. When internal pins are in the system, they do not see the same CDM discharges, and the transient voltage overshoot is not an issue. In summary, reducing both HBM and CDM IC levels while safely meeting all IC handling and manufacturing requirements would have no influence on system level protection for all pins that do not interface with the outside world. Even in the case of external pins, an efficient and systematic procedure can be followed using the transmission line pulse information on these pins without any particular regard to HBM and CDM IC levels. This is a more effective method for system design than relying on specified HBM and CDM IC levels. This method, known as "System-Efficient ESD Design" (SEED), is described in detail in Section 6.6. Demanding artificially high HBM and CDM IC levels for system protection of the external pins can potentially backfire as the internal ESD clamp can severely interfere with operation of the external clamp. An example of this is illustrated in Section 6.7.

6.1.4 Impact of Latch-Up Sensitivity on System Level IEC ESD Stress

As mentioned earlier, increased latch-up sensitivity can impact on the system robustness. This applies strongly to external pins but can also affect internal pins. If a soft failure from system ESD testing is seen on internal pins it could be related to EMI phenomenon and indicate a need for improved shielding of the IC on the board. But if a soft failure is seen at the external pin, the effect could be coming from the fast rise time of the initial spike of the system level IEC ESD waveform. The powered test often exposes this sensitivity. Close attention to the System-Efficient ESD Design is required in these cases.

On a final note, looking forward, system protection issues have to be addressed with the new realistic ESD levels kept in proper perspective. These are described in detail along with a co-design methodology for achieving effective integrated system protection design in the following sections.

6.2 Secondary Effects of System Level ESD Stress - Impact on Internal Pins during System Stress

While internal pins should not be directly affected by system level stress, they can still be stressed during a system level ESD event due to secondary effects such as electro-magnetic coupling from directly stressed external pins and/or discharge from the case to the circuit board. These secondary risks are described and methods for reducing exposure to them are discussed in the following paragraphs.

A system level ESD pulse injected onto the external ports of a system can either capacitively or inductively couple to PCB traces neighboring the forced trace, inducing a voltage spike on the neighboring traces and any internal pins connected to these traces. The overvoltage spike can disturb the IC function in various ways:

1. It can induce a malfunction of the IC due to a misinterpreted signal.
2. It can cause surges on the supply traces leading to an upset of the IC.
3. It can cause latch-up of the circuit attached to the IC pin (e.g. the on-chip protection element).

All three effects correlate to the magnitude of the overvoltage spike appearing at the IC pin. Since the induced voltage spike is only caused by the sharply rising or falling part of the forced primary pulse, the energy that is electromagnetically coupled to any neighboring trace is typically too low to cause physical damage to internal pins connected to that trace. In extreme cases where no board level measures are taken, physical damage has been observed at higher levels of forced pulses. However, functional problems at lower stress levels are observed even when no physical damage has occurred yet. These functional problems at lower stress levels almost always precede physical damage when the stress level is increased step-by-step.

To control induced voltage spikes on internal pins, board designers must be aware of and work to mitigate electromagnetic coupling between external and internal traces during board design. For example, care must be taken to limit the maximum length the trace of an external pin runs unshielded parallel to the trace of an internal pin. Passive board elements may also be used to damp the voltage spike on both the forced trace and the coupled victim trace(s) next to it. Note that on-chip ESD protection optimized for HBM or CDM stress does not contribute to the reduction of these perturbation effects as they are designed for the protection of the unpowered device. To the contrary, a highly efficient on-chip ESD protection is usually more prone to latch-up which enhances the susceptibility to functional failures.

If there is a direct discharge inside the case to a PCB trace, e.g. arcing from metallic parts of the case, internal pins connected to that trace may be damaged. The mechanism is distinctly different from electromagnetic field coupling and can only be controlled by distance rules and selection of appropriate materials. The case design must provide low impedance from any point on the case to the point where it is connected to ground. If the impedance through the case is too high, arcing to the board at the board attach points or at locations where the case is physically very close to the board may occur. Should this happen, charge may be injected directly into a internal interconnect, leading to destruction of its circuits.

Whenever discharge inside the case cannot be completely eliminated, the pins connected to the endangered traces need to be considered as exposed or 'external' pins with appropriate system level ESD protection. The achieved HBM level of the IC has only marginal impact on the system robustness in these types of failures due to the very high energy of the system level ESD discharge pulse. Note that here the design effort for the system and board design will not be different between ICs qualified for 1 kV or 2 kV device level HBM.

6.3 Full IEC Protection On-chip Design Strategies

6.3.1 On-chip System Protection Design

Protecting an IC against system level pulses using only on-chip protection measures is a difficult task for a number of reasons:

- The shape of the incoming pulse is largely unknown as the board surroundings greatly influence the pulse.
- Due to radiation and coupling effects, it is difficult to establish exactly which pins will be stressed – multiple pins can be stressed simultaneously.
- During system level tests, the IC can be in a powered state, possibly leading to latch-up failures.
- Other ESD tests only take destructive failure into account, while for system ESD, system disruption needs also to be considered as failure mechanism, depending on the application requirements.
- At the time of IC manufacturing , the off-chip circuit is most often unknown.

These elements complicate the on-chip protection strategy. In general, the unknown pulse shape and the unknown amount of energy per pin make it impossible to guarantee a certain amount of system level robustness by design of on-chip measures. This design is often done without a worst case analysis and thus may lead to either an overdesigned or insufficient design. To design for an ideal 8 kV system level IEC ESD pulse applied to the IC directly is much more straight-forward from the perspective of the ESD design engineer, as the energy and pulse shape are better known. Unfortunately, this would be insufficient for the final system level test. The standardized IEC test is applied at the system level. Therefore, an IC design which did not consider the influence of the system environment would not lead to the intended system level ESD robustness.

Some ESD protection strategies focus closely on specific ESD stress waveforms like HBM. ICs with such a strategy can be inefficient for system level IEC stress due to the unknown pulse shape (even if scaled up to handle the energy). For instance, the pulse duration during system level ESD can be much longer than during regular HBM. It could be assumed that the pulse rise time should be equal or slower than the system level standard waveform, but this is not guaranteed. Due to switching (e.g. of a snapback clamp), a local parallel (parasitic) capacitance can discharge at rates higher than the system level standard, depending on the size of the capacitor, and the impedance seen during discharge.

In case of system level stress, typically the ground line/plate takes a lot of current. Since the supply voltage level of an IC is referenced outside the IC (e.g. in the battery), this current influences the supply voltage seen by the IC: it can become smaller for positive stress, but larger for negative stress. This means that it should be expected that the power clamp will more likely trigger during negative system level stress than positive stress. The power clamp must be able to survive this stress and recover fast enough from a latched state. In some cases where system upset is not allowed, this poses great difficulties in designing an efficient power clamp.

In Table 6, the merits and demerits of typical on-chip ESD protection approaches are compared in terms of trigger speed, latch-up and long duration pulses.

Table 6 – Comparison of typical on-chip protection concepts regarding IEC stress

	Zener	Snapback		bigFET
		SCR based	ggNMOS/Bipolar based	
Trigger	+ ok	+ Can be adjusted	+ Can be adjusted	- Depends on external capacitance
Clamping	+ No latch-up concerns - Too high resistive/huge area	- Latch-up must be avoided + small area possible	Medium area	+ less latch-up critical - Very large (especially if only needed for 1 pin)
Long stress pulse duration	+ ok	+ ok	+ok	- May switch off before stress is finished.

The protection device trigger scheme must be able to handle a broad spectrum of transient pulses, from very fast, to extremely slow. The supply lines are often stabilized with board-level capacitances such that the voltage rise time can be very slow, though the voltage can still rise above the critical level. For transient (rise time) triggered clamps, this can be a critical issue.

During clamping, latch-up must be avoided. Therefore, an approach with a holding voltage below Vdd poses a risk, since the power supply would then inject large currents into the system as the protection device would remain on. The duration of the event is also unknown. Techniques where the clamp's turn off is defined by an RC scheme after a fixed time can therefore be considered risky.

Zeners can shunt system level stresses well, but can consume significant silicon area. They remain fairly popular in some high voltage processes, but are hardly used in advanced CMOS anymore where minimal protection device area is needed.

SCRs are widely popular as ESD protection structures because of their high current capabilities per unit area, low capacitance, and effective clamping behavior (low holding voltage, low on-resistance). They can be used as local clamp and/or as power clamp. SCRs have a number of important advantages for use as system level protection devices. Due to their small area per failure current, it is relatively easy to scale up these devices to dissipate large energies. In many cases the stress needs to be dissipated to the lowest impedance pad, which in most cases is ground due to the many bond pads connected to Vss. Since the SCR can be placed locally, the current can be redirected to ground without passing the supply line. This is an important advantage to avoid system upset.

The drawback of SCRs is their low holding voltage. Thus careful engineering must be applied to avoid latch-up. This can be done either by increasing the holding voltage, or by increasing the trigger/holding current.

Bipolar/ggNMOS based circuits have good properties for dissipating system level stress as well. In general triggering, clamping and turn off behavior can be designed to cover a wide range of specifications. The optimization of these circuits and devices is not easy though, and is often very process dependent. Multi finger triggering issues can be dependent on the applied pulse; trigger voltage engineering is often necessary. For very large energies, the area consumption can be significant.

BigFET are the more risky approach. The trigger voltage is not guaranteed if the higher frequencies are filtered from the ESD pulse. This can result in damage even for a low level pulse. The capacitance seen by the IC is the most critical factor: for small capacitances, the ESD pulse reaching the IC has sufficiently high frequency components to trigger the bigFET clamp. For medium capacitances, the bigFET might not trigger. For large capacitances, the voltage rise will be too small to endanger the core. The risk for latch-up is minimal if guard banding is done correctly. It is however possible that the bigFET clamp will draw some additional leakage current for a small period of time, as the trigger circuit might weakly bias the bigFET gate.

If only one pin needs system level protection, increasing the bigFET for the full 8 kV IEC specification is a significant overdesign. Concerning turn off behavior: if the energy at low frequencies is too high, the clamp's behavior will have increased impedance (i.e. turned off) while significant energy remains in the pulse, again causing damage.

Special care should be taken with high voltage technologies, as the latch-up threat is even more prominent. For high voltage applications, the system level stress is a very hazardous test, since it is extremely difficult to shunt high current at a holding voltage level high enough to avoid any latch-up issue. This often results in very large area clamps. Also, more spacing is needed from the protection device to core circuitry for high voltage technologies, to protect against latch-up to nearby core circuitry. Specific bipolar circuits are commonly used, but they are often very process specific.

6.3.2 Interaction Between On-chip Protection and IEC Waveforms

This section lists various types of devices and their reaction to IEC pulses.

6.3.2.1 Devices that May Not Trigger Properly During 1st Pulse

If the response time of the PCB diode is too slow, or if there is no such protection at all, the on-chip ESD concept has to shunt an initial current peak and clamp the voltage seen by the sensitive nodes. The amplitude of this initial peak is strongly influenced by on-board passives. Usually only a minor part of this initial current spike will reach the IC. Nevertheless, on-chip protection design needs to take into account this part of the system level ESD stress which can be experienced by the external pins of the IC.

The initial rise time described in the IEC standard is fast (0.6 ns – 1.0 ns), but is still slower compared to the CDM specification (~100-200 ps). Taking into account however that the IEC peak current is a factor of three to five times higher compared to CDM specification of 500-1000 V (dependent on package size and characteristics), the dI/dts from both models can be very comparable. However, package and board parasitics, e.g. bond-wire inductance or inductive trace coupling, can slow down the pulse significantly. Therefore, the on-chip protection schemes must be designed for the resulting wide range of dI/dt. The failure modes due to the non-triggering of the on-board diode during the first transient resemble CDM fails of IO circuitry. High levels of CDM protection at the external pins, able to handle 5-10 A of CDM current, will reduce the risk of this type of failure. However, this may not be practical; especially for large packaged IC devices with high speed serial link (HSS) IO designs [2].

It should also be noted, that only the CDM discharge path to the local ground or the local supply line is relevant for shunting the 1st transient of a system level ESD pulses. IC level CDM testing often addresses different failure mechanisms which are located inside the circuitry of the IC and as such are system level irrelevant.

6.3.2.2 Device Designs that Require Energy Absorption from the 2nd Pulse

If the quasi-static clamping behavior of the on-board diode is insufficient or there is no such protection at all, the on-chip ESD concept has to shunt a large current causing significant energy dissipation in the device. This is ~10 times the energy of the typical HBM protection levels. Accordingly, the area of the protection devices has to scale up by a factor of ~10, increasing the parasitic capacitance of the IO circuit in order to shunt the full system level pulse on chip. On-board protection elements are capable of shunting large stress currents at lower capacitive loads and are beneficial for the overall performance of the system. In an optimum concept, the on-board diode and the on-chip ESD protection respectively act like the primary and secondary stages of a typical input protection scheme, where the current carrying capabilities of both branches are balanced by the serial impedance. Figure 26 illustrates how the TVS is properly isolated by on-board impedance from the IC's ESD clamp.

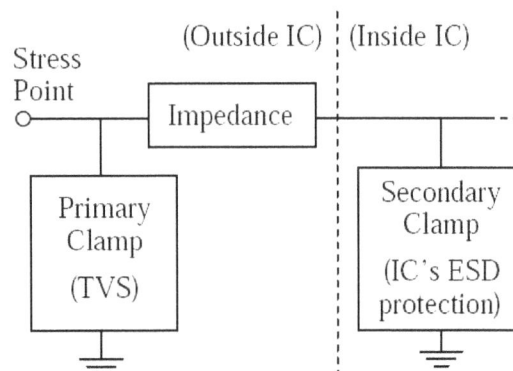

Figure 26: Two stage system level ESD protection by matched primary (on-board TVS) and secondary (IC ESD protection) clamps [3]

6.3.2.3 Devices that Require Latch-Up Immunity from the IEC Pulse

IC and system latch-up immunity depends on how well protection structures can filter different noise and stress pulses. ICs can see mainly two major kinds of high frequency disturbances in a system: direct conducted or radiated current pulses due to ESD and induced radiated noise which is emitted from internal RF lines. Conducted waveforms depend on the source circuit and pulse transmission line impedances and can have high current and voltage amplitudes in the sub GHz range. Radiated noise typically has main frequencies spread from MHz to several GHz range and has relatively low current and voltage amplitudes compared to lines with direct pulse injection. An example of the noise levels are shown in Figure 27 and Table 7. Capacitively coupled noise has mainly high frequency components remaining as the capacitance acts like a high pass filter. On-board and on-chip protection structures have to be able to filter both these types of high amplitude stress pulses and withstand high frequency interferences without disturbing product operation.

Product ESD/EMC and latch-up immunity is typically validated according to the IEC 61000-4-2 standard by using both contact and air discharges. These stress pulses create both conducted and radiated disturbances in signal and power lines and may lead to damages, system latch-up or product operational failures. Operational failures, such as software resets, are the most common failure symptoms. The level of internal disturbances depends mainly on the system topology, including mechanical design and PCB layout. Latch-up or reset-sensitive components should be protected or located in a layout in such a way that maintains the immunity level. For example, disturbances can be measured with near-field probes if the target area is accessible during the stress. However, very often measurements are challenging to obtain without affecting system configuration and immunity problems are solved just by trying out different designs.

A more sophisticated way to solve immunity challenges is to use simulation. For example, 3D *Time Domain Transient* simulation can be used to calculate induced peak voltages, currents and waveform frequencies with selected stress waveforms [4]. Simulations are accurate if the product physical design, connections and materials are known in detail. Two example cases are shown in Figure 27 and Table 7 where the internal interference frequencies and peak voltages are simulated when a product metal cover is stressed with 1 kV and 8 kV IEC contact pulses. This information can be used to optimize system design either by preventing the interference with mechanical changes (better shield design or grounding), by adding filter components in the PCB layout, or by specifying an appropriate on-chip protection. The pulse resonance frequency is design-specific and the residual voltage depends on the impedance and location of the signal line.

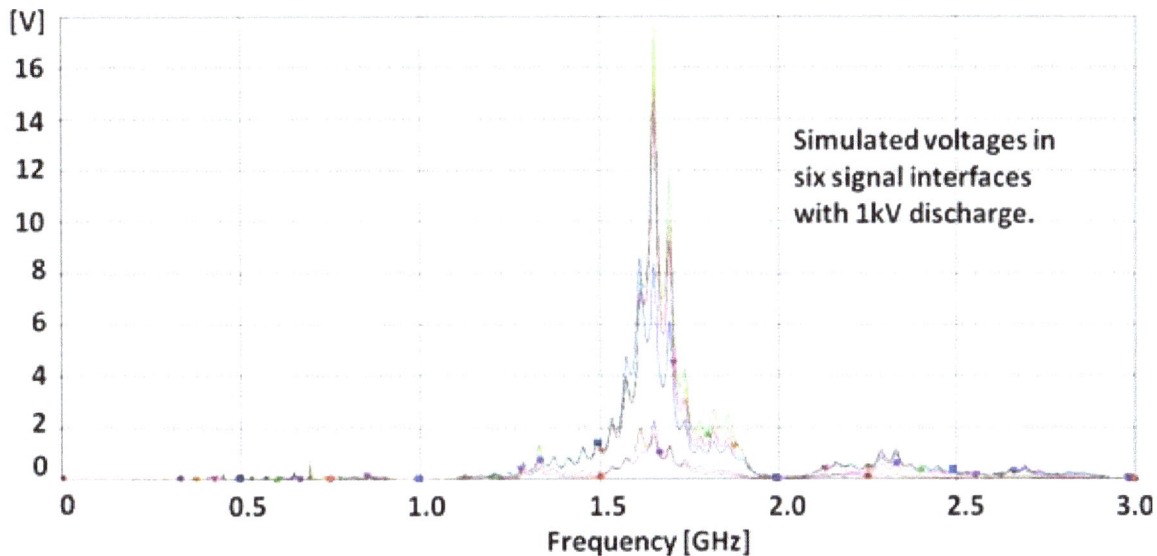

Figure 27: Time domain transient simulation of internal disturbances with 1 kV IEC pulse.
Depending on the components used and the operation voltage for example 10 V maximum induced noise level can be set as a target. More care in ESD/EMC design is needed if a signal line has higher varying voltages and the noise frequency is close to system clock.

Table 7. An example of simulated peak voltages and frequencies with different mechanical designs.

Peak voltages scaled to 8kV discharge and the related ringing frequency				
Cases	GND (0.1Ω)	Conn (10Ω)	CLK (10kΩ)	DATA (10kΩ)
CASE1	0.012V @ 2.2GHz	5.4V @ 1.8GHz	41V @ 1.9GHz	22V @ 1.8GHz
CASE2	0.005V @ 3.2GHz	0.84V @ 1.9GHz	15V @ 2.08GHz	6.3V @ 2.0GHz
CASE3	0.003V @ 2.1GHz	1.5V @ 2.0GHz	15V @ 2.1GHz	11V @ 2.3GHz

6.3.3 Discussion of Pro & Cons of Full On-chip IEC Protection

Based on the preceding information, it is possible to design an IC itself to be robust against an IEC ESD event; however, in the context of robust system level design, the practicality and effectiveness of such an approach must be considered. The typical motivation of equipment manufacturers for pushing IEC protection requirements onto ICs is the reduction of system level costs and area. Consequently, the practicality and effectiveness of IC-level IEC protection should be gauged in terms of its ability to reduce the overall system level protection measures and costs.

- The primary benefit of designing external pins of an IC to be IEC-robust is the IC itself will be more impervious to IEC-induced physical damage and operational upset, i.e. soft failures.
- The primary problems with designing external pins of an IC to be IEC-robust are:
    1) The added IC-level costs

2) The misconception that a single IC can guarantee IEC robustness of the overall system

An IEC-robust IC can eliminate the need for on-board ESD protection devices directly at an interface; however, the costs associated with IC-level IEC robustness – even for only a small set of external pins (typically two to six pins) – should be calculated to determine if overall system cost and performance are optimized. As mentioned previously, without information regarding the overall system design (e.g. what other components will be placed along with the IC) and how the system will be IEC tested, an IC would need to be designed to handle the worst-case IEC stress. Under such an assumption, IC-levels costs are impacted by two main factors.

1) Increased IC area: having to conduct the tens of amperes from an IEC discharge can increase IC area by as much as 30% due to the following determinants (in turn the increased silicon area can then drive the need for a larger package).

   a. For IEC-rated pins, ESD protection area will increase to withstand the higher peak currents of the IEC ESD event compared to HBM and CDM events; in general, an order of magnitude area increase can be expected for 8 kV contact protection compared to 2 kV HBM protection (Figures 28 and 29).

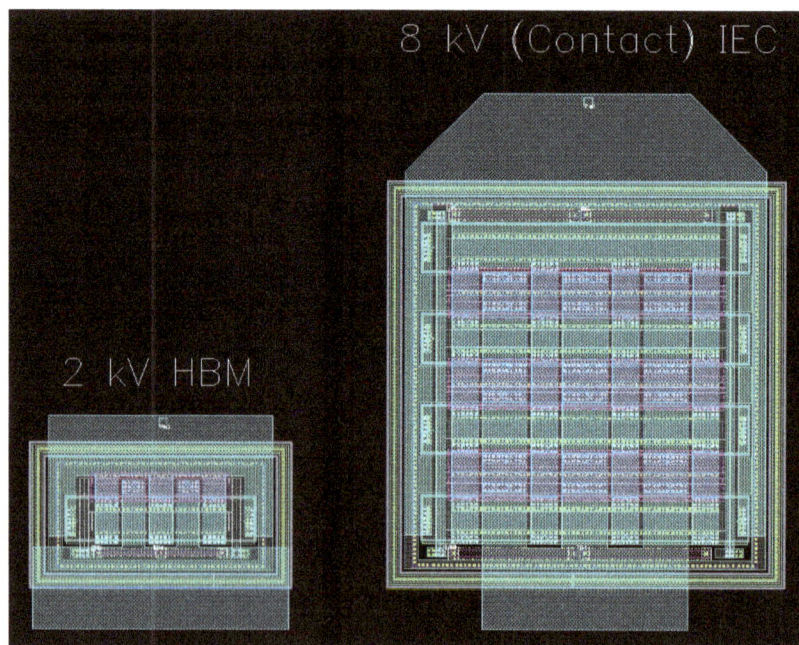

Figure 28: Comparison of 10 V ESD cell size in 1.0 um BiCMOS technology: 2.0 kV HBM version versus 8.0 kV (contact) IEC version.

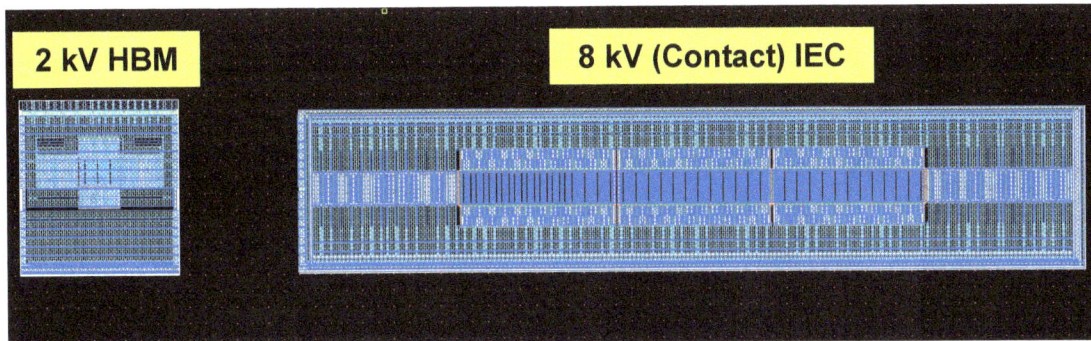

Figure 29: Comparison of ESD cell size in 65 nm CMOS technology: 2.0 kV HBM version versus 8.0 kV (contact) IEC version.

IC silicon technologies are typically optimized for lateral transistor performance; discrete (on-board) IEC protection technologies are specially optimized for vertical current flow. The resulting improvements in active conduction area and thermal impedance mean that discrete IEC protection devices can typically provide greater IEC performance with smaller active area and, consequently, with less degradation of system operation.

b. IC-level metals required to carry the current from IEC discharges will need to be widened by roughly an order of magnitude (comparing 8 kV contact protection to 2 kV HBM protection) to lower resistance and maintain reliable current densities.

c. The higher level of substrate injection that will occur during an IEC discharge drives more stringent guard-ring and inter-device spacing rules to avoid IEC-induced latch-up [5]. Unlike the area impact of ESD protection and its associated metal routing, which is localized to the IEC-rated pins (typically two to six pins), the area impact of IEC LU protection affects the entire IC, making it the most significant source of the area increase.

2) Increase in design time: having to contend with conducted and radiated energies from an IEC discharge increases design complexity. This added design complexity can directly translate to a 2x – 3x increase in the overall design time required to produce a working IC that passes the IEC requirement.

a. Due to IEC testing involving powered-up operation, an IC's functional and reset circuitry must be designed to account for the stress effects. Due to possible coupling effects, the possible impact to circuit design must be considered for all IC pins. As an example, an audio amplifier product in a 1.0 um BiCMOS technology required two additional metal-level design revisions to eliminate LU and functional issues which occurred during IEC testing.

b. As mentioned previously in this section, the possible variations in the stress waveform reaching the IC and the nature of powered-up stress testing restrict the choices for ESD protection on both the external and internal pins. ESD protection that would be optimum for HBM/CDM protection may not be usable due to system level ESD protection requirements; the resulting protection choice can then lead to more stringent protection implementation rules. As an example, an industrial transceiver product in a 0.25 um BiCMOS technology required three all-level design passes to successfully implement protection compatible with functionality requirements, HBM/CDM protection requirements and IEC protection requirements.

As with a single on-board ESD protection device, a single IC – independent of its IEC robustness – will have little effect on the propagation of the radiated and conducted energies throughout a system if some level of effort and cost is not budgeted for proper system design. While an IEC-robust IC can eliminate the need for on-board ESD protection devices directly at an interface, proper enclosure design, proper system functional design, proper overall system protection design and proper circuit-board layout, are all still required to eliminate the occurrences of both system level soft failures and physical damage.

While IEC-robust ICs are relatively new, IEC-robust systems are not. It is known that IEC-robust systems can be built successfully using ICs with typical HBM robustness ratings from 500 V to several kV. It is also apparent that a system can fail IEC testing if an IEC-robust IC is improperly relied upon for protection within a system. A resulting conclusion is that proper system level protection measures are still required along with IEC-robust ICs to ensure overall system level IEC robustness. IC-level IEC protection can be a part of robust system design; however, for IC-level IEC protection to be optimally practical and effective, equipment manufacturers will need to work with their suppliers regarding aspects of their system design and IEC test methodologies.

6.4 Common On-board System level ESD Protection Approaches

6.4.1 General Aspects of System Level Protection

PCB protection strategies depend largely on product physical design and operational requirements. The primary system level ESD protection relies heavily on EMC design of both the system and the PCB that keep most of the ESD/EMC energy outside of the system. With a good cover design, the on-board protection can focus on external connections as well as on the cover hole and seam areas which may leak ESD energy inside. Non-grounded and badly grounded metal structures with high impedance to ground should also be avoided between the covers and electronics as those can leak secondary sparks and high frequency EM noise inside. The PCB ground has to be the first on-board structure to which the residual pulse is directed as the ground spreads and attenuates pulse energies to a safe level. Air gaps between the electronics and covers can also prevent ESD or limit discharge energies, as shown in Figure 30.

External connectors and antennas are typically the most challenging parts from ESD protection point of view. Connectors have signal lines open to unknown stress pulses, and on-board protection methods have to be set to attenuate and guide stress pulses to the electrical ground of the product. At the same time the design must limit electromagnetic disturbances to a level the product can withstand without major faults in operation. Protection must also limit the residual voltage and current to a level which can be handled by the IC's internal ESD protection structures (marked with red dots in Figure 30).

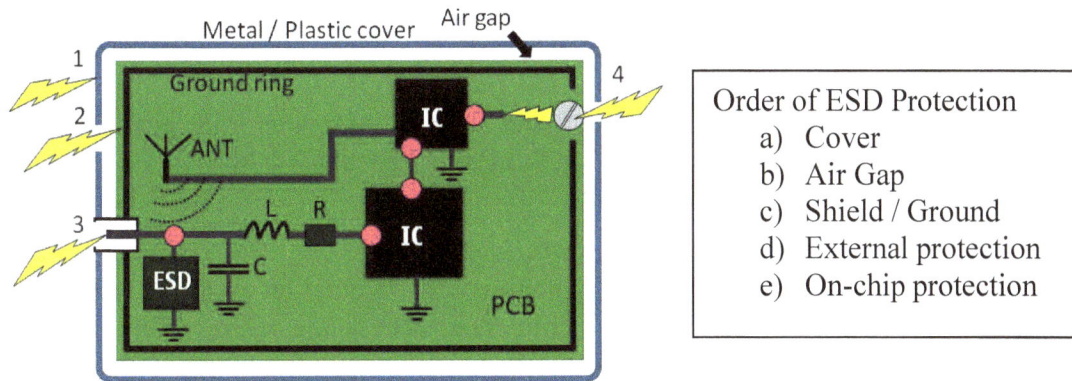

Figure 30: System ESD protection depends on product physical protection (covers), shields and groundings, on-chip and external signal protection and signal integrity targets.

6.4.2 On-board ESD Protection Designs

There are several different methods available to build up an on-board ESD/EMC protection system. The protection can be built with basic passive components or with specified ESD/EMC filters. On-board ESD protection has typically the next common design targets:

- Low capacitive load (especially when used with >100 MHz signal lines)
- Low dynamic resistance after turn on, to drain the ESD current and to keep the residual potential low
- Application specific trigger voltage (e.g. at 1 mA level) and low clamping voltage at relevant current levels of 15 - 30 A (e.g. after 30 ns)
- Low leakage current (especially with portable devices, demand is often <10 nA)
- Sufficient turn on speed to protect against IEC 61000-4-2 pulses
- Capability to withstand multiple ESD pulses with low impedance and return to a normal high impedance state immediately after stress
- Ability to provide both ESD and also EMI protection (depending on the need)
- Ability to withstand IEC 61000-4-2 pulses up to 8 kV contact and 15 kV air
- Ablity to withstand many IEC 61000-4-2 pulses without degradation
- Capability to protect against positive and negative pulses
- Frequency response of the tested circuitry without signal integrity problems while protection component works together with the on-chip protection design
- Easy protection to implement from PCB design point of view
- Small in physical size and low cost

Many of these requirements are mutually exclusive and the protection design must be selected based on PCB level design targets. For example, smaller dynamic resistance may increase protection component capacitance and limit component usage with high speed RF signal lines. The use of ESD protection often requires extra compensation as the impedance match and signal integrity have to be kept along the signal transmission line. This increases component count, requires more space on the PCB and increases product cost.

System designers make the ESD protection component selection based on the parameters described above. However, certain electrical parameters have a major effect on the selection.

- The capacitance of the component can be a major limiting parameter with high speed signal lines. Capacitance should be below 1 pF when frequency exceeds 500 MHz.

- The dynamic resistance strongly affects the residual voltage and current ($V_{clamping} = V_{trigger} + R_{dynamic} \times current$). For example a diode with low R_{dyn}=3 Ω and V_{trig}=7 V can create about 100 V residual peak potential with a 30 A (equivalent to 8 kV contact level) IEC 61000-4-2 peak current pulse. With higher dynamic resistance the residual voltage can be easily hundreds of volts.

- The trigger voltage depends not only on the system operation voltage but also on the inter modulation distortion and harmonics. For example, the coupled voltage amplitude in a 200 Ω trace can be over 15 V when the product has a transmitting antenna close to the PCB.

- The leakage current can be the main limitation, especially with battery operated products where it is often limited e.g. below 10 nA.

The residual potential is typically reported as a voltage left over on a test board when the protection device is used to filter the IEC 61000-4-2 pulse. This voltage waveform is not the same as the one used for IC HBM and CDM validation, and direct comparison between those ESD model current levels and potentials should not be made. Unfortunately, there is no information commonly available about on-chip protection triggering levels which could be used to choose external protection components. The residual potential may also be tested on a board with different impedance than the PCB where the protection component is going to be used. In addition, the protected IC can also be a factor on the residual pulse shape. Some examples of residual pulses are shown in Figure 31 and in Section 6.5.2.

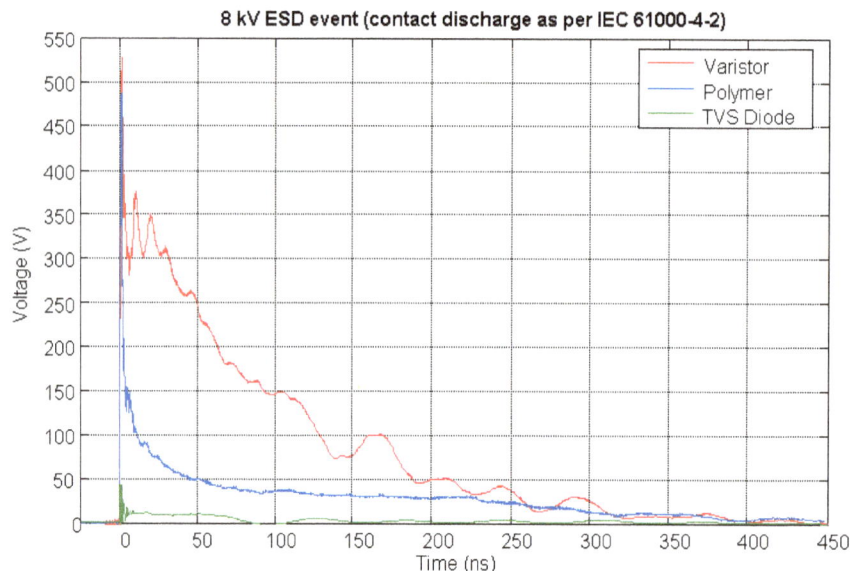

Figure 31: Typical residual potentials for various on-board protection components.

Protection components are also evolving and there are already devices available specifying advanced capacitance, dynamic resistance and leakage current parameters. Impedance matched protection components are also available and can be chosen depending on the case-specific targets. Some typical electrical parameters for diode, varistor / suppressor, polymer and spark gap protection components are presented in Table 8. Some example protection designs are shown in Figures 32 and 33.

Table 8. Typical parameters for SMD ESD protection components which are used to protect low voltage medium & high speed data lines.

	R_{dyn}	VBR @1mA	Clamping	Capacitance (@1MHz)	Ileak	Turn-on	ESD withstand	Linearity
TVS Diode (C > 2 pF)	typ. 0.3..1 Ω	6...20 V	10...30 V(1)	complete range	1 nA	<1 ns	excellent (2)	depends on application
TVS Diode (C < 2 pF)	typ. 1..1.5 Ω	6...20 V	20...40 V	<< 0.5 pF	1 nA	<1 ns	excellent (2)	depends on application
Varistors (C > 2 pF)	> 20 Ω	30 ...300 V	> 100 V	complete range	< 10 nA	< 40 ns	Ok (3)	Ok
Varistors (C < 2 pF)	> 20 Ω	50 ... 300 V	> 200 V	< 2 pF	< 10 nA		limited (3)	Ok
Polymers	< 1 Ω	100 .. 600 V	20 ... 100 V	< 0.5 pF	100 nA	< 10 ns	Ok (3)	Ok
Spark gaps	> 30 Ω	> 250 V	> 200 V	< 0.1 pF	< 1 nA	> 5 ns	Ok	Ok

(1) at 30 ns according to IEC 61000-4-2.
(2) multistrike capability beyond IEC 61000-4-2 without degradation effects.
(3) leakage current may stay high after a single pulse. Some components have a good recovery, but not all.

- TVS Diodes typically have lower capacitance and higher ESD multistrike absorption capability than multilayer varistors. Once the ESD strike is absorbed by the TVS diode, the protection device returns to its high-impedance state very quickly.
- Varistors typically have a trigger voltage over 50 V, clamping voltages over 100 V and a dynamic resistance over 20 Ω after turn on. The capacitance of a varistor can be below 1 pF. Some varistors have significant leakage currents after ESD stress.

- Polymer devices typically have lower capacitances (<0.5 pF). The triggering voltage can be considerably higher than the clamping voltage and polymer components may also have a delay before they return to their high impedance condition after stress. Polymer devices also have lower endurance to ESD strikes.

- Spark gaps have a very low leakage current and very low capacitance, but the triggering voltage is typically more than 250 V. The residual voltage and current will stay high due to the high dynamic spark resistance.

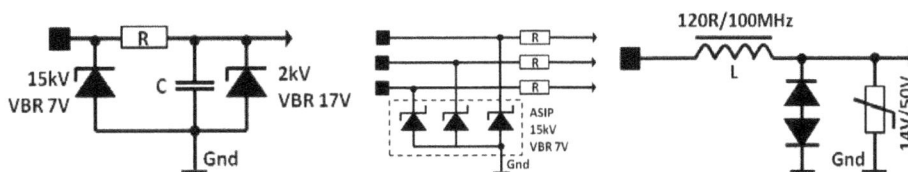

Figure 32: Two level diode, ASIP™ and a choke-varistor-diode designs for IEC 61000-4-2 15 kV air discharge protection.

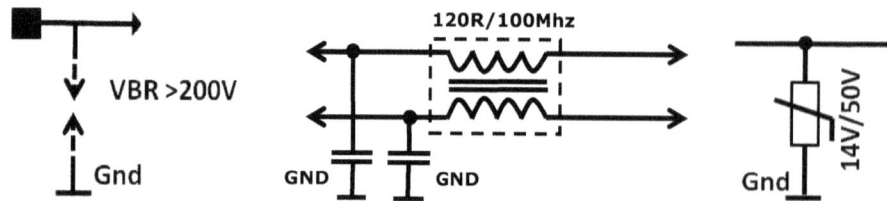

Figure 33: Spark gap, common mode filter and varistor protection.

Passive components can also provide protection against ESD. A series resistor, capacitor to GND or a common mode filter in a discharge path improves system ESD/EMI withstand but may limit signal quality, increase component count and may increase power consumption. Most signal lines have only limited frequency range where the data is transferred and passive components can be used for example to build up a specific LC filter with a low attenuation in the signal frequency. However, there can be space and cost limitations when multiple discrete components are needed for several signal lines. Small SMD passive components do not provide good protection against direct 8 kV IEC pulses but they can be used to protect against secondary pulses.

6.4.3 PCB Parasitic Components

The main purpose of discrete parasitic components, such as common mode filters, capacitors and ferrite beads, is to attenuate noise currents during ESD events. These components are typically not used as primary ESD protection components in a system design, but are located close to a possible ESD stress point in a layout. For example a resistor-capacitor pair or a diode with a known parasitic capacitance and a resistor is used to filter selected noise frequencies.

Parasitic components together with good mechanical protection can be the easiest and lowest cost ESD protection design. Product mechanics can block the major portion of ESD pulses and parasitic components will provide all required protection against residual waveforms and EMC noise. For example a memory card socket is typically accessible in a system and has to withstand contact and air discharge IEC 61000-4-2 pulses. When the socket is made of metal and is grounded to a PCB ground, the spark energy will not flow deep into the card contact pins. However, there will be some induced noise on pins and traces, therefore filter components are needed to prevent EMC disturbances.

6.4.4 'Realistic' Situations in the System

Product shape, the product's position on the test bench and transmission lines can all have an effect on the ESD pulse waveform when it moves through the stressed product. An IEC waveform has a pulse rise time specified from 0.6 to 1 ns when it is measured with a current-sensing transducer [6]. However, extra inductance and impedance mismatch in the discharge path will slow down, reflect and attenuate the current pulse when a real product is stressed with contact discharge. A rise time faster than 1ns with the main frequency of the ESD pulse is rare at the IC pin. Typically main pulse frequencies are below 500 MHz.

The second and third common discharge types are CDE or CBE. One example of fast cable and/or board discharge is shown in Figure 34. If the discharge moves through balanced impedances the waveform can keep its original shape and the component can see the stress according to the original pulse. This is more or less a special case and can exist only when an impedance matched cable or part such as an antenna is connected to a product.

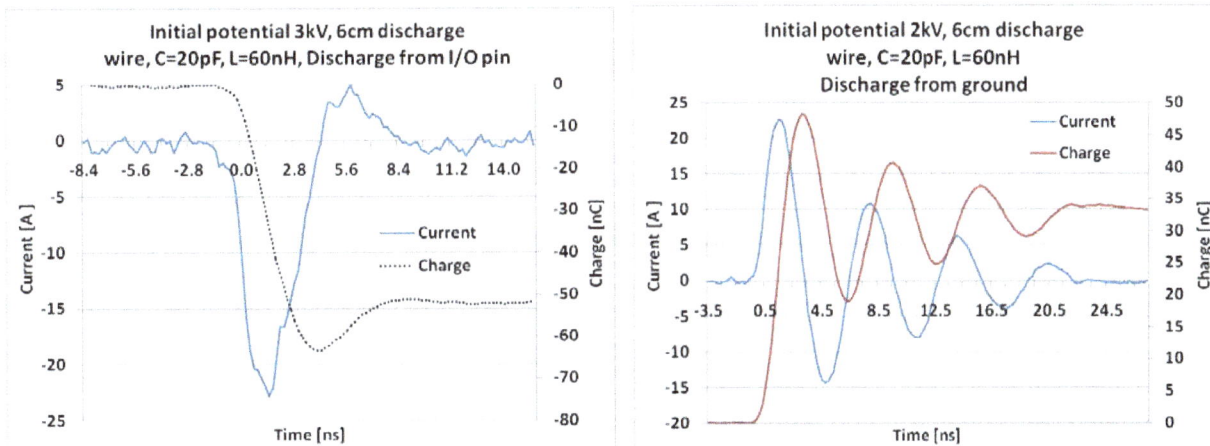

Figure 34: CCE/CBE discharge waveforms from an IO pin and a ground plane.

On-board protection components must trigger before on-chip protection, leaving only the residual current pulse to be handled by the on-chip protection. The residual stress waveform that components would see during IEC system level testing or during a CBE/CDE discharge is very different if compared to an IC HBM and CDM event at the pin. HBM has a time domain in the same range, but the discharge energy is not distributed in a similar way as occurs when the component is part of a system. In a system all IC IOs are terminated to ground, while supply voltages, signal lines and parasitic components vary depending on the PCB design.

A CDM pulse, in comparison, has an extremely fast rise time which typically does not occur when the IC is part of the system. CDM ESD levels are also reported as a voltage withstand levels only. Since package capacitance and effective peak current are not communicated; it is difficult to draw a conclusion on the robustness in terms of ESD current. Also, ICs must withstand high frequency EMC noise in a system, sometimes a fast CDM optimized IO protection may help to handle high frequency EMC noise in a system. However, as these high frequency signals have typically low current and voltage amplitudes (see Section 6.3.2.3), a high current withstand level, which is expected in the CDM validation of a large package device, is in most cases irrelevant.

In summary, both HBM and CDM qualification passing levels of ICs are not very relevant parameters from a system designer point of view and do not provide relevant information for system level ESD/EMC design. Also, on-board protection elements must have the specifications of residual voltage waveforms, dynamic resistance and other important parameters. However, this information varies from supplier to supplier. A System-Efficient ESD Design can only be performed if on-chip and on-board protection component characterization are linked to each other, for example, by using appropriate high current IV data valid for a few nanoseconds to 100 ns pulse duration.

6.4.5 Potential Competition Between on-chip HBM and On-Board System Level Protection

The on-board protection design should take into account the on-chip ESD design window. In other words, the maximum residual voltage over the on-board protection voltage clamping level must be smaller than the ESD failure voltage of the IC pin being protected. Additional margin can be created by the on-chip protection. As the on-chip protection draws current during the ESD event, an additional voltage drop is created over the board parasitics and package bond wires. This voltage drop is distributed over the on-board protection but not over the IC. Although there is no perfect correlation due to the numerous reasons already described, the HBM protection qualities give a reasonable estimate of the on-chip protection capabilities to handle this residual part of the ESD stress.

The time response of the on-board protection can be too slow for the fast rising edge of the system level pulse, depending on the strength of the damping effect of the board parasitics on the system level pulse. The on-chip protection then acts as a first protection.

These calculations are hard to make in the IC design phase, as most of these system related parameters are unknown. Different on-board protection elements can be tested on the final board in a trial and error approach and the best version can be selected. Good on-chip ESD protection can help to increase the number of usable on-board protection devices. Inappropriate on-chip ESD protection might make it impossible to find such a solution. The quality of the on-chip ESD protection is determined by the maximum voltage and current which the discharge path involving the IC pins can withstand. This can be extracted by electrical analysis methods as described in the following section (6.5) and is NOT the HBM withstand voltage!

6.5 Advanced Characterization of ICs for Achieving System Level ESD

6.5.1 IO Characterization of IC

For an effective integrated system protection design, the transient behavior of the ESD clamp at the IO pin must be first understood. TLP is commonly applied for extracting the IV characteristics at high current densities, as any forced DC current at those currents would immediately destroy the device under test. A TLP is a square current pulse generated by discharging a transmission line as shown in Figure 35. Transmission line pulses of 100 ns duration can be related to the energy of an IC level HBM stress pulse, which gives a certain preference to this waveform.

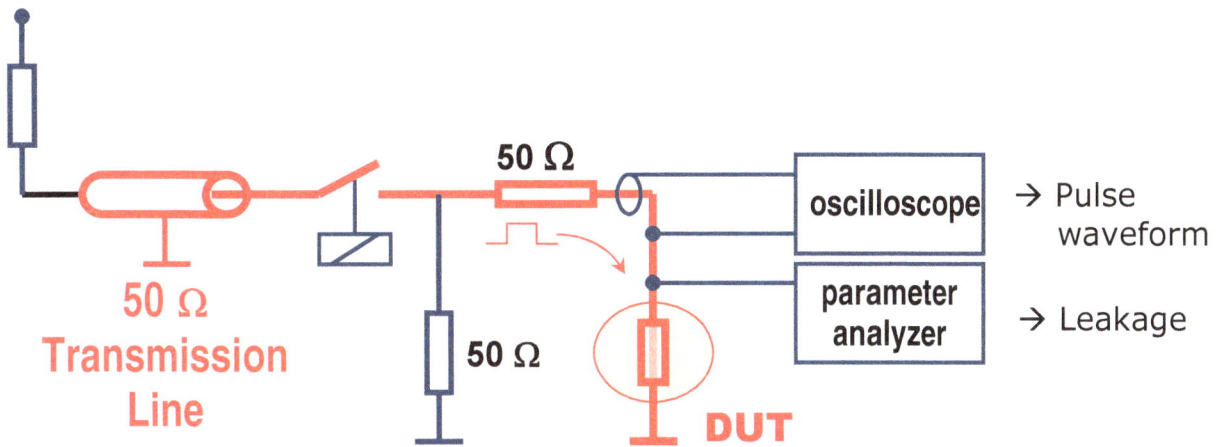

Figure 35: Typical set-up of a Transmission Line Pulse

The current and voltage values are extracted as mean values in the late phase of the pulse, where usually a plateau appears in the waveform (Figure 36).

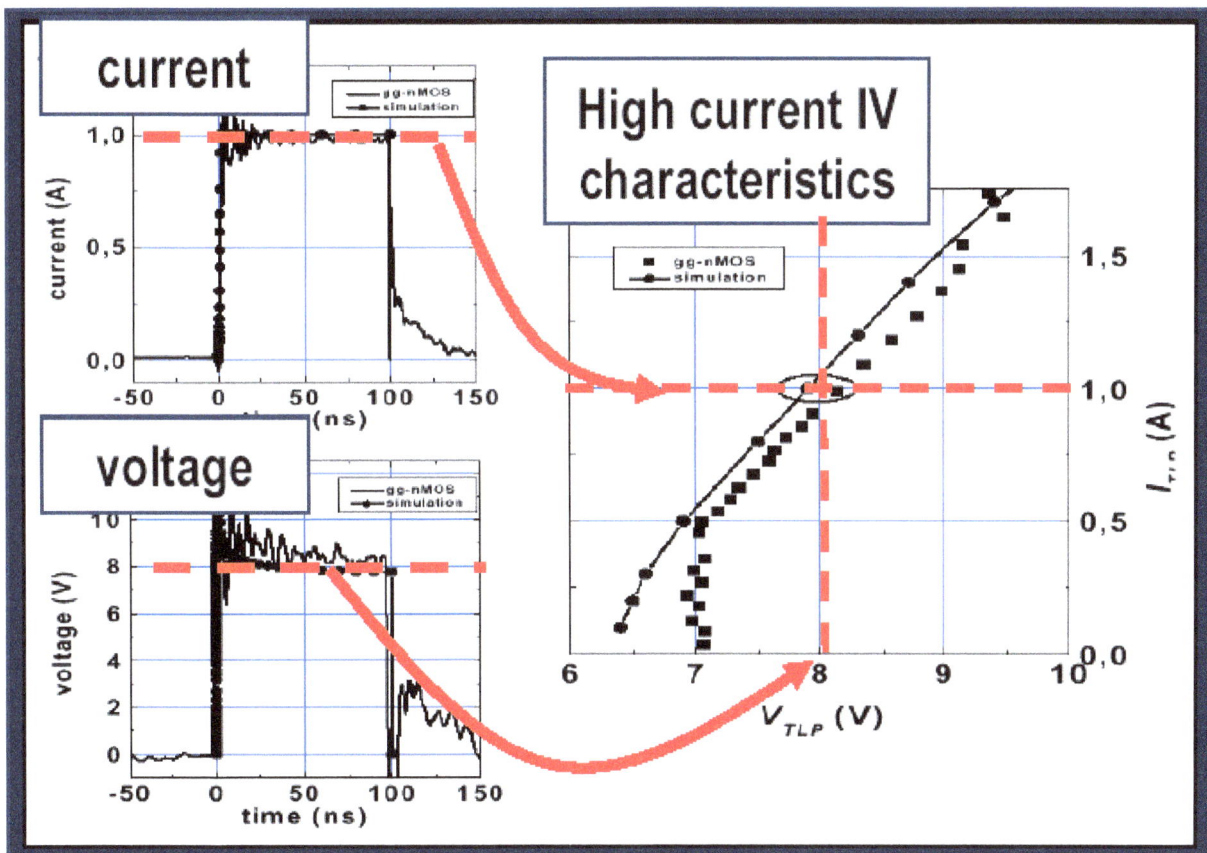

Figure 36: Typical 100 ns TLP I-V characterization for an IC pin.

To assess the possible impact of the initial high frequency (fast slope) contributions of the IEC pulse, very fast (VF) TLP with ns-wide square pulses can be used to sense the transient behaviour of the stressed IO circuit or protection diode. This is especially relevant as fast CDM-like pulses usually trigger different mechanisms, which occur at different levels compared to HBM or 100 ns TLP.

6.5.2 Characterization for PCB Clamps

The Residual Pulse (RP) results from the limited clamping behavior of the (purple) on-board protection diode (Figure 37). The applied pulse is shunted to the respective rail (typically ground) via an on-board shunt device. A voltage drop appears across the shunt device (e.g. on-board protection diode), which leads to a 'residual' voltage pulse sensed by the protected circuit, e.g. the IO circuit of the IC (green).

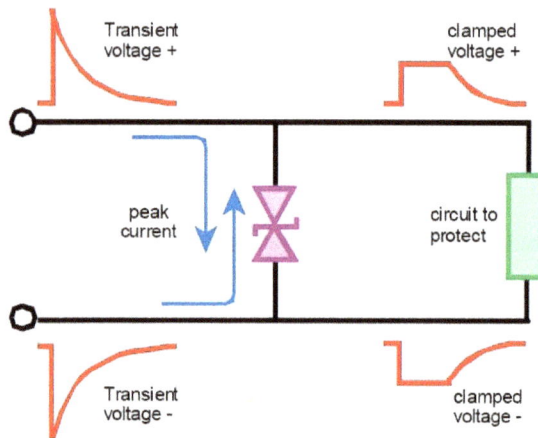

Figure 37: Residual pulse (waveform on the righ hand side) resulting from the finite clamping voltage of PCB diode (purple). Ideal turn-on behavior is assumed.

Typical clamping behavior of various on-board protection diodes is shown in Figure 38. TVS diodes are usually superior in their clamping behavior when compared to varistors for the same application. Figure 38 a) shows the capability to clamp long pulses. However, the transient behavior at turn-on in the first few ns can be even more critical. High voltage overshoots are detected for varistors exceeding the plateau value by more than 100 V as shown in Figure 38 b).

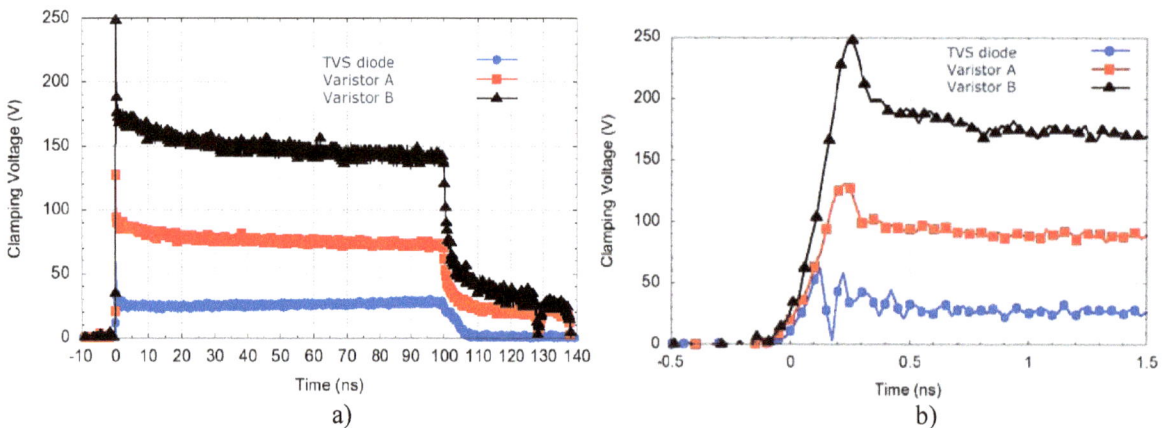

a)   b)

Figure 38: Waveform of various PCB diodes using a)100 ns TLP and b) very fast TLP

A worst case residual pulse can be extracted from the waveforms found in Figure 38 by taking into account the resistance of the PCB trace connected to the IC pin and the voltage drop across the IO circuit.

6.5.3   Characterization of PCB

The purpose of characterizing a PCB regarding system level ESD is to extract a sufficiently detailed RLC network, which allows a simulation of the board that includes the passive components and the ESD voltage clamping elements under system level ESD stress. Commonly an IEC 61000-4-2 pulse waveform is considered. While the initial fast spike might require including even very low parasitic contributions of inductance and resistance to correctly simulate gigahertz transmission, in many practical cases (e.g. automotive engine control unit boards) this initial spike can be neglected due to the large capacitance available on board or in the IC package which damps fast components. 3D field simulators enable the extraction of capacitance and inductance for a given board layout. Cross coupling phenomena caused by mutual inductance and capacitance can be extracted as well. For the sake of simplicity and reduction of the simulation nodes, only the self-inductance and the capacitance to ground will typically be used in a first overall optimization step.

Alternatively, analytical models can be applied to model the various parts of the wiring of the PCB [7, 8].

6.6   System-Efficient ESD Design (SEED) – An Optimized IEC Protection Co-Design for External Pins

6.6.1   Benefits for PCB and IC Designer

When an IEC 61000-4-2 discharge is applied to an external pin on the system board, the main ESD current Ip flows directly to ground through the on-board TVS, but some current Ic will enter a connected IC and flow via its internal protection to ground (see Figure 39).

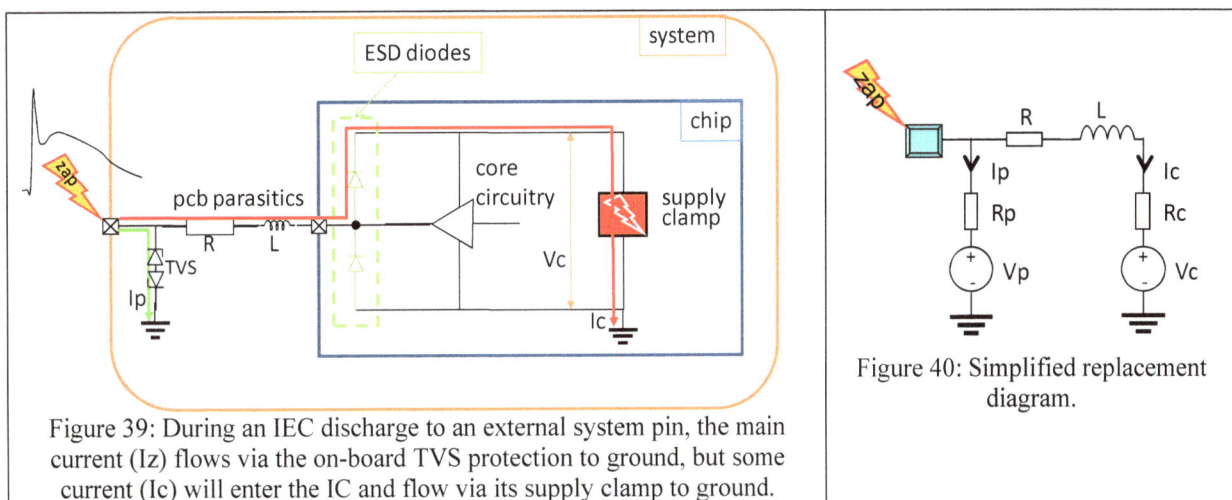

Figure 39: During an IEC discharge to an external system pin, the main current (Iz) flows via the on-board TVS protection to ground, but some current (Ic) will enter the IC and flow via its supply clamp to ground.

Figure 40: Simplified replacement diagram.

In order to prevent damage to the IC, it is important to assess the amount of current Ic which may enter the IC and the associated (over-)voltage Vc across the connected IC circuitry. The calculation may be simplified by approximating the TVS and the on-chip protection by diode-like devices, characterized by their on-voltage Von and on-resistance Ron. A simplified replacement diagram is shown in Figure 40. If all parameters of the devices in Figure 40 are known, straightforward application of Kirchhoff's Current Law will yield the current distribution (Ip, Ic). The resistance R and inductance L of the wiring, both on-board and on-chip, play important roles. Particularly for wire-bonded chips, L may be significant, and the additional impedance may reduce the current from the first peak entering the chip to a negligible value. In that case, the IEC robustness may be estimated by considering the current distribution in the second peak only. Since the timescale of the second peak is very close to the HBM timescale (around 100 ns), the relevant I-V characteristics of all ICs may be estimated using a 100 ns TLP test.

6.6.2 Example of SEED Design Using 100 ns TLP Data

As an example, a 5 V audio output designed in 65 nm CMOS and placed on a board is planned to be protected by an on-board 6 V TVS diode. Figure 41 shows the I-V characteristic of the audio output to VDD (forward biased ESD up-diode). Figure 42 depicts an I-V characteristic of the TVS diode. Both characteristics have been measured separately by means of 100 ns TLP.

The 65 nm chip is wire-bonded inside its package. For an estimated wire inductance of about 1 nH, the additional impedance $Z = \omega L$ equals about 12 Ω for 0.5 ns (first peak) and about 0.2 Ω for 30 ns (second peak), which justifies neglecting the first peak.

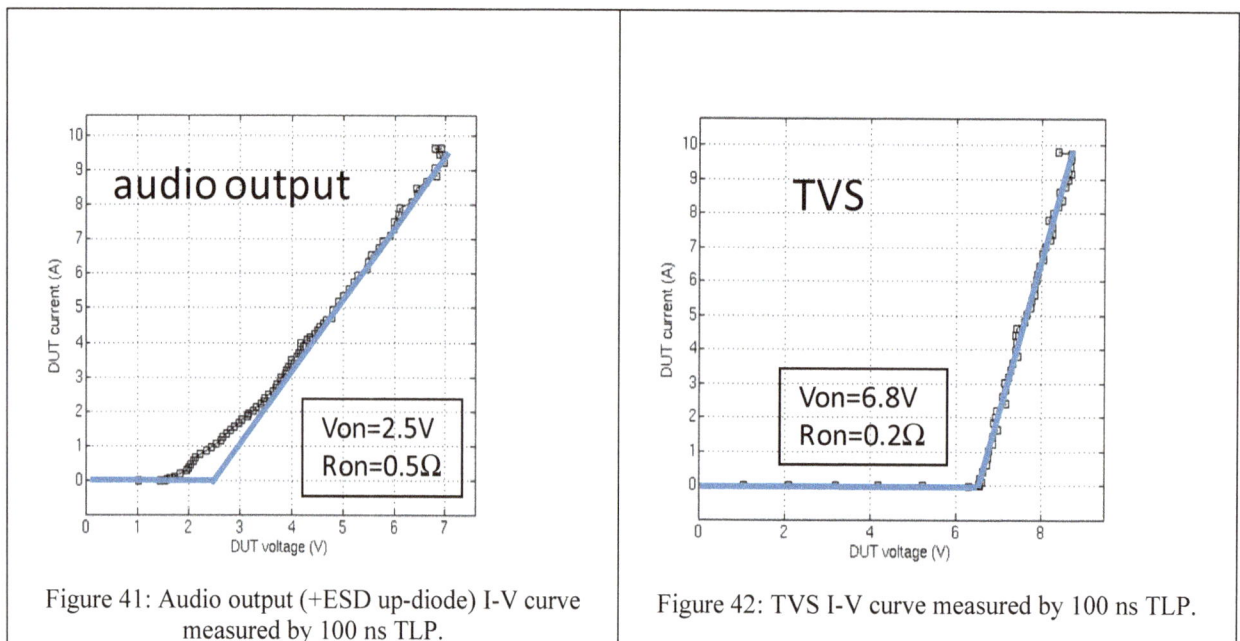

Figure 41: Audio output (+ESD up-diode) I-V curve measured by 100 ns TLP.

Figure 42: TVS I-V curve measured by 100 ns TLP.

Using the device parameters above, the current into the on-chip supply clamp and the ensuing clamp voltage may be calculated for any given on-board resistance R. Figure 43 shows Vc and Ic assuming no additional board resistance R exists. In that case the 'operating point' of the on-chip supply clamp during an IEC discharge (Vc, Ic) turns out to be outside the safe operating area (SOA) for the output domain on chip, which is defined by the design target for the clamp. The SOA is bounded by the maximum ESD current for which the clamp is designed, e.g. Imax = 4 A, and the maximum voltage on the core circuitry before oxide damage will occur, e.g. Vmax = 11 V (e.g. oxide break of thick oxide transistors).

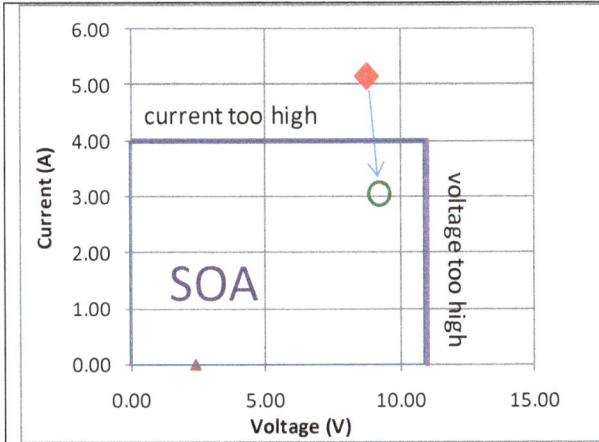

| Figure 43: Increasing the series resistance on-board by 1 Ω moves the output 'operating point' (Vc, Ic) during an 8 kV IEC discharge into the SOA. | Figure 44: Choosing a different TVS with Von= 6 V and Ron=0.1 Ω moves the output 'operating point' (Vc, Ic) during an 8 kV IEC discharge into the SOA. |

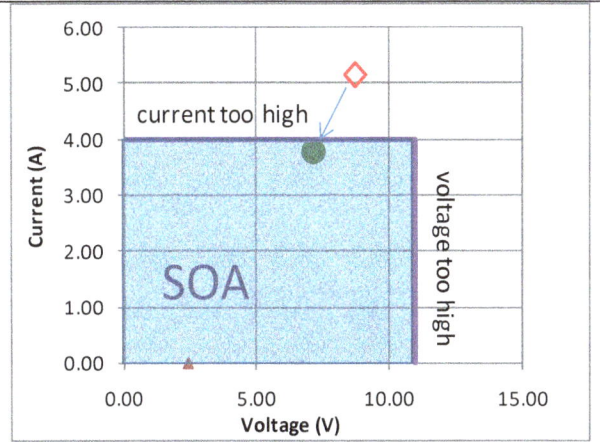

By increasing the on-board resistance R, the current into the clamp supply clamp may be reduced. However, in the case of an audio output, the required efficiency of the power stage usually does not allow increasing the output impedance. An alternative solution is to find another TVS which has either a lower on-voltage or a lower on-resistance. Figure 44 shows that by using a TVS with Von = 6 V and Ron = 0.1 Ω, the output (Vc, Ic) point moves into the SOA.

Note that the example shows a situation with limited options for co-design of the on-chip ESD protection and the on-board system level protection since the output impedance of the audio needs to be low-ohmic. The on-chip protection is low-ohmic as well by design. Therefore, in this situation the only solution is to find an on-board TVS with the proper Von and Ron. Von cannot be lower than the maximum operating voltage of the output (in normal operation) plus some margin. So, the proper solution is to find a TVS with a Von as low as possible and a sufficiently low Ron.

6.6.3 Design Verification Using VFTLP Data

In the previous example, the impact of the first peak in an IEC 61000-4-2 discharge has been neglected in a first approximation. In order to verify this approximation is valid for this design case, the response of the system to a fast initial pulse on the order of the first peak (about 0.7 ns rise time) needs to be tested. A 100 ns TLP tester is too slow for this purpose (rise time of about 10 ns), but a very-fast TLP system with a 2 ns pulse width / 500 ps or faster rise time would be well suited for this test.

6.6.4 A Generic Design Methodology for External Pins (SEED)

From the design example in Section 6.6.2 it is clear that the design of on-chip ESD protection and on-board system level protection cannot be performed separately, but co-design is needed to account for the interaction of the on-chip and on-board protection. The approach illustrated by the example may be generalized into a generic design methodology for each external pin. This concept is called System-Efficient ESD Design (SEED) and comprises the following generic steps (see Figure 45 for an illustration):

SEED Concept Details

1. **IC Supplier provides Transmission Line Pulse (TLP) data on the External Pin**

2. **Board Designer characterizes the Transient Voltage Pulse (TVP) to determine the Residual Pulse Stress (RPS) data (Voltage Vs. Time)**

3. **Board components are adjusted to balance the RPS data to the TLP data**

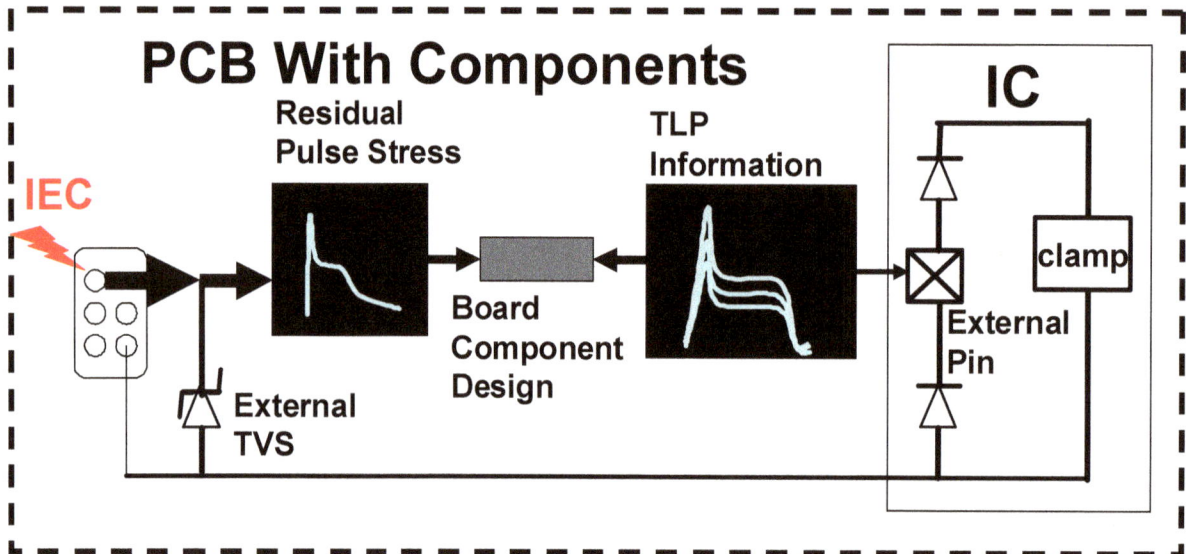

Figure 45: System-Efficient ESD Design (SEED) design methodology

1. For the domain which contains the external pin, design the on-chip protection such that the chip meets the usual 1 kV HBM target as well as a 250 V CDM target, without more than typical margin. This is to ensure that the current path through the on-chip protection circuitry is not lower-ohmic than needed to meet the HBM and CDM targets.

2. For any given pin to be protected, determine the SOA for the on-chip domain connected to the external pin, including both ESD protection and circuitry to be protected. Usually, the current capability of the supply clamp (in case of rail-based protection) will determine the current capability. The maximum ESD voltage will usually be determined by the breakdown voltage of the gate oxide of the core circuitry to be protected.

3. Select a TVS for the external pin with the lowest breakdown voltage which is above the maximum normal operating voltage of the pin plus some sensible margin, e.g. for a 5 V pin, the maximum operating voltage is 5.5 V, so the minimum TVS breakdown voltage should be 6 V.

4. Characterize the external pin under ESD conditions, e.g. by means of 100 ns TLP from pin to ground. Characterize the TVS in the same way.

5. Estimate the resistance and inductance in the path from TVS to chip pin. Determine the current distribution for an 8 kV contact discharge between TVS and external pin.

6. If the (Vc, Ic) operating point of the external pin is outside the SOA, try modifying any of the following parameters to move the operating point inside the SOA:

a. Increase series resistance or inductance of the path between TVS and pin. This is usually not a problem for inputs, but the maximum series resistance for outputs, especially high-power outputs, may be limited.
b. Find a TVS with a lower on-resistance. There will probably be a trade-off between lower on-resistance/capacitance and price.
c. Increase Von of the on-chip protection. This will not always be possible. Usually, the on-chip protection will use library cells which cannot easily be adapted for each new project.
d. Increase voltage level at which damage occurs in the circuit (e.g. by using thick oxide transistors).

6.7 Examples for System Level ESD Protection Design of Typical Interfaces / Ports

6.7.1 USB Designs and Trends

In the USB 2.0 system, the specifications [9] detail the overvoltage conditions as well as the nominal IO voltage range for VSSP of 0 V to VDDP of 3.3 V. An AC stress, which models the reflections at the other termination to support USB 1.1 backward compatibility, spans from -1 V to 4.6 V at 6 MHz with 4 to 20 ns rise and fall time. The short-circuit stress withstand condition requires the circuit to withstand 5.25 V or 0 V DC applied at the pad for 24 hours, which models a short-circuit of the IO pads to bus voltages. In addition, the USB 2.0 IO should support full-speed signaling at 12 Mb/s with a voltage level from 0 to 3.3 V and the high-speed signaling at 480 Mb/s between 0 V and 400 mV. These operational constraints lead to a limitation in the ESD protection concept, e.g. no diode to VDD is allowed. Also the acceptable capacitive load of the ESD protection is limited to a few pF.

In typical usage, cable discharge events are the most relevant and severe ESD threat for the USB 2.0 IO ports. In CDE events, the discharge occurs directly into the pin whereas the system level ESD stress according to IEC 61000-4-2 [6] is not applied directly on the USB pin but to the nearby chassis or connector shield. A typical CDE discharge shows a square-like current waveform overlaid by an initial peak (Figure 46). In an even more critical field event an additional (large) current peak can occur at the end of the pulse due to charge stored on the connecting device.

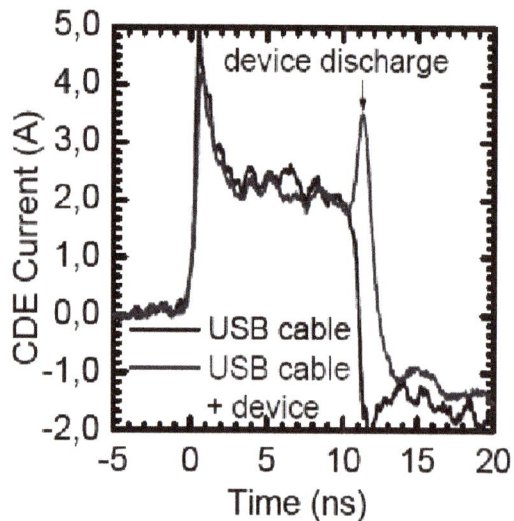

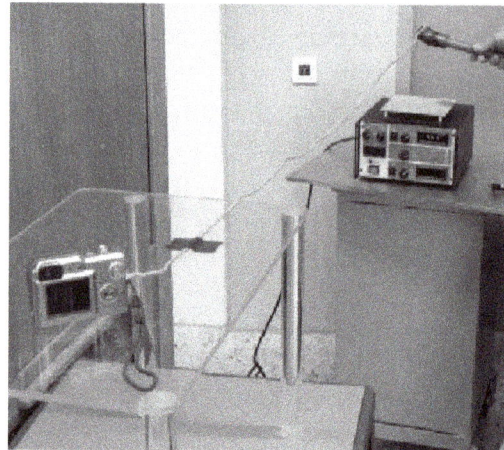

Figure 46: Cable discharge events for USB interfaces [10]

While on-chip only system level ESD protection is feasible for USB interfaces in special cases [11], a combined on-board and on-chip protection scheme is more flexible and better fulfills any EMC constraints. However, on-board and on-chip protection needs to be matched as described in Section 6.6.1, where most of the IEC current is drawn by the on-board shunt element (Figure 47). The relevant parameter is the voltage at fail Vt2, which is not necessarily correlated to the IC level HBM robustness that scales with the failure current parameter It2 (Figure 48). Only a high current IV characterization, acquired by TLP analysis, reveals the matching properties of the on-chip and on-board protection circuitry. In this case any additional serial resistance between the PCB clamping element and IC pin may not be allowed, due to the performance requirements of high speed USB.

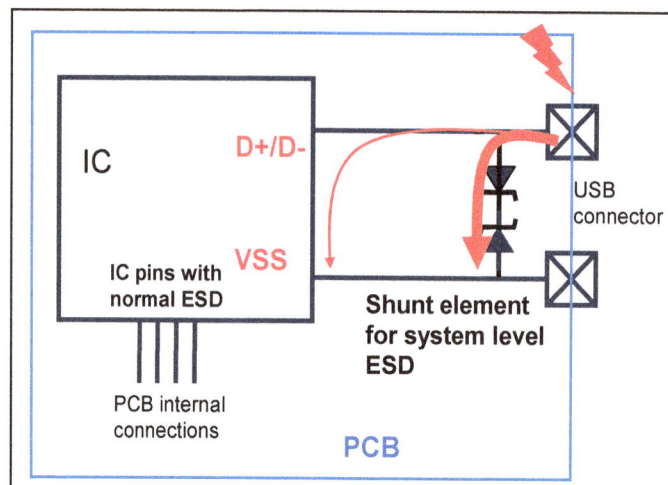

Figure 47: Current paths through on-board PCB and IO circuit

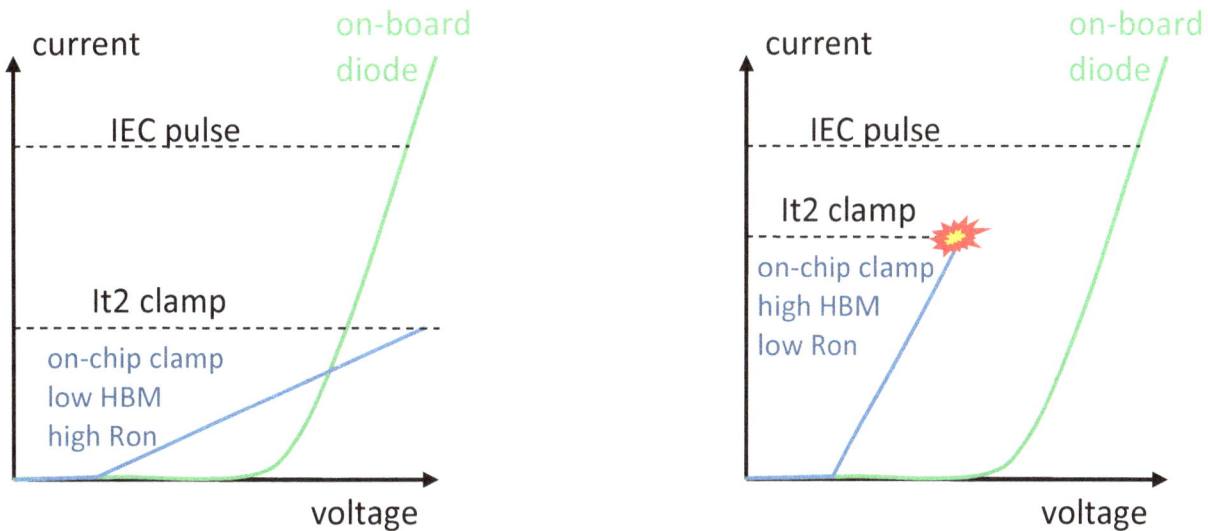

Passes IEC
(On Board Diode takes current)

Fails IEC
(IC takes current)

Figure 48: Matching (left) and mismatch (right) of on-board and on-chip protection IV characteristics. To avoid damage by a parallel current path through protected USB circuitry on the IC, the clamping voltage of the on-board diode Vclamp must be lower than the voltage Vt2 at which the IO (on-chip protection) is damaged.

It should also be noted that these harsh requirements are only applicable to USB connectors of the system (e.g. a mobile system). Inter-chip USB lines do not have this critical exposure.

6.7.2 CAN Interfaces

CAN interfaces have to survive high levels of system level ESD events. CAN ICs are tested with an IEC gun while mounted on small test boards [12]. The mandatory inductance of the common mode choke will damp any initial spike of the IEC pulse (Figure 49). The dominating failure mechanism is due to Joule heating resulting from the broad peak of the IEC waveform.

In the typical application, self-protection of the IC pins against IEC pulses is expected. In this case the placement of additional on-board ESD protection elements (ESD1/2/3) is very restricted, and it is highly recommended that SEED design concepts discussed in this section be used by the OEM and supplier. For example, due to overvoltage requirements of the CAN busses, no diode to VDD is allowed.

The required system ESD specific on-chip protection devices have to satisfy stress levels which exceed common IC ESD protection levels by more than an order of magnitude, Typically, more than 10 times the on-chip layout area, compared to a standard solution, is required to achieve the elevated ESD robustness level.

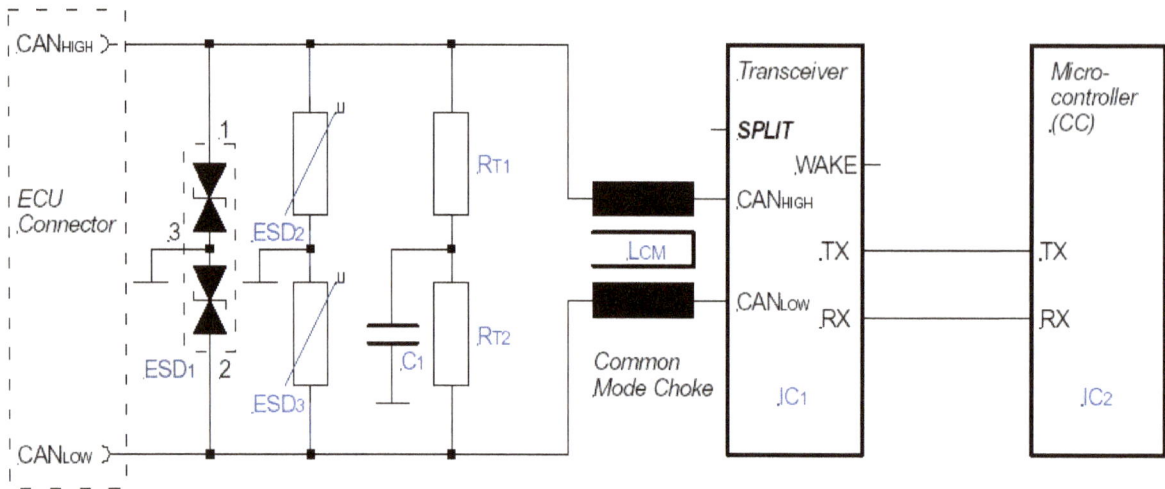

Figure 49: Typical system ESD protection of CANinterfaces [13]. Self-protection of the IC pins, CANlow and CANhigh, is assumed. ESD 1/2/3 can only be placed in exceptions.

6.7.3 Antenna Port Design

The antenna port, e.g. of a mobile device (Figure 50), is not only extremely critical to performance but it also needs to satisfy high system level ESD robustness requirements [14]:

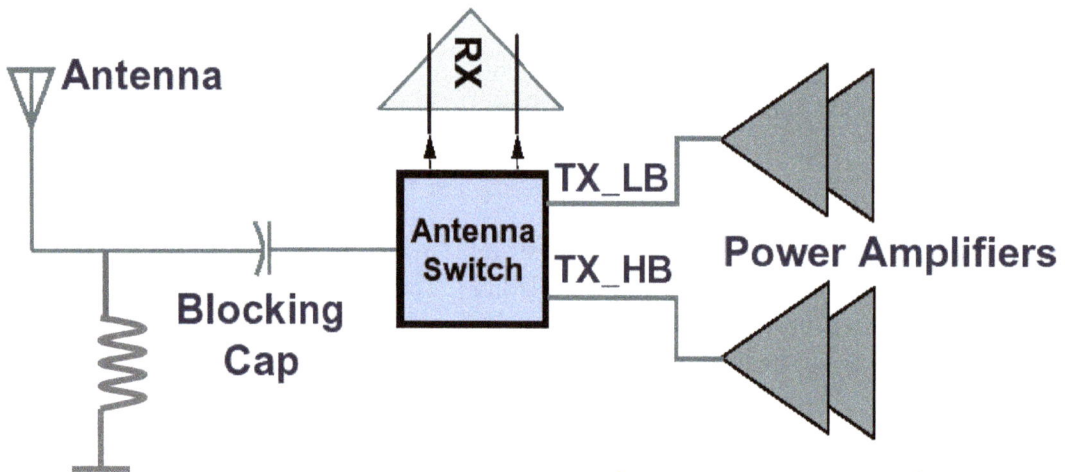

Figure 50: Typical antenna port configuration circuit

Antenna ports apply ac decoupling with a blocking cap. Therefore only limited energy can be transmitted to the chip. Most of the pulse energy is shunted to ground by the inductance of the matching network. However, in a certain frequency range, ESD pulse energy can be transmitted to the pin of the antenna switch. This could be sufficient to damage the sensitive antenna switch or even generate sufficient overvoltage at the receiver (RX) or transmitter (TX) pins to damage SAW filters connected to them. The residual voltage at these ports (RX, TX) is commonly required to be less than 150 V peak-to-peak.

The conventional system level ESD protection approach is to design a RLC network with a sufficiently wideband inductance from antenna port to GND which minimizes the transmission of energy from the IEC pulse into the protected antenna switch. Either an additional PCB clamping device or on-chip protection attached to the antenna IO can shunt the residual pulse efficiently. A cost efficient SEED protection design methodology will be achieved by an on-board protection component integrating filter network and ESD protection diodes.

6.8 Conclusion

6.8.1 Discussion of System Cost for Various Design Strategies

The following list of case histories illustrates that in the past there was no single best solution regarding system level protection design:

1. **DSP product - failed 8 kV air discharge in a system**: The customer insisted on a system level fix, which caused a production delay of 2 months. The engineering effort spent was 3 man-months and the fix involved adding a new component to the board.

2. **DSP product - failed 15 kV contact discharge**: Multiple customers required a fix, which caused a production delay of 3 months. The engineering effort spent was 4 man-months and the fix involved adding a new component to the board. The issue caused a business loss of $1.5 M.

3. **On-chip system level protection** was developed in a 0.13 μm technology, which involved about 4000 μm^2 of additional die area to protect one pin. The total capacitance loading including the metal was 400 fF.

4. **PMU product - failed system level test:** During a system level test the 2 kV HBM protection on the die was destroyed by either CDE or EOS. The solution involved adding TVS protection on-board for some customers and / or improved control for other customers.

5. **Mixed signal product - hot plugging failure:** A hot plugging (powered system connected to powered cable) issue was solved by using shorter cables. In addition the PCB was re-designed to improve system level ESD performance.

6. **Wireless communication product - system failed when using cheap on-board TVS:** The customer discovered that the board worked fine with expensive on-board TVS parts, but with less expensive diodes, system level fails occurred. Reason for the discrepancy was investigated, which involved TLP testing of TVS parts and product pins in order to establish the maximum current flowing into the IC for a given TVS. Based on this data, it was possible to advise the customer to use cheaper diodes on certain pins, while still requiring the expensive TVS for a single pin, which saved 95% of extra cost. The engineering effort was about 2 weeks.

7. **Wireless communication product - On-chip system level protection for FM pin**: On-chip protection for a FM TX/RX antenna pin was developed to allow the customer to save on external components. During system development IEC testing, failures occurred at 7 kV contact discharge. Analysis showed that the TX output buffer was destroyed since it took more current during a system level discharge than originally anticipated. After

redesign of the metal / via connectivity to the on-chip protection, the issue disappeared. The engineering effort was 4 man-months.

8. **Computer type product:** This product had 12 discrete on-board protection components to protect the display driver IC on a motherboard. The driver IC had only moderate on-chip protection, and protection components were used for assembly purposes to limit ESD stress from the charged display. The driver IC was stressed, without protection components, with the same discharge waveform that the charged display was able to produce during assembly phase. The calculated stress level was estimated to be below the driver components withstand level. There were no failures in the final tests, and 11 on-board protection components were removed from the design. One protection component was still needed to prevent system resets during IEC 61000-4-2 validation. The engineering effort was ½ man-months.

Changing to System-Efficient ESD Design, a co-design approach including both on-board and on-chip protection, allows a systematic optimization of system level ESD protection. When deciding about an optimum solution, various aspects have to be considered:

1. **Cost**
The obvious advantage of using on-chip protection is to reduce the number of individual on-board TVS, of which there may be many dozens on a single PCB. The price of these parts various widely, from less than $0.01 to >$0.3 per piece. In addition, the external protection requires extra cost for assembly, area on the board, etc.

Although the size of the on-chip protection is independent of the technology, the price of the chip nevertheless goes down with each new technology. So, for the OEM there is an apparent price drop of the product with each new technology. The price of external parts on the board, on the other hand, remains at the same level, independent of the IC technology.

When looking only at manufacturing cost, the use of additional die area will always be preferred than using external protection components. However, R&D cost due to IC respins might be significantly higher. Also these respins can lead to a major delay in time-to-market of the system.

2. **RF-Disturbance**
Particularly in RF applications, it is of paramount importance to contain (electromagnetic) disturbances as much as possible at their point of origin, i.e. as close as possible to the gun. By placing the protection on-chip, however, the full ESD current has to enter the IC by default. This increases the risk of internal RF disturbances significantly, possibly to the point where the system level performances can no longer be met. Furthermore, on-chip shielding is very costly in terms of area.

3. **Flexibility**
On-board protection may be a lot more flexible than on-chip protection. The cost involved in changing a particular protection design on-chip (e.g. mask costs) may be prohibitive. On-board components in combination with board parasitics including track resistance and inductance may be used to build effective filters which block part of the ESD current. Such filters can be relatively easily optimized by re-routing tracks, changing track length, etc.

4. **Debugging**
Failure analysis at the board level can be done with regular lab tools. On-chip failure analysis

is far more labor/cost intensive. Also, board level modifications are easier to implement, allowing new ideas to be tried out.

There is a major difference in the ESD protection design between high value & low volume and low value & high volume products. High value products can have all needed protection components on a layout, and protection methods are mainly limited by the available space and signal integrity targets. Low value products have the same limitations in space and signal quality, but have also major limitations on the design and material costs. For example, one on-board protection component purchase and assembly can cost 2 cents/piece. When 1 million low value products are made and each product requires 10 devices the protection expenses are >$200k. This represents a significant percent of the product sales price. With a proper mechanical design, layout optimization and good on-chip protection, the number of on-board protection components can be limited and the product can still pass the IEC 8 kV target.

6.8.2 Effort and Benefit of System-Efficient ESD Design (SEED)

The co-design methodology SEED requires a common approach for the protection design, with collaboration from both the IC and system designer. System designers have to specify which IOs require higher withstand levels in a system and then request IC operational parameters with similar I-V curve based specifications. These specifications are used for ESD validation by the IC suppliers. The system designer can use this data with calculation and simulation tools to optimize the on-board design to fulfill the final system requirements. These requirements can be IEC based or some other selected stress events. The primary benefit is that SEED enables a systematic step by step approach for ESD robust system design. SEED reduces costly trial and error rounds and decreases the wrestling between system and IC suppliers over on-chip protection requirements.

6.8.3 Next Steps Required by IC Suppliers, PCB ESD Diode Suppliers, Board and System Designers

The presented extraction and simulation approach enables a systematic development of an optimum and cost-efficient system level ESD protection. To apply this SEED method to regular product design requires certain preparation steps in the industry.

1. Suppliers of ICs and on-board ESD protection ESD elements need to align on a common standard for characterizing the on-board protection elements and the external pins of the protected ICs. A highly accurate TLP analysis up to 30 A is recommended for characterization of on-board protection elements. TLP IV characteristics of external pins have to be provided by the IC suppliers. Also, the transient turn-on behaviour of the on-board protection elements and IO circuitry have to be determined over a wider range of currents (mA to several 10 A), which may require very fast TLP (VFTLP) testing. The extracted parameters and models have to be provided in a standardized way, which allows the system manufacturer to use this input from various suppliers of his board components.

2. The system manufacturer should adjust the requirements catalogue for the IC and the protection devices accordingly. The reliance on HBM robustness of the IC pins must be abandoned and must be replaced by the request for detailed TLP characterization of the system relevant pins of the IC. The selection of on-board protection devices is then based on

the quality of the IC and on-board ESD protection element IV characteristics and the transient behavior of the protection elements up to 30 A (for an 8 kV contact level IEC ESD event).

3. Finally, the board designer must have the capability to use these data in the simulation environment and to perform IEC pulse simulations.

By using such a procedure, the risk and the required effort due to any late changes in the system design can be minimized while the full performance of the semiconductor devices can be exploited.

References

[1] Industry Council on ESD Target Levels. A) White Paper 1: A Case for Lowering Component Level HBM/MM ESD Specifications and Requirements," August 2007, at www.esda.org;

[2] Industry Council on ESD Target Levels. "White Paper 2: A Case for Lowering Component Level CDM ESD Specifications and Requirements," Revision 2, April 2010, at www.esda.org.

[3] S. Marum, C.Duvvury, J. Park, A. Chadwick, A. Jahanzeb; "Protection Circuits From the Transient Voltage Suppressor's residual Pulse During IEC 61000-4-2 Stress", Proc. EOS/ESD Symposium 2009 (2009), p.377-385.

[4] T. Reinvuo, T. Tarvainen, T. Viheriäkoski, P. Tamminen; "Electrostatic Discharge Measurement and Simulation of a Charged Power Amplifier Board", Proc. of the European Microwave Conf., p. 292-294, 2009.

[5] D. Wang, S. Marum, W. Kemper, D. McLain; „ System event triggered latch-up in IC chips: Test issues and chip level protection design", Proc. EOS/ESD Symposium 2006 (2006), pp. 1-7.

[6] IEC 61000-4-2, 'Electromagnetic compatibility (EMC) – Part 4-2: Testing and measurement techniques – Electrostatic discharge immunity test', Ed. 2.0, 2008.

[7] B. Arndt, F. z. Nieden, Y. Cao, F. Mueller,J. Edenhofer, S. Frei; "Simulationsbasierte Analyse von ESD Schutzelementen auf Systemebene"; Tagungsband 11. ESD Forum 2009 (2009), p 155- 164.

[8] B. Arndt, F. zur Nieden,Y. Cao, F. Mueller, J. Edenhofer, S. Frei; "Simulation Based Analysis of ESD Protection Elements on System level", presented at IEW 2007, Tutzing, 2010.

[9] Universal Serial Bus 2.0 Specifications, http://www.usb.org/

[10] W. Stadler, T. Brodbeck, R. Gärtner, H. Goßner; "Cable Discharges into Communication Interfaces", Proc. EOSESD Symposium 2006 (2006), p.144-151.

[11] M-J Kim, G. Langguth, K. Esmark, H. Gossner, T. Lee; "High Voltage Tolerant Fail-safe ESD Protection for USB 2.0- Compliant I/O Circuits", presented at IEW 2007, Lake Tahoe, 2007.

[12] LIN Conformance Test Specification LIN EMC Test Specification", Version 1.0; August 1, 2004

[13] C. Schanze, G. Linn, M. Wriedt, L. Claus, B. Körber; "Hardware Requirements for LIN, CAN and FlexRay Interfaces in Automotive Applications", v1.1, 2009-12-2.

[14] N. Peachey, S. Muthukrishnan, K. Muhonen, "Protection of Mobile Phone Antenna Ports Against System Level ESD Stresses", presented at IEW 2009, Lake Tahoe, 2009.

Chapter 7: Summary, Conclusions and Outlook

7.0   Summary

In this paper, we have attempted to present the first comprehensive analysis of system level ESD issues including ESD related system failures and design for system robustness. The discussion throughout shows that designing for system level ESD involves bridging the misconceptions between system designers (OEMs) and semiconductor component (IC) providers. We have drawn on the expertise of both OEMs and IC designers to address respective misconceptions and propose a system level ESD analysis and design methodology that increases system level ESD robustness while simultaneously reducing IC-level ESD design difficulty.

We have shown a clear distinction between physical failures and soft failures. Soft failures appear exclusively during powered conditions and are deferred to Part II of White Paper 3. Physical failures, with few exceptions, are related to external pins that are directly exposed to ESD stress, as proven by analysis of field returns and qualification test results. We have demonstrated that both design and robustness evaluations of these external pins have to follow a different methodology than standard ESD qualification. This is especially important since, contrary to the prevalent assumption in the industry, HBM and CDM testing do not provide sufficient information for system robust design.

This paper builds a framework for successful system level ESD protection using the following key concepts:

- ESD test specification requirements of system providers must be clearly understood as a separate domain from IC level ESD specifications. IC level ESD specifications should not be used as a basis for system level requirements.
- Understanding of the ESD failure and upset mechanisms is critical to recognizing their relevance for robust protection design and for correlating them to the IC specifications.
- Responsibility <u>must</u> be shared between system designers and IC providers for proper system level ESD protection.

From these concepts, we have introduced a new methodology described as "System-Efficient ESD Design" (SEED) that promotes a common OEM/IC provider understanding of correct system level ESD needs. The key objective has been the development of a framework for communicating IC / system level circuit information so that best practice ESD protection and controls can be co-developed and properly shared.

7.1   Conclusions

From this extensive collaboration between the IC providers and the system builders, this paper has been able to establish some key conclusions in Part I of the white paper. Our intention has been to remove misconceptions about system ESD design and requirements and at the same time, to present a fully comprehensive view of system level ESD protection design. We can conclude the following:

- Component ESD requirements are critical for IC production and handling, but requiring them to be much higher than the necessary safe levels can have a direct impact on circuit speed and consequently on system performance itself.

- As previously established in White Paper 1 and White Paper 2, and now more importantly confirmed in the present White Paper 3, artificially high HBM and CDM requirement for individual ICs either do not correlate to a robust system ESD performance or do not necessarily add value when designing for better system protection.

- By the same token, components/ICs just passing a certain level of any stress test (such as IEC, HMM, etc. as qualification goals) do not guarantee robustness of the complete system. More work is needed in this area to establish the true nature of system ESD events.

- A good design strategy for system protection requires a clear definition and understanding of *external* versus *internal* IC pins and those conditions under which stress to external pins can couple to internal pins. Only after establishing these distinctions can system design methodology and the process for it be properly and efficiently communicated and practiced.

- Placing large area protection clamps directly on-chip for an IC external pin may not ensure robust system ESD, and the approach will have many disadvantages in implementation for both the IC supplier and the system designer. A better strategy relies on external clamps as much as possible and also involves an understanding of the interaction with the IC pins' internal clamps.

- In order to offer a better and more interactive approach, the System-Efficient ESD Design (SEED) strategy has been introduced. This method uses a TLP based <u>characterization</u> of on-board protection diodes and on-chip protection circuits to co-design on-chip and on-board protection circuits.

- SEED is proposed as a superior design methodology to optimize system cost vs. performance and to reduce overall R&D effort.

- By following the SEED approach, some new efficient ESD systems have already been demonstrated.

7.2 Outlook

The overall system ESD protection can be more complicated when considering both hard and soft failures. The so called soft failures may involve complex EMC/EMI effects and also some Transient Latchup (TLU) phenomenon. The latter TLU effect could come from the technology development along with the system application, and thus requires thorough understanding. In Part II of this White Paper, the Industry Council will address a comprehensive system level ESD design strategy in more detail, using the information from Part I. This information will be used to establish recommendations for IC and system level manufacturers regarding proper protection / controls and best practice ESD tests which can be used to properly assess ESD and related EMI performance of system level tests. This is intended to better define the IC manufacturer / system level OEM ESD relationship and responsibilities.

Revision History

Revision	Changes	Date of Release
1.0	Initial Release	Oct 2010

White Paper 3
System Level ESD
Part II: Implementation of Effective ESD Robust Designs

Industry Council on ESD Target Levels

March 2019

Revision 2.0

The Industry Council on ESD
http://www.esdindustrycouncil.org/ic/en/

The Electrostatic Discharge Association
http://www.esda.org/

JEDEC – Under Publication JEP162
http://www.jedec.org/

Abstract

This document (White Paper 3 Part II) is the second of two Electrostatic Discharge (ESD) Industry Council white papers dealing with System Level ESD.

In Part I, the misconceptions common in the understanding of system level ESD between supplier and original equipment manufacturer (OEM) were identified, and a novel ESD component / system co-design approach called system efficient ESD design (SEED) was described. The SEED approach is a comprehensive ESD design strategy for system interfaces to prevent hard (permanent) failures. In Part II we expand this comprehensive analysis of system ESD understanding to categorize all known system ESD failure types, and describe new detection techniques, models, and improvements in design for system robustness. Part II also expands this SEED co-design approach to include additional hard / soft failure cases internal to the system.

Part II begins with an overview of system ESD stress application methods and introduces new system diagnosis methods to detect weak ESD failure areas leading to hard or soft failures, and provides a "cost vs. performance vs. robustness" analysis of present-day state-of-the-art EMC/EMI design prevention methods that have been developed to prevent system level ESD failure. It follows with an expansion of SEED failure classifications to cover a combination of hard (permanent) and/or soft (resettable) system failures and stresses which could cause these errors, and describes cases where the SEED co-design approach can be expanded to provide additional benefits to system ESD design. System design simulation tools are described in the context of their potential improvements to simulating system level ESD stress and failure modes. Application-specific industry system ESD test methods are then described in the context of their ability to reveal hard and soft failure modes from actual system deployment. Finally, a technology roadmap of the system design components is described, including IC technology and related circuit speeds, automotive electronics, packaging technology, system / board interconnect technology and ESD protection materials, illustrating continuing challenges for system ESD design improvement.

About the Industry Council on ESD Target Levels

The Council was initiated in 2006 after several major U.S., European, and Asian semiconductor companies joined to determine and recommend ESD target levels. The Council now consists of representatives from active full member companies and numerous associate members from various support companies. The total membership represents IC suppliers, contract manufacturers (CMs), electronic system manufacturers, OEMs, ESD tester manufacturers, ESD consultants and ESD IP companies. In terms of semiconductor market leaders, the member IC manufacturing companies represent 8 of the top 10 companies, and 12 of the top 20 companies as reported in the EE Times issue of November 9, 2010.

Core Contributing Members	Core Contributing Members
Robert Ashton, ON Semiconductor	Guido Notermans, ST-Ericsson
Jon Barth, Barth Electronics	Nate Peachey, RFMD
Patrice Besse, Freescale	Ghery Pettit, Intel
Stephane Bertonnaud, Texas Instruments	David Pommereneke, Missouri University of Science & Technology
Brett Carn, Intel Corporation	Wolfgang Reinprecht, AustriaMicroSystems
Ann Concannon, Texas Instruments	Alan Righter, Analog Devices
Jeff Dunnihoo, Pragma Design	Theo Smedes, NXP Semiconductors
Charvaka Duvvury, Texas Instruments (Chairman) c-duvvury@ti.com	Teruo Suzuki, Fujitsu
	Pasi Tamminen, Nokia (* - Advisory board also)
David Eppes, AMD	Matti Uusimaki, Nokia
Howard Gan, SMIC	Benjamin Van Camp, SOFICS
Reinhold Gärtner, Infineon Technologies	Vesselin Vassilev, Novoroll
Robert Gauthier, IBM	Terry Welsher, Dangelmayer Associates
Horst Gieser, FhG IZM	Joost Willemen, Infineon Technologies
Harald Gossner, Intel Corporation (Chairman) harald.gossner@intel.com	
	Advisory Board*
LeoG Henry, ESD/TLP Consulting	Fred Bahrenburg, Dell
Masamitsu Honda, Impulse Physics Laboratory	Heiko Dudek, Cadence
Michael Hopkins, Amber Precision Instruments	Bob Dutton, Stanford University
Hiroyasu Ishizuka, Renesas Technology	Johannes Edenhofer, Continental
Satoshi Isofuku, Tokyo Electronics Trading	Stephen Frei, Technical University of Dortmund
Marty Johnson, National Semiconductor	Frederic Lefon, Valeo
John Kinnear, IBM	Christian Lippert, Audi
David Klein, Freescale Semiconductor	Patrice Pelissou, EADS
Timothy Maloney, Intel Corporation	Wolfgang Pfaff, Bosch
Tom Meuse, Thermo Fisher Scientific	Tuomas Reinvuo, Nokia
James Miller, Freescale Semiconductor	Marc Sevoz, EADS
	Werner Storbeck, Volkswagen
	Wolfgang Wilkening, Bosch
	Rick Wong, Cisco

Associate Members	Associate Members
Stephen Beebe, Global Foundries	Soon-Jae Kwon, Samsung
Jonathan Brodsky, Texas Instruments	Brian Langley , ORACLE
Michael Chaine, Micron Technology	Frederic Lafon, Valeo
Freon Chen, VIA Technologies	Moon Lee, SemTech
Tim Cheung, RIM	Markus Mergens, QPX
Ted Dangelmayer, Dangelmayer Associates	Roger Peirce, Simco
Ramon del Carmen, Amkor	Donna Robinson-Hahn, Fairchild Semiconductor
Stephanie Dournelle, ST Microelectronics	Jeff Salisbury, Flextronics
Melanie Etherton, Freescale Semiconductor	Jay Skolnik, Skolnik-Tech
Yasuhiro Fukuda, OKI Engineering	Jeremy Smallwood, Electrostatic Solutions Inc.
Kelvin Hsueh, UMC	Jeremy Smith, Silicon Labs
Natarajan Mahadeva Iyer, Global Foundries	Arnold Steinman, Electronics Workshop
Larry Johnson, LSI Corporation	David E. Swenson, Affinity Static Control Consulting
Peter de Jong, Synopsys	Nobuyuki Wakai, Toshiba
Jae-Hyok Ko, Samsung	Jon Williamson, IDT
ChangSu Kim, Samsung	Michael Wu, TSMC
Han-Gu Kim, Samsung	MyoungJu.Yu, Amkor
Marcus Koh, Everfeed	

Acknowledgments:

The Industry Council would like to thank all the authors, reviewers, and specialists who shared a great deal of their expertise, time, and dedication to complete this document. The authors of Appendix B would like to thank several individuals for their insight into the future roadmap, Chris Barr (TI) and Christopher Opoczynski (TI) for projections on universal serial bus (USB) and high definition multimedia interface (HDMI) and Andrew Marshall (TI) for an overview on optical interconnects. Wolfgang Stadler (Intel) is acknowledged for his contributions to the transient latch-up discussion in Chapter 4. We would especially like to thank our OEM advisors from Audi, Bosch, Cadence, Continental, Cisco, Dell, EADS, Nokia, Technical University of Dortmund, Valeo and Volkswagen who gave us valuable direction. The Industry Council would also like to thank the Electrostatic Discharge Association (ESDA) for facilitating the face-to-face meetings of the Council.

Editors:
Brett Carn, Intel Corporation
Terry Welsher, Dangelmayer Associates

Authors:
Robert Ashton, ON Semiconductor
Jon Barth, Barth Electronics
Patrice Besse, Freescale Semiconductor
Jeff Dunnihoo, Pragma Designs, Inc
Charvaka Duvvury, Texas Instruments
David Eppes, AMD
Harald Gossner, Intel Corporation
Leo G. Henry, ESD/TLP Consulting
Masamitsu Honda, Impulse Physics Laboratory
Mike Hopkins, Amber Precision Instruments
Marty Johnson, Texas Instruments
David Johnsson, Intel Corporation
David Klein, Freescale Semiconductor
Tom Meuse, Thermo Fischer Scientific
Tim Maloney, Intel Corporation
Guido Notermans, ST-Ericsson
Ghery Pettit, Intel Corporation
David Pommerenke, Missouri University of Science & Technology
Alan Righter, Analog Devices
Wolfgang Reinprecht, Austria Microsystems
Pasi Tamminen, Nokia
Matti Uusimaki, Nokia
Benjamin Van Camp, SOFICS
Vesselin Vassilev, Novorell
Joost Willemen, Infineon Technologies

Mission Statement

The Industry Council on ESD Target Levels was founded on its original mission to review the ESD robustness requirements of modern IC products to allow safe handling and mounting in an ESD protected area. While accommodating both the capability of the manufacturing sites and the constraints posed by downscaled process technologies on practical protection designs, the Council provides a consolidated recommendation for future ESD target levels. The Council Members and Associates promote these recommended targets for adoption as company goals. Being an independent institution, the Council presents the results and supportive data to all interested standardization bodies.

In response to the growing prevalence of system level ESD issues, the Council has now expanded its mission to directly address one of the most critical underlying problems: insufficient communication and coordination between system designers (OEMs) and their IC providers. A key goal is to demonstrate and widely communicate that future success in building ESD robust systems will depend on adopting a consolidated approach to system level ESD design. To ensure a broad range of perspectives the Council has expanded its roster of Members and Associates to include OEMs as well as experts in system level ESD design and test.

Preface

While IC level ESD design and the necessary protection levels are well understood, system ESD protection strategy and design efficiency have only been dealt with in an ad hoc manner. This is most obvious when we realize that a consolidated approach to system level ESD design between system manufacturers and chip suppliers has been rare. This White Paper discusses these issues in the open for the first time, and offers new and relevant insight for the development of efficient system level ESD design. This effort has been divided into two parts. In WP3 Part I, we identified the misconceptions common in the understanding of system level ESD. In this document (Part II) we will explore realistic system ESD protection requirements and strategies. We would also like to note that in Part I we addressed direct stress effects on external and internal pins of an IC while this document investigates the intricate effects of inter-chip pin coupling. This document is intended to be useful for both chip suppliers and OEMs/ODMs. As a final note, we would like to clarify that this document addresses system level ESD issues only, not electrical overstress (EOS) issues unless they manifest from a system failure.

Disclaimers

The Industry Council on ESD Target Levels is not affiliated with any standardization body and is not a working group associated with JEDEC, ESDA, JEITA, IEC, or AEC.

This document was compiled by recognized ESD experts from numerous semiconductor supplier companies, contract manufacturers and OEMs. The data represents information collected for the specific analysis presented here; no specific components or systems are identified.

The Industry Council, while providing this information, does not assume any liability or obligations for parties who do not follow proper ESD control measures.

Table of Contents

Glossary of Terms

AAMI	Association for the Advancement of Medical Instrumentation
AC	alternating current
ADC	analog digital converter
AEC	Automotive Electronics Council
AMR	absolute maximum rating
ANSI	American National Standards Institute
ASIC	application specific integrated circuit
ASTM	American Society for Testing and Materials
ATE	automated test equipment
BCD	bipolar-CMOS-DMOS
BOM	bill of materials
CAD	computer aided design
CAN	controller area network
CBE	charged board event
CCL	capacitive coupled latch-up
CDE	cable discharge event
CDM	charged-device model
CISPR	Comité International Spécial des Perturbations Radioélectriques
CMOS	complimentary metal-oxide-semiconductor
CMF	common mode filter
CRC	cyclic redundancy check
DC	direct current
DDR	double data rate
DIP	dual inline package
DMOS	double-diffused metal-oxide-semiconductor
DSP	digital signal processing
DVI	digital visual interface
DUT	device under test
ECU	electronic control unit
EDA	electronic design automation
EEPROM	electrically erasable programmable read only memory
EFT	electrical fast transients
EM	electromagnetic
EMMI	emission microscopy
EOS	electrical overstress
ESD	electrostatic discharge
ESDA	Electrostatic Discharge Association; ESD Association
ETSI	European Telecommunication Standards Institute
EUT	equipment under test
FB	ferrite bead
FDA	Food and Drug Administration
FET	field effect transistor
FFT	fast Fourier transform
GND	negative voltage supply in digital logic, neutral voltage supply in analog logic
GPIO	general purpose input/output

HBM	human body model
HDL	hardware description language
HDMI	high definition multimedia interface
HMM	human metal model
IBIS	input/output buffer information specification
IC	integrated circuit
ID	identification
IO	input/output
IEC	International Electrotechnical Commission
IR	infra-red
ISO	International Organization of Standards
IT	information technology
ITE	information technology equipment
I-V	current/voltage
JEDEC	Joint Electronic Devices Engineering Council
JEITA	Japan Electronics and Information Technology Industries Association
JTAG	joint test action group
LCD	liquid crystal display
LED	light emitting diode
LIN	local interconnect network
MCM	multi-chip module
MDD	medical device directive
MEM	micro-electro-mechanical
MID	molded interconnection device
MM	machine model
OEM	original equipment manufacturer
ODM	original design manufacturer
OTP	one time programmable
PC	personal computer
PCB	printed circuit board
PCI	peripheral component interconnect
PHY	physical layer interface
PLL	phase lock loop
PN	p-type/n-type junction
PWB	printed wiring board
R2R	roll-to-roll
RAM	random access memory
RC	resistor capacitor network
RF	radio frequency
RLC	resistor inductor capacitor network
RTOS	real time operating system
RX	receiver
SAE	Society of Automotive Engineers
SAW	surface acoustical wave
SCR	silicon controlled rectifier
SERDES	serializer/deserializer
SIM	subscriber identity module
SiP	system-in-package

SMA	sub-miniature version A
SMT	surface mount technology
SOA	safe operating area
SoC	system-on-chip
SoP	system-on-package
SP	standard practice
SPD	surge protected device
SPICE	simulation program with integrated circuit emphasis
TCAD	technology computer-aided design
TIVA	thermally induced voltage alteration
TLP	transmission line pulse
TLU	transient latch-up
TSV	through silicon via
TV	television
TVS	transient voltage suppression
TX	transmitter
USB	universal serial bus
UV	ultra violet
VDD	positive voltage supply
VFTLP	very fast transmission line pulse
VHDL	Verilog hardware description language
VSS	negative voltage supply
WP	white paper
WSP	wafer scale package
XTAL	crystal oscillator

Definitions

Crosstalk: Any phenomenon by which a signal transmitted on one circuit or channel of a transmission system creates an undesired effect in another circuit or channel. This phenomenon is usually caused by undesired capacitive, inductive, or conductive coupling from one circuit, part of a circuit, or channel, to another.

EMC: electromagnetic compatibility – The condition which prevails when telecommunications (communication-electronic) equipment is collectively performing its individual designed functions in a common electromagnetic environment without causing or suffering unacceptable degradation due to electromagnetic interference to or from other equipment/systems in the same environment (MIL-STD-463A).

EMI: electromagnetic interference - Electromagnetic emissions (radiated or conducted) which may cause harmful interference to communications services or other electronic devices.

ESD Design Window: The ESD protection design space for meeting a specific ESD target level while maintaining the required IO performance parameters (such as leakage, capacitance, noise, etc.) at each subsequent advanced technology node.

External Pin (interface pin): An external pin is one which at the board/card level is exposed to potential ESD threats from the outside world.

FR-4 (or FR4): A grade designation assigned to glass-reinforced epoxy laminate sheets, tubes, rods and printed circuit boards.

Hard Failure: Failure of a system due to physical damage to a system component which can only be repaired by the physical repair or replacement of the damaged component.

IEC-Robustness: The capability of a product to withstand the required IEC ESD-specification tests and still be fully functional.

IEC ESD event: An ESD stress as defined in IEC 61000-4-2.

Immunity: The ability of an electronic device to function properly in its electromagnetic environment.

Internal Pin (non-interface pin): An internal pin is one which is exposed to ESD threats typically only during IC manufacturing or during a crosstalk situation at the system level.

Residual Pulse: The resulting voltage/current (after system level ESD protection devices) seen by an IC component from an IEC stress waveform.

Second breakdown trigger current - I_{t2}: The current point at which a transistor enters its second breakdown region under ESD pulse conditions and is irreversibly damaged.

<u>SEED:</u> System-Efficient ESD Design - Co-design methodology of on-board and on-chip ESD protection to achieve system level ESD robustness.

<u>Susceptibility, conducted:</u> A measure of the interference signal current or voltage required on power, control and signal leads to cause an undesirable response or degradation of performance (MIL-STD-463A).

<u>Susceptibility, electromagnetic:</u> The degree to which an equipment, subsystem or system evidences undesired responses caused by electromagnetic radiation to which it is exposed (MIL-STD-463A).

<u>Susceptibility, radiated:</u> A measure of the radiated interference field required to cause equipment degradation (MIL-STD-463A).

<u>Susceptibility threshold:</u> The signal level at which the test sample exhibits a minimum discernible undesirable response (MIL-STD-463A).

<u>System level ESD Robustness:</u> The capability of a product to withstand the required IEC ESD-specification tests and still be fully functional.

<u>Soft Failure:</u> Failure of a system, not due to physical damage, in which the system can be returned to a functional state without the repair or replacement of a component. Return to a functional state may or may not require operator intervention. Operator intervention may include rebooting or power cycling. Soft Failures can involve software issues and software fixes but in the context of this document they are primarily due to ESD events injecting unwanted signals into the system which place the system into a state in which it does not function as intended.

<u>VHDL-AMS:</u> Verilog-AMS is a derivative of the Verilog hardware description language. It includes analog and mixed-signal extensions (AMS) in order to define the behavior of analog and mixed-signal systems. It extends the event-based simulator loops of Verilog / System Verilog / VHDL, by a continuous-time simulator, which solves the differential equations in analog-domain. Both domains are coupled: analog events can trigger digital actions and vice versa.

Executive Summary

Overview

White Paper 3 Part II, while establishing the complex nature of system level ESD, proposes that an **efficient ESD design can only be achieved when the interaction of the various components under ESD conditions are analyzed at the system level**. This objective requires an appropriate characterization of the components and a methodology to assess the entire system using simulation data. This is applicable to system failures of different categories (such as hard, soft, and electromagnetic interference (EMI)). This type of systematic approach is long overdue and represents an advanced design approach which replaces the misconception, as discussed in detail in White Paper 3 Part I, that a system will be sufficiently robust if all components exceed a certain ESD level.

In the first step, a method for categorizing the failure types has been introduced. An advanced characterization and simulation approach is discussed through examples. However, a full design flow cannot be established without **a common effort across the electronic industry involving IC suppliers, suppliers of discrete protection components and original equipment manufacturers (OEMs) as well as tool vendors**. This paper identifies existing tools with both simulations and scanning techniques that are applicable for this purpose and calls out fields for further development.

Equally important is the notion that **efficient system ESD design can ideally be achieved by improved communication between the IC supplier, the OEM and the system builder**. As technologies advance even further, and as systems become more complex under various applications, this shared responsibility is expected to gradually shift more towards system design expertise.

Understanding Component to System ESD

Towards achieving the goals mentioned above, it is first important to decouple the component ESD requirements from system level ESD design. White Papers 1 and 2 established that component electrostatic discharge (ESD) levels can be safely reduced to practical levels with basic ESD control methods that are mandatory in every production area. We have also established that these ESD target levels enable fabrication of integrated circuits (ICs) with on-time delivery (in billions of units) for electronic systems in consumer applications with high circuit performance. The general perception has been that component ESD (for example, human body model (HBM)) is a prerequisite for good system level ESD robustness. But this misconception once again needs to be clarified, as shown below in Figure 1, **system level ESD and component ESD are not correlated with each other.**

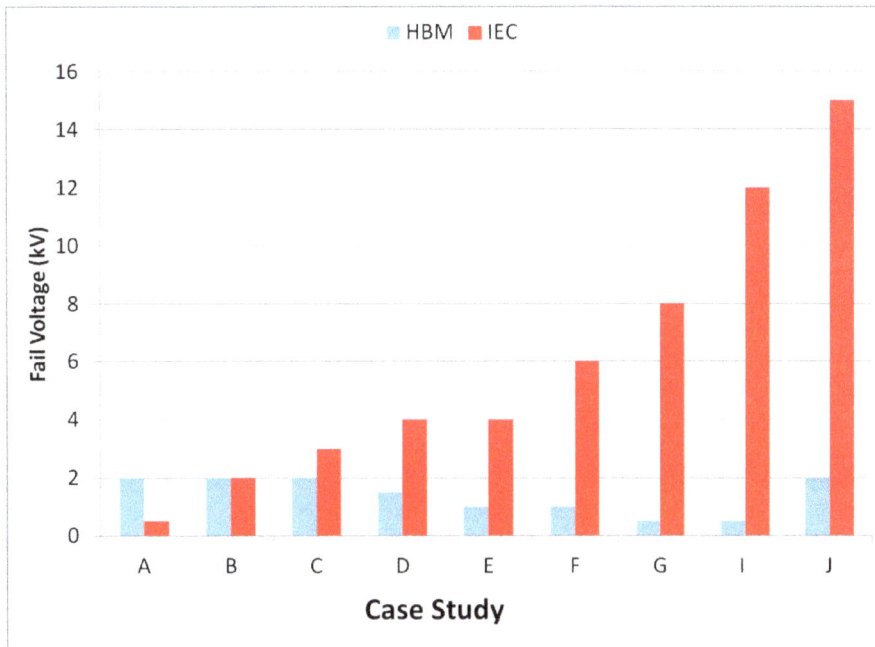

Figure 1: Comparison of IC level and system level ESD failure threshold of various systems (A-J) showing that HBM protection is not related to System level ESD robustness

In fact, at the system level, ESD robustness is a much more complex issue requiring a deeper understanding to address the ESD protection requirements for electronic systems such as laptops, cell phones, printers and home computers. These system complexities come about as a result of protecting the external interfaces, such as the universal serial bus (USB), to the outside world. Such systems, after encountering the more severe ESD pulses defined by the IEC standard, can lead to hard or soft failures. As introduced in White Paper 3 Part I, the basic version of system-efficient ESD design (SEED) addresses **hard failures** related to IC pins with a direct external interface, **soft failures,** which are more frequently reported, are challenging to understand and overcome. In this latter case, addressing soft failures requires an extension of the SEED approach to other failure mechanisms that include latch-up and EMI effects. In this document, the steps to categorize the different failure mechanisms, and the appropriate characterization and simulation methodologies, are identified through various forms of **Advanced SEED**.

Communication and Strategy

This white paper documents a rigorous approach to describing **the challenges related to all categories of system level ESD failures** that can arise from energy injection due to the IEC contact pulse stress, as well as from electromagnetic compatibility (EMC) and EMI effects. To classify these fails and to provide a common terminology, three categories of fails have been introduced:

- SEED Category 1 (physical damage due to pulse energy)
- SEED Category 2 (damage or interference of function due to transient latch-up)
- SEED Category 3 (interference of function by noise or bursts on supply net and signal lines)

Understanding these different categories of failures is an important part of addressing the appropriate solutions. Thus, one main objective is to close the existing communication gap between OEMs and IC providers by involving the expertise of both OEMs and system design experts. As a result, the completion of this second phase of White Paper 3 required the participation and contributions from world class experts on the art of system level ESD phenomena and protection techniques.

One of the challenges which remains elusive is the trade-off between cost, performance, robustness, and time-to-market. This white paper also addresses these issues, bringing forth a dialogue between the IC supplier, customer, and the system designer.

Implementation of Advanced Tools

White Paper 3 Part II specifically covers in detail **an overview of system ESD stress application methods, system diagnostic techniques to detect hard or soft failures, and the application of tools for susceptibility scanning.** For example, as illustrated in Figure 2, these types of advanced tools can be used to differentiate the characteristics of products and enable proper system protection methodology.

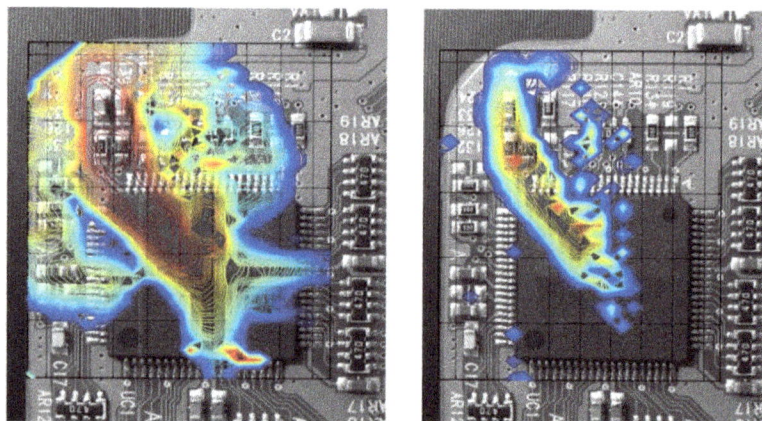

Figure 2: Susceptibility scanning using pulse techniques on Product A (left) and Product B (right)
(Courtesy of Amber Precision Instruments)

Along the lines of communication and interaction, IC suppliers and system designers can share their knowledge of tools and their applications. For example, suppliers would provide a single definition, high quality model of their input/output (IO). Then OEMs would use analytical tools to integrate the IC's IO models into their system models for system level stress analysis. These tools will not reach their full potential unless the data they collect can be used within the standard design flow for an electronic system. For these to be effective, model files such as input/output buffer information specification (IBIS) and simulation programs with integrated circuit emphasis (SPICE), which describe the electrical properties of components, need to be enhanced to describe component behavior in the ESD range. The general framework for such modeling has been defined through IEC 61433-6 and the detailed models around the integrated circuit itself, such as discussed above, are currently in definition based on an industry wide investigation performed by

ESDA WG26 on system level ESD modeling. Suppliers of components, ranging from integrated circuits to ESD protection components also need to characterize their products in the appropriate ESD range.

Impact from the Technology Roadmap

Finally, we bring into focus that IC technology and related circuit speeds will increasingly have an impact on system designs. This roadmap will cover market segments ranging from information technology (IT), communications, automotive electronics, IC package technology development to advances in board and assembly technologies. **The cost of ESD must be considered along with all development and innovation; the industry must decide who should bear the cost of ESD design and its pressure on production schedules and time to market**

System level ESD will continue to be a challenge in the future. The dilemma of meeting technology demands for speed and performance will inevitably require either the development of more effective shielding or innovation of novel on-board protection solutions.

Conclusions

In summary, this white paper has pointed out the necessary framework required for comprehensive improvement in the first time success rate of ESD robust system designs. A number of helpful tools and techniques already exist. However, standardization between IC suppliers, PCB protection device suppliers and system designers requires common models, common methods and compatible simulation tools in order to meet the goal of better ESD design capability. **None of this will occur without a great deal of communication between EDA tool vendors, component suppliers and system designers.**

Finally, this document has provided a major step forward in identifying and understanding the technical issues. An important message to remember is that **the current focus on component ESD performance must shift towards improving system level ESD performance. That is, while minimum component ESD levels provide for safe component handling, the bulk of future research and development efforts should be directed towards system level ESD to reach a day when a high first pass success rate in system level ESD design is straightforward.**

Chapter 1: Introduction & Purpose

Robert Ashton, ON Semiconductor
Charvaka Duvvury, Texas Instruments

1.0 History of the Industry Council on ESD Target Levels

The Industry Council on ESD Target Levels (Council) was formed in 2006 to address a growing disconnect between original equipment manufacturers (OEMs) and manufacturers of integrated circuits (ICs) with regard to electrostatic discharge (ESD). The competitive environment requires that OEMs create products with higher performance in ever smaller form factors while maintaining and improving reliability and offering the products at competitive prices. This progress was made possible by imposing the following improvements in ICs; higher levels of functionality, higher speeds, smaller component size and lower power. That is, the burden of higher performance has been transferred to the IC suppliers, who must achieve these improvements through enhancements in their chip designs. IC manufacturers have been able to make these improvements by employing new IC technologies with smaller feature sizes to improve functionality and speed and lower working voltages to prevent wear-out of the circuit and maintain reasonable power levels.

OEMs of course need these improvements, but they do not want them at the cost of reliability and yield. OEMs therefore maintained the same quality and reliability requirements on ICs produced in the latest technology as they had for a more mature product. One of these requirements was for the ESD robustness of ICs. 2 kV of human body model (HBM) robustness had long been a *de facto* standard and 500 V of charged device model (CDM) robustness was becoming a similar requirement. Also, there was an unintended, albeit perceived important, requirement that the ICs must simultaneously meet 200 V for the machine model (MM). However, starting around the 90 nm technology node, IC manufacturers were finding it increasingly difficult, and sometimes impossible, to consistently meet these requirements. The smaller feature sizes, lower operating voltages and high speed signal requirements of ICs were shrinking the design window available for ESD protection. IC manufacturers, however, knew that 2 kV HBM and 500 V CDM levels were not needed. HBM and CDM tests are performed to ensure that ICs can survive manufacture in an ESD controlled manufacturing facility. Experience has shown that with basic ESD control procedures, ICs with far lower HBM and CDM levels could be handled. Furthermore, without basic ESD controls, robustness far beyond 2 kV HBM and 500 V CDM will not survive. Attempting to meet 2 kV HBM and 500 V CDM was therefore costing the industry millions of dollars in redesigns and delays to market without any benefit to the industry.

The Council was formed to show the electronics industry that lowering ESD levels from 2000 V HBM and 500 V CDM was not only possible without sacrificing quality and reliability, but also the right thing to do. "White Paper 1: A Case for Lowering Component Level HBM/MM ESD Specifications and Requirements" [1] demonstrated why it was becoming more difficult to design high levels of HBM performance in advanced technologies and presented data that proved lower HBM levels do not lead to higher levels of failure. White Paper 1 also showed that designing for HBM would guarantee adequate levels of intrinsic MM performance and that specifically testing for MM was not necessary. "White Paper 2: A Case for Lowering Component Level CDM ESD Specifications and Requirements" [2] made a similar case for CDM levels below 500 V by presenting explanations of why high levels of CDM performance were not necessary and presented data to back up the arguments.

White Paper 1 (WP1) and White Paper 2 (WP2) were influential in changing the industry's mind about the level of ESD robustness needed to allow high yield manufacturing of systems. OEM's had a lingering concern however; what will lower component ESD levels do to system reliability? It has been assumed by many that the first step in producing ESD robust systems is to choose components with high HBM and CDM levels. The assumption that higher HBM and CDM levels will automatically produce ESD robust systems seems logical, but members of the Council knew this assumption was not true and that it was blocking the adoption of more reasonable HBM and CDM levels. The Council decided to address the problem with a third white paper to demonstrate the lack of correlation between component level ESD and system level ESD, and to address the wider question of how to effectively design for system level ESD robustness. Although *ad hoc* system level ESD designs have been done in practice, the important message of the White Paper 3 (WP3) Part I was to remove the misconceptions about system level ESD.

System level ESD is a very broad subject and differs substantially from HBM and CDM ESD. HBM and CDM tests only look for physical damage to unpowered ICs. System level ESD is most often performed on a powered and functioning system, and failures include not only physical failures but system upset. Due to the broad subject of system level ESD, WP3 has been divided into two parts. Part I focused on physical failures and the lack of correlation between component level ESD (HBM and CDM) and system level ESD. Part I also introduced the (basic) system-efficient ESD design (SEED) strategy for designing ESD robust systems. In this document, WP3 Part II, we address the broader topics of system level ESD design and system upset by an advanced SEED concept. For those not familiar with WP3 Part I a summary has been placed in Appendix D. The remainder of this chapter will preview the content of WP3 Part II.

We would also like to reference a recently published white paper by JEITA which addresses system level ESD related tests of ICs as used by the Japanese Industry [3]. Tests based on 200 pF/0 Ohm and 200 pF/100 Ohm RC discharge circuits are discussed to characterize transient latch-up robustness of powered ICs. The JEITA white paper focuses on this specific test procedure while this white paper discusses the general design approach.

1.1 White Paper 3 Part II: Summary of Chapters/Appendices

Chapter 2 looks at how ESD affects systems, beginning with how these events are coupled to internal system circuits and devices, followed by the system response (both hard and soft failures) and concluding with analysis methods used to determine root causes of upsets and failures. Included are methods of categorizing failures and coupling modes to internal circuits. Also described are classic trouble-shooting methods for solving ESD upset problems plus new technologies capable of identifying sensitive circuit areas and components.

Chapter 3 examines the advantages and limitations of ESD/EMI design methodologies as they are presently practiced. State-of-the-art ESD/EMI design methodologies span the continuum from comprehensive theoretical 3D modeling of Maxwell's equations for an entire system to the most basic trial and error bench tests, and numerous advanced practical and theoretical tricks-of-the-trade in between. Utilizing one or more of these methods in the concept, design, and prototype phases, a designer can observe and improve the extent and effectiveness of any element or flow in a process intended to balance ESD/EMI robustness, system performance, and cost.

Chapter 4 summarizes the SEED approach. In the SEED approach the path of the ESD stress is analyzed and predictions of system response are made using the properties of the printed circuit board and components on the board. Not all ESD paths are the same, however. ESD stress can range from high energy stresses, which are directly coupled from a system IO connector to a sensitive pin on an integrated circuit causing physical damage, to a much lower energy stress that is capacitively coupled to sensitive circuit nodes and results in system upset, but no physical damage. This chapter categorizes the types of ESD stress which a circuit can be exposed to and shows how different approaches and circuit properties will be needed to analyze the different stresses and prevent system failures.

Chapter 5 describes the models and analytical tools needed to support the SEED concept. The first requirement is a standard model for the description of integrated circuit IOs. After describing the requirements for a standard model, existing models are examined that may be extended or adapted for SEED requirements. Next, simulation tools that use these models are also examined for their adaption to SEED. This is followed by an examination of the limitations of both software and hardware analytic tools. Finally, an example using available models and tools is presented.

Chapter 6 presents a summary of the white paper and final conclusions. The white paper shows that there are tools available, such as transmission line pulse (TLP), which can study component properties in the ESD time and current domain as well as scanning tools which can measure the propagation of ESD events through a system and locate sensitive circuit nodes. The white paper categorizes the paths through which ESD stress can enter a system and outlines how to analyze and design for these types of stress. To use this increased understanding, design tools need to be upgraded to handle the information from the extended characterization tools and to perform the analysis needed to quantify the behavior of the paths through which ESD stress enters a system. The models used to describe components need to be modified to extend their useful range into the current, voltage and time domain of ESD events. Finally, component manufacturers need to characterize their products outside the range of normal operation so that system level ESD design can be done based on known component properties. This can only be done with extensive dialog between system designers, measurement equipment manufacturers, electronic design automation (EDA) tool vendors, model file developers and component suppliers. However, this paper does define the important issues and provides a starting point and significant first steps in this dialog.

Appendix A addresses the typical stress conditions and design solutions used in various fields of electronic industry ranging from mobile phones, medical devices, computers, consumer electronics, and automotive components to avionics. It provides an overview of test standards to be applicable beyond IEC 61000-4-2 and discusses the impact of the multiple stress requirements on the design solutions. The applicability of the previously introduced SEED concept is analyzed for the various types of systems.

Appendix B addresses the changing technologies from different points of view that are slated to have an impact on the design capability of system level ESD protection. The main objective of this appendix is to review the integrated circuit technology roadmaps and trends with consideration to the system level ESD performance of information technology (IT), communications, and automotive electronics. The various issues that are critical for these trends include IC package development into more complex 3-D structures, new advances in board assembly techniques, emerging optical interconnects as applied to IC connections, and the promising development of novel suppressive materials such as polymers with embedded nano-particles that can potentially offer low-capacitive and efficient solutions at both the package level and the board level. In its overall essence the appendix assesses the impact of rapidly advancing technologies on system level ESD requirements and designs. While not specifically recommending any future system level ESD specs, the appendix points to where the difficulties could arise from scaled down technologies, higher speed circuit designs, and broader analog and high voltage applications. The appendix also attempts to include some comments on the demands for resources and the cost for system ESD designs from the perspectives of both IC suppliers and system builders.

Appendix C describes how ESD threats to systems include both a fast component, from the ESD gun's initial current spike, followed by a slow component, with lower current but with considerably longer duration. The appendix explains how the current paths for the two components can be very different, can cause different types of failures and require different solutions. The section ends with diagnostic suggestions for locating susceptible areas within a system or circuit board.

References

[1]   Industry Council on ESD Target Levels, "White Paper 1: A Case for Lowering Component Level HBM/MM ESD Specifications and Requirements," August 2007, at www.esda.org or JEDEC publication JEP155, "Recommended ESD Target Levels for HBM/MM Qualification", www.jedec.org

[2]   Industry Council on ESD Target Levels. "White Paper 2: A Case for Lowering Component Level CDM ESD Specifications and Requirements," Revision 2, April 2010, at www.esda.org or JEDEC publication JEP157, "Recommended ESD-CDM Target Levels", www.jedec.org

[3]   JEITA document EDR-4709, 2012

Chapter 2: Overview of ESD Stressing and System Response

Jon Barth, Barth Electronics
David Eppes, AMD
Leo G. Henry, ESD & TLP Consulting
Mike Hopkins, Amber Precision Instruments
Marty Johnson, Texas Instruments
David Pommerenke, Missouri University of Science & Technology

2.0   Introduction

The purpose of Chapter 2 is to look at ESD system level testing as it is done today as well as the complications (including coupling issues), failure modes, and trouble-shooting methods used to determine root cause of failures. "Soft" failures, where no obvious component damage occurs, and "hard" failures, where component damage *is* apparent, are discussed in some detail. Considerable attention is placed on electromagnetic coupling modes and the propagation of electrostatic discharges in a system. New evolving methods such as susceptibility scanning, software tools and system specific test boards to monitor the health of a system are all discussed.

2.1   Presently Used Stress Tests for ESD

There are presently only a few basic stress tests used in the industry for testing systems for the effects of electrostatic discharges. These are:

- **IEC 61000-4-2** and its derivatives for compliance testing of virtually all consumer, telecommunications, medical and light industrial electronics.
- **ISO 10605** for automotive electronics.
- **DO-160** for commercial avionics.
- **CDE (cable discharge event).** There are various corporate standards in use, and at this writing, Electrostatic Discharge Association (ESDA) is close to issuing a draft standard practice to help standardize test methods for CDE.

The following sections summarize these tests. For a more detailed discussion see Chapter 2 of White Paper 3 Part I [1].

2.1.1   IEC 61000-4-2

IEC 61000-4-2 applies a standard test methodology to equipment, systems, subsystems and peripherals that could be exposed to ESD events in the final installation. The environmental conditions (humidity, flooring materials, low-conductivity carpets, grounding methods) may influence the intensity of the discharge, which in turn affects the susceptibility to damage of the systems to an ESD event, and for the purposes of compliance testing, this standard closely defines the conditions under which testing is to be done. IEC 61000-4-2 specifies a current pulse that is quite different than those used for testing integrated circuits for ESD during assembly and packaging and is not correlative.

2.1.2 ISO 10605

ISO Standard 10605 is used to characterize the ESD sensitivity of electronic subsystems in road vehicles. This standard draws on the IEC 61000-4-2 for most of its procedures and physical set-ups, but adds three additional current impulse variants for conditions specific to the automotive environment. It applies to discharges in the following cases:

- ESD in assembly
- ESD caused by service staff
- ESD caused by occupants

ISO 10605 applies to all types of road vehicles regardless of the propulsion system (such as spark-ignition engine, diesel engine, electric motor) and for some cases requires testing to 25 kV.

2.1.3 DO-160 Section 25

This document is used to test airborne equipment (avionics) for the effects of static discharges from human contact. It is based on IEC 61000-4-2 and uses the same ESD test pulse. It consists of a 15 kV air discharge test applied to points on operating avionics that are accessible to people.

2.1.4 CDE Testing

This test is designed to determine the susceptibility of electronic products to electrostatic discharges that occur when a cable, such as Ethernet or universal serial bus (USB), are connected. If a charge differential exists between the cable and the product being connected, a discharge will occur to the affected IO pins. A proposed ESDA document provides test methods using both the ESD gun and a modified TLP system.

2.2 Definitions

2.2.1 Hard vs. Soft Failures in Systems

Both hard and soft failures in an electronic system can be induced by transient events from outside the system, such as noise transients on connecting cables and electrostatic discharges, but also by over-voltages due to mis-operation or improper connection of the system. In any case, the system has ceased to operate as intended.

A "hard" failure is one where the outside event has caused damage to a component that will require a service technician to replace the affected component. Examples of hard failures include damage to gate oxide, resistors and metallization in the signal path which could have been as a result of latch-up or overvoltage. These will all be discussed in more detail later in this chapter.

A "soft" failure is one where the outside event can have several effects. Using some definitions paraphrased from IEC documents, one can demonstrate these situations:

- System automatically recovers with no loss of data and no operator intervention. This upset is completely transparent to the operator. An example might be a change to a command or data stream that is caught and corrected by system software.

- System malfunctions in a way that the operator notices but operation continues. Examples might be interference to or momentary loss or distortion of a display, keystrokes that are temporarily corrupted or not recognized, or static or noise in an audio system.
- System malfunctions in a way that requires some recovery action be taken by the operator. Examples might be system lock-up, display becomes blank, controls or keyboard ceases to function. Typically, this kind of upset or malfunction requires powering down the system and re-booting.

In the above scenarios data could be corrupted or lost without being apparent to the operator. Under IEC definitions, loss of data is never allowed when testing a product.

2.2.2 Sub-Assemblies and Modules

The definition of a system is generally considered to be a completed, self-contained product such as a cell phone or lap top computer; however, for testing purposes, a system can also be a sub-assembly or module such as a video card, engine control module (in the automotive industry), or other device which can only function as part of a larger self-contained or distributed system. Sub-assemblies such as these (except automotive) are now included under the European Union's EMC Directive and must be tested as systems with similar failure criteria.

Testing and evaluation of sub-assemblies and modules present an obvious problem: the device being tested needs to be operating in a larger system to properly test for susceptibility. Ideally, it should be tested in the actual system where it will reside, but this isn't always possible. Take for example the manufacturers of sound cards, video cards, memory modules and other devices that could be assembled into a wide range of systems. These manufacturers may elect to test their product in a typical installation – a computer with known susceptibility characteristics – or in a test jig. It's up to the manufacturer to decide the best way to test the product to insure compliance, and often a system manufacturer, who is going to purchase and use the sub-assembly in question, may require testing in a specific way.

2.3 Coupling of ESD into Systems and Circuits

2.3.1 How Does ESD Couple into a System and Affect a Specific Component During System Level Testing?

Depending on the requirements of the test standard, system level tests typically require discharges to system enclosures and controls and may include some externally exposed pins. IEC simulators produce currents and transient fields and the resulting current pulses have been described as having both CDM and HBM like characteristics. Furthermore, they induce strong electromagnetic fields, which are not necessarily associated with the discharge current; including those fields that are also excited by the discharge relay used for contact mode testing (i.e., the ESD current is directly injected into the test object through metallic contact between the tester and the device under test (DUT), as opposed to an air discharge between the tester and DUT). Annex D of the IEC 61000-4-2 standard (2009) [2] provides good insight and further details the effect of different sensitivities and coupling mechanisms [3].

During discharges to system enclosures the injected current pulse spreads at close to the speed of light over the metallic surface of a metallic enclosure. This occurs because the current must satisfy the tangential electric field boundary condition; Etan = 0. This can only be fulfilled by the creation of a current on the surface. At locations on the enclosure where the current has to divert due to the presence of slots or openings, fields will couple from the outside to the inside of the enclosure causing transient electromagnetic fields inside the enclosure ranging from very small to very large (kV/m). Furthermore, the current pulses will reflect on edges of the enclosure and couple to attached cables, predominately in common mode, which can then be converted to differential mode currents and possibly cause system upset.

The current densities flowing on the inside of the enclosure, its cables and PCBs are associated with strong electromagnetic fields. The electric fields can couple capacitively while the magnetic fields will induce voltages in any loop they can penetrate. The net effect of the dynamically changing electric/magnetic field strengths and localized field are induced voltages and currents. The induced voltages are often pulses of < 2 ns in width. Typical coupling scenarios are:

- A current flowing on the outside of a personal computer (PC) close to the peripheral component interconnect (PCI) slots of the enclosure will couple from the outside to the inside as the PCI slots are not well connected. Once the fields enter the inside, they can couple to any board, such as a graphics card, upsetting that board.

- The current from one PCB is coupled to another PCB connected by a flex circuit cable. The flex circuit cable does not confine the fields of the wanted signals well, and as a consequence the common mode current of the ESD on the flex cable can couple easily to signal traces.

- A current is injected into an audio trace, which is robust against ESD. This trace is routed inside the PCB parallel to a reset line. The coupling between the audio trace and the reset line causes noise pulses on the reset line; upsetting the system.

- A charged cable is plugged into a USB port. The ground contacts first and a large current is injected into the shell of the USB connector. This will cause a voltage drop between the USB connector shell and the chassis (as the connection is not perfect, but it forms a small inductance). The voltage between the shell and the chassis is visible on both the outside and the inside of the enclosure, and will force a current on the inside of the enclosure. Furthermore, although the shell connected first, a voltage is induced on the differential pair, which may upset a USB hub.

The fields can also interact with the direct current (DC) voltage supplies. While this is likely to lead to upsets if an IC's internal power nets receive large amounts of current from an ESD event, it is unlikely to upset power planes on circuit boards. The high decoupling capacitance and low impedance of typical board supply planes tend to limit the induced voltage, consequently, upset or damage is seldom seen by ESD injection to circuit board power planes. If power plane or ground plane injection causes ESD upsets, the mechanism is not the result of DC voltage fluctuations, but rather due to the transient fields of the ESD event coupling into other nets or directly into ICs.

For test standards requiring discharges directed at exposed pins (such as USB ports) or plugged in fixtures that simulate system events like cable discharge, the current pulse will travel from the connector to board traces which direct charge to the component pin. ESD injected into an input or output pin will typically be diverted to the internal positive voltage supply (VDD) and negative voltage supply (VSS) net within the IC. However, the internal voltage transient can cause soft failures or hard damage. The effect of both signal and supply transients will be discussed in more depth later in the chapter.

Depending on board layout, trace length/thickness, signal shielding, and the passive/active components connected to a board trace or plane, the resultant reactance of the signal trace will attenuate rise time, duration and peak current as seen by the targeted IO or supply pins. This ultimately affects what happens to the charge when it gets into a system.

2.3.2 What Happens When the Charge Gets into the System?

The hybrid CDM and HBM characteristics of IEC simulator pulses results in high peak currents and pulse durations on the order of 60 ns at the point of charge injection. The degree of ESD pulse attenuation/degradation before arriving at the part will determine how the on die IO and power supply protection reacts. Attenuation of a pulse that would otherwise exceed the built-in protection capability could help protect a part from damage. However, if the perceived pulse characteristics fall outside of the component protection design window, the attenuation could also inhibit the response of pin level protection circuits as well as the resistor/capacitor (RC) triggered supply clamp. The net effect of inhibited ESD protection response could be failure at voltage levels lower than those seen by a component during component testing.

Signal attenuation can affect the way internal supply rail power clamps engage to dissipate the charge that is dumped onto the rail when the ESD diodes turn on. Power clamps often use triggers that expect a certain voltage rise time to engage. If clamps engage quickly and their current shunting capacity is not exceeded, the component should be protected. If the clamps turn on too slow or not at all, overvoltage damage of transistors connected to the supply rail can occur.

Because component protection is designed for an unpowered component, the response of a component when power is applied in a system and IO pins are transitioning can be much different. In some cases, the response can be inhibited because the ESD protection circuits see voltage on output pins and DC bias on internal supplies. These voltages affect the threshold at which protection schemes, like double diode [4], turn on and consequently the ability of the pin level protection circuits to shunt current. ESD power clamps that expect to see ESD pulses with the part unpowered now see pulses on top of a pre-existing DC bias. This voltage starting point, along with the possibility of a change in the time constant of an ESD pulse due to a reactance contribution from the voltage regulator, can retard the response of the power clamps.

The scenarios described above do not take into account external signal/power protection that may reside on the board to protect from system events. Transient voltage suppression (TVS) components protect a component by shunting charge before reaching the component. However if the TVS component does not fully shunt the signal, and the rise time of the residual IEC gun pulse is not fast enough to engage the on die supply clamps, overvoltage damage can occur as described above.

Coupling to signal or power lines adjacent to a targeted signal line can also lead to failure in signals or supplies that are not stressed directly. These lines may not have the same degree of protection as lines leading to the outside world. Failure of these lines will depend on the degree of coupling and protection on the coupled IO or supply.

2.3.3 Recoverable or "Soft" System Failure Modes

Many of the IO and power supply overvoltage/transient issues that cause hard failures at a given voltage may cause recoverable or soft failures at a much lower voltage, causing a system upset or a change of state in the system. Typically, the upset levels are much lower than the destruction levels. Signal or voltage supply rail (such as VDD or VSS) pulse induced transients is one mechanism that could lead to a loss of state of a system component, requiring user intervention to reset the system. Transient latch-up due to ESD pulse induced power glitches, or locally induced overvoltage of a power domain or cluster of transistors in a discharge path is another mechanism that could lead to a loss of state of a component/system.

2.3.4 System Degradation

In theory, so called "soft damage" to a transistor could lead to system performance degradation, which could be considered a system level failure if the degradation is self-detected by the system. An example would be soft oxide damage to an output driver transistor of a stressed pin which affected the robustness of the transistor. The driver could continue to operate satisfactorily at low frequencies, but higher frequencies or low supply voltage could lead to failure. Soft oxide damage due to voltage transients on voltage supply rails could also lead to degradation of internal circuit paths that ultimately affect performance as well. Passive components on system boards can also be damaged to the point that performance is degraded. Examples are surface mount chip capacitors and resistors used for filtering or terminations. Over voltage/current damage can change the effective value of these components to the point the circuits malfunction if they continue to degrade after initially being electrically damaged. This is sometimes referred to as a "latent" failure: one which occurs at a later time due to damage from a transient at an earlier time.

2.3.5 Hard System Level Failure Modes

Unlike a functional powered up system which can put a component in many states (and consequently induce many different types of failures), the state of a standalone component prior to the initiation of component testing is constant. System level testing can induce hard failures that would appear similar to those seen in component testing. There are extra failure modes that can occur in system testing that don't occur in component testing, but some of the more common component testing failure symptoms that could be seen in system level testing includes [5]:

a) Damage to gate oxide and/or channel damage to drivers and receivers attached to an IO signal.
b) Damage to primary signal path ESD protection circuits such as ESD diodes whose purpose is to protect the signal path. This occurs when the robustness of the protection circuit is exceeded.
c) Damage to resistors and metallization in the primary signal path.
d) Overvoltage damage to transistors which reside in power domain(s) and absorb charge routed to them by the ESD protection circuits. This occurs when the protection circuits (diodes, clamps, etc.) for the power domain fail to respond fast enough or their robustness is exceeded.
e) Latch-up

Because component protection is designed for an unpowered component, the response of a component when power is applied and IO pins are changing state is much different. In some cases, the response can be inhibited. This can be due to the following causes:

a) ESD protection circuits on IOs, such as protection diodes between the pin and internal VDD rails, are designed to shunt to unpowered rails. Voltage on the rail affects the effective voltage threshold of the protection diodes. For a diode with the anode connected to the pin and the cathode to VDD, the voltage threshold increases with bias. Elevated thresholds result in higher pin voltages during discharge, which typically increases the voltage seen by downstream circuits during a discharge.

b) dV/dt (for example, RC) triggered ESD overvoltage internal supply clamps that expect to see ESD pulses with the part in an unpowered condition now see pulses that start from a powered condition. This pre-existing voltage, along with the possibility of a change in the RC trigger time constant due to a reactance contribution from the voltage regulator, can retard the response of a clamp trigger circuit.

Aside from traditional component level ESD damage and damage due to inhibited ESD protection, localized voltage differentials which are not seen in standard operation could lead to latch-up of transistors. This can cause hard damage to a component if the latch-up condition is sustained. In theory, signal corruption due to an injected pulse could lead to a loss of state and signal "contention", where transistors are simultaneously driving opposing signals that lead to electrical overstress and permanent damage if sustained.

2.4   Troubleshooting to Determine the Cause of Failures

2.4.1   Hard Failures

Classic methods of troubleshooting electronic products work well in the case of hard failures. Faulty sub-systems, boards and finally components are systematically isolated by measurements of power supplies and signal tracing. In the case of hard failure of a component, damage is often detected with methods such as photon emission microscopy (such as emission microscopy (EMMI)) and/or laser induced circuit perturbation (such as thermally induced voltage alteration (TIVA)) that highlight circuit damage. These are well understood and documented techniques [6], and as such, won't be dealt with in this paper.

2.4.2   Soft Failures

Classic troubleshooting methods don't work well on soft failures simply because there is no physical evidence to find. Either the system recovered on its own through error correction routines or re-booting the system corrected the error. Soft failures are often found during compliance testing for ESD (as well during other types of immunity testing) but they also occur in field at the customers' site. In either case, the root cause of the failure needs to be determined and corrections made to the system to prevent further upsets.

When a soft failure is detected, the method of troubleshooting is often trial-and-error. This can be effective when the person doing the troubleshooting is familiar with the system, but if the failure occurs in the field or at an outside compliance lab that may not be the case. Even for a soft failure being investigated by someone familiar with the system architecture, the real root cause may still not be resolved although the symptom may be eliminated. Filtering, shielding or the addition of protective components can be effective at removing the symptom, but the root cause might actually be a power bus running too close to a sensitive control line. Adding components may resolve the soft failure issue, but the real root cause -- the design problem -- remains un-detected.

One method of localizing a soft failure problem that is sometimes used is injection of a localized magnetic field or other disturbance in the vicinity of the suspect circuit. By moving the probe around the circuit and watching for upset it may be possible to determine the area of a board likely to be the problem. This method has recently been automated to allow very precise scanning and provide a 3-dimensional picture of a circuit board's weak spots and will be described in detail in the following section.

2.5   New Technologies for Determining Root Cause of Failures

2.5.1   Susceptibility Scanning

As discussed above, once a system fails due to a system level ESD event, isolating the root cause can follow different paths. For soft failures the most promising tool is local magnetic or electric field injection using susceptibility scanning. This can be performed by hand as previously mentioned but it is very difficult to obtain significant precision or generate a visual picture of the boards' sensitive locations. When done automatically [7, 8], these difficulties are removed and the ability to generate data for later analysis is achieved [9].

The objective of susceptibility scanning is to identify locations, ICs, modules and electrical nets that exhibit the same failure symptoms that were observed during system level testing. If those locations are found within the system, it is likely that during system level testing energy is being coupled from the outside to those locations, thereby causing the failure.

To identify these sensitive locations a locally strong field needs to be created that resembles the noise that might be coupled from the outside ESD generator to the suspect board or circuit. The methods are detailed in [7, 8]. In brief, a transmission line pulser having rise times of < 1 ns is connected to magnetic and/or electric field probes. These field probes are moved close to the ICs, nets or modules in question and pulses are applied by the TLP, which induces the field. The system reaction is observed. By varying the probe type, size and applied voltage, locally sensitive regions can be identified. Further, the results can be visualized as susceptibility sensitivity maps as shown below in Figures 3 & 4.

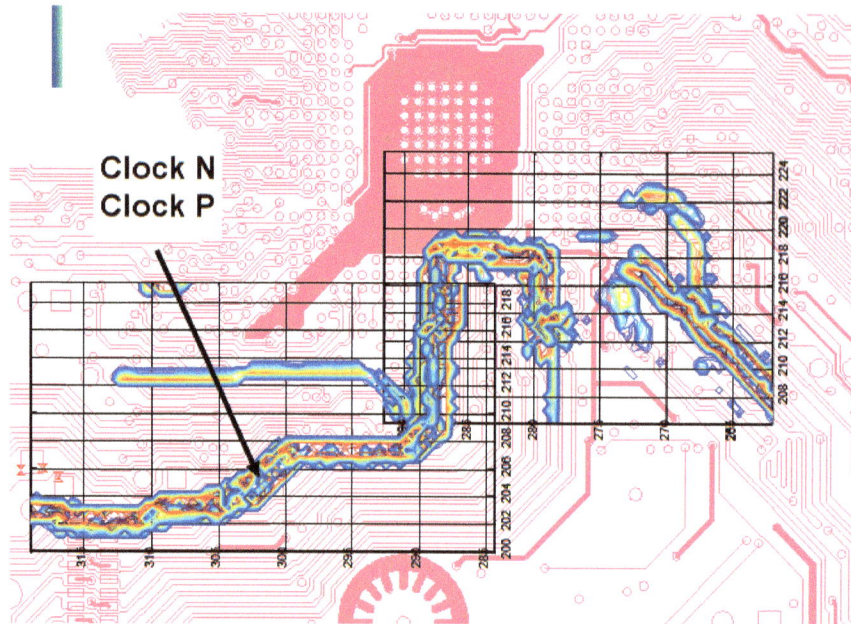

Figure 3: Result of susceptibility scanning: Sensitive differential clock on a motherboard [7]
(Courtesy of Amber Precision Instruments)

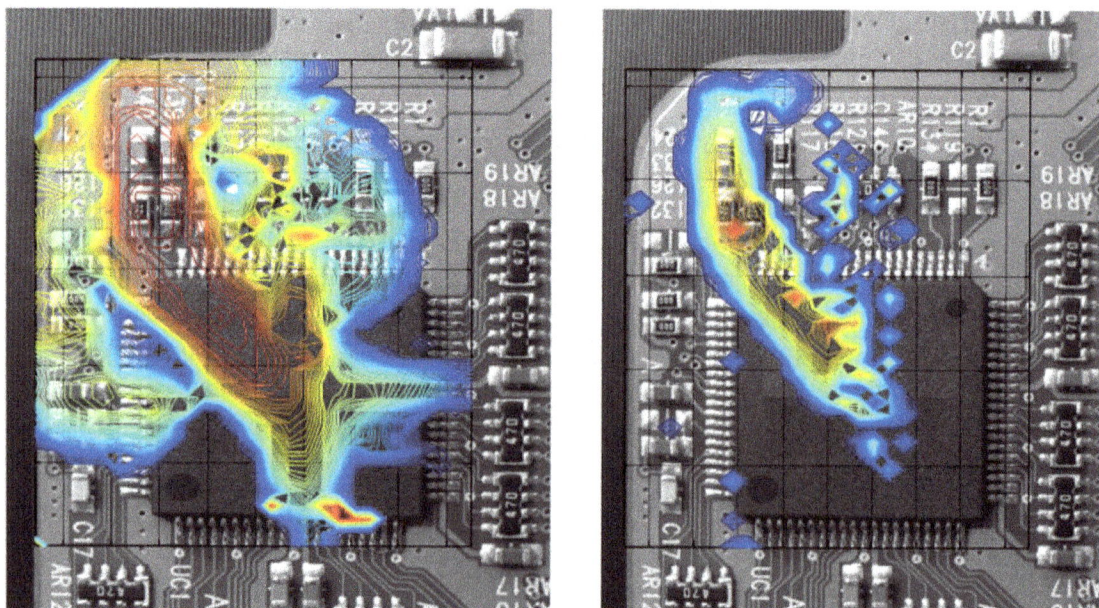

Figure 4: Result of susceptibility scanning: Comparison of two functional identical ICs from different vendors. The color grade indicates the TLP charge upset voltage with blue being 4 kV, red being 1 kV [7]
(Courtesy of Amber Precision Instruments)

2.5.2 New Software Methods

In addition to susceptibility scanning, software based methods can be used to uncover the root cause of ESD failures. These methods often require having pre-existing diagnostic tools built into the system that are targeted for debug. For system upset, where some system functionality is retained, software tools that collect the contents of die data registers can reveal status register bits, error codes, data loss or data corruption. All of this will provide insight into the nature of the failure. Other software tools include error handling routines that collect system response data for diagnosis but again, some system functionality must be retained to run the software diagnostic [10,11].

If system functionality is lost but components are accessible through special test modes such as JTAG (Joint Test Action Group) ports, it can provide another means of accessing the data registers. Factory debug and test modes can also be employed if they are available with appropriate documentation from the component manufacturer. Ultimately, for these tools to work, the system analyst needs to have a good working knowledge of the system and the information that can be extracted from the components.

2.5.3 System Specific Test Boards

System specific debug test boards are boards that can be embedded in a system to track and save system status for recall in the event of an ESD induced fault. Dedicated system monitor chips that report system status could also be used. Simple methods include fault indicator lights or display screen messages/codes that can point to unusual power consumption conditions by a component or else report the status of a component pin that only comes on during a specific fault condition.

When a failing component is identified and diagnostic information is gathered, component debug tools can be employed to isolate a particular location on a component. If the failing component can be exercised in the system to the point of demonstrating the mode of failure, various diagnostic tools can be employed including optical and infrared emission cameras and laser based scanning tools to pinpoint damage sites on silicon. Complications associated with use of these tools include special die preparation (die thinning) to support the optical nature of these tools. Software and/or system functionality may need to be modified to increase the acquisition rate of the optical signal. If the component cannot be operated in the system, it may be necessary to use special test hardware in conjunction with knowledge of the component to stimulate the failing condition. The complexity of the hardware can range from sophisticated automated test equipment (ATE) testers to curve tracers or DC power supplies.

Once a failure location is isolated, the next step is to understand the implications of the location of failure and, in some cases, the mechanisms seen in the physical analysis. Software based design tools that can correlate physical locations to circuit schematics are particularly useful for this analysis. The schematics can then be evaluated through analog circuit simulators such as simulation program with integrated circuit emphasis (SPICE) or digital simulators such as Verilog to confirm that the physical failure symptoms match both the analog response to ESD stimulus and the system behavioral response to the ESD upset.

As the response to a system failure becomes better understood, examination of physical die layout may be appropriate:

- If latch-up is suspected, the layout can be examined for latch-up design sensitivity.
- If damage to gate oxide is suspected, the circuit path that results in overvoltage during the stress event must be understood.
- If metal shorts or opens are discovered, the current carrying capacity of the metal routing at the affected location can be compared to the current expected during the system stress event.

2.6   Summary

It is important to recognize that both hard and soft failures occur as a result of ESD events in the test lab and in the field, and both need to be addressed. Component manufacturers rate components for their ability to withstand a certain voltage during handling, but an ESD event to a system producing only a few volts on an input pin of that component may cause a system to re-set unexpectedly. Fortunately, new software analysis methods and test instruments are becoming available to enable us to see what's happening as the result of an ESD event. This enables both the component and system manufactures to make good decisions in design that will save money and benefit everyone.

References

[1]   Industry Council on ESD Target Levels, "White Paper 3 System Level ESD Part I: Common Misconceptions and Recommended Basic Approaches," December 2010, at www.esda.org or JEDEC publication JEP161, "System Level ESD Part I: Common Misconceptions and Recommended Basic Approaches", www.jedec.org

[2]   IEC 61000-4-2, 'Electromagnetic compatibility (EMC) – Part 4-2: Testing and measurement techniques – Electrostatic discharge immunity test', Ed. 2.0, 2008.

[3]   Jayong Koo; Qing C,ai; Kai Wang; Maas, J.; Takahashi, T.; Martwick, A.; Pommerenke, D., "Correlation Between EUT Failure Levels and ESD Generator Parameters", IEEE Trans. EMC. Vol.50, Issue 4, Nov. 2008 pp. 794 - 801

[4]   Industry Council on ESD Target Levels. "White Paper 2: A Case for Lowering Component Level CDM ESD Specifications and Requirements," Revision 2, April 2010, at www.esda.org or JEDEC publication JEP157, "Recommended ESD-CDM Target Levels", www.jedec.org

[5]   Anderson, W., Eppes, D, Beebe, S. "Metal and Silicon Burnout Failures from CDM ESD Testing", Proc. EOS/ES Symp., Sept. 2009, pp. 2A.6-1 - 2A.6-8.

[6]   Voldman, Steven H., "ESD Failure Mechanisms and Models", John Wiley & Sons, Ltd, 2009, pp. 35-37.

[7]   Muchaidze, G., Jayong Koo, Qing Cai, Tun Li, Lijun Han, Martwick, A., Kai Wang, Jin Min, Drewniak, J.L., Pommerenke, D., "Susceptibility Scanning as a Failure Analysis Tool for System-Level Electrostatic Discharge (ESD) Problems", IEEE Trans. EMC, May 2008, Vol 50, No. 2, pp. 268-276

[8]   Kai Wang; Jayong Koo; Muchaidze, G.; Pommerenke, D.J. "ESD susceptibility characterization of an EUT by using 3D ESD scanning system", IEEE Int. Symp. on EMC, 8-12 Aug. 2005, pp. 350-355, Vol.2

[9]   Zhen Li, David Pommerenke, "Electrostatic Discharge Induced Upset Probability Database for Transient Field Coupling to Integrated Circuits", presently under review (2012) IEEE Transactions on EMC, contact davidjp@mst.edu

[10]   Maheshwari, P.; Tianqi Li; Jong-Sung Lee; Byong-Su Seol; Sedigh, S.; Pommerenke, D.; , "Software-Based Analysis of the Effects of Electrostatic Discharge on Embedded Systems", Computer Software and Applications Conference (COMPSAC), 2011 IEEE 35th Annual , pp. 436-441, July 2011

[11]   Maheshwari, P.; Byong-Su Seol; Jong-Sung Lee; Jae-Deok Lim; Sedigh, S.; Pommerenke, D.; , "Software-Based Instrumentation for Localization of Faults Caused by Electrostatic Discharge," High-Assurance Systems Engineering (HASE), 2011 IEEE 13th International Symposium, pp.333-339, Nov. 2011

Chapter 3: State-of-the-Art ESD/EMI Co-design

Patrice Besse, Freescale Semiconductor
Jeff Dunnihoo, Pragma Design, Inc.
Masamitsu Honda, Impulse Physics Laboratory
David Pommerenke, Missouri University of Science & Technology
Pasi Tamminen, Nokia
Matti Uusimaki, Nokia

3.0   Introduction

The threat to a design posed by ESD is essentially defined by the term "Electrostatic Discharge," that is, the movement of an accumulated positive or negative static charge from the user environment to (or from) a component. The intensity of discharge that the component can withstand without damage or functional degradation defines its robustness and affects its cost.

Existing state-of-the-art ESD/EMI design methodologies span the continuum from advanced theoretical 3D modeling of Maxwell's equations for a system to the most basic trial and error tests, and all of the limited practical and theoretical implementations in between. A matrix of these approaches is presented in Table 1.

Table 1: Practical and theoretical co-design methodologies for ESD/EMI robustness

		Methodology Type	
		Practical	Theoretical
Complexity & Cost	Basic	Designer's experience, lessons learned, design reuse	Datasheet comparisons
	Advanced	Component qualification testing	SPICE modeling of ESD injection
	Comprehensive	Rigorous head-to-head system level iterative testing	Full 3D Maxwell Field-Solver Simulations

The overall design methodology is ultimately chosen to help the designer compromise on an optimum fixed point in the "tradeoff gamut" between three generally conflicting goals: (1) ESD/EMI robustness and susceptibility resistance, (2) signal integrity and functional performance (bandwidth, data-rate, minimum emissions, etc.), and (3) cost and/or time to market as depicted in Figure 5.

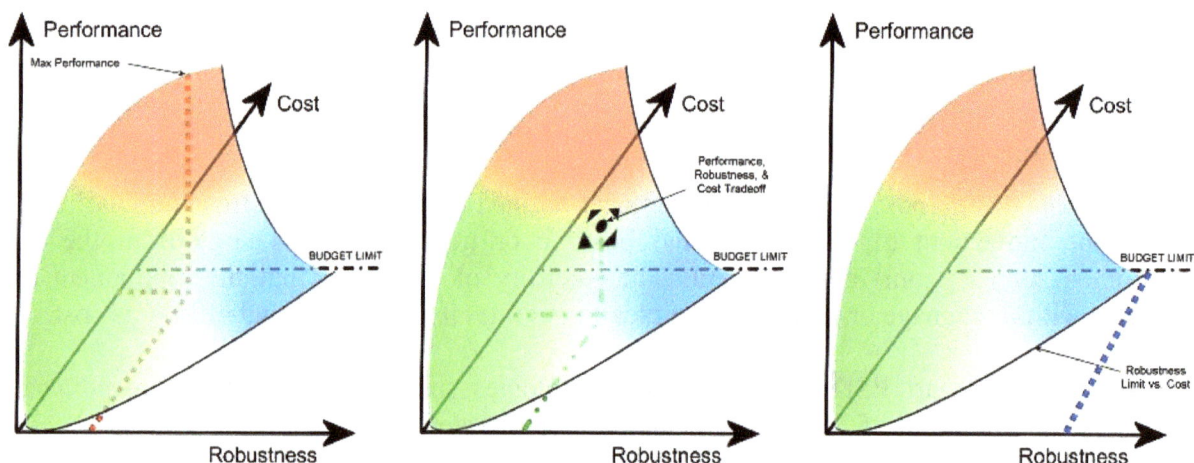

Figure 5: ESD/EMI Co-Design Tradeoff Gamut

This concurrent shepherding of sometimes cooperative but often conflicting design goals during product design and development is referred to here as the "ESD/EMI Co-Design" problem.

Not surprisingly, many of the protection strategies for ESD are tightly related to EMC robustness and EMI suppression methods. This chapter examines the advantages and limitations of existing and emerging methodologies in the art with respect to their effectiveness in arriving at the optimum robustness/performance/cost tradeoff for a given application. Obviously, the survey of such an extensive field is necessarily limited in its depth, but specific recommended system design guidelines are proposed.

From the beginning of the design process, critical exposed and susceptible nets must be identified and assigned reasonable and sufficient robustness levels, including the excess robustness protection margin desired. Aggressor pulse entry points or entry paths (whether conducted or induced) for hard and soft failures should be outlined as described in Chapter 2. Component selection, layout and placement should be considered with regard to ambient and direct ESD/EMI strikes and disturbances and the effect on signal integrity. System partitioning, grounding, clamping, shielding and return path shaping are also considered, along with the problems of potentially hidden multiple/secondary discharge points within the system.

By managing these issues in the concept, prototype, and design phases, a designer can observe and improve the extent and effectiveness of any element or flow in a process intended to balance ESD/EMI robustness, performance, and cost with existing and emerging methodologies. The remainder of this chapter will examine common examples of co-design as presently practiced, from the most basic rules of thumb, through advanced practical and theoretical methods, to comprehensive modeling of systems.

3.1 The Basics

As with any endeavor, a strong foundation or solid beginning can provide the most "Bang-for-the-Buck" in robustness vs. cost. Adherence to a few basic ESD/EMI principles and concepts throughout the functional design is often all that is needed to design an ESD robust system.

3.1.1 Shielding

The purpose of connecting a lightning rod and ground cable on a building is to route and shunt a potential lightning strike safely to ground in a reasonably controlled manner, away from flammable interior and structural components. Beyond encasing the building in a continuous Faraday Cage, the uncertainty and complexity of a strike makes the safe and cost-effective routing of the strike around delicate structures very difficult.

Consistent with ESD strikes on electronic systems, the primary goal for robustness design is to anticipate ESD entry points, and try to plan for effective return paths away from sensitive components.

While managing these primary effects, the designer also must be aware of potential indirect secondary effects of strikes or zaps, such as transient latch-up, or secondary discharges as described in Chapter 4.

The first and generally most effective method of protection is to encapsulate the entire system in a conductive enclosure, but as system input and output ports often do not permit complete enclosure, it will be necessary to add some holes in the enclosure for buttons, displays and connectors. These holes then become potential entry points for system level ESD strikes.

When these entry points are identified on or near susceptible nets (whether directly injected or indirectly coupled as discussed in Chapter 2), then discrete and/or on-chip protection circuits are often used in an attempt to catch and shunt ESD energy from the circuit back into the shield, and these methods are discussed in more detail below.

Good implementation choices made about how and where to implement limited shielding often provide the first and only protection needed for a sufficiently ESD-robust system. For example, certain systems naturally are constrained to certain standard IO connector requirements. A television (TV) or a monitor may require a high definition multimedia interface (HDMI) or digital visual interface (DVI) input. But whereas the bare HDMI connector presents tightly packed IO pins with interleaved ground pins and an enclosed ground shield as shown in Figure 6, the 0.100" (2.54 mm) pitch of the DVI pins exposes sensitive IO lines to potential unprotected strikes.

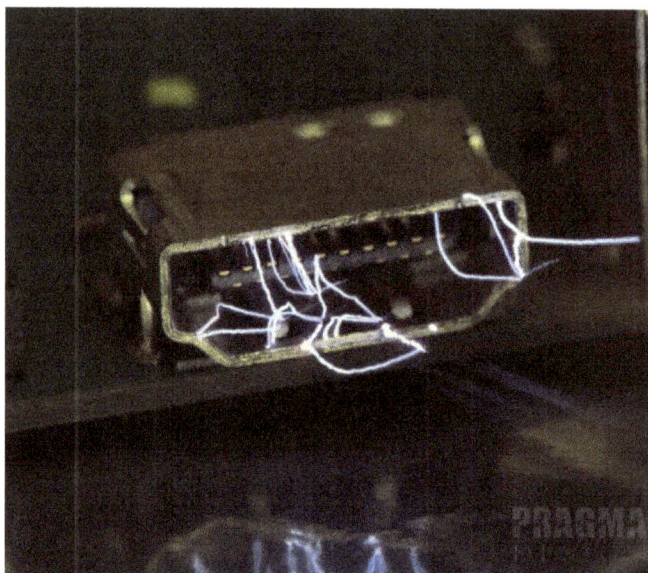

Figure 6: Multiple air-gap strikes to shield and pins as an entry point or aperture

A system is as vulnerable as the first component to fail during an ESD strike. But the mere selection of an interface type may also introduce some engineered "ESD robustness" embedded in the basic design of the interface (such as the connector grounding design mentioned above). Thus basic knowledge of typical IO susceptibilities may provide a large portion of the solution.

3.1.2 Beyond Shielding

Signals into and out of a system must be designed to handle the potential of unwanted ESD entry through those points. Once such a potential is identified, the design engineer must provide a preferential path to dissipate the strike(s). The following are the basic points to consider from the ESD perspective:

- Good Return: Provide good ESD ground return paths for connector shells and grounds, and TVS or other ESD attenuation components.
- Short Path: Clamp and divert ESD charge as close to the entry point as possible. Place TVS components or other attenuators as close to the entry point as possible to limit the total loop area and direct coupling of the aggressor pulse into safe paths.
- Hidden Impediments: Minimize secondary parasitics in the return path (for connectors, grounds, and clamping components). Consider component package parasitics and PCB plating/stackup considerations which may affect overall system parasitics.

For a more detailed conceptual analysis of guiding ESD current from the entry point out of the system, see Appendix C.

3.1.3 Component Selection

Any given system board will be populated with components of varying individual ESD robustness and protection. The same set of components implemented in different ways on the board, or within the system, may result in different levels of system ESD robustness.

While system specifications such as IEC 61000-4-2 are often applied to component evaluation boards as the "System-Under-Test," the results of these tests cannot be directly compared to system level testing of complete, end-user ready systems. Component suppliers are limited in their ability to accurately characterize the system level robustness of their particular component since the choice of other interconnected components, connectors, routing and enclosures is out of their control and up to the system designer. Even if a component supplier provides a "reference design" circuit with a particularly well-characterized TVS component combination, a poorly designed enclosure can dramatically affect overall system robustness.

For example, consider an ESD stress to an IO pin of an integrated circuit. If the IO is intended to be connected directly to a system IO, such as a USB or HDMI pin, it would seem that stressing the component's IO pin versus the component's ground pin would correlate directly to the IO pin's ESD performance in a system. The current path for a component ESD test may, however, be totally different from the current path during a system level ESD test. In a component ESD test, such as HBM, one pin combination (other pin combinations would look at IO to power rail or other IO) tests between the IO and ground, all current entering the IO pin must exit from the ground pin(s). In a system ESD test the current may all enter into the IO pin but the return path to the ESD source may include one or more power pins and have very little current exiting the integrated circuit through the ground pin(s).

Selection of components and optimal layout for the system may be the only degrees of freedom that a board or subsystem designer may reasonably have at his disposal. Determining component robustness from manufacturer specifications is not always straightforward. For example, consider the two TVS clamping components below in Table 2:

Table 2: Sample attempt at comparing and contrasting two TVS datasheets

	Component A	Component B
ESD "Rating"	10 kV ???	8 kV IEC 61000-4-2
R_{DYN}	0.8-1.2 Ohms *IEC 61000-4-5 (8/20us)	0.6-0.7 Ohms *TLP 100ns
V_{CLAMP}	18 V *IEC 61000-4-5	240 V peak ???
Strikes	10 kV - 10 positive / 10 negative 12 kV - 1 positive / 1 negative	8 kV - 1000 positive / 1000 negative

The 10 kV ESD withstand rating of the TVS protection component "A" on its evaluation board initially appears better than "B's" 8 kV except that the test condition is not clear. But in any event, this only attempts to describe what the TVS component will withstand on its own, and tells nothing in particular about how well it can protect other components (For more information on the difficulties associated with testing components individually at the system level see Section 2.2.2 of Chapter 2 of White Paper 3, Part I [1], and Section B.7 in the Appendix of this paper).

The dynamic resistance is interesting in that it suggests that Component B might shunt more current from a node under protection than Component A, but upon further inspection we may find that the dynamic resistances are specified for different test conditions, test pulses, and stress levels. In the third row of Table 2, the "clamping voltage" is provided, although with a very low surge current (slow risetime, low bandwidth) in one, and a very high peak transient (fast risetime, high bandwidth) measurement setup in the other. In the end, the designer is left with essentially no meaningful comparison or contrast between the two sets of seemingly similar TVS parameters.

The last line of Table 2 reflects the attempts of some vendors to characterize the robustness with repeated strikes at a particular or multiple rated levels. These may imply a guarantee of survivability, in that either their component specifications will not change after repeated strikes, or that while the physical characteristics of the component may change dramatically within limits after repeated strikes, the absolute limits of the component specification will not be violated. While there is a maturing body of literature on repetitive damage in semiconductor and polymer clamp components, there presently is no accepted industry standard for the characterization or definition of "multi-strike" robustness or margin [2].

Even when components from the same vendor are considered under the same test conditions, the specification data may not provide meaningful comparative data for proper selection, especially when protection components are interacting with other components competing for ESD currents in a net and the sharing (and real-time behavior) becomes much more complex; see Figures 7a and 7b for typical examples.

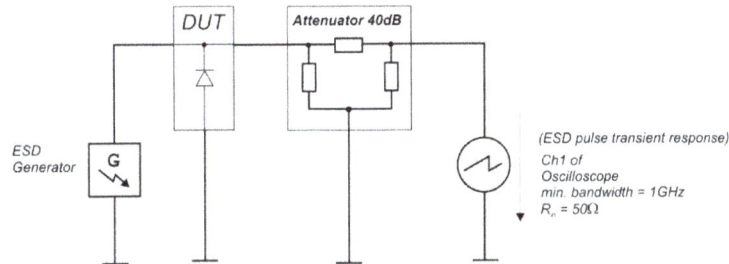

Symbol	Parameter	Conditions	Min	Typ	Max	Unit
$V_{CL(ch)trt(pos)}$	positive transient channel clamping voltage	V_{ESD} = 8 kV per IEC 61000-4-2; voltage 30 ns after trigger	[1] -	8	-	V

Figure 7a: Example of a well specified ESD test condition, but with arbitrary test parameter reported.

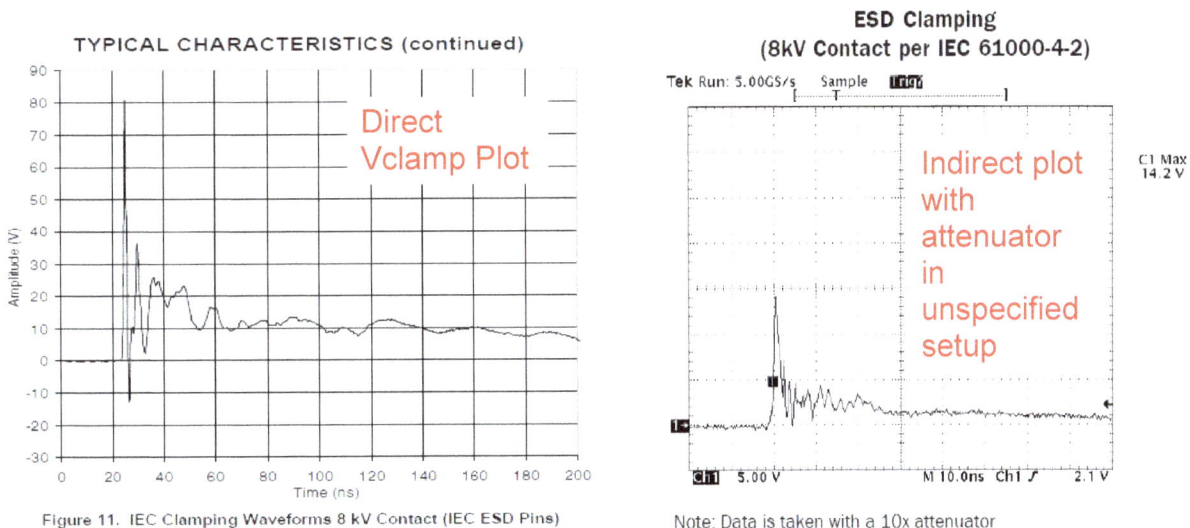

Figure 11. IEC Clamping Waveforms 8 kV Contact (IEC ESD Pins)

Note: Data is taken with a 10x attenuator

Figure 7b: Competing examples of unspecified ESD test conditions providing peak clamping values also affected by unreported measurement bandwidth and test conditions

The result is that design reuse becomes an important safety line for designers from one product generation to another in the realistic climate of incomplete ESD comparison data available to designers today. Lessons learned about replacing Component A with Component B may encourage the designer to stick with a known design, and render him less willing to consider a cost reduction of New-and-Improved Component C. Without a reliable methodology to qualify the performance in-house in specific systems, the designer may be forced to forgo an otherwise margin-improving cost reduction.

3.1.4 ESD/EMI Budget Strategy

For critical IO nets, such as high-speed serial interfaces or radio frequency (RF) antenna connections, the system designer may be faced with a passive parasitic budget as well as a bill of materials (BOM) pricing budget. As illustrated in Figure 8, similar challenges exist for EMI filter design and TVS clamp selection where there is a direct external interface with signal integrity restrictions on filter and protection components.

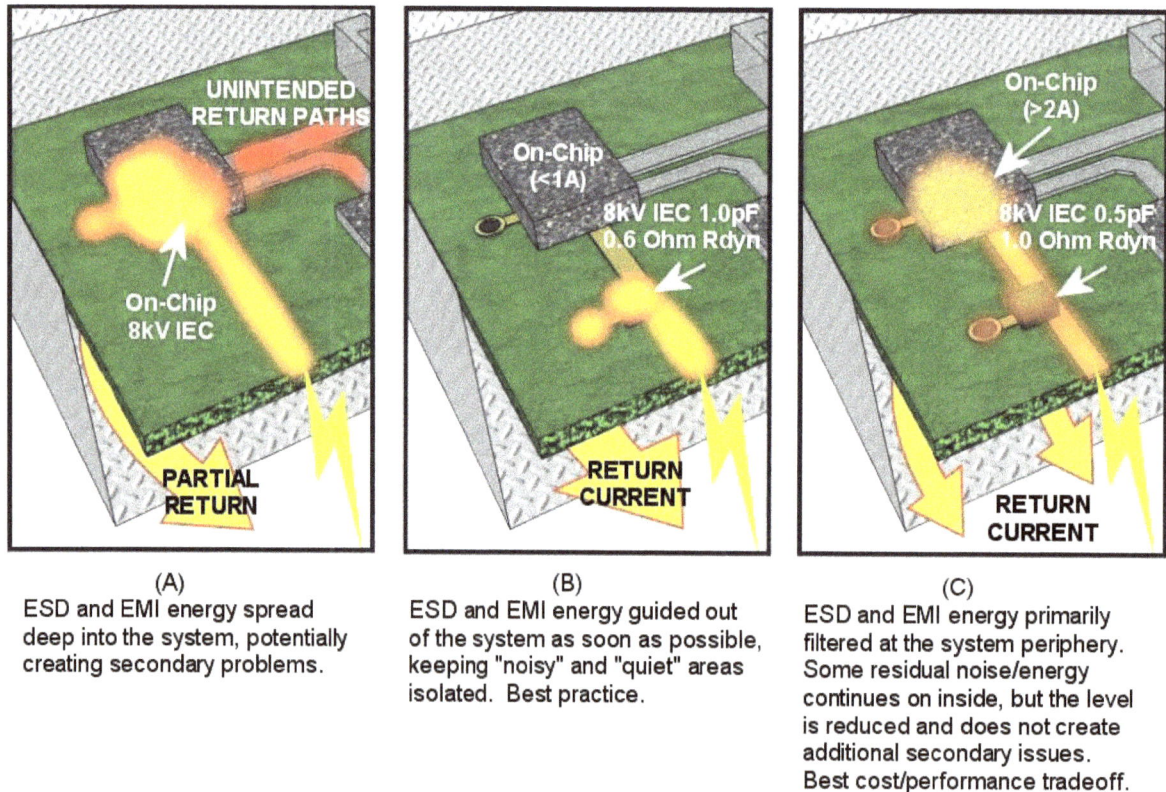

(A)	(B)	(C)
ESD and EMI energy spread deep into the system, potentially creating secondary problems.	ESD and EMI energy guided out of the system as soon as possible, keeping "noisy" and "quiet" areas isolated. Best practice.	ESD and EMI energy primarily filtered at the system periphery. Some residual noise/energy continues on inside, but the level is reduced and does not create additional secondary issues. Best cost/performance tradeoff.

Figure 8: A qualitative illustration of RF ESD design issues with respect to ESD/EMI budget

In very critical design cases, minimizing the "ESD shunt budget" for each component may become a consideration as a way to attain a more desirable position on the Robustness/Performance/Cost tradeoff gamut. Since parasitic capacitance tends to increase with increased robustness in a TVS, a "weaker" TVS component may be chosen to obtain a lower parasitic capacitance that fits within the signal integrity design constraints. In order to maintain equivalent system robustness levels with the weaker TVS component, the component under protection will be required to shunt more of the "ESD shunt budget", otherwise the achievable overall robustness will be reduced.

For instance, a high-speed serial bus may allow for as little as 1 pF of additional parasitic capacitance on each IO line to pass compliance tests. If that port may be subjected to an 8 kV IEC contact strike, but the physical layer interface (PHY) chip can only tolerate a 2 A peak residual current, then there likely must be some additional external protection added to meet the system level robustness goal. Rather than choose an 8 kV, 1 pF, 0.6 Ohm TVS component that can attenuate the majority of the system level pulse, the system designer may need to choose an 8 kV, 0.5 pF, 1.0 Ohm component that stresses the application specific integrated circuit (ASIC) closer to its maximum limits (Note again, as in Table 2, that while both TVS components might be "rated" at 8 kV, their effect on overall robustness and signal integrity in the system may be radically different due to their dynamic clamping resistance (Rdyn) and other parameters).

In this manner the "ESD shunt budget" is shared more efficiently between the two components (see Figure 8C). It is important to understand that this very critically allocated ESD budget should be contemplated in the most restricted circumstances. If the ESD pulse penetrates deep into the system (See Figure 8A) then secondary upset effects are more likely, meaning that there could be a secondary spark inside the system or at least the ESD return current on the ground plane that will generate additional noise effects.

This is analogous to a total distributed insertion loss budget for an inline EMI filter, where two filter components in series both contribute to the total budget. While the two filters could share the attenuation equally or in any arbitrary fraction, the majority of the EMI attenuation should occur in the elements closest to the connector to avoid "polluting" the interior of the system with interference.

Likewise, the ESD protection should be implemented as in Figure 8B, where TVS and filters are placed as close to the IO port as possible and the ESD is returned out of the board as soon as possible.

Both the EMI and ESD budgets are thus optimized by minimizing the overall design margins for the system. This potentially makes the system more vulnerable to both EMI and ESD; therefore this method should be utilized in only the most critical situations.

3.1.5 Equipment Ground

The ESD charge must be stored or returned in a manner that does not affect the operation of the system. This aggressor charge may be safely conducted along the perimeter of a metal product case to an alternating current (AC) mains ground line, or it may charge the volume between a laptop and a tabletop as shown in Figure 9. But it may also charge an IO input gate capacitance beyond its breakdown voltage, or it may be transformed into enough heat by an ESD clamp to permanently damage it.

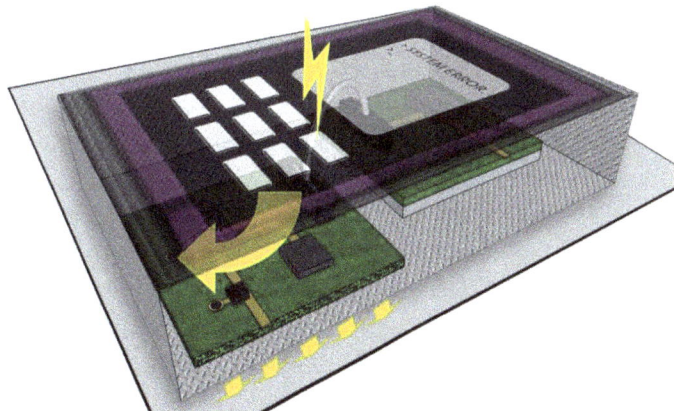

Figure 9: Discharge path into a keypad, through the system to a local ground, then through the case on the system return path (of a generic device as shown here)

Again, the overall ability to gracefully transfer or transform ESD charge without damage or functional degradation depends on a good system level path from the system to the outside world. What must be considered are the discharge paths from the system entrance point, to the different local ground nodes, to the PCB ground, to the case ground and to the reference ground of the system level test.

3.1.6 Quick Fixes

Even the best trouble-free legacy designs may suddenly exhibit an unexpected susceptibility out of the designer's control due to user application environment changes, unavoidable supply chain substitutions, end-of-life component revisions, or component variation interactions not anticipated during system design.

In this case, the immediate engineering problems for the designer are to "fix-what-we-have", and fix it fast! Unfortunately, this problem is resolved through the same design-for-robustness exercise applied during normal product development and design, except that the designer is now constrained to the existing board and components already committed to the supply chain. In this case, an "add-on" component may be the only hope to avoid complete redesign of the system board, if not the entire system. Again, the more information the designer has on failure and suspected "upset vector," the more accurately he can focus his solution options.

Some options include:

1. Enhancing shielding/grounding with copper tape or improved bonding of conductors, fasteners, product covers, cases and gaskets or other components.
2. Populating "depop" TVS footprints with additional clamping components if the pads were included initially as an insurance policy.
3. Upgrading the components to more robust (potentially more expensive) versions in the same footprint
4. Software recovery methods (catching and correcting ESD upset vectors, see Section 3.2.8)
5. Absolute last resort: Reduce qualification ESD level. (Often not possible for a given application.)

3.1.7 Information Available to the Designer

As described above, little helpful system ESD protection information is provided by component suppliers. System designers may consider external shielding as a first step. While placing a roll of copper tape on top of a laptop is not going to degrade the system ESD robustness, it is the skillful application of such tape/shielding, in the right amount and location, which can actually improve the robustness. Likewise, reactively adding TVS clamps on an IO may not help with the problem at all, and ironically may even make some system ESD problems worse.

System test qualification standards are universally accepted (and required for regulatory purposes, see Appendix A for examples) for various applications. However, as shown in Table 2 previously, a uniform test and measurement reporting regimen is not commonly adhered to by component vendors for _components_ that comprise those systems.

3.2 Advanced System ESD Protection Methods

3.2.1 Primary Goals of ESD/EMI Co-design Today

The goal of ESD/EMI co-design in its present form is to primarily meet minimum ESD and EMI susceptibility and robustness qualification levels without affecting either the cost or the performance of the end product. Indeed, while best practice recommends extensive in-depth prediction and analysis of ESD performance, in reality many engineering groups include ESD as part of a hit-or-miss qualification result. In all fairness, the complexity of this analysis, and the lack of accurate, inexpensive system level tools to do this analysis makes it almost impossible to justify more than a cursory basic approach to ESD robustness design for some low-cost, quick-to-market products.

Regardless of the pre- and post-design attention paid to ESD robustness, three key elements are considered in the "tradeoff gamut."

1) Sufficient Signal Integrity and Functionality
2) Adequate ESD/EMI Robustness and Compliance
3) Cost and Time to Market

3.2.1.1   Sufficient Signal Integrity and Functionality

The requirement for IEC 61000-4-2 based testing is to test it on a functional system. If the system is not functional to begin with, then IEC testing has no meaning.

Extending the extremely trivial example from the beginning of the chapter, if it becomes necessary to seal off all IO ports with copper tape in order to pass IEC testing, then your system results aren't meaningful. However, adding robust, high parasitic-capacitance clamps and/or series resistance to IO lines which limits the bandwidth or signal levels below the required performance is effectively the same as deleting the port from the system altogether.

Thus meeting the new performance and functional specifications are the first priorities in any design.

3.2.1.2   Adequate ESD/EMI Robustness and Compliance

EMI compatibility is not the primary focus of this paper, but it certainly overlaps the scope, as ESD immunity in many instances reflects a reduction in susceptibility of electromagnetic interference from electrostatic discharge. Very often, solutions for attenuating EMI interference (received or emitted) also result in signal attenuation at ESD frequencies.

The appropriate target for system level ESD is often set by the industry for a given class of applications through a plenary or composite standard (See Appendix A for examples). Sometimes, a corporate system ESD level is mandated for competitive or historical issues [3].

3.2.1.3   Cost and Time to Market

Less ESD protection cost usually means higher profit margins. Over-engineering the system ESD may be fine if it adds little to the total costs. Of course this is true for all costs and time, not just the product BOM. The engineering analysis cost for modeling, simulations and human resources for best-performance with optimum robustness may be much more expensive than simply relying on a known, slightly more expensive component.

Time to Market is directly and indirectly related to cost and sales/profit. One year of ESD simulation and co-design for a system on a 6 month design cycle to product release is unacceptable. Reducing expectations or missing product launches can impact entire corporate product perception or product reputations. So additional BOM costs and committed ESD development time must always be considered against schedule slips and field quality levels.

3.2.2 Desired Results vs. Actual Process Reality

State of the art ESD system design often relies on a core competence individual (or team) who can quickly analyze and recognize both ESD problems on a schematic or layout and also recover from the disaster quickly with "first aid" in the lab. The most effective tool in this kit is an understanding of ESD events, ESD damage, and ESD mitigation strategies and structures. This specialist may have responsibility on the design team, or may be a full-time ESD/EMI engineer in the corner office, or he/she may be found at a component vendor that can help spot the system problem or avert potential problems.

A further reality is that testing resources and equipment may not always be as readily available to the designer as a debugger or design consultant. These testing resources may not even be resident within the company but could be limited to an external test house. These testing resources may also be at a vendor or even integrated into a customer acceptance team. In this case, it is even more important for the designer and the ESD expert to communicate effectively with industry accepted and well defined ESD fault analysis and recovery concepts.

Examples of specific ESD-induced problems without obvious origination points include:

- Indirect test modes triggered by ESD noise superimposed on signals, op-code corruption due to glitched prefetched instructions.
- Air discharge induced into the case, PCB grounding and return path ground-bouncing, leading to in-band susceptibility issues.
- Direct injection into keypad IOs or USB thumb drive case through plastic enclosure seams.
- System component EOS failure where the root cause analysis from manufacturing line is determined to be the result of ESD-induced latch-up failure.

Each of these examples represents the end of a root cause analysis phase of investigation and the beginning of a transition to remedial design revisions to prevent the problem.

Any combination of these potential problems may have been lurking near the "robustness margin" surface and became a failure by chance, despite the best efforts of an ESD/EMI co-design methodology. But at the point of emergence, these problems must be identified and assigned to the appropriate resolution team. This team must not only identify the missed energy enclosure opening, but also the failure mechanism and hopefully remedial options, even if they cannot be immediately implemented.

This can result in a point of recovery, or a decisive breakdown in the development success. Pushing back on a component vendor or the manufacturing department, or even on operator error, may be justified in some cases. Either the failure criteria are too severe, or the robustness is insufficient for the application and the system designer must decide to either improve the product or relax the specifications. On the other hand, the design constraints may very well be such that there is no acceptable solution for a given target and the yield loss or product vulnerability must be tolerated.

3.2.3 Design Reuse

Reusing layouts and known components can help propagate the techniques and lessons learned in previous designs, or even result in accidental success. Conversely, changing everything from design to design won't necessarily make the design more or less robust, but reduces the confidence level that performance observed in previous systems is relevant to the new design. A different system layout, or the same system layout with a new revision of silicon, may provide radically different ESD results.

3.2.4 Revision Decisions

New revisions of silicon can unexpectedly impact the robustness of the system, even when all the specifications appear to be "improved." For example, an improved, faster clamping on-chip ESD protection circuit in an IO for an ASIC may trigger before an old reliable external TVS component and actually fail sooner than the TVS component when combined in the system.

Changes in silicon are obviously inevitable and usually desirable from both the cost and performance side of the "tradeoff gamut," and they often offer improvements in the robustness corner as well. But each deviation from a previous design correspondingly reduces confidence that the new design will (a) provide a similar level of robustness and (b) be responsive to tweaks and enhancements in the same fashion as the previous design.

3.2.5 Post-Design Iterative Improvement

After the new design prototypes are received, the moment of truth for any design is the test and validation phase of the predictions and simulations of the design phase. Obviously, the _cost margins_ for the given BOM are easily calculated and well understood, and the _functional performance margins_ of the system can be tested, but the third vertex on the tradeoff gamut, _ESD robustness **margin**_ must also be assessed.

IEC system level testing helps ensure that a given sample system can survive a given strike level such as 4 kV contact (Level 2). But while one can measure bit error rates or eye-diagram margin on data links, IEC testing is pass/fail, and does not provide a statistical margin assessment of the robustness.

3.2.6 Robustness Margin Fine Tuning

Unfortunately, there is not a well-defined definition for the ESD robustness margin, much less an industry consensus for how to measure it. In general, it is thought that if a system passes 6 kV, then it could be said that it has 2 kV of margin above a 4 kV target. This is not as strictly informative or meaningful as it may seem.

For example, this borderline system may very well pass 4 kV once, but then fail at 4 kV if retested due to multiple factors; including number of strikes (variation in cumulative damage) or test/gun variability. The same system can vary by as much as 2 kV or more due to these issues [4].

Some vendors specify robustness beyond the defined standard test method to guarantee their component for multiple or even thousands of repetitive strikes. The intent is to provide a confidence level that cumulative damage (due to silicon filament formation for example) will not threaten the component performance with a fraction of the number of rated multiple strikes.

Additionally, the BOM may have many qualified supplier and version options for the same footprints, and these may impact the final cost substantially. So in addition to validating the absolute system robustness level for a given BOM selection and price, it is also helpful to quantify the relative robustness margin improvement for cost increments of different component choices.

3.2.7 Head to Head Component Comparisons

For example, if a board has a TVS footprint that will support two or more components with varying specifications, which component gives better ESD results for the price? If a system passes 4 kV with a 3 cent TVS part and a 1 cent TVS part in the same socket, then is the 1 cent part always a better choice? Alternately, if the system passes 6 kV with the 3 cent part and only passes 4 kV with the 1 cent part, is the extra cost worthwhile?

Building and evaluating a statistically significant number of systems each with different TVS components is usually cost prohibitive and impossible due to resource and time constraints, so often, this problem cannot be easily resolved.

One way that this is addressed is with low level head-to-head testing (same system with the only change being different TVS products) as described above. Additionally, testing from individual components (often supplied by the vendor under favorable conditions) is compared to results from other vendors. None of these methods are exhaustive, but can often identify an improved component selection option that can still meet the budget. This then becomes a hard-won starting point for the next design.

Sometimes the most critical comparison and contrast cannot be made under identical test conditions. For example, a typical Shunt TVS component may need to be contrasted with an improved Series-Shunt protection component as shown in Figure 10. Obviously, the 3-pin device cannot be directly tested on a system board designed for a 2-pin TVS device.

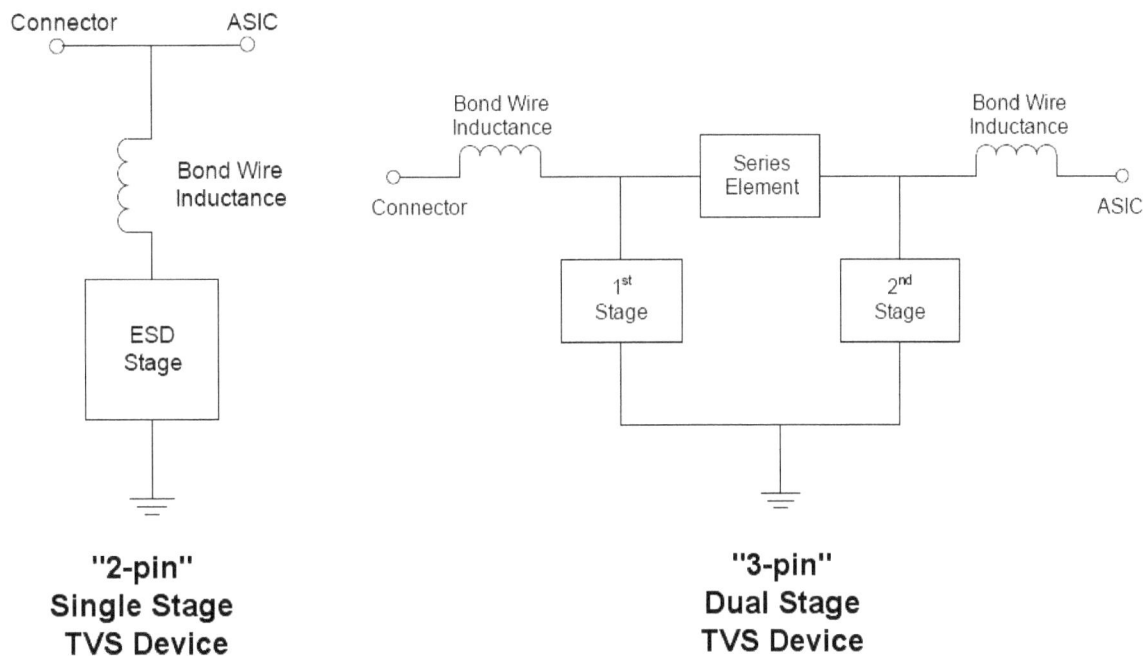

Figure 10: 2-pin "Shunt" vs. 3-pin "Series Shunt" TVS Clamps

The introduction of an alternative test board revision provides a concrete "what if" platform to evaluate. However, since the PCB must change slightly for different pinouts, the results cannot be assumed to indicate a strict accordance with the component's parameters only. PCB layout choices (such as vias, routings) will play a part in the end result and must be considered as well.

To extend this methodology to a quantitative improvement analysis, various methodologies described in Chapter 5 are being used to directly measure the in-system performance of components during actual ESD strikes. This comprises multiple samples of either complete systems or sub-assemblies being scanned during the strike, and then evaluating the evident current paths and levels. This is a very resource intensive exercise but may immediately pinpoint unexpected anomalies, which of course is the domain of ESD strikes.

3.2.8   Software Recovery Methods

In the case of readily repeatable soft failures and upsets, another remedy available to the system designer may be software or firmware patches and routines which can detect and correct the ESD upset condition. Normally, a digital system must assume that it is coherent, and that registers and state machines will not randomly lose their mind. But when ESD/EMC immunity is considered during a system upset, an external effect can invalidate many of the assumptions software may make about the rational machine on which it is running.

This is an immense subject area, which may be as complex as the software architecture of the system or as simple as a configuration bit setting. Additionally, this option may be severely limited by a lack of control over the software for the system. If an original design manufacturer (ODM) supplies an industry standard hardware platform intended to support multiple software stacks, software recovery methods may be impractical.

Consider the case where a codec is upset by ESD and merely loses its data link. An existing error correction algorithm may already remedy the ESD induced error as if it were any other signal integrity issue or "bit error" contemplated in the design. On the other hand, if the same functional block locks up but is not damaged and the rest of the system is stable, it may be possible for the driver software to recognize the upset state and seamlessly reset and reconfigure the block for operation, making the system operation appear uninterrupted at the user level. In such cases, a few lines of code may mean the difference between a pass or fail during ESD qualification.

In other cases, where the entire system resets, it may be necessary to add a discovery algorithm during the soft reset boot code that is able to recognize the difference between a normal warm boot and an asynchronous, ESD induced soft reset or upset. There is no set method to implement this mechanism, but often registers which are known to be upset to a certain state can give clues to the boot monitor as to what may have just happened and what recovery actions may be necessary. Watchdog timers and periodic interrupts, as well as dedicated system monitor microcontrollers which are already used in the system, may be adapted to help with ESD event recovery tasks. Below are some examples of "best practices" for software recovery or software robustness techniques in ESD/EMC co-design:

1. In low level configuration software or firmware, make sure that all unused IO ports are set as outputs, not inputs. This will reduce the possibility that floating inputs will accidentally collect some unexpected status inside the register. In the initial design, it may not cause any harm, but may impact subsequent software versions which may read that register or bit, making it difficult to analyse ESD susceptibility later on in the field.

2. For edge triggered interrupts that initiate critical functions, such a shutdown operation, it is a good practice to have a secondary check for the line before executing the operation. This would be analogous to the operation of a soft-power button debounce circuit. The action is edge-triggered, but the state of the button is validated only after the debounce period where the "ringing" of the mechanical contacts dies out. An ESD induced glitch may occur deep within the system registers or memory and if there is a check method to validate the action, then the system robustness can be improved by ignoring and recovering erroneous transient events by executing one or more "second opinions" about the validity of a state before acting.

3. The examples above describe ESD/EMC immunity. However, similar techniques are related to EMI emissions management for low level operations which need detailed controlling and planning. For example, if a system has internal and external memory in use, there may be a high difference in power consumption between the two blocks. Accessing each block for a relatively long time before alternating to the other one may create long periods of high and low power consumption. In a battery operated component, these alternating periods may create long voltage drops, which may complicate the supply

design. But if the high current and low current memory accesses can instead be rapidly interleaved without affecting the component performance, then current pulses may be minimized and these EMI inducing transients may be easier to mitigate with less expensive bypass capacitors and filtering. Other functional blocks may also be considered, including interleaving high current functions such that the system does not transition rapidly from "all on" to "all off", but instead keeps a consistent allocation of load versus time.

These three examples vary in the type of transient robustness or immunity being addressed, but share similarities in the use of software to mitigate the issue. The concepts are usually straightforward, but the implementation depends on the complexity of the system software. Direct access and control over system states can be easily handled in a small microcontroller based component or even an advanced real time operating system (RTOS). But virtual machines and multitasking operating systems may impede the ability to actively monitor sequential system states to detect transient events on a discrete time scale, limiting the opportunities for software recovery methods.

3.3   Comprehensive Co-Design Methodologies

On a long enough product cycle with sufficient product value at risk, 3D modeling of ESD/EMI design becomes viable. A one-shot deep space probe simply doesn't allow for an iterative product development cycle. "Lessons learned" may result in an inoperative probe. In such cases, the aggressor discharge strike might not even be reproducible in a development lab, and the actual environment may only be created and tested in a simulation.

Another case is when the product has a very complex mechanical and circuit design and has limited options for on-board protection. This is especially true in the RF area which often requires 3D simulations to optimize the RF performance and ESD robustness.

Over time as computing power increases, the simulation algorithms improve, and costs decline, this "brute force" methodology may become the obvious choice for any designer.

Of course, any simulation is only as valid as the models and conditions applied, but many areas of design utilize this methodology exclusively. Nodal circuit simulation environments like SPICE are mature and widely available, but while models of ESD simulators are well known and published, and the circuit elements are well characterized in the frequency domain for EMI and signal integrity analysis, there are few resources for out-of-the-box high-current/fast-transient ESD models of TVS, ASIC and even passive capacitors and inductors/filters.

Simulation is thus the foremost shortcoming of the present-day state of the art in achieving ideal ESD/EMI co-design, and these factors are expanded upon more fully in Chapter 5.

3.4   Conclusion

The chosen methodology from Table 2 is ultimately selected by the designer as a compromise on an optimum fixed point in a "tradeoff gamut" between three generally conflicting goals: (1)

ESD/EMI robustness and susceptibility resistance, (2) signal integrity and functional performance (bandwidth, data-rate, minimum emissions for example), and (3) cost and time to market.

The distinction between Basic, Advanced and Comprehensive methodologies in this chapter is defined by the complexity and effort required of the method as well as the ability of the underlying model to predict robustness of the system. In addition to this depth of complexity, the breadth of design methods is primarily divided into practical analysis and theoretical simulation.

Complexity does not always imply improved robustness, however, and in some cases the most basic shielding design of a product enclosure can deliver the highest system level robustness for essentially no additional product cost or development time. Alternatively, expensive simulation environments utilizing component models which are inaccurate in the ESD domain may indicate the need for costly and unnecessary additional components, redesign delays, and expensive silicon spins, and result in a less robust system than expected. This chapter explored the prevalence and cost-effectiveness of the different methods as well as the ESD performance of the results that can be expected.

Within each method, critical exposed and susceptible nets must be identified and assigned reasonable and sufficient robustness levels. Aggressor pulse entry (conducted, induced) points for hard- and soft failures should be outlined. Component selection, layout and placement should be considered with regard to ambient and direct ESD/EMI strikes and disturbances and the effect on signal integrity. System partitioning, grounding, clamping, shielding and return path shaping are also considered, along with the problems of hidden potential multiple/secondary discharge points within the system.

By managing these issues in the concept, design, and prototype phases, a designer can observe and improve the extent and effectiveness of any element or flow in a process intended to balance ESD/EMI robustness, performance, and cost with existing and emerging methodologies.

References

[1]   Industry Council on ESD Target Levels, "White Paper 3 System Level ESD Part I: Common Misconceptions and Recommended Basic Approaches," December 2010, at www.esda.org or JEDEC publication JEP161, "System Level ESD Part I: Common Misconceptions and Recommended Basic Approaches", www.jedec.org
[2]   Laasch, Ingo et al, "Latent Damage due to multiple ESD Discharges," EOS/ESD Symposium, 2009, pp. 1-6
[3]   Hua, Vanessa, "Sparks over static: An age-old problem still vexes makers of tech devices," San Francisco Chronicle, August 27, 2001
[4]   Frei, S.; Wu, W.; Hilger, U.; Edenhofer, J.; Stecher, M.: "Packaging and Handling Test according to ISO 10605: How comparable are tests with different ESD generators?", ESD-Forum 2007, Munich, Germany, 2007

Chapter 4: Reference Methodologies for IC/System Protection Co-Design

Harald Gossner, Intel Corporation
Guido Notermans, ST-Ericsson
David Pommerenke, Missouri University of Science & Technology
Benjamin Van Camp, SOFICS

4.0   Introduction

SEED was introduced in White Paper 3 Part I as a method enabling board designers to develop an ESD protection concept based on ESD specific models for all elements involved in the ESD discharge path. Here in Part II, a 'divide and conquer' approach is chosen, where the various failure mechanisms are classified and the appropriate design and testing methodology for each category is described. Many detailed aspects of this methodology are still to be developed. The purpose of this description is to give a common base for discussion and to provide a setting for direction.

The first step of analysis is to determine the discharge path tree to be considered. The path can be traced from the entry point of the pulse. There are various ways to force the IEC gun stress pulse to the system. A system consists of a case, the PCB including external ports and the electronic components thereon. Usually a contact mode IEC 61000-4-2 discharge is applied to all metal housing parts. The IEC 61000-4-2 air discharge is applied to insulating cases or parts of it as described in Chapter 1. In the latter case, pulse energy can be coupled to the board either by electromagnetic fields or by sparking to some metal parts. The metal parts guide the waves and both fields and currents can reach the board. From that point of view, air discharge and contact modes are similar. The focus of the SEED concept is on measures which are implemented on the PCB. Mechanical design aspects of the system regarding shielding and avoidance of secondary sparking are not treated by the SEED method.

We will first start by introducing a series of SEED categories through examples which will be categorized by the impact ESD has at the system level. An example of an ESD event which can be covered by the SEED approach is a discharge to metal pieces which are galvanically connected to the wiring of the PCB. This can be a PCB port which is hit by a contact discharge (like a cable discharge event), a portion of an audio or USB jack which picks up energy from a nearby discharge to the ground shield or a part of the metal connection to a discrete component of the system (like an ear piece or a keypad of a mobile) where the discharge can occur through gaps in the case. The effect of the system level ESD event on the pin(s) that are galvanically connected to the metal which is zapped will be called a SEED Category 1a event; it addresses physical damage from discharges to external pins.

From the entry point the energy can spread along the direct metal connection to an IC pin. The energy is shunted through PCB protection diodes and IO circuitry (including on chip ESD protection) to VSS and VDD nets of the IC and PCB (See Figure 11). Category 1b is a slightly different case: it involves other paths which can lead to damage to pins which pick up a high energy pulse without being directly (galvanically) connected to the outside world.

Alternatively, part of the charge might be forced into the substrate of the IC at any pin, which can cause latch-up events or upset errors. This type of path is assigned to Category 2.

Parallel to the current in the galvanic network connected to the entry point, a small part of the energy might be coupled to neighboring wire lines of the PCB by mutual inductance and capacitance or can be picked up from electromagnetic (EM) radiation by PCB traces acting as an antenna. The induced pulses can travel in these parallel metal networks and be shunted at IO pins or dedicated board shunt elements like TVS diodes to VDD and VSS nets. Such an induced pulse of low energy is not able to create damage directly due to its dissipated energy, but it might cause an upset or other malfunction of the system. Also, as VSS and VDD nets on an IC or even on a PCB might not represent an ideal sink, the pulse energy can penetrate at low energy levels throughout the whole system as supply noise. Additionally, disturbances of the signal lines at this low level of energy can degrade signal integrity. These disturbances can even pass a first IC and be transferred to another IC which it is electrically coupled to. All these mechanisms, related to rather low energy levels of the stimuli initiating the fails, are part of Category 3.

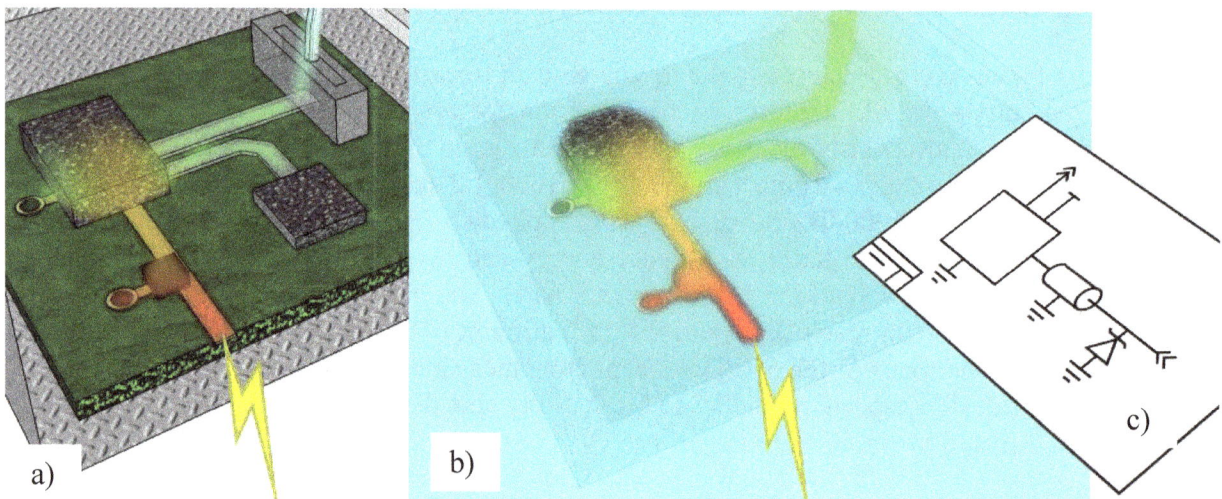

Figure 11: Qualitative illustration of system ESD discharge along PCB lines and typical protection concept: a) PCB b) highlighted discharge paths and c) related circuit block diagram

As discussed in Chapter 3, applying an IEC ESD pulse to an encapsulated (end)-system, to the PCB only or even to single ICs, usually leads to very different discharge paths and failure mechanisms. There is a need for sub-component characterization to achieve a cost efficient solution for the final system while maintaining short design cycles. SEED provides a very good direction for optimization of ESD protection of certain PCB paths. This design solution can actually be verified by system level ESD tests on a PCB level for specific paths or by other stimuli of high energy pulses like TLP. This intermediate test is a valuable step in the system development process and provides a good basis for the qualification of the final system, if design measures for shielding and prevention from secondary discharge are taken into account properly. At this point it should also be mentioned that the SEED concept can be equally applied in cases where the main ESD discharge path leads through the IO circuit, such as in the case of no TVS diode protection. Though having no TVS protection is discouraged due to soft fail risks and RF performance constraints it might be a viable solution for some applications.

4.1   Approaches – Categories

4.1.1   Categories – Definitions

In summary, the previous discussions can be rolled up into the following SEED categories:

SEED Category 1:
   a) External pin experiences hard failure due to a direct ESD zap (failure root cause: high pulse energy at exposed line)
   b) Non-external pin experiences a hard failure due to an indirect ESD zap (failure root cause: high transferred pulse energy to non-exposed lines)

SEED Category 2: Pin experiences a transient latch-up event which can lead to either a hard or soft failure (failure root cause: current injection into the substrate which is too high)

SEED Category 3: Describes protection of an IC experiencing soft failure due to low amplitude transient bursts in the system during an ESD zap (for example, this may be caused by degraded signal integrity of an exposed line or cross-talk to a neighboring line or supply noise).

Note that for Category 2, if the part is powered up, the energy delivered by the supply can be the source of the damage. For example, if the supply dumps a current which is too high for a longer duration in the low ohmic path created by the latch-up event, damage will occur. For SEED Category 3, failures can only be seen if the system level pulse is applied when the component is powered up.

4.1.2   Category 1

The SEED Category 1 approach focuses on physical damage and higher energies and is not really intended to reduce soft fails. When the energy of an IEC 61000-4-2 discharge is coupled into a PCB line connected to an (external) pin, the main ESD current typically flows directly to ground through the on-board TVS. Nevertheless, some residual current will enter the connected IC pin and flow through its internal protection to ground (see Figure 12).

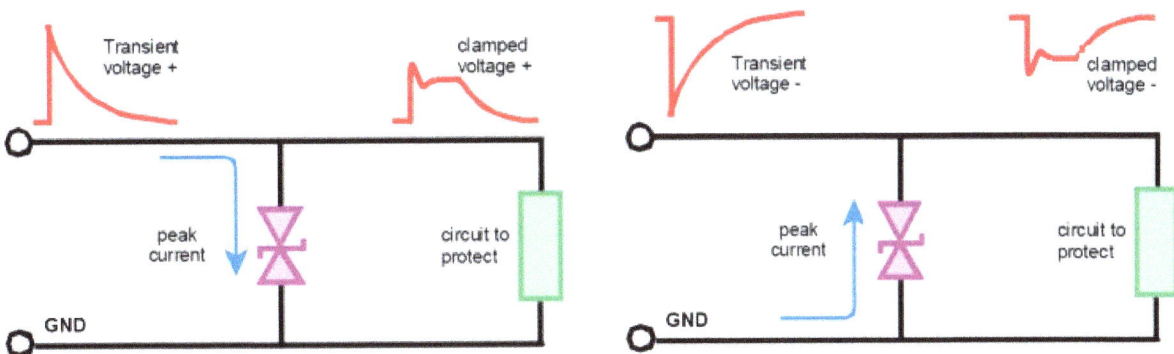

Figure 12: During an IEC discharge to an external system pin, the main current flows through the on-board TVS protection to ground, but some current will enter the IC and flow through its ESD clamp to ground.

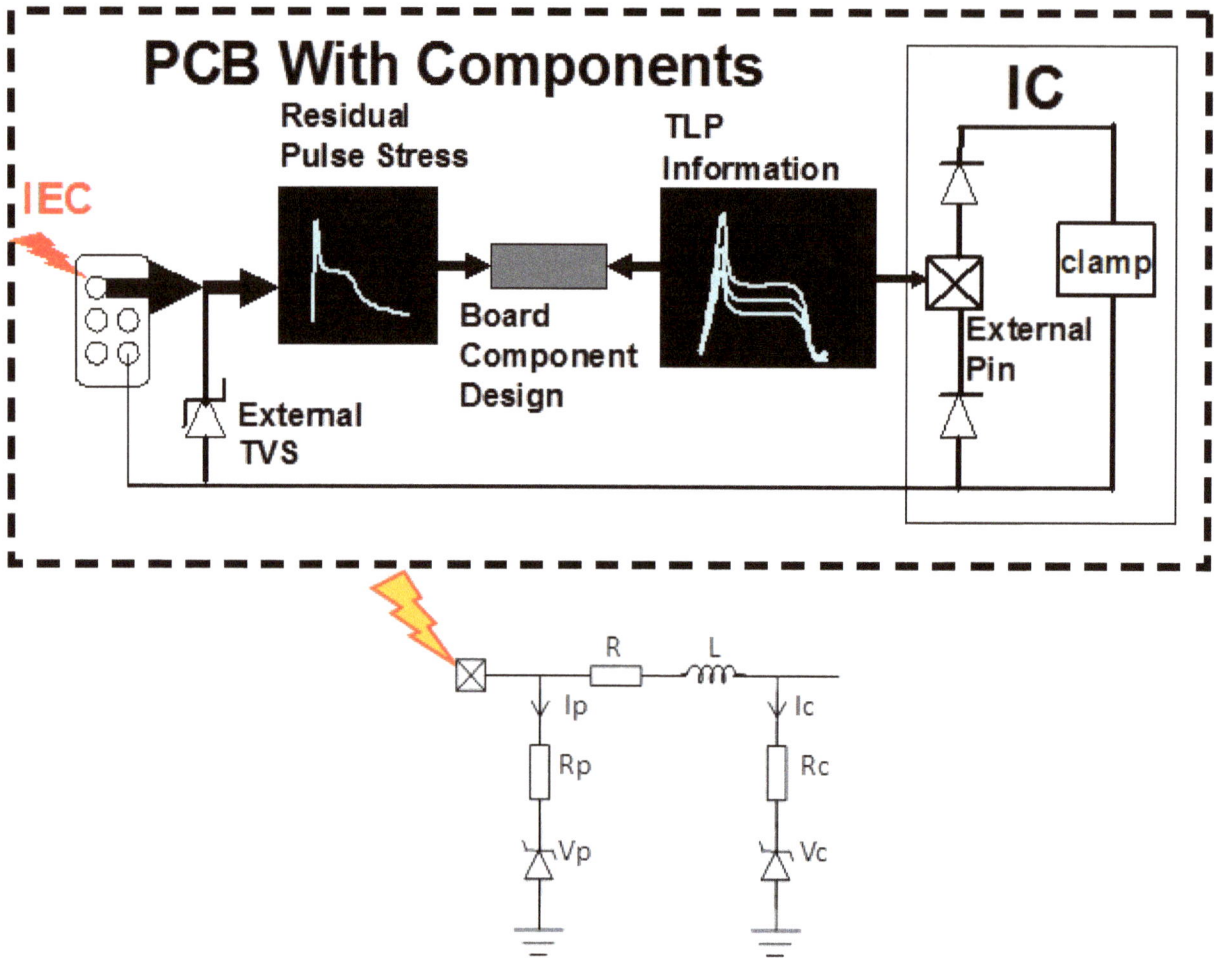

Figure 13: Simplified replacement diagram showing the essential components of a basic SEED concept [26]

In order to prevent damage to the IC, off-chip protection is often added as shown in Figure 13. It is important to assess the amount of current (Ic) which may enter the IC and the associated (over-) voltage (Vc) across the connected IC circuitry.

To analyze the residual pulse and the optimization of the PCB protection, three basic SEED approaches of increasing complexity might be chosen depending on the system conditions. If the protection can be configured by fast and low ohmic breakdown elements like some TVS diodes and high ohmic resistors (in the > 10 Ohm regime), a quasi-stationary model allowing a back-of-the-envelope estimation of the required parameters can be applied.

In most practical cases, complex impedance plays a major role involving inductance of wiring and discrete capacitors. This requires a more detailed transient approach. Still, a transient model of the IC pin itself is not always needed due to the suppression of fast transients in the PCB network [1, 2]. Characterization of the IC and simulation effort can significantly be reduced by this simplification.

However, in cases where a matched network without clamping elements allows fast transients to travel to the IC pin without significant damping, transient models of all elements of the path including the IC IO circuit have to be considered.

In case of a quasi-stationary model the calculation may be simplified by approximating the TVS and the on-chip protection by ideal zener components, characterized by their on-voltage (Von) and on-resistance (Ron). A simplified replacement diagram is shown in Figure 13. If all parameters of the components in Figure 13 are known, a straightforward application of Kirchhoff's Current Law will yield the current distribution (Ip, Ic).

This approach does not reflect the actual behavior of the system of TVS diodes and IO on chip protection, as these are competing elements with characteristic transient responses. It is important to realize that the on-chip protection might need to take the first transient, such that the combined effectiveness of on- and off-chip protection needs careful consideration. If the on-chip protection shows snapback behavior, it is possible that the on-chip protection takes all the current, before the off-chip protection triggers. For the RC-bigFET approach, the current conduction capability is time-dependent, with potential triggering and/or turn off issues. However, for the sake of a PCB design decision, a worst case scenario requires a robust protection concept. In the case that the IO circuit gets into the low ohmic state first, turn on of the TVS diode might be an issue. The PCB design has to account for this and the assumption of a low ohmic/high current I-V characteristic is appropriate. Still, the assumption of an effective turn-on of the TVS diode without delay is an optimistic one. In real cases there will be an initial overshoot, which may already damage the IC. Therefore, the methodology of comparing quasi-stationary I-V characteristics only applies to very fast triggering components, like TVS diodes, with small overshoot plus the use of filtering elements on board between the TVS and IO which blocks fast transients in the sub-ns regime. PCB traces may serve this purpose. Note that if the system level test is performed with the component in a working mode (supply lines charged), the characterization of the ESD cell must be performed with a biased supply line in order to give relevant data.

The resistance R and inductance L of the wiring, both on-board and on-chip, play an important role. The total inductance between the shunt elements, comprised of trace, interconnect, and bond wire inductance, may reduce the current from the first peak entering the chip to a negligible value. In that case, the IEC robustness may be estimated by considering the current distribution in the second peak only. Since the timescale of the second peak is very close to the HBM timescale (around 100 ns), the relevant I-V characteristics of all components may be estimated using a standard TLP test.

4.1.2.1 Safe Operating Area (SOA)

Using the simplified basic SEED model depicted in Figure 13, calibrated with 100 ns TLP characterization data, the residual current into the IC and the ensuing pin voltage may be calculated for any given on-board resistance R (R_{on}). Figure 14 shows Vc and Ic assuming no additional board resistance R exists. In that case the 'operating point' of the on-chip supply clamp during a system level ESD discharge (Vc, Ic) turns out to be outside the SOA for the output domain on chip, which is defined by the design target for the clamp. The SOA is bounded by the maximum ESD current for which the clamp is designed, such as Imax = 4 A TLP (~6 kV HBM voltage), and the maximum voltage on the core circuitry before oxide damage will occur, such as Vmax = 11 V (oxide breakdown of a thick oxide transistor).

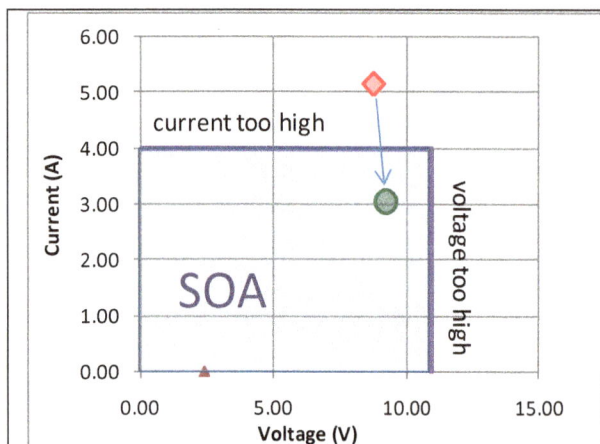

| Figure 14: Increasing the series resistance on-board by 1 Ω moves the output 'operating point' (Vc, Ic) during an 8 kV IEC discharge into the SOA [26]. | Figure 15: Choosing a different TVS with Von= 6 V and R_{on}=0.1 Ω moves the output 'operating point' (Vc, Ic) during an 8 kV IEC discharge into the SOA [26]. |

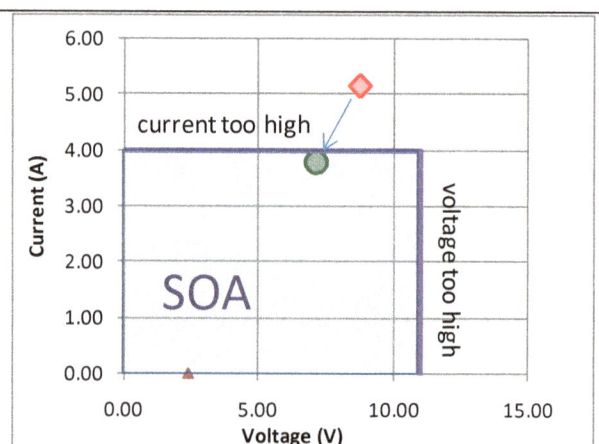

The SOA approach allows assessment of possible improvement measures. For instance, by increasing the on-board resistance R, the current into the supply clamp may be reduced. However, in many cases, such as the case of an audio output, the required efficiency of the power stage usually does not allow increasing the output impedance. An alternative solution is to find another TVS which has either a lower on-voltage or a lower on-resistance. Figure 15 shows that by using a TVS with V_{on} = 6 V and R_{on} = 0.1 Ω, the output (Vc, Ic) point moves into the SOA. Increasing the holding voltage of the on-chip protection or designing the IC to allow a larger current to flow through the on-chip protection, are alternatives.

V_{on} cannot be lower than the maximum operating voltage of the output (in normal operation) plus some margin. So, the easiest solution is to find a TVS with a V_{on} as low as possible and a sufficiently low R_{on}. To cover transient behavior due to impedance of the PCB network, package and protection elements, an extraction of resistor/inductor/capacitor (RLC) models of these elements at high currents is required. One example is the behavior of ferrite beads, which are widely used to suppress high frequency surges reaching sensitive pins. It has been shown that their damping behavior is strongly dependent on current amplitude and needs to be extracted such as through high current TLP pulsing [1, 2].

The simulation approach also includes transient models representing the turn-on behavior of the PCB protection elements at fast pulse rise times (or at high frequencies). Preferably these should be delivered by the component supplier. However, if needed, the characterization can also be done by the PCB designer using TLP equipment with a fast pulse rise time (typically < 1 ns). Finite turn-on time and package inductance of the protection elements such as PCB voltage clamps (i.e., if it is a varistor, but also capacitors) lead to voltage overshoots and will strongly impact the effectiveness of the ESD protection strategy. It is important that the transient response of any off-chip protection is characterized for these fast pulse slopes. Regular TLP measurements have a rise time of 10 ns and pulse duration of 100 ns. Typically the voltage shown in the TLP I-V curve is an average of the voltage measured between 75 ns and 85 ns, that is, near the end of the TLP pulse. It therefore lacks information of what the voltage is during the first part of the pulse. Furthermore, a rise time of 10 ns is too slow to characterize the overshoot during very fast ESD events. Some TLP systems allow changing the setting to faster rise times. To correlate with the fast transients of an IEC event, a rise time of 400 ps is a good typical value. The extraction of the voltage waveform as the component responds to a fast risetime pulse is the basis for an effective analytical design methodology.

The RLC network of the PCB and package can be extracted from standard 2.5D or 3D EM simulators solving Maxwellian equations. 3D simulations are especially slow, however, and only a few paths can be analyzed in a reasonable time frame in real life.

IC pin models need to reproduce the breakdown characteristic and the on-resistance of the discharge path through the IC as well as the I_{t2} where a transistor enters its second breakdown region and is irreversibly damaged. Transient turn-on behavior of the IC IO circuit is not relevant as long as the network damping is significant (Note that in some cases lumped capacitors might create non-negligible transient effects) [1, 2]. This reduces the effort in characterization and, even more importantly, avoids convergence problems and misinterpreted current distribution due to competition between the IC circuit and a PCB protection diode. Accepting the IO IC protection in the on-state usually results in a worst case condition for the current distribution and allows safe guidance for the optimization of the protection circuit. If transient effects have influence on the triggering of the on-chip protection (as explained above), these need to be taken into account.

A possible flow of a transient simulation of the PCB optimization for system level ESD protection is described below:

1. Extract the netlist of board (RLCk or S-parameter) from the PCB layout data
2. Import the electrical board netlist to a network simulator program
3. Add discrete PCB elements (TVS, resistors, inductors, ferrites, capacitors) to predefined nodes of the board schematics
4. Add IC pins to the board schematics
5. Bind high current models to IC pins and discrete PCB elements
6. Perform a simulation using ESD stimuli
7. Assess fail criteria at IC pins (and discretes)
8. Select better discrete device (TVS, R, L,...) if necessary
9. Optimize placement of discrete elements (go to 1.)

4.1.2.2 Design Verification Using Very Fast Transmission Line Pulse (VFTLP) Data

In the previous examples, the specific transient response of an IC IO circuit to initial overshoots of the residual pulse passing though the PCB network has been neglected. In most PCB networks this is a valid approach, as the overshoot is reduced by the damping of the PCB network. However, in the case where the fast pulse passes the PCB network and generates a large overshoot at the IC IO, the described methodology doesn't meet the real situation. Even though the short duration of the overshoot generally shifts the failure current to higher values, damage can still occur. If the circuit doesn't turn on in time, the increased voltage may lead to a physical fail.

To extract the ESD behavior realistically, a VFTLP characterization of the IC IO circuit provides the relevant information needed. A VFTLP system with a 2 ns pulse width and 400 ps rise time is well suited for (typical) characterization of the IO circuit in terms of turn-on, on-resistance and I_{t2} of short pulses with very fast rise times. This might be applicable for antenna ports for example.

To include the results of VFTLP measurements for specific PCB traces, the following procedure is given as an example. It is assumed that the off-chip protection is chosen according to the principles discussed previously. VFTLP will enable extraction of the voltage overshoot of the off-chip protection, as well as the I-V curve for the on-chip protection. Using Kirchoff's laws, one can check whether the current flowing through the IC is sufficiently low enough to avoid damage similar to the basic SEED as discussed in Section 4.1.2.1. If not sufficiently low, the impedance of the connection can be modified to lower the current peak through the IC. Other measures include faster off chip protection, additional series resistance in the connection, better on-chip ESD protection or parallel capacitance.

General trends for the choice of capacitance and inductance are:

- Higher parallel capacitance lowers voltage.
- Higher series impedance/inductance (such as in IC packaging lead wire) reduces current

The interaction between two or more components on board can cause residual pulses to vary dramatically when the pulse energy is shared between different types of clamps over non-negligible electrical distances. For example, a polymer component may clamp an IEC pulse at 30 ns very well when tested on its own. The initial voltage peak helps to trigger the component and shunt the current effectively. However, if it is placed in a circuit parallel to a very fast silicon shunt component which has a lower clamp voltage and lower failure threshold, then the silicon component may inhibit the polymer component from triggering, and the "protected" IC will fail at unexpectedly low system levels.

In this case the "DUT" becomes the entire system of nodes on the IO line under inspection. The VFTLP or any chosen stress pulse has to be applied to the actual system or equivalent evaluation board with all TVS and auxiliary bypass components installed.

Another transient aspect is the turn-off behavior of the RC-triggered on-chip and off-chip protection. If sufficient energy is left in the late phase of the pulse when either protection component turns off, damage can occur. Though this effect is not very common, it is still important to understand the turn-off behavior of these protection components.

4.1.3 Category 1b

Category 1b, protection of a <u>non-external pin</u> which experiences a <u>hard failure</u> due to <u>an indirect ESD zap</u> (failure root cause: high transferred pulse energy to non-exposed lines), is described below.

There are well reported entry paths for such indirect zaps. One way they can occur is by sparking from nearby connector pins or from parts of the housing to PCB traces connected to these pins, or breakdown of isolation between neighboring PCB lines or package balls. Also, discharge through gaps in the case can lead to stress of non-external pins. In very specific cases, like flex cables connecting modules in the system, very strong electromagnetic coupling can result in a destructive fail at a non-external pin.

One approach is to add off chip protection to lines that are *likely* to see a high energy induction. Capacitors and filter elements can be used as well. Effective capacitors have to be dimensioned sufficiently large to avoid acting as *temporary* storage components, which then release charge in very short timeframes. The term 'likely' is hard to define, as it requires PCB and system analysis. Of course some examples are:

- Traces that connect to flex cables or other board-to-board connectors
- Traces connecting to user interface devices like a liquid crystal display (LCD) or keypad
- Traces that have long parallel runs with high noise traces (like running parallel to an audio line which connects to the outside)
- Reset lines that are routed over many boards or on board edges
- Power Good lines that go to a power supply
- JTAG lines often cause trouble
- LCD connection lines as they connect to flex cables
- All lines that connect to user interface devices like knobs

Pins in Category 1b need the same protection approach as the pins in Category 1a, although the pulses used for IC characterization might be slightly different, as explained in the examples below.

The suspicion that the transfer of relevant (destructive) energy by inductive or capacitive coupling between PCB traces has been proven to be very unlikely by recent investigations [3]. In this case the waveform of the induced pulse is typically short (up to a few ns), but can reach high current levels (up to 2-3 A) [3, 4]. It is assumed the pulse is capacitively/inductively coupled into the line, such that lowering the capacitance between the lines shortens the pulse and lowers the amplitude. The need for good design in this respect is higher for pins with low CDM withstand levels for example.

The response time of off-chip protection can be of importance. As in Category 1a, VFTLP like measurements (as well as any TLP measurement with fast rise times of ~400 ps) should provide sufficient information. Note that the rise time and pulse duration of the event seen by the pin is largely influenced by the amount of coupling between the exposed (aggressor) and non-exposed (victim) lines, such that careful board design can help to reduce the requirement for fast and highly robust on-chip ESD protection.

Further system level ESD design considerations are similar to those for EMC design; the goal for both is to limit the amount of current/voltage/energy transferred to non-exposed lines. This prevention of ESD stress from reaching the IC is addressed by the advanced SEED concept and can be assumed to be a common design practice for EMC robustness.

4.1.4 Category 2

Protection of any pin experiencing transient latch-up which can lead to either a hard or soft failure (failure root cause: current injection into a substrate which is too high) is described below.

Overstress to any pin of an IC can result in latch-up. Overstress includes surges above the power supply voltage, below GND to a supply line, and current injection into an IO pin.

The phenomenon "latch-up" and a qualification methodology for latch-up robust products is defined reasonably well in JEDEC standard JESD78D of 2011 [5]. Historically, latch-up covers product fails which result from the occurrence of low-ohmic paths, caused by overstress, which triggers a parasitic pnpn or npnp clamping component. To account for this, all ICs have to pass JESD78D latch-up testing during qualification.

The stress defined in JESD78D has a rise time (t_r) of 5 µs–5 ms and pulse duration of $2 \times t_r$ to 1 s. Compared to typical ESD events on an IC or system level, these pulses are very slow, thus, JESD78D is often referred to as a "static" latch-up test. However, it is well known from several examples that transient latch-up (TLU) can occur if only a small part of the IEC energy pulse reaches the IC. TLU can occur in products during IEC testing or even in the field although they passed the JEDEC latch-up qualification test. This effect and its physical root cause have been intensively discussed in literature [6-24]. The reasons for the fails are surges which are (much) faster than the pulses defined in JESD78D. Such surges can occur during system level stress. Working Group 5.4 of ESDA is presently reviewing this topic and plans to release new recommendations as a Technical Report (TR) toward the end of 2012.

At the time of this white paper writing, the WG5.4 TR covers about 20 different examples of transient latch-up fails from business applications of digital circuits/processors, mobile devices, automotive, high-voltage circuits, and analog circuits (audio and power management). Those examples are now categorized with respect to a possible test methodology to reproduce the fail. Basically, there are two classes of TLU events. In the first class, the latch-up event can be reproduced by JEDEC JESD78D testing with increased compliance values. Of course, it is not straightforward to just increase the compliance values of JESD78D without changing the pulse width, as it may result in EOS damage of the IC. In the second class, transient latch-up can only be reproduced by transients with rise times significantly shorter than the JEDEC JESD78D surge. Thus, for both classes a pulsed methodology based on square pulses could possibly work; discussion has started. The final goal of the ongoing work is to determine the appropriate emulation of stress and failure signature by TLU testing.

Once the IC has been characterized up to a certain TLU current threshold on relevant IO pins, this information can be used to detect critical excess current, when injection of a system level ESD pulse to the PCB network is simulated. The simulation has to take into account inductive or capacitive coupling between PCB traces, which can spread ESD pulse energy to various parts of the PCB network attached to IC IOs. PCB design measures as part of an advanced SEED co-design concept must be taken to limit the current at IOs to a safe level below the characterized TLU threshold.

4.1.5 Category 3

Category 3, describes <u>protection of an IC</u> experiencing <u>soft failure</u> due to <u>low amplitude transient bursts</u> in the system during an ESD zap (for example, this may be caused by degraded signal integrity of an exposed line or cross-talk to a neighboring line or supply noise).

Failures of Category 3 are diverse and could be triggered by a very low fraction of the ESD pulse. The underlying causes of soft failures are often not known. Often times, indirectly coupled ESD currents are ringing signals having ring frequencies from tens of MHz to a few GHz. Not only is the ringing frequency determined by resonances of the enclosures, wiring and PCBs but also by the spectral content of the specific ESD generator model used. Each test point might have its own resonant coupling path and each ESD generator model has its own spectral content leading to larger variations if the same DUT is tested for soft failures by different ESD generators [25].

However, one can categorize soft failures by different criterion:

1) <u>In-band / out-of-band with respect to voltage.</u> For an in-band error the noise voltage needs to be larger than VSS and less than VDD, or, for differential signals, within the allowed common mode and differential mode voltage swing range; thus, within the normal operating range of the input, otherwise it is called out of band. In general, in-band errors (with respect to voltage) add noise and cause signal integrity violations, but the voltages stay within normal operating limits. For an output, the current forced into the output needs to be less than the maximal allowed current, and the voltage at the output must be maintained within the normal range of voltages. Most in-band errors are caused by voltage changes that allow noise to be confused with legal data. If, for example, the ESD causes the common mode level of a differential mode signal to swing up beyond the maximum common mode range of the differential input, then this would be considered an out-of-band soft failure with respect to voltage.

   If the common mode would swing below VSS or above VDD, then the ESD protection circuit can inject currents into VSS, VDD or the substrate.

2) <u>In-band / out-of-band with respect to pulse width.</u> If the intended minimal pulse width is, for example, 2 ns and an ESD pulse with 2 ns width arrives at the receiver, the receiver will confuse the ESD induced voltage for valid data. Such a pulse is considered as an in-band signal with respect to pulse width. However, if the same receiver is able to react to a 200 ps wide pulse, although the fastest system signal would have > 2 ns pulse width, then the 200 ps receiver can react to the 200 ps pulse. Such a pulse is considered as an out-of-band signal with respect to pulse width, as its width is narrower than the narrowest intentional pulse within the system. These types of errors are very common for reset and other status lines, as the input buffers are often much faster than they need to be. Together with long traces or poor routing of status lines over connectors, for example, strong coupling paths are formed between the ESD pulse and the receiver, causing the receiver to react to a very narrow pulse. Low pass filters at the IC input can help to improve the situation.

3) <u>Local vs. distant errors</u>. A local error is caused by changes in the IO buffer which receives the ESD; a distant error is caused by changes far away from the IO buffer that received the ESD. For example, if a negative pulse opens a p-type/n-type junction (PN) diode and leads to an injection of charge carriers into the substrate, and this current disturbs a crystal oscillator (XTAL) input pin at a different IO, then this would be considered a distant error. The same is true for a positive pulse injected into an output which forces current into VDD. This current leads to voltage drops within the VDD system and can cause an error at a level translator, or a phase lock loop (PLL) away from the output.

4) <u>Amplified / non-amplified soft failures</u>. Amplified soft failures involve transient latch-up or the trigger of power clamps, while non-amplified soft failures are caused by voltage changes, IR drop, or cross coupling without triggering high current devices. Ringing pulses can lead to fast transient latch-up. A fast transient latch-up can have many different consequences, from increased current consumption with no direct effect on functionality, to increased current consumption with soft failures, to destruction of the IC. Another example of an amplified soft failure is the trigger of a power clamp. This can again have multiple consequences. If the holding voltage is higher than VDD, the power clamp will recover after some time. During this time, the ICs logic can be disturbed, but the IC will survive. If the holding voltage is below VDD the IC will try to pull VDD down. Depending on the power supply this can destroy the IC.

It should be noted that fails in Category 3 are often resolved by modification of software. This is also possible to a certain extent for Category 2 fails. The hardware aspects are addressed by the advanced SEED approach.

4.1.6 Pass-Through Effects

While in most cases the semiconductor component suffers from a hard or soft fail once its pin is hit by the ESD residual pulse, in some cases the part only acts as transfer gate into another network, where a subsequent element might fail. A typical example is the front end module attached to an antenna where a portion of the ESD energy from the antenna pin interface can be transferred to a surface acoustical wave (SAW) filter, which can only sustain a certain overvoltage. To account for this, a characterization of the 'transfer' or better 'blocking' capability of the first component is required. This is applicable to only a few, mostly well-known components, and is not covered in the SEED characterization above.

4.1.7 Summary

In conclusion, the failure mechanism and protection methodology of failures in Category 1 are addressed by the basic SEED concept as introduced by White Paper 3, Part I [26]. Failures of Category 2 and 3 need different analysis and protection methods which require an advanced SEED approach. This comprises the characterization of the sensitivity of the IC against low energy level pulses. This is partially overlapping with EMC considerations, but is not identical, as the typical effect addressed herein appears in the time domain, while EMC analysis is typically done in the frequency domain.

4.2 Examples

4.2.1 Category 1a

4.2.1.1 Voltage Waveform Extraction

Voltage waveform extraction can be done with VFTLP measurements. As the incident and reflected waveforms do not overlap, the waveforms contain all information necessary to extract the overshoot information. The key elements of the VFTLP which make it useable for this purpose are:

1. The ability to produce fast rise time pulses
2. The ability to measure the response accurately

Some portion of the fast transients can reach the component when the transmission line between the source and component matches. One example of fast transients can be seen in Figure 16. The IEC contact discharge current waveform is measured just before an IC component on a small size printed wiring board (PWB). The current moves through an 8 cm long trace through the component into the board ground. The PWB is placed on an IEC test bench close to the electrical ground and is grounded through a high impedance ground cable. The fast transients will stress the component when the PWB is capacitively coupled to the plane of the IEC bench electrical ground and the rest of the charge will dissipate through the grounding cable within the next 500 ns.

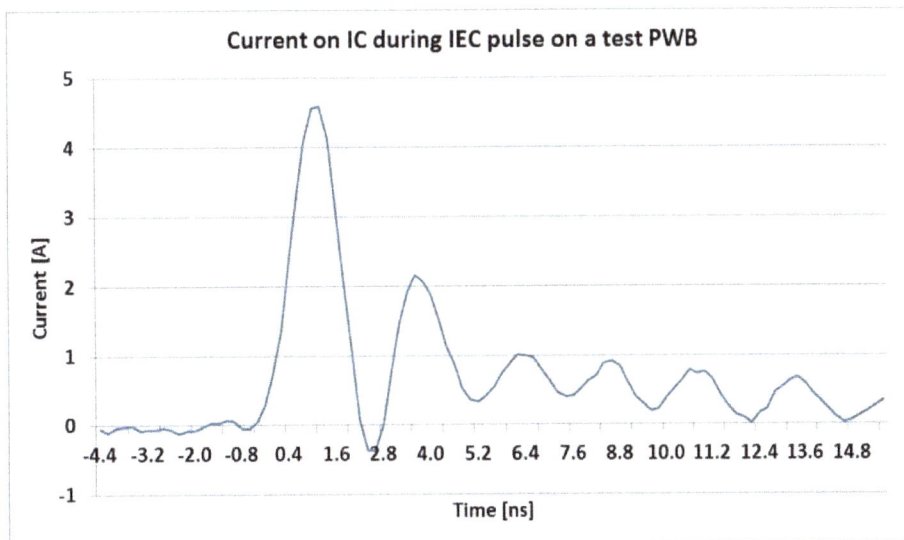

Figure 16: IEC contact discharge current waveform is measured just before an IC component on a small size PWB
[26]

4.2.2 Category 1b

4.2.2.1 CDM as a Robustness Indicator

Today, there is no accurate information available about the robustness of an IC pin for Category 1b failures. As the expected waveform entering the pin is somewhat CDM like, it is tempting to use the CDM value as a predictor for this case as well, especially if the CDM peak current level is available. For Category 1b, it is estimated that about 3 A (typically) should be shunted, which is normally below the peak current of the CDM pass level. If not, this pin should be considered critical, and the addition of off chip protection is recommended.

Using CDM as a predictor for Category 1b might be tempting, but it comes with many pitfalls:

1. CDM is a one pin test, while in system level stress two or more pins are involved.
2. Pulse rise time in a CDM tester is always faster than when the component is soldered on a PWB with additional board parasitics.
3. As said, in the system level case, the IC can be powered up, which can lead to destructive or non-destructive transient latch-up. This type of failure is not covered in the CDM case.
4. The CDM data does not provide any information on whether or not the IC remains functional during or after the event. Rebooting might be required, even if the IC passes the highest level of CDM.
5. The CDM level is dependent on the IC and package size; larger components see more ESD current for the same CDM level. This is not the case for the system level event. Therefore, from a system level point of view, for the same CDM performance, a large component is more robust as compared to a small component though the above notes about transient latch-up and functionality must be considered as well.
6. One more thought on polarity - If we look at positive and negative polarity and compare CDM to pin injection into an IC that is mounted to a board, the current on the pin might be the same, but the current on the other pins is not the same, as the current will exit the component mainly on VDD or VSS, but possibly also on other pins if the VDD or VSS connection has resistance (1 Ohm is enough) or inductance (a few nH is enough). This complication is not present in the CDM data.

Therefore, if no other information is available, using CDM data might be pragmatic, but it is a better approach to extract the correct information using, for instance, a VFTLP system.

4.2.3 Category 2

There are no standardized tests for transient latch-up. Some tests used in the field are listed below merely as examples.

Note that for all tests, a relevant supply should be used to power up the component, as parameters such as source impedance and response time are critical.

Next, Sections 4.2.3.1 through 4.2.3.3 will give some examples of how TLU is dealt with in the industry; similar techniques and characterization methods might have merit for avoiding failures during IEC 61000-4-2 as well, though none of them are generally accepted as solving the issue of a TLU triggered by an IEC 61000-4-2 pulse.

4.2.3.1 Triggering of Supply Clamp during ESD System Level Tests on a Form-Factor Board in a Mobile Application

This example addresses a TLU trigger induced by a system level stress. A baseband IC in a 65 nm complimentary metal-oxide-semiconductor (CMOS) process latches up during an IEC 61000-4-2 ESD system level stress of the entire mobile form factor. The physical root cause for the latch-up was triggering the power clamp in one domain. Electrically, the latch-up could be detected by increased power consumption on the particular power supply. For a complete discussion, see Technical Report TR-5.4-5 of the ESDA, which should be released by the end of 2012.

The baseband product passed the (static) latch-up test according to JEDEC JESD78D up to 150 mA of injected current at a temperature of 85 °C. In the JEDEC latch-up test, the compliance was set to 1.5 times the absolute maximum ratings (AMR) according to the JEDEC JESD78D spec. If higher compliance values are used, the transiently triggered latch-up could be reproduced by the JEDEC JESD78D stress. In Figure 17, the DC I-V characteristics of the power clamp are shown. The power clamp's holding voltage is significantly below the supply voltage, the clamp's trigger voltage, however, is above the levels tested in the LU overvoltage test. To identify the latching silicon controlled rectifier (SCR), the IC was analyzed by EMMI. As shown in Figure 18, the EMMI spot was localized over the power clamp.

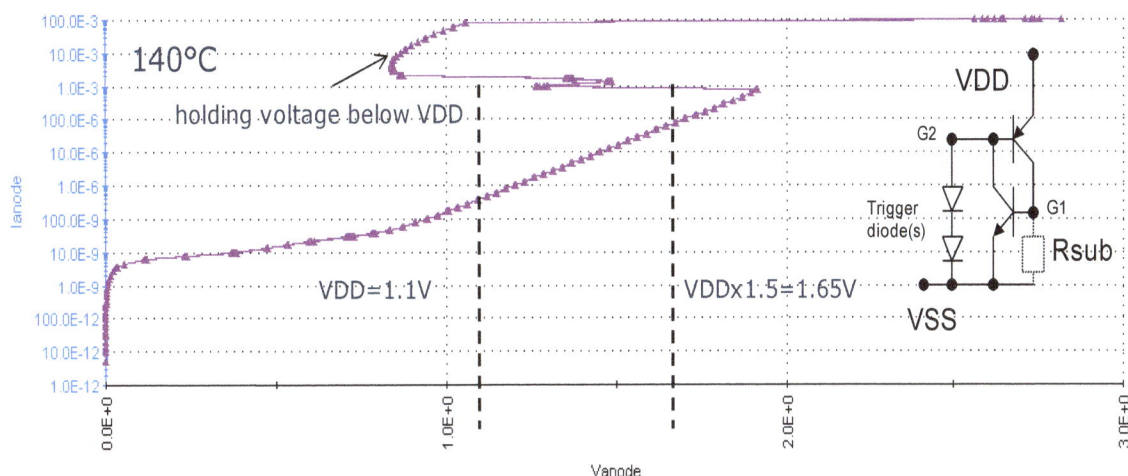

Figure 17: DC I-V characteristics of the diode triggered SCR power clamp. The clamp has the holding voltage below VDD; this posts a potential issue for LU/TLU. It is, however, not sensitive to slew rates, because it does not contain any breakdown elements or RC trigger circuitry

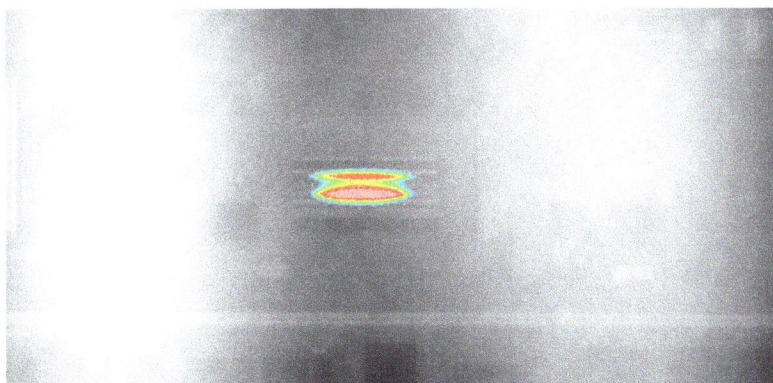

Figure 18: EMMI snapshot of the power clamp after EOS stress. The emission spot indicates that the component has latched (courtesy of TR-5.4-5)

Consequently, the latch-up phenomenon observed in that baseband product is not a "true" TLU phenomenon, because it is not dependent on the latch-up pulse slew-rate or duration. The SCR is triggered due to overvoltage and associated current injection. Obviously, a system level stress can induce quite high current amplitudes into the IC; at least considerably higher current levels than defined in the JEDEC JESD78D qualification tests. However, using such high currents far beyond the AMR in static latch-up tests would very likely cause EOS or gate oxide damage, which of course is not the focus of latch-up testing.

Transient TLU pulses are used in this example, in order to avoid overstress failure. The parameters of the transients (rise time, pulse duration ...) do not play any significant role as long as the pulse duration is short enough to avoid overstress and the current amplitude is high enough.

4.2.3.2   CDE triggering TLU

It is not surprising that fast transients, which occur during CDE or ESD system level surges according to IEC 61000-4-2, can potentially trigger latch-up.

CDE using cables with lengths of 1.5 m, 10 m, 20 m, and 40 m were applied on an IO pin of a specially designed IO test chip at 100 °C ambient temperature. Only trigger currents with negative polarity were applied and the values of the trigger currents are given in absolute numbers.

No latch-up was observed for the shortest cable. For the longer cables, increased power consumption at the voltage supply VDDP1 indicated latch-up for charging voltages between 240 V and 580 V; the corresponding waveforms are depicted in Figure 19 [9].

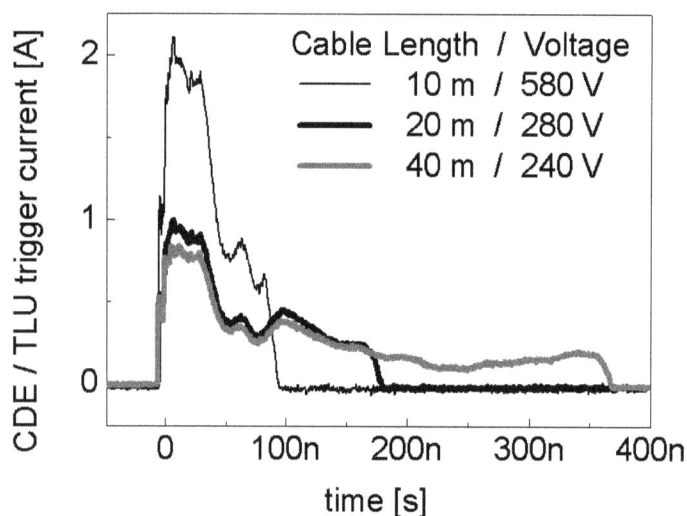

Figure 19: CDE waveform for 10 m, 20 m, and 40 m long cables after they are charged with voltages corresponding to the CDE / TLU trigger threshold of TC2_IO1 [9]

No latch-up at the second voltage supply (VDDP) could be induced by CDE. This observation is surprising because the results of other TLU tests estimate a lower threshold level at VDDP (see Figure 20). Again, crowding of the substrate current might favor the triggering of latch-up at voltage supply VDDP1.

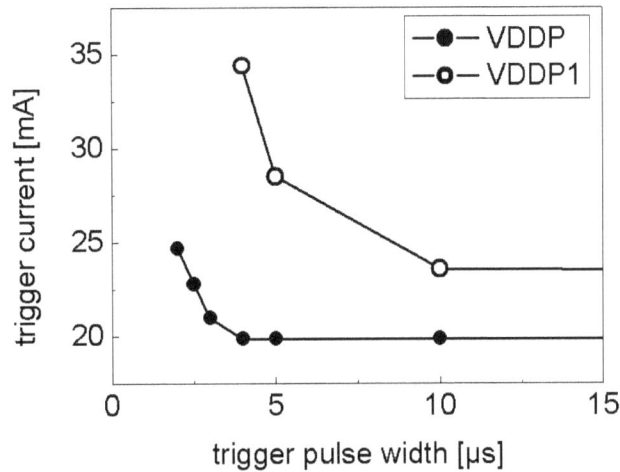

Figure 20: TLU threshold trigger current (negative polarity) of IO1 pin of TC2 versus duration of the trigger pulse (81104A, t_r = 10 ns) at 100 °C ambient temperature (courtesy of TR-5.4-5).

External latch-up can be triggered even by comparably short cables (cable length = 10 m). The CDE trigger current threshold can be two orders of magnitude higher compared to the "static" latch-up trigger level. Thus, the TLU sensitivity in terms of injected current amplitude is significantly relaxed compared to static latch-up. However, during TLU events high voltage amplitudes can appear. This may be one of the reasons why components can fail during CDE stress even if they pass JEDEC JESD78D. The CDE trigger can be well reproduced by TLP. This has to be considered carefully in upcoming standardization activities.

4.2.3.2   Transient Latch-up Tested through Capacitive Discharge

CCL (capacitive coupled latch-up, also referred to as ESD induced latch-up) is a test which is popular in the Middle East and Japan. A bipolar HV pulse is applied to an IC under powered up conditions. Though there is some merit in using this test to predict the ICs susceptibility for transient latch-up, there is some mismatch in the peak current vs. energy in the bipolar pulse versus the assumed field case. A 200 V pulse using a MM generator has a peak current of 3 A, which corresponds reasonably well with the expected peak current of the field case (though this consideration better fits Category 2), but the time duration of a bipolar stress pulse is much larger, such that applied energy is too high, which might lead to a destructive failure which is not transient latch-up related. In general, the applied voltage for the CCL test should not exceed the MM capability of the IC. In [27] it is explained that the use of such a bipolar stress is useful, but the test in this reference uses a waveform of a lower frequency (using a 20 Ohm source) than the MM stress. In [10], a comparison is made between 0 Ohm (MM) and 20 Ohm pulses for characterizing the transient latch-up susceptibility.

4.2.3 Category 3

Though many Category 3 soft failures have been solved, there are few very well documented cases showing the analysis and the solution.

In depth or trial and error methods can be used for ESD soft failure analysis. The trial and error methods usually incorporate:

- Changing the assumed current path through shielding or "ground" connections and observing changes in the system behavior. For example, the improvement of the shell to chassis connection if an ESD strike to a USB cable causes errors in a USB connection.
- Adding filters such as ferrite beads and capacitors to traces that are suspected to carry the disturbing signal to an IC. For example, if the fast changing electric field couples into the key pad of a cell phone and wakes the phone from sleep mode (as the phone mistakes the induced currents in the high impedance keypad circuitry as a user action) then adding capacitors to the traces will slow the response to a point that the charges induced by the rapidly changing E-field are insufficient to overcome the threshold voltages. Another example is adding resistor/capacitor (RC) filters, like 1 kOhm/100 pF, to a reset line to avoid unwanted responses to a narrow pulse induced by the magnetic field of an ESD event.

In depth analysis methods may utilize a more systematic approach:

- Conducting system level ESD testing such that repeatable results are achieved (avoidance of air discharge, verification whether secondary ESD occurs, usage of a sufficiently large number of pulses, test for repeatability, determination of failure threshold by increased test levels in small increments, and tests of multiple units.) [28].
- Near field susceptibility scanning to identify sensitive nets, modules and ICs which show the same failure mechanism as observed during system level testing [29].
- After sensitive nets have been identified, direct injection through small capacitors can be used to quantify the sensitivity of a net. During this injection the injected current is usually measured using a current clamp.
- Current spread scanning to try to visualize the coupling paths [30].
- Quantification of the coupling path to the net by substituting the ESD generator by a network analyzer driven ESD generator and probing the sensitive net [31].
- Numerical modeling of the system on block level to determine the current densities and transient voltages between metal parts during and ESD [32].
- Tracing the root cause from the software side by analyzing register information. [33-34].

In summary, the methodology associated with systematic and relevant characterization of components addressing category 3 fails has not yet reached the level of sophistication associated with the methodology for category 1 and 2 failures. To advance beyond the empirical trial and error approach as the standard used by many companies will require a significant development effort.

References

[1] D. Johnson, H. Gossner; "Study of System ESD Co-design of a realistic Mobile Board", Proceedings of 33rd EOS/ESD Symposium (2011), pp. 360-369

[2] G. Notermans, S. Bychikhin, D. Pogany, D. Johnsson, D. Maksimovic, "HMM–TLP correlation for system-efficient ESD design", Microelectronics Reliability, 2011

[3] zur Nieden, F.; Scheier, S.; Arndt, B.; Frei, S.: „Gefährdung von integrierten Schaltungen durch Überkopplungen zwischen Leiterbahnen auf Platinen", ESD-Forum 2011, Munich, Germany, 2011

[4] S. Poon, T. Maloney; "Shielded Cable Discharge Induces Current on Interior Signal Lines", Proc. 29th EOS/ESD Symposium (2007), pp. 311-317

[5] JEDEC Standard "IC Latch-Up Test"; JESD78D; November 2011.

[6] Ming-Dou Ker, Sheng-Fu Hsu. "Physical Mechanism and Device Simulation on Transient-Induced Latch-up in CMOS ICs Under System-Level ESD Test" IEEE Transactions on Electron Devices, vol. 52, no. 8, August 2005, pp. 1821-1831

[7] Farzan Farbiz and Elyse Rosenbaum. "Understanding Transient Latch-up Hazards and the Impact of Guard Rings", International Reliability Physics Symposium, 2010, pp. 466-473

[8] K. Domanski, B. Poltorak, S. Bargstaedt-Franke, W. Stadler, W. Bala. "Physical fundamentals of external transient latch-up and corrective actions", Microelectronics Reliability 46 (2006) pp. 689–701

[9] Brodbeck, T.; Stadler, W.; Baumann, C.; Esmark, K.; Domanski, K; "Triggering of Transient Latch-up (TLU) by System Level ESD", EOS/ESD Symposium 2010, pp. 1-10

[10] I. Morgan, C. Hatchard, M. Mahanpour, "Transient Latch-Up Using an Improved Bi-polar Trigger", Proc. 21st EOS/ESD Symposium (1999), pp. 190-202

[11] S. Bargsteadt-Franke a, W. Stadler, K. Esmark, M. Streibl, K. Domanski, H. Gieser, H. Wolf, W. Bala. "Transient latch-up: experimental analysis and device simulation", Microelectronics Reliability 45 (2005) pp. 297–304

[12] Wang, Dening; Marum, Steve; Kemper, Wolfgang; McLain, David; "System event triggered latch-up in IC chips: test issues and chip level protection design", EOS/ESD Symposium 2006, pp. 1-7

[13] D. Bonferta, H. Giesera, H. Wolfa, M. Frankb, A. Konradb, J. Schulzc, "Transient-Induced latch-up test setup for Wafer-Level and Package Level", Microelectronics Reliability 46 (2006) pp. 1629–1633

[14] R. Llido a, J. Gomez a, V. Goubier a, N. Froidevaux a, L. Dufayard a, G. Haller a, V. Pouget b, D. Lewis b. "Photoelectric Laser Stimulation applied to Latch-Up phenomenon and localization of parasitic transistors in an industrial failure analysis laboratory", Microelectronics Reliability 2011

[15] J. Quincke, "Novel Test Structures for the Investigation of the Efficiency of Guard rings Used for IO-Latch-up Prevention", ICMTS 1990, pp. 35-39

[16] S. Liao, C. Niou, K. Chien, A. Guo, W. Dong, C. Huang, "New Observance and Analysis of Various Guard-Ring Structures on Latch-Up Hardness by Elackside Photo Emission Image", IRPS 2003, pp. 92-98

[17] Ming-Dou Ker, Wen-Yu Lo, and Chung-Yu Wu, "New Experimental Methodology to Extract Compact Layout Rules for Latch-up Prevention in Bulk CMOS ICs", Custom Integrated Circuits, Proceedings of the IEE 1999, pp. 143-146

[18] Ming-Dou Ker, Member, IEEE, and Chung-Yu Wu, Member, IEEE, "Modeling the Positive-Feedback Regenerative Process of CMOS Latch-up by a Positive Transient Pole Method-Part I: Theoretical Derivation", IEEE Transactions on Electron Devices, vol. 42, no. 6, June 1995, pp. 1149-1155

[19] D. Estreich, R. Dutton, "Modeling Latch-Up in CMOS Integrated Circuits", IEEE Transactions on Computer-Aided Design of Integrated Circuits and Systems, vol. cad-1, no, 4, October 1982, pp. 157-162

[20] Felipe Coyotl Mixcoatl, Alfonso Torres Jacome, "Latch-up prevention by using guard ring structures in a 0.8 µm bulk CMOS process", 2004

[21] B. L. Gregory and B. D. Shafer, "Latch-up in CMOS Integrated Circuits" IEEE Transactions on Nuclear Science, Vol 20, 1973, pp. 293-299

[22] Gensuke Goto, Hiromasa Takahishi and Tetsuo Nakamura, "Latch-Up Immunity Against Noise Pulses in a CMOS Double Well Structure", 1983 Electron Devices Meeting, pp. 168-171

[23] K. Domanski, M. Heer, K. Esmark, D. Pogany, W. Stadler, E. Gornik. "External (transient) Latch-up Phenomenon Investigated by Optical Mapping (TIM) Technique", EOS/ESD Symposium 2007, pp. 6A.3-1 – 6A.3-10

[24] Ming-Dou Ker, Senior Member, IEEE, and Sheng-Fu Hsu, Student Member, IEEE, "Component-Level Measurement for Transient-Induced Latch-up in CMOS ICs Under System-Level ESD Considerations", IEEE Transactions On Device And Materials Reliability, vol. 6, no. 3, September 2006, pp. 461-472

[25] Jayong Koo; Qing C,ai; Kai Wang; Maas, J.; Takahashi, T.; Martwick, A.; Pommerenke, D., "Correlation Between EUT Failure Levels and ESD Generator Parameters", IEEE Trans. EMC. Vol.50, Issue 4, Nov. 2008 pp. 794 - 801

[26] Industry Council on ESD Target Levels, "White Paper 3 System Level ESD Part I: Common Misconceptions and Recommended Basic Approaches," December 2010, at www.esda.org or JEDEC publication JEP161, "System Level ESD Part I: Common Misconceptions and Recommended Basic Approaches", www.jedec.org

[27] M. Kelly, "Developing a Transient Induced Latch-up Standard for Testing Integrated Circuits", Proc. 21st EOS/ESD Symp. (1999), pp. 178–189

[28] Xiao, J.; Pommerenke, D.; Drewniak, J. L.; Shumiya, H.; Maeshima, J.; Yamada, T.; Araki, K.;"Model of Secondary ESD for a Portable Electronic Product", IEEE Trans. EMC, Oct. 2011, pp. 1-10

[29] Muchaidze, G., Jayong Koo, Qing Cai, Tun Li, Lijun Han, Martwick, A., Kai Wang, Jin Min, Drewniak, J.L., Pommerenke, D., "Susceptibility Scanning as a Failure Analysis Tool for System-Level Electrostatic Discharge (ESD) Problems", IEEE Trans. EMC, May 2008, Vol 50, No. 2, pp. 268-276

[30] Wei Huang; Dunnihoo, J.; Pommerenke, D.; "Effects of TVS integration on system level ESD robustness.", EOS/ESD Symposium 2010, pp. 1-6

[31] Jayong Koo, Qing Cai, Muchaidze, G. Martwick, A.Kai Wang, Pommerenke, D, "Frequency-Domain Measurement Method for the Analysis of ESD Generators and Coupling", IEEE Trans. EMC, Aug. 2007, Vol. 49, No.3, pp. 504-511

[32] Liu, D.; Nandy, A.; Zhou, F.; Huang, W.; Xiao, J.; Seol, B.; Lee, J.; Fan, J.; Pommerenke, D. "Full-Wave Simulation of an Electrostatic Discharge Generator Discharging in Air-Discharge Mode Into a Product", IEEE Trans. EMC, Vol.99, 2010, pp. 28-37

[33] Maheshwari, P.; Tianqi Li; Jong-Sung Lee; Byong-Su Seol; Sedigh, S.; Pommerenke, D.; "Software-Based Analysis of the Effects of Electrostatic Discharge on Embedded Systems" 35th Annual Computer Software and Applications Conference (COMPSAC), 2011, pp. 436-441

[34] Maheshwari, Pratik; Seol, Byong-Su; Lee, Jong-Sung; Lim, Jae-Deok; Sedigh, Sahra; Pommerenke, David; "Software-Based Instrumentation for Localization of Faults Caused by Electrostatic Discharge", 13th Ins. Symposium on High-Assurance Systems Engineering (HASE), 2011, pp. 333-339

Chapter 5: Standard Model and Analytical Tool Needs To Support SEED

Patrice Besse, Freescale
Jeffery Dunnihoo, Pragma
Harald Gossner, Intel Corporation
David Johnsson, Intel Corporation
David Klein, Freescale Semiconductor
Timothy Maloney, Intel Corporation
David Pommerenke, Missouri University of Science & Technology

5.0   Introduction

In chapter 4, the SEED methodology was divided into three categories depending on the different kinds of stress a component in a system can experience beyond the normal operating parameters of the component. For that approach to be successful, component suppliers and OEMs must agree on a standard model definition that includes the fast transient, high current response of a component's IO (Category 1), the fast transient, high current response of a component's IO and the circuits connected or in proximity to it (Category 2), and finally, in order to support Category 3, the model must move beyond single IO responses and become system aware, taking into account low level transient stress due to crosstalk between traces on the board or radiated EM fields. Having such a standard model allows:

- suppliers to focus on providing a single definition, high quality model of their IC. This model must be able to accurately describe Category 1 type stress response, but should also be extensible to Category 2 and, eventually, Category 3 type stresses.
- OEMs to use analytical tools to integrate the IC IO models into their system models and carry out system level stress analysis.
- EDA vendors to have a single definition to integrate into their existing or new simulation products.

With these points in mind, this chapter describes methods for characterizing component response to system level ESD stress with the aim of creating a model of that response. Available signal integrity and functionality focused models are generally characterized only within normal operating parameters for the device. Typically these models do not accurately describe the behavior of the component when it is exposed to extreme voltages and currents over the short duration of an ESD event. We will show how correct characterization and modeling can be used to optimize new designs for ESD robustness as well as debug and understand these failures in existing designs. Also, because current modeling and simulation tools are limited, methods for examining existing full system designs are presented.

This chapter is divided into three sections. The first section introduces the standard model and component pin characterization required to support the Category 1 design approach. The next section expands the standard model to encompass the requirements of the Category 2 approach. Further discussion of the component characterization follows for the additional level of required detail. In the final section, Category 3 analysis recognizes that the nodal analysis focus of the first two sections may not fully capture the system level response to ESD stress. This section examines what to do when software analysis proves insufficient and hardware analysis of a prototype or finished system is necessary. Category 3 analysis moves beyond nodal analysis into full 3D field solver analysis based on overall system level susceptibility to spatial E- and H-fields. The use of 3D EM scanners and antennas to determine both the susceptibility of a system to induced fields and the strength of those fields in a system are described. Finally, we describe how, in the near future, this model/system structure could be extended into the EM simulation realm by extracting a field susceptibility model from a 3D EM scan and running full-wave 3D simulations to predict not only hard failures, but perhaps difficult to analyze soft failures as well.

5.1 Component Characterization and Model Requirements to Support SEED Category 1

The level of detail required by different OEMs varies with the level of sensitivity of the components used in their systems and the system operating environment. The Category 1 standard model must provide for simulations that range from simple pass / fail analysis to full current, voltage and frequency response.

5.1.1 Standard Model Needs to Enable SEED Category 1 Level Analysis

To enable Category 1 basic nodal analysis, the standard model must:

- accurately describe the characteristic current and voltage response of each component pin to direct and indirect ESD stress, both with the component powered and unpowered.
- accurately describe voltage and current levels at which hard failures occur.
- extend from a simple pass/fail type simulation for a given pulse shape to the full pin response to the given ESD stress.

To do this, the model must contain the basic elements that describe the ESD behavior of a component. These are:

1. Description of each interface pin's inductance, resistance, and capacitance.
2. Description of the voltage and current response to each pin when exposed to ESD level stress events. This could be an equation that curve fits to the pin's behavior when tested. It could also be a set of data tables that presents the reponse of each pin to voltage, current and the frequency and duration of these stimuli. Multiple equations or tables may be necessary to describe the behavior of the pin when the component is powered, unpowered, or in a particular logic state.
3. Description of the failure criteria. This could be as simple as a pair of voltage and current values beyond which failure is expected to occur. Or, because failure is often dependent on the duration and rate of change of the ESD pulse, this might be described as points in the data table or equation surface beyond which characterization data says either an oxide will break down or the current density will exceed the pin's current handling ability and thermal runaway or metal fusing will occur.
4. Description of the behavior states each pin transitions through as it is exposed to ESD stress levels. Behavioral modeling allows the model to account for the cumulative effects of multiple stress events and other effects that a look-up table or curve fit model would be unable to provide.

5.1.1.1 Existing Model Standards that could be adapted to include the SEED Standard Model

There are many different models available in the industry, ranging from high level hardware description languages (HDLs) that describe the behavior of circuits and systems of circuits to SPICE compact models that describe the behavior of individual components such as transistors and resistors. There are also technology computer-aided design (TCAD) models, but because they are based on fundamental physics, they would not be suitable for system level modeling given presently available computers.

Traditionally, an IC designer would often use SPICE type models to simulate the continuous electrical or analog behavior of their IC. They might also use event driven HDLs to describe the logic behavior of the IC. A system designer would be most familiar with high level HDL models that describe the overall system behavior using an IO Buffer Information Specification (IBIS) or similar model to ensure electrical and signaling compatibility between the various components in their system. Relatively recent extensions of HDLs have allowed them to simulate both event driven logic behavior and continuous time analog behavior. Coupling between the two types of simulation allows digital or logic events to trigger analog events and vice versa.

The model that will best support SEED Category 1 will need to work at both a high level (pass/fail) for overall system robustness simulation and at a low level (I-V response) to allow for optimization of the system by providing detailed information which can be used to determine the necessary upstream protection and filtering components. Therefore, a successful model will need to encapsulate both high level behavioral modeling with the ability to dive into the details low level models afford. For this reason, HDL languages that provide behavioral modeling with analog and mixed signal capability should prove useful.

Behavioral modeling offers many advantages. Computing time is only a few seconds with a good convergence. Intellectual property is kept proprietary as no design architecture or description is required.

Behavioral modeling can be used to describe the transient behavior of an integrated circuit submitted to external stress such as an ESD event. It can be useful to simulate high injection phenomena on a nanosecond time scale, where limitations of existing compact models often appear. Depending on the complexity of the circuit, the model can be built to simulate the response of an entire IC when a pin is stressed with an ESD event or to describe only a part of the circuit such as the ESD protection [1].

Usually behavioral models for ESD are based on TLP measurements (100 ns). Main transient parameters are extracted from the TLP curve and reported in equations to describe the I-V response of the pin under ESD stress. Behavioral models are particularly good at modeling non-linear behaviors such as strong snapback in the case of SCR protection mechanisms. Different techniques exist to improve simulation convergence in the VHDL-AMS language [2].

Of course there are limits. Behavioral models can be particularly difficult to create if different biasing conditions exist. Resonance or dv/dt effects can also be missed.

Recently several papers have reported a good correlation between the level of robustness predicted by behavioral models and the level observed when a component was stressed with a TLP zap or a human metal model (HMM) zap using a 330 Ohm gun discharge. The correlation was validated when the models predicted the IC failure, which was due to energy overstress leading to permanent damage [3, 4]. In this case, failure criteria based on time to failure measurements (energy dissipation capability of the DUT depending on the stress duration) can be defined [5] and successfully used to predict damage after ESD system level stress. This kind of simulation can be performed with the addition of external protections and PCB models.

One option which makes this information available for current simulators is to map it to IBIS multi-lingual models. The ultimate objective is to facilitate a reference methodology for capturing and merging ESD simulations, models and parameters into industry standard signal integrity and functional simulation workflows. ESD pulse generator models can be used with "ESD ready" component models in existing simulator environments to predict overall system robustness and pinpoint potential susceptible problem areas.

There are currently three methods to support ESD modeling in common simulation environments:

1. **Use environment variable watching:** All simulators have some type of watchpoint environment variables that can be set for system-level nodes. The simulator would flag a environment variable when stress levels beyond the SOA are detected. The drawback to this method is that the system designer must manually calculate/assess what these limits actually should be at the system level. This requires extensive knowledge and background in ESD simulations and interactions, in addition to both considerable expertise in the simulator environment and full-system level functional and signal integrity simulations. Contrary to typical workflow, this requires manually assigning accurate system nodal I/V limits extracted from individual component characterizations. This places a considerable time and error-prone burden on the system designer.

2. **Virtual SPICE signals:** In this technique, a virtual circuit with nodes that are not exported to pins creates a virtual "FAIL" signal when component ESD_SOA stress limits are reached. This is supported commonly across platforms and would be supplied independently by component vendors inside their functional SPICE model. This is probably the most accurate, immediate and widely supportable option at this time. The drawback of this method is that this will require model support from the component vendors which is not commonly in place.

3. **HDL error messages:** This is a HDL abstraction of the SPICE method above where a Verilog-A or VHDL-AMS model referenced by an IBIS model provides stateful monitoring of I/V levels and "prints" an error message during simulation when ESD_SOA parameters are exceeded. Beyond what a SPICE model would provide, some stateful information could be handled in HDL about cumulative damage or stress level effects. The problem with this approach is that at present many lower-cost license and simulator packages do not incorporate analog HDL functionality.

As noted, all of these methods have limitations. The best solution will be one that is accurate and still easy to incorparate into the existing tool chain for the end user: the system designer. Looking forward, the best long term solution may be to extend the IBIS standard to include optional ESD specific (model specification) parameters. This will involve adding appropriate ESD SOA characterization parameters as defined by workgroups and presented to IBIS for adoption. In this case, the simulation vendors could incorporate specific ESD analysis capabilities. This would reduce the burden on the system designer by automatically propagating appropriate watchpoint levels throughout the netlist at each instance of an "ESD ready" model.

Eventual implementation could be as simple as importing the ESD SOA parameters as nodal current/voltage limits for a pass/fail ESD test when a standard ESD pulse model is applied to the system. This modeling proposal will require a concerted effort from both the component vendor and system design communities to settle on the appropriate parameters and characterization methods that can then be presented to IBIS for adoption.

At present, there is no accepted standard system level (IEC 61000-4-2) aggressor simulation model, nor does the standard lend itself to a common model that fits all the tests that have grown out of the standard. Such a standard system level aggressor model should also be pursued as an integral part of the modelling proposal mentioned above. In the meantime, while good models should provide rational results for a range of aggressor models, vendors should carefully identify which aggressor model or model(s) are verified with their component models [6-8].

Currently ESDA WG26 on system level ESD modeling, building on a general framework outlined by IEC 61433-6, is drafting a technical report which recommends standard model variants for SEED. These include PCB diode as well as IO pin models. Their applicability is proven by round robin validation tests using test boards for LIN, D LATCH and USB3 components.

5.1.1.2 Cumulative Multi-Strike and Other Extensions

An ESD event as understood by the basic SEED methodology does not exclude dynamic failure criteria or other real-world approximations of cumulative ESD damage from subsequent ESD events. A simulation with fixed failure criteria will not address the case where a component forms filaments and maintains good leakage and clamping performance for nine strikes but fails on the tenth. Some research [1, 9, 10] has been done to model the dynamic characteristics of components as noted above, but this could be extended during the simulation to include dynamic derating of the performance and parasitic characteristics of a component to more accurately portray the SOA with respect to subsequent ESD events.

A mixed-mode model (with event driven behavioral states) could provide many important enhancements to understanding and potentially predicting accurate system susceptibility and failure modes. In the most basic usage, a state-driven model might simply encapsulate a "run-time" state where the model presents signal-integrity and functional performance, and a "clamping" model which is selected when voltages and currents enter the ESD domain. Alternately, a complete analog model of a functional buffer and clamps may cover inside and outside the run-time "SOA". However, when the ESD SOA is exceeded in the model, the simulation advances to a "fail" state where the component becomes a short or open, for example.

Cumulative damage and soft-failures at the system level could be simulated in more advanced models with the same approach. For cumulative damage, each strike may advance a "damage state" in the model which presents different leakage and SOA limits resulting from previously applied stresses as shown in Figure 21. For example, in the case where a TVS component is rated for 1000 strikes at 8 kV IEC and 10 strikes at 16 kV, it may be possible to model the specified limits from these datasheet parameters. It may also be possible to model them more accurately as cumulative energy dissipated and/or peak power applied, or any number of unique damage mechanisms.

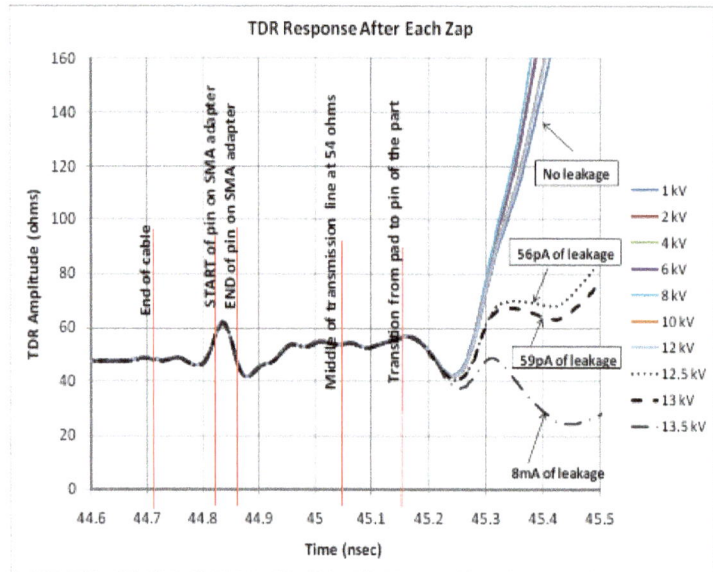

(a) An example of TDR data used to create model leakage and damage tables.

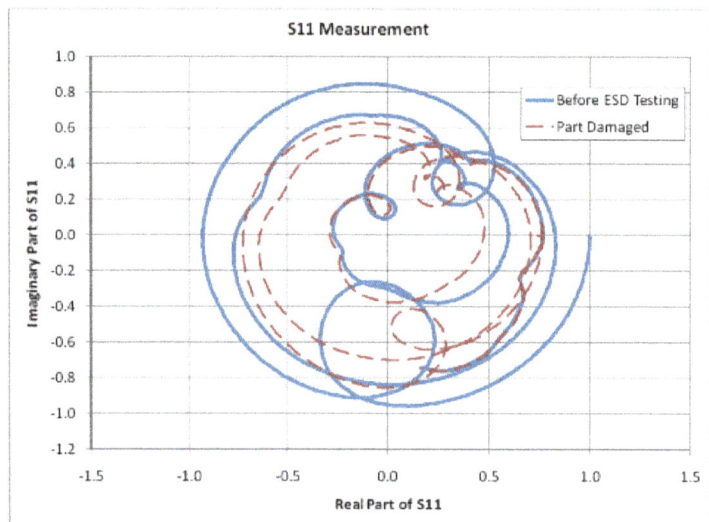

(b) An example of measurement data to calibrate the failure criteria tables.

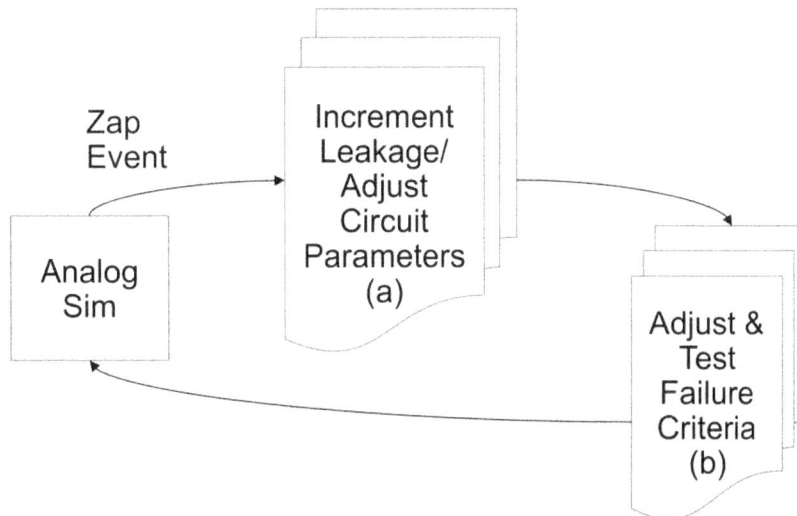

(c) Resulting mixed mode behavioral model.
Figure 21: Conceptual State Diagram of a Cumulative ESD Damage Device Mode [11]

The IBIS model container provides a potential comprehensive transport vehicle for transmitting this advanced information more accurately from the chip designer to the system developer. The market can then exploit advances in the research and adopt them within a common (optional) transport mechanism which vendors can support with minimal accuracy (basic static SOA with conservative guard-banding for multiple-strikes) or extended accuracy allowing, more aggressive ratings (with dynamic SOAs that describe a distribution of ESD robustness and susceptibility.)

5.1.2 Component Pin Characterization Description

The most widely used tool for analytical ESD characterization is TLP. A TLP is based on a transmission line which is charged to a high voltage. The line is discharged through a mechanical switch, which gives rise to a rectangle-shaped pulse. The pulse width is defined by the length of the transmission line. The rise time of the pulse depends on the spark formation speed in the switch, which is usually in the range of 100 ps. The ability to form such steep pulses at voltage levels of several kilovolts makes the TLP an ideal instrument for measuring components in the ESD regime. The approximately square-shape of the pulse allows the component to settle into a quasi-stationary state where the voltage and current over the component can be measured. The well-defined coaxial environment is also of advantage since RF metrology methods can be applied to measure the turn-on response of a component in the sub-ns regime. The most common TLP systems are discussed below and Table 3 shows the recommended parameter settings for system level characterization.

Table 3: Recommendations for a TLP system suitable for system level characterization

Pulse amplitude	At least 30 A
Rise Time	0.5 ns & 10 ns
Pulse widths	2.5 ns & 100 ns

Standard TLP

The most commonly used TLP pulse width is 100 ns since it corresponds well to an HBM discharge. A rise time of 10 ns is usually selected for the same reason. The system creates the I-V curve by averaging every applied pulse in the window at the end of the pulse when the component has settled to a quasi-stationary state and any ringing due to inductance in the connection path has decayed. The component probe setup is often not optimized for high frequencies, which makes it hard to study fast transient effects.

Very Fast TLP (VFTLP)

The term very fast TLP (VFTLP) commonly refers to a TLP system with a pulse width below 10ns and a rise time below 1 ns. A VFTLP system uses advanced RF probing methods to capture the transient turn on response of a component. Since VFTLP pulses are much shorter in duration, components can be measured to higher pulse levels without thermal destruction. This addresses other failure mechanisms relevant for the CDM regime. VFTLP is also suitable for investigating the susceptibility of a component to the first peak of an IEC 61000-4-2 pulse.

Combined TLP/VFTLP systems

Differentiation between standard TLP and VFTLP systems is only historical. Today several test systems can deliver any combination of pulse width and rise time, and with a suitable probe setup, the transient response can be measured for both short and long pulses. This is a great advantage since only a single system is necessary to characterize both transient turn-on effects and thermal destruction.

5.1.2.1   Characterization of ICs

In order to use the SEED methodology, the I-V response of the relevant IC pins exposed at the system level must be known. It is also necessary to know how much current an IC pin can handle without being damaged. Even if an IC is not specified for more than 1 kV HBM, it is inevitable that external pins will be exposed to higher stress in application. System designers often follow a trial-and-error approach, which results in designs where several pins may be exposed to currents in the range of 10 A. IC manufacturers may be reluctant to rate external pins to such high current levels, but it would be a great advantage for both the IC manufacturer and the system designers if such data would be made available.

The current robustness of an IC pin depends on the applied pulse shape, which can vary greatly depending on which pulse form is applied. The IEC 61000-4-2 pulse has a time constant around 50 ns. However, the pulse appearing at the IC usually looks very different from the IEC shape due to the impedance transformation that takes place at the external protection component. The pulse through the IC has lower amplitude, but often longer time duration. A pulse width of 100 ns TLP is a good approximation for a residual system-level pulse and easily created.

Measurement configuration

There usually are several paths for the ESD current to flow through an IC. Thus, the response of a certain pin can vary depending on how the other pins are terminated and the state of the system. An example is shown in Figure 22(a), where the I-V curve is presented for a typical system-level ESD relevant pin in different configurations. The curve "PIN vs. VSS" shows the measured I-V response between the pin and the GND pin of the same domain. However, when the IC is measured on a PCB, the curve "PIN vs. PCB GND, off state" is obtained. Note that the voltage of the pin is almost half, which would lead to higher current through the IC in a co-design configuration. The reason for the lower voltage is the parallel ESD current path through the VDD pin (Figure 22(b)). The relatively high quality (ceramic) capacitors placed between VDD and VSS have lower impedance in the ESD region. Thus the Vdd bus can be considered equipotential to ground for an ESD pulse. With the system powered on, the higher potential of the VDD bus will change the pin response. This is shown in the curve "PIN vs. PCB GND, on state".

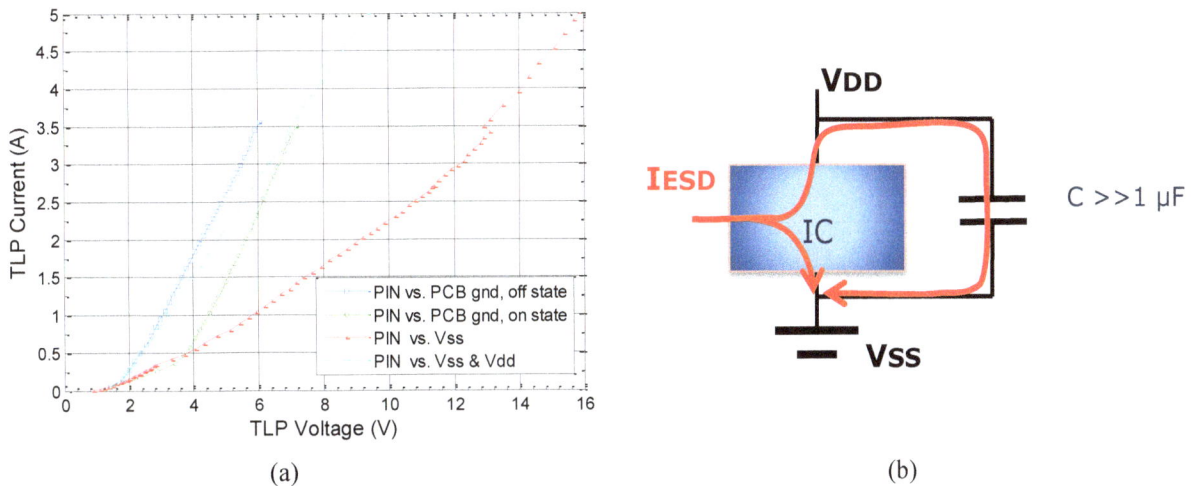

(a)   (b)

Figure 22: a) I-V curves for a typical external pin measured in different configuration. b) Schematic figure of the current flow through an IC with second discharge path over VDD.

The previous example shows that ideally an IC should be characterized in a configuration similar to its final application. However, it is tedious work to test every pin on a product PCB, since the IC must be replaced after every destructive pulse. It is also necessary to remove external protection elements from the PCB to obtain the component response alone.

Figure 22(a) also shows the IC measured with both VSS and VDD grounded. This curve is very similar to the curve obtained with the IC mounted on a system PCB. Since this curve can be considered "worst case" for a co-design, this configuration can be used to systematically test an IC. An advantage of this configuration for parts where all the internal grounds are shorted together in the package is that the IC can be measured in a 3-pin setup, either in a socket or with needle probes touching down on the IC bumps/pins. To make such a simplification, the ESD paths of the IC must be carefully analyzed to pick the correct pins to be grounded during the test.

5.1.2.2   Characterization of External Protection Components

The external protection component is the first barrier that keeps ESD pulses away from the main IC. Its turn-on speed is crucial for achieving high system robustness against ESD. Even if a certain voltage overshoot can be tolerated, it is a good practice to use components with as fast a turn-on speed as possible. A TLP system with a rise time on the order of 0.5 ns is ideal to measure the turn-on characteristics of a TVS. To connect the package, a test board like the one presented in Figure 23 can be used. This test board incorporates a 50 Ohm micro-strip line connected on both sides with sub-miniature version A (SMA) connectors. The 1 kOhm resistor is used as a voltage divider to give a scaled voltage at the port "Pulse Sense". The resistor can also be omitted to directly measure the residual voltage arriving at the pulse sense port. In this case pulse withstanding attenuators must be placed in series with the oscilloscope input to protect it from damage.

An alternative setup is presented in Figure 24. This figure shows a TVS contacted with Cascade RF probes in a Kelvin configuration: one probe is used to deliver the pulse and the second to sense the voltage. This setup achieves the best transient resolution of the measurement, but needs expensive equipment such as a probe station with micro-positioners and RF probes. An example of voltage waveforms captured with the setup as shown in Figure 24 is shown in Figure 25.

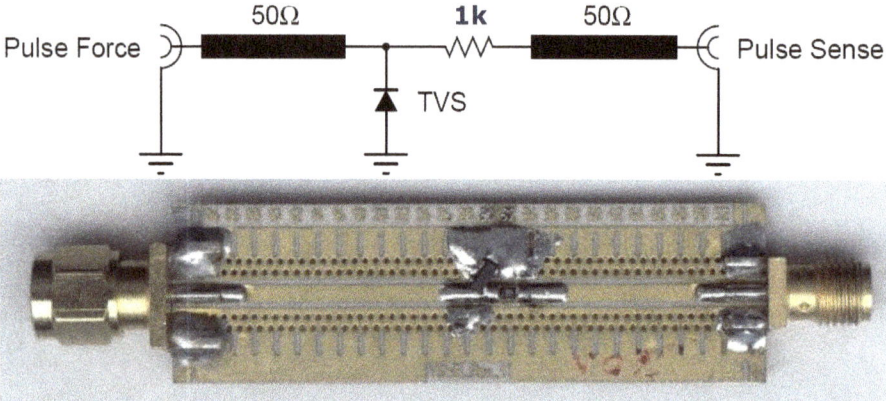

Figure 23: Measurement configuration with evaluation board

Figure 24: Measurement configuration with RF probes

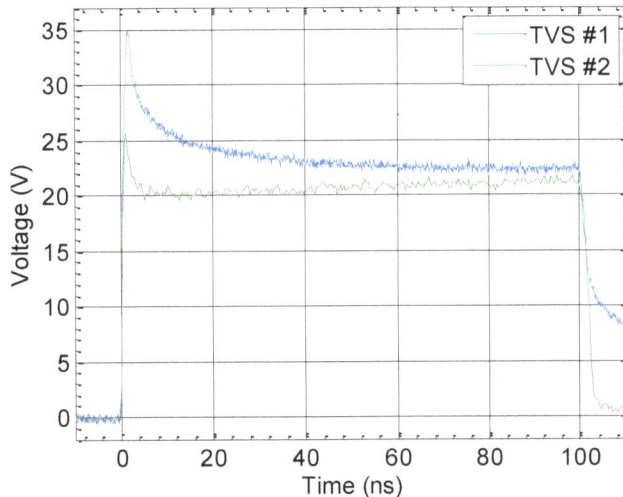

Figure 25: Example of voltage waveforms from two TVS diodes with different turn-on time. The TLP pulse has current level of 10 A and a rise time of 0.6 ns. The waveform with large overshoot is the response of a low capacitance diode. This overshoot is caused by the diode's strong conductivity modulation.

5.2   Component Characterization and Model Requirements to Support SEED Category 2

Characterizing and modeling a pin of a component is relatively straightforward when all the circuitry that can be considered connected to that pin is not dependent on the state or mode of operation of the component. Examples of these sorts of IOs are general purpose input/outputs (GPIOs) and serializer/deserializer (SERDES) interfaces. However, there are pins where the functionality and exposure of internal circuits are dependent on the system state or mode of operation. An example is a multifunctional analog pin which exposes different internal circuits to the pin depending on the mode of the component. For successful characterization and subsequent modeling of these pins, extensive testing of the IO may be necessary to determine the most susceptible (worst case) mode of operation for that specific pin. As a result, while the Category 1 models could be lumped by type, to successfully support Category 2, each pin may need to be separately modeled for its response to low energy pulsing. It is of interest to determine the pulse shape that leads to either physical damage due to latch-up or functional upset due to transient latch-up caused by injection of current into the substrate.

5.2.1   Standard Model for Category 2 Stress

Changes necessary to the standard model in order to accommodate Category 2 type stress are fairly minimal. A behavioral model that is able to model accumulated stress damage and/or accelerated triggering due to very fast edge rates can also accommodate the descriptions of pulse shapes (rise and fall time, duration, amplitude, etc.) that can lead to latch-up or system upset. The larger problem is characterizing the component to determine not only the SOA of the pin, but also to ensure that the pin's mode of operation during characterization is indeed the worst case for low pulse induced hard or soft failure.

5.2.2   Category 2 Component Pin Susceptibility Characterization

The immunity of IOs against injection of substrate current is typically probed by standard JESD78 latch-up qualification of ICs. The standard requires an injection level of 100 mA. The pulse applied ranges in the 10 ms region. While this is a useful basis for assessing the general robustness of the IC, it might not be sufficient if short pulses of up to Ampere amplitude occur during a system level ESD event. Simply increasing the current level is usually not appropriate as 1) other failure mechanisms might be triggered due to thermal overheating and 2) rise time dependent mechanisms are neglected. Here transient latch-up testing based TLP can offer an appropriate way of characterization.

The TLP testing requirements are similar to those used in Section 5.1.2 with the following changes:

- Testing should be done under powered conditions, having realistic on board power decoupling for achieving VI curves relevant to system level design.
- Damage parameters, such as increased leakage, need to be monitored.

These modifications are necessary to detect hard failures. To characterize soft failures, further work is necessary:

- A significantly more complex test setup is needed, as peripheral ICs may be required to operate the DUT. Additional software may be necessary to cycle each pin through its possible operating modes to determine the worst case mode during testing. Also, additional components and software may necessary to detect operational changes.
- For meaningful soft failure characterization, the TLP pulse width of the injected noise should be varied from ns to µs. Such testing can be performed automatically by injecting pulses from a transmission line pulser through a small value capacitor or a high impedance resistor. An automatic scanning system can position the injection probe from pin to pin while varying the pulse strength and shape and observing the functional behavior of the IC. In special cases, narrow, possibly bipolar pulses need to be considered.

The characterization is applicable for any system external or internal IO. Supply pins are excluded. Due to the ambiguity of some supply pins, for instance pins attached to internally generated voltage, such pins should be characterized like IO pins if substrate current injection can occur through forward biased junctions attached to this pin.

5.3 System Characterization and Model Requirements to Support SEED Category 3

The modeling and characterization up till now have focused on how pins of components react to direct or induced stress. While this is very important to understand and model, the environment in which the component will be placed is also important in determining whether that component will be susceptible to damage in that system. For example, a component with low damage thresholds may be successfully used in a system where the surrounding components, PCB design and enclosure or case all cooperate to shield it from damage. At the same time, a component with a high damage threshold can easily be destroyed if it is exposed during a system level ESD event to stress beyond its SOA. For these reasons, characterization and modeling of the system are necessary.

An appropriate characterization method is not yet defined and poses a challenging task as such an approach must work for a wide range of systems. One option is to apply TLP pulses to modules, traces, and system ports, and couple the resulting transient EM field into various parts of the system. This can probe the susceptibility of the system. Another option is to map system susceptibility by injecting noise from an EM scanning head. These options will be discussed in detail below.

Ideally, modeling and simulation of system response during a ESD strike would be carried out using full wave tools and models. However, full wave simulation is not possible at present because to model most systems in detail will result in too many variables and unknowns to be practical. However, one can model a system using a full wave block level approach. In this approach, the main metallic parts, like the enclosure and batteries or main cable connections, are modeled while the details of the PCB and components mounted on it are ignored. Excitation by an ESD generator can be included in the model. This model will provide the current densities induced on PCBs, connecting cables and the chassis. Further, it will provide an estimate of the field strengths inside the system and the current paths. This information can be used to estimate the risk of direct coupling to ICs and used as stimulus for further, more detailed simulations which might determine the coupling into a single layer flex circuit relative to a dual layer flex circuit.

At this point, let's review the tools presently available for ESD analysis [12]:

TLP for I-V characterization:

A variety of TLP concepts, such as direct current and voltage measurements exist for the characterization of the transient I-V curve of components. These can be IOs or power pins on an IC. TLP can also be used to measure the non-linearity of capacitors and ferrite beads in addition to the breakdown limits of resistors and capacitors

TLP for susceptibility scanning: [13]

In a similar fashion, a TLP can be used to couple to modules, traces and pins through the transient field. The TLP may need up to 8 kV of charge voltage to provide sufficient field strength if small probes are used. Typical rise times are < 500 ps. The TLP is coupled to the circuit using probes magnetic field probes ranging from 50 mm to 0.5 mm or similar electric field probes. These probes induce currents (E-field) or induce voltages (H-field). The induced pulses follow the time derivative of the TLP rising and falling edge, thus, they are usually less than 2 ns wide. This is not atypical for system level ESD signals that are indirectly coupled onto traces, such as at board to board connectors. Field coupled noise can be directly injected onto a trace using a small (1 pF) capacitor to connect the TLP to the trace. In this case the injected current can be measured, (for example, with a small current transducer such as one from the Tektronix CT-series). In order to directly inject current onto a trace, the scanning system should be equipped with a probe that contains either a small capacitor (such as 1 pF) or a higher value resistor (such as 1 kOhm). A probe so equipped allows connection of the TLP output to the trace. The displacement current from the local ground of the probe to the ground of the PCB is the high frequency component of the return current. The low frequency component of the return current flows through a wire connection to the PCB. Now the system can touch the probe to different nets and increase the injected pulse until an error is observed. A record should be kept of the current at which this error occurs and the observed failure phenomenon.

Susceptibility scanning:

In susceptibility scanning, a probe is usually moved automatically from position to position and noise is coupled to the traces, modules or ICs while observing the system response. At each location the soft-failure threshold is detected creating a map of the susceptibility [13].

In the remainder of this section, system level characterization and analysis will be discussed, starting with the basics of measuring crosstalk on a PCB board. Next, the discussion will briefly introduce and describe a way to observe the radiated fields inside a system. This is followed by a brief explanation of how these methods, along with susceptibility scanning, introduced in Chapter 2, can be used to debug a system. Finally, present short-comings and future directions for both modeling systems and software are discussed.

5.3.1   Investigation of PCB Crosstalk

Crosstalk between PCB lines can be investigated by TLP or IEC gun stress of test boards with aggressor and victim lines in worst case arrangement such as long parallel lines of narrowly spaced distance. An example schematic and test board is shown in Figure 26. This type of investigation is shown for a typical automotive board [14]. The induced energy and the waveform at the victim line strongly depends on the matching of the impedance and can be modified by serial resistance and shunt elements like TVS diodes and IC IO protection. All this needs to be considered when setting up the test. Recent simulations have shown that an 8 kV IEC discharge to a controller area network (CAN) line leads to significant peak currents of several amps in a neighboring line attached to a typical microcontroller IO [15]. However, due to the very short length (< 5 ns) of the induced pulse, the energy coupled to the victim line is small compared to a 1 kV HBM pulse. Thus, thermal damage is very unlikely, but CDM type damage as well as transient latch-up can be failure mechanisms of relevance.

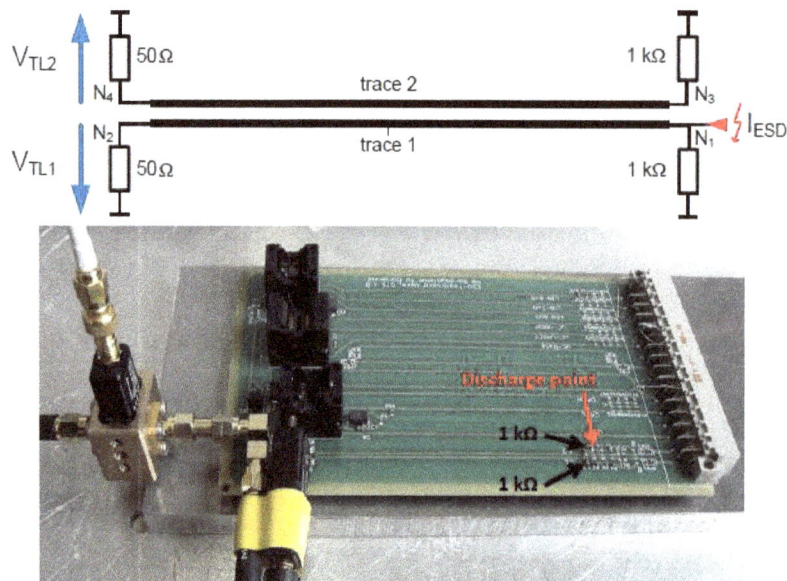

Figure 26: Crosstalk experiment for an automotive PCB [14]

5.3.2 Description of a Method for Measuring Radiated Fields inside a System Case

As discussed in Chapter 2, the concept of signal upset or soft failure refers to the kind of system crash due to signal integrity problems that can occur when a chassis is ESD pulsed according to the IEC spec. Most often this is due to electric or magnetic (E or H) fields injected into the system, despite chassis grounding efforts. Pulsed fields can be picked up by board wiring, loops, and other structures (such as heat sinks) inside the chassis, influencing signal integrity.

While the EM scanner, discussed above, examines the impact of local fields on component and system functionality, the actual IEC test or ESD event will inject fields globally into the chassis and influence all components at once. This is not an easy situation to analyze, so it is natural to want to find local sensitivities to local fields with the EM scanner. As a complement to the EM scanner, it is also useful to know the overall level and time-dependent behavior of field injection into the chassis. This is because the actual injected fields during ESD will usually not be localized, though it is these injected fields that produce many failures. A measure of the magnitude and time dependence of the internal chassis E and H fields could reveal whether field injection is high, low, or "reasonable" for the product. EM scanner results using local fields should be more meaningful in light of such measurements of global field injection.

If care is taken, internal field detection is possible using various methods of installing a transmission line and readout connections into the product [16, 17]. Micro-strip lines could be built into the PCB, but could also be retrofitted onto the chassis of a completed product. A demonstration setup showing these principles is shown in Figure 27.

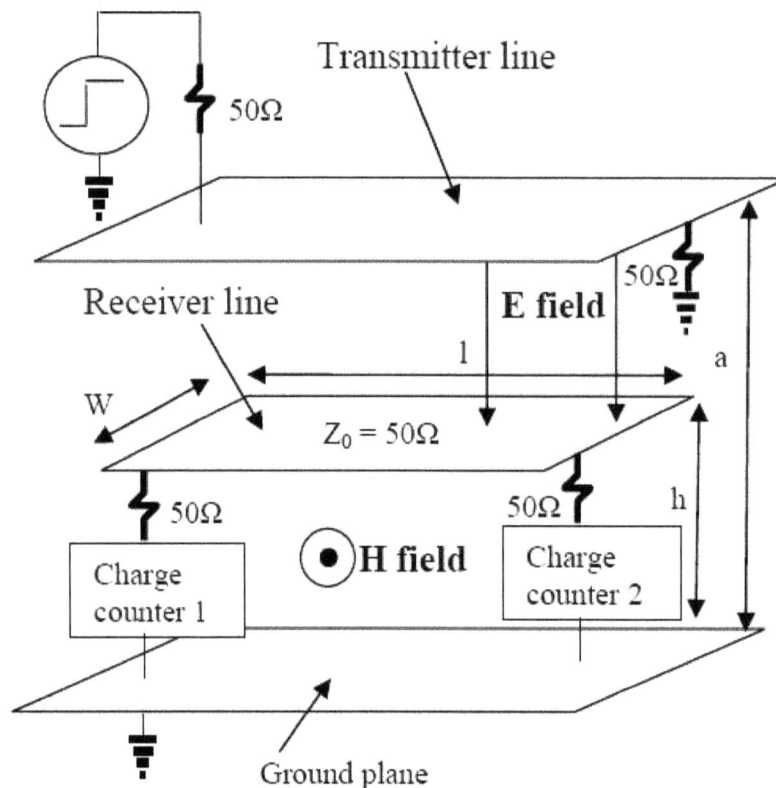

Figure 27: Transient electric (*E*) and magnetic (*H*) fields as described in [16]. Two transmission lines with characteristic impedance of 50 Ω are used, one as the receiver (middle line) and the other as a transmitter (top line).

Thus a micro-strip or similar line on a board can be terminated with 50 Ohms at both ends, and connected on one or both sides in order to read the fields (or, more accurately, the field derivatives) as described in [16, 17]. E-field and H-field strength on a given segment can be distinguished by noting the sum and difference of the signals on the two "charge counter" outputs. This kind of detector senses the time derivative of the field and the principles are well covered in review articles, for example [18].

If, as often happens, the product of interest is complete and there is no opportunity to design field detectors into the product from scratch, one can consider installing a micro-strip-like transmission line on the inside panel of the chassis. If space allows, a nearly ideal method is to use common 300 Ohm "twin lead" transmission line with 1 mm wires separated by 7 mm and embedded in plastic; this is familiar TV antenna wire. A cross-section is shown in Figure 28.

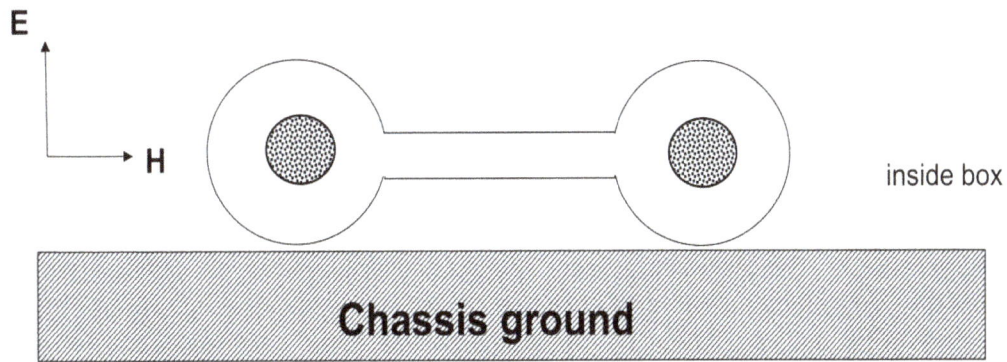

Figure 28: Cross section of twin lead wire (1 mm wires separated by 7 mm, embedded in plastic and sitting about 1 mm off the metal ground plane) mounted inside a chassis for use as an E-H field detector as in Figure 27.

Each of the two single wire lines above the ground plane (i.e., mirror plane) is equivalent to an almost perfect 100 Ohm parallel-wire line, so the two in parallel make a good 50 Ohm line with respect to the ground plane. Figure 29 shows a top view of such a detector wired to BNC connectors from the inside of a chassis. This arrangement has been shown to work very well as a field detector, and is briefly discussed in [16].

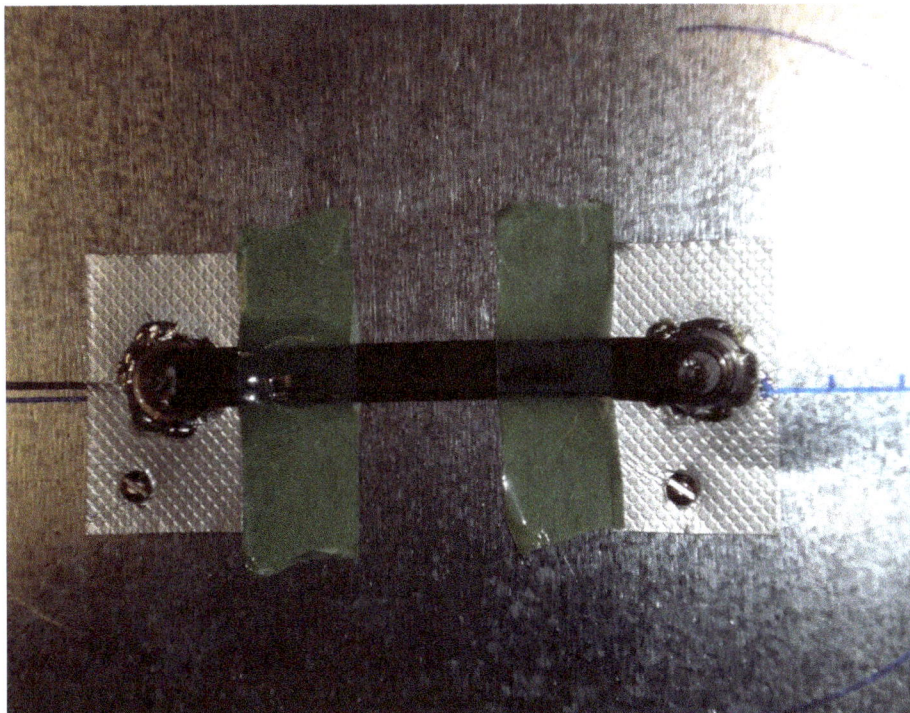

Figure 29: Top view of twin lead 50 Ohm T-line field detector, wired to BNC panel mounts.

An example of the use of this field detector follows. A new personal computer (PC) reference motherboard was installed into a standard oversize metal PC chassis but promptly failed at an unexpectedly low IEC ESD voltage when pulsed externally. The twin lead field detector was installed at a convenient place in the middle of the chassis as shown in Figure 29 and signals read out to 50 Ohm scope inputs. As expected, E and H field derivatives (determined through the sum and difference of the channels) were about the same in this location. But a fast Fourier transform (FFT) of the + and − 6 kV induced waveform yielded an interesting feature -- a sharply peaked resonance at 537 MHz (points were 50 MHz apart so the nearest were 488 and 588 MHz, both over 50% of the strength observed at 537 MHz). This box resonance just happened to be close to the 2nd harmonic of one of the IO clock frequencies (266 MHz). Apparently, the interior fields injected by the IEC pulser drove the clock transition board signal lines in the wrong direction every other cycle for as long as the pulse lasted. This sort of weakness would be hard to pick up with the EM scanner, as the chassis has to be opened and the local signal is usually a pulse. The field detector thus complements other methods of determining system ESD failure by serving as an in situ detector of exactly the kind of induced signals that could cause system upset.

In addition to the twin lead detector, it is clear that other single conductor or parallel wire schemes above a ground plane can be arranged to interface with convenient 50 Ohm outputs. For example, SMA connectors can be used for a more compact fit into a tight chassis. Today's products are a little tighter than yesterday's PC boxes, but lines made with copper tape on polyethylene tape are very compact and just need a carefully built connector interface.

5.3.3 Explanation of how these Methods can be used to Empirically Debug a System

Faced with the difficulty of a system failing a system level test, a variety of methods are suggested to identify the root cause of the failure. These methods approach the problem from three sides: system level testing, software, and near field susceptibility scanning. Please note that this list of methods is not exhaustive, other methods that are different or better may be available.

5.3.3.1 System Level Testing

System level testing cannot be reproduced well, yet measurements can be taken to obtain the most reliable information. A short list of considerations follows:

- Measure the discharge current during system level testing (such as in the ESD generator ground strap using an F-65 current clamp or similar) to test if secondary ESD has occurred. Secondary ESD will show up as a second pulse delayed by nanoseconds to microseconds after the first pulse and happens if a non grounded piece of metal is charged by the ESD causing a spark inside the product. All two-wire components are in danger of having secondary ESD. The secondary gap discharge can have a much higher current peak value and shorter rise time than the primary ESD which caused the secondary ESD. This, and the fact that secondary ESDs are often very close to the electronics of the product, emphasizes the need for controlling secondary ESDs [19].
- Test for threshold not for limits. One should not only determine if the product passes at 2 kV and fails at 4 kV, but determine the threshold of failure and its repeatability
- Apply a sufficient number of pulses. It is recommended to use a few hundred pulses in contact mode at each test point using 10 pulses a second. Of course, this recommendation may not be suitable for each product, as the standard assumes that the product can return to its baseline condition after each discharge. The baseline condition relates both to the removal of charge and to a baseline software status in case the ESD has triggered a recovery or a self correcting cyclic redundancy check (CRC) error.
- Observe the nature of the failure and report carefully (an example of careful reporting; "striping observed in the system display and then the display turns white").

- To test if a system is sensitive to the initial pulse of the ESD generator in contact mode or to the slower discharge of the RC network, replace the ground strap with a resistive wire having around 10 kOhm resistance. This way a charge return is still possible, however, the second pulse will have a very small magnitude. Of course, this only works with battery powered ESD generators.
- It is known that soft failures are often dependent on the ESD generator used [20], as each ESD generator will create a different spectral composition of its transient fields, especially at greater than 300 MHz. It is impossible to predict which test point will react more sensitively to which ESD generator, and it is usually not possible to identify the most severe ESD generator in general, as soft failures also depend on the spectral sensitivity of the DUT which varies at each test point. However, using two different ESD generators in contact mode might give an indication of the frequency range that could cause the ESD soft failure. If both results differ greatly (> 1.5x) then it is more likely that the soft failure is caused by the high frequency content of the ESD generator's current and transient fields, as ESD generators differ more in the high frequency range.

5.3.3.2   Software

Software [21, 22] can be a great tool for identifying the root cause. However, this usually requires access to the firmware and having a firmware engineer working in close cooperation with the test labs. Internal register information often documents IO failures very well. Further, bus analyzers can provide an insight into IO related errors (provided the bus analyzer is robust). If firmware is written well, it will provide clues about unexpected events by allowing access to internal registers or a bus analyzer if one is present; if it is written badly, it will just hang up or reset without reporting. In most cases, one will need to have firmware design guidelines that encourage the firmware designers to include reporting methods that can identify the root cause of ESD soft failures.

5.3.3.3   ESD Near Field Scanning

As mentioned in other parts of this white paper, ESD near field scanning injects voltages and currents into the system locally using field coupling or direct injection. If the board is scanned while observing for soft failures, one can identify the sensitive regions, nets, and ICs using near field scanning. Once those regions are identified, re-routing, filtering or software changes can be used to increase the system's robustness.

5.3.4   Software and Hardware Tool Limitations and Future Directions

The SEED methodology applies the nodal simulation concept of SOA current and voltage limits to an idealized circuit topology. In the future it could be expanded to address E and H field susceptibility limits through a full wave system simulation of an ESD strike. There are immediate computational and analytical limitations to this extension, just as there are limits to the utility and accuracy of the nodal analysis methods.

There are obvious directions in which the SEED methodology concepts could be extended. Full wave simulation of ESD pulse energy applied to the system level enclosure (Figure 30a) extends the nodal counterpart of HMM ESD circuit simulation models. 3D characterization with the susceptibility scanning methods (Figure 30b) described in Chapter 2 would be analogous to the TLP nodal characterization of components.

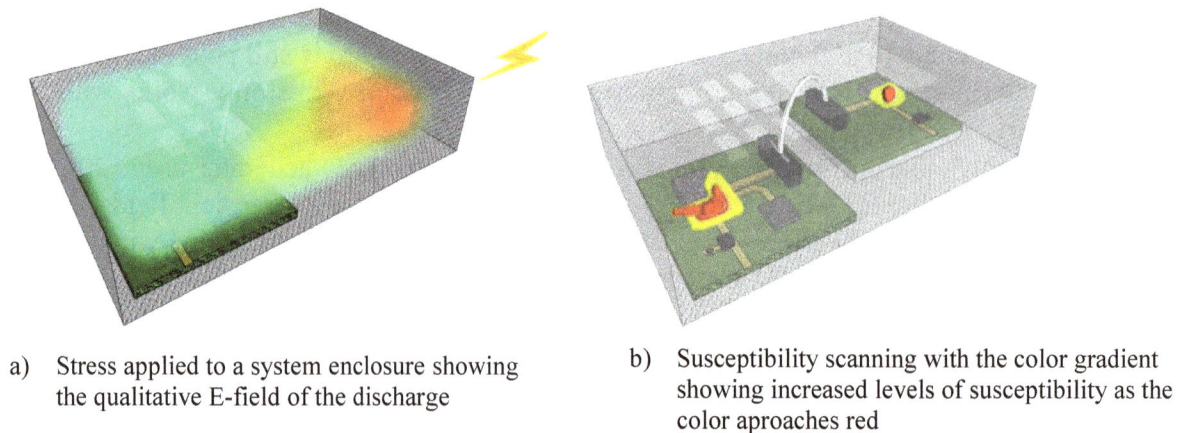

a) Stress applied to a system enclosure showing the qualitative E-field of the discharge

b) Susceptibility scanning with the color gradient showing increased levels of susceptibility as the color aproaches red

Figure 30: Examples of extensions to the SEED methodology

Overlaying these representations and using the susceptibility scans as a field-induced failure criteria mask against the expected field distribution of a given strike could form the basis of a complete virtual 3D component-to-system analysis before the first prototype is assembled. For example, in the simple representation of Figures 30a and 30b, two susceptibility "hotspots" are analyzed. In this particular case, shown in Figure 31, the fields induced affect the component on the right (red) and do not upset the component on the left (blue). Remedial design methods described in Chapter 3 can be carefully applied to mitigate this condition, minimizing the cost of the preventative measures as well as potentially cutting an entire prototype iteration from the design cycle. In this manner it may be possible to fully anticipate both hard and soft failures in simulation, as long as many of the practical and analytical limitations are overcome, such as the difficulty of creating a "component level" susceptibility map on an appropriate vendor evaluation board which is meaningful when integrated into an arbitrary system enclosure.

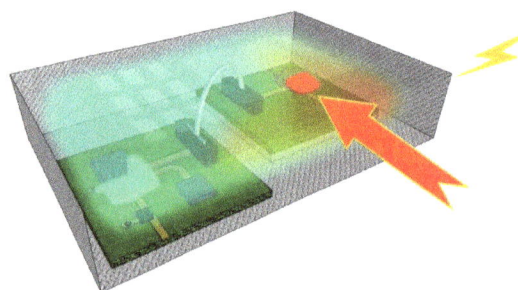

Figure 31: An example of how overlaying full wave simulation and susceptibility scanning can highlight a problem spot

At the moment, this type of analysis is beyond the capabilities of most ESD research, not to mention actual system designers. While the choice of an IBIS container can provide a platform for future expansion, the industry would welcome advanced research which can suggest an equivalent common platform for System Efficient 3D ESD Design [23].

Tools which are presently not available but within reach in the next 5 years are:

- Complete IC characterization and qualification methodologies for modules and ICs. While many sub-steps are available, there is no comprehensive method to completely characterize ICs and modules for ESD soft-failure robustness.
- Rule checker for enclosures. ESD soft failure and damage countermeasures depend on shielding, or on guiding the discharge current so that the current and associated fields do not cause unacceptable coupling. While full wave simulation can determine the current paths and fields, it is still a rather complicated method and requires very many model verification runs (we estimate that 90% of all full wave simulations are performed to verify model assumptions or simplifications). A rule checking tool that is able to import mechanical geometry files and estimate currents to flag weaknesses would reduce or eliminate the need for a full wave simulation.
- Rule checker for PCB for susceptibility. Knowing the sensitivity of the IC pins and using a matrix that connects the sensitivity of the more sensative pins with the likelihood of noise coupling (from an external IO, board to board connecting traces, long on board traces, and short on board traces) it should be possible to estimate the overall susceptibility of the PCB to coupling upset. Additional information can be extracted from the general enclosure structure and from the partitioning in different PCBs and their connections. Existing tools can extract electromagnetic PCB characteristics so that models can be created that include conducted coupling and estimated field coupling. The later might require a data base that estimates the field to net coupling by the field confinement of the structure (for example, a single layer flex has a very low field confinement). The needed field strength information can be taken from a full wave simulation at a highly simplified level or by the enclosure rule checker mentioned above. At best this will estimate the sensitivity and at least it can flag risky design choices.
- Software based analysis. It is often difficult to determine the root cause of ESD soft failures without information from the inside of the IC. This inside information can be obtained through software by monitoring things like register values and errors on internal buses.
- Simulation suite including standardized models.

5.4   Conclusion

In this chapter, we presented several methods for characterizing components and proposed models to support the SEED methodology. Using presently available models and simulation tools, ESD behavioral models can be built and simulation of components can be carried out to debug hard failures (Category 1) of components in a system and to optimize a system under design.

Next, we investigated extending these models to cover the much harder to find and correct collateral damage and soft failure or system upsets often found in Category 2.

For both categories, TLP is suggested as a powerful tool for characterization of IC pins and PCB discrete elements like TVS diodes or ferrites.

Finally, we addressed Category 3. We recognized that the present simulation tools use primarily nodal type analysis and any system is much more complex. 3D system models were discussed that would allow modeling of complex coupling due to H and E fields present in real designs.

These involve using scanners to map susceptibility of components in the system. In closing, we gave ideas for future tools and models that could be used to simulate and predict system behavior from proposed board design and component selection.

References

[1]   Weiying Li, Yu Tian, Linpeng Wei, Chai Gill, Weil Mao, Chuanzheng Wang, "A Scalable Verilog-A Modeling Method for ESD Protection Device", 2010 EOS/ESD Symposium, pp. 1-10

[2]   Patrice Besse, Fréderic Lafon, Nicolas Monnereau, Fabrice Caignet, Jean-Philippe Laine, Alain Salles, Sophie Rigour, Marise Bafleur, Nicolas Nolhier , David Trémouilles, "ESD system level characterization and modeling methods applied to a LIN transceiver", 2011 EOS/ESD Symposium, pp. 329-337

[3]   T. Smedes, J. van Zwol, G. de Raad, T. Brodbeck, H. Wolf, "Relations between system level ESD and (vf) TLP", EOS/ESD Symposium 2006. pp. 136-143

[4]   Patrice Besse, Jean-Philippe laine, Alain Salles, Mike Baird, "Correlation Between System Level and TLP tests Applied to Stand-alone ESD Protections and Commercial products", 2010 EOS/ESD Symposium, pp. 151-156

[5]   Yiqun Cao, Ulrich Glaser, Joost Willemen, Filippo Magrini, Michael Mayerhofer, Stephan Frei, Matthias Stecher, "ESD Simulation with Wunsch-Bell based Behavior Modeling Methodology", 2011 EOS/ESD Symposium, pp. 187-195

[6]   K.Wang et al., "Numerical modeling of electrostatic discharge generators," in IEEE Trans. Electromagnetic Compatibility, vol. 45, no. 2, pp. 258- 271, May 2003

[7]   S. Caniggia and F. Maradei, "Circuit and Numerical Modeling of Electrostatic Discharge Generators," IEEE Trans. Ind. Appl., vol. 42, no. 6, pp. 1350-1357, Nov. 2006

[8]   R. Mertens et al., "A Flexible Simulation Model for System Level ESD Stresses with Application to ESD Design and Troubleshooting," in Proceedings of EOS/ESD Symposium, 2012, pp. 1-6

[9]   N. Monnereau; F. Caignet; N. Nolhier; D. Trémouilles, "Investinating the probability of susceptibility failure within ESD system level consideration", EOS/ESD Symposium 2011, September 11-16, 2011, pp. 1-6.

[10]   N. Monnereau; F. Caignet; N. Nolhier; D. Trémouilles; M. Bafleur, "Behavioral-modeling methodology to predict electrostatic-discharge susceptibility failures at system level: an IBIS improvement", Electromagnetic Compatibility (EMC) Europe, York, Uk, September 26-30, 2011, pp. 457-463.

[11]   K. Muhonen, J. Dunnihoo, E. Grund, N. Peachey, A. Brankov, "Failure Detection with HMM Waveforms", EOS/ESD Symposium 2009, 5B.3, pp. 1-9.

[12]   Jiang Xiao, D. Pommerenke, Fan Zhou, J. L.Drewniak, H. Shumiya, T. Yamada, K. Araki, "Model for ESD LCD upset of a portable product", 2010 IEEE International Symposium on Electromagnetic Compatability (EMC) , pp. 354-358

[13]   Muchaidze, G., Jayong Koo, Qing Cai, Tun Li, Lijun Han, Martwick, A., Kai Wang, Jin Min, Drewniak, J.L., Pommerenke, D., "Susceptibility Scanning as a Failure Analysis Tool for System-Level Electrostatic Discharge (ESD) Problems", IEEE Trans. EMC, May 2008, Vol 50, No. 2, pp. 268-276

[14]   zur Nieden, F.; Scheier, S.; Arndt, B.; Frei, S.: *"Gefährdung von integrierten Schaltungen durch Überkopplungen zwischen Leiterbahnen auf Platinen"*, ESD-Forum 2011, Munich, Germany, 2011

[15]   zur Nieden, F; private communication

[16]   T. W. Chen, T. J. Maloney, and B. Chou, "Detecting E and H Fields with Microstrip Transmission Lines", 2008 EMC Symposium, August 2008, pp. 1-6

[17]   T. J. Maloney, A. Martwick, K. Wang, "Measuring Electric and Magnetic Field", US Patent application, filed Aug. 30, 2007. Issued as US Patent 7,750,629, July 6, 2010.

[18]   C. E. Baum, E. L. Breen, J. C. Giles, J. O'Neill, and G. D. Sower, "Sensors for Electromagnetic Pulse Measurements Both Inside and Away from Nuclear Source Regions", IEEE Trans. Electromagnetic Compatibility, Vol. EMC-20, No. 1, pp. 22-35, February 1978.

[19]   Xiao, J.; Pommerenke, D.; Drewniak, J. L.; Shumiya, H.; Maeshima, J.; Yamada, T.; Araki, K.; , "Model of Secondary ESD for a Portable Electronic Product," Electromagnetic Compatibility, IEEE Transactions on , 2011, pp.1-10

[20]   Jayong Koo; Qing C,ai; Kai Wang; Maas, J.; Takahashi, T.; Martwick, A.; Pommerenke, D., "Correlation Between EUT Failure Levels and ESD Generator Parameters", IEEE Trans. EMC. Vol.50, Issue 4, Nov. 2008 Page(s):794 – 801

[21]   Maheshwari, P.; Tianqi Li; Jong-Sung Lee; Byong-Su Seol; Sedigh, S.; Pommerenke, D.; , "Software-Based Analysis of the Effects of Electrostatic Discharge on Embedded Systems,", 2011 IEEE 35th Annual Computer Software and Applications Conference (COMPSAC), pp.436-441, 18-22 July 2011

[22]   Maheshwari, P.; Byong-Su Seol; Jong-Sung Lee; Jae-Deok Lim; Sedigh, S.; Pommerenke, D.; , "Software-Based Instrumentation for Localization of Faults Caused by Electrostatic Discharge,", 2011 IEEE 13th International Symposium on High-Assurance Systems Engineering (HASE), pp.333-339, 10-12 Nov. 2011

[23]   Liu, Dazhao; Nandy, A.; Zhou, F.; Huang, W.; Xiao, J.; Seol, B.; Lee, J.; Fan, J.; Pommerenke, D. "Full-Wave Simulation of an Electrostatic Discharge Generator Discharging in Air-Discharge Mode Into a Product " IEEE Trans. EMC, Vol.99, 2010, pp. 28-37

Chapter 6: Summary and Conclusions

Robert Ashton, ON Semiconductor
Charvaka Duvvury, Texas Instruments
Alan Righter, Analog Devices

6.0   Introduction

This white paper has sought to capture the state of the art in system level ESD test and design and plot a course which will lead to improved system level ESD robustness while reducing the costs and effort to obtain this robustness. While today there is considerable knowledge of system level ESD, it is far from a mature science, and system level ESD design is often done with rules of thumb, trial and error and a combination of good and bad luck. Some of the design methods established in this paper seeks to minimize the often *ad hoc* approach to achieve system ESD robustness. This final chapter of the white paper will summarize what has been learned in the previous chapters and outline the advances which need to be made in the field.

6.1   Overview of ESD Stressing and System Response - Chapter 2

There are a number of system level ESD tests for different classes of products and applications. Most tests are based on the IEC 61000-4-2 ESD standard. The common elements of the test methods are to stress the completed systems with an ESD pulse source (ESD gun) using a combination of contact and air discharge directly to the system and indirectly to the coupling planes adjacent to the system under test. The tests look for both hard, physical, failures and soft failures in which system operation is interrupted. These tests are obviously very effective, since most electrical systems on the market have sufficient ESD robustness that the average system users are unaware of the threat from ESD.

The challenge today is when systems fail the ESD tests, usually during system qualification. Determining the path of the ESD stress of a hard failure is relatively straightforward since there is the physical damage of the failure. Diagnosing soft failures is much more difficult. In recent years there have been significant advances. There are now scanning technologies which can be used for diagnostics. These technologies can be classified into two general methods. In the first method, the ESD stress which leads to a failure can be applied repeatedly to a circuit board; probes sensitive to voltage or current scan the board surface to measure the path of the stress though the board. In the second method, probes which can inject electric or magnetic fields are scanned over the surface of a functioning circuit board to find where on the circuit board there is sensitivity to the applied fields.

These methods can be very effective in finding susceptible locations on a circuit, but they can be very labor and capital intensive. Scanning techniques are also most effective on designs with a known ESD or EMC weakness. It would be much more effective to be able to design a system to be robust the first time, rather than have to debug a failing system. Improvements in system design techniques are needed to avoid the need for debugging failures.

6.2 State-of-the-Art ESD/EMI Co-Design - Chapter 3

ESD/EMI design is similar to most design problems, requiring a balance between often conflicting goals. In ESD/EMI design, the conflicting goals are performance, robustness and cost. ESD/EMI design consists of identifying exposed and susceptible internal nets, evaluating the pulse entry point and selecting the components and layouts that will create a robust design. The design processes used to meet these goals will vary based on the complexity of the design, the experience of the designers and the available design resources. ESD/EMI design styles can be divided into practical and theoretical approaches. Each approach can range from very basic to comprehensive as illustrated in Table 4, (reproduced from Chapter 3).

Table 4: Practical and theoretical co-design methodologies for ESD/EMI robustness

		Methodology Type	
		Practical	Theoretical
Complexity & Cost	Basic	Designer's experience, lessons learned, design reuse	Datasheet comparisons
	Advanced	Component qualification testing	SPICE modeling of ESD injection
	Comprehensive	Rigorous head-to-head system level iterative testing	Full 3D Maxwell Field-Solver Simulations

As computer power has increased and electrical components and systems have become more complex, there has been a movement in design toward the use of more theoretical computer based simulations during the design process. The progress toward a theoretical approach is hampered in the ESD/EMI field by a lack of fundamental information available to the designer. This exists not just at the Full 3D Maxwell Field-Solver level but also at the SPICE modeling and even at the datasheet comparison level.

At the datasheet level, different manufacturers specify the protection properties of TVS components in very different ways. ESD survival levels are often quoted as an IEC 61000-4-2 voltage while others quote a voltage but do not reference the standard used for the test. Dynamic on-resistance of the components in the on state are often measured with different methods such as TLP or an IEC 61000-4-5 8/20 µs pulse; however the resistance measured from one method to the other is likely to differ dramatically. Most datasheets, however, do not specify a dynamic resistance at all. The datasheets for the integrated circuits requiring protection are even less helpful. Some integrated circuit datasheets do quote IEC 61000-4-2 survival levels for some interface IO pins. Integrated circuit datasheets often specify ESD robustness levels for HBM and CDM as well. The HBM and CDM tests are relevant for component handling during system manufacture; however, they are very misleading for predicting system level ESD performance. The relevant information of turn-on voltage or breakdown voltages and dynamic on-resistances are never quoted on datasheets and are often not known by the manufacturer of the circuit.

Attempting to do SPICE modeling is even more problematic. SPICE models are seldom available from manufacturers for TVS components. SPICE models of the properties of an integrated circuit appropriate for use in a system level ESD simulation are even rarer. A critical need in the ESD design environment is learning the properties of electrical components which are important during a system level ESD event. It is also important that the form of this information be similar from supplier to supplier so that meaningful comparisons can be made. The component information must also be in a form that is appropriate for the tools being used. A SPICE model may not be much help to a designer looking for a simple datasheet comparison but is perfect for a designer wanting to do a circuit simulation.

A standardized specification for data, characterizing electronic components in the system ESD range of currents, voltages and stress duration, is a high priority. The format that such a specification should take depends critically on the tools that will use this information. There may need to be more than one format. The most basic format would be standardized datasheet parameters. Next would be a specification for high current SPICE models for TVS components and integrated circuits appropriate for circuit modeling of system level ESD events. The same would apply to passive components in the ESD mitigation path like ferrites. The appropriate form needed for 3D field simulations may need to be developed as this method of simulation matures.

6.3   Reference Methodologies for IC/System - Chapter 4

In White Paper 3 Part 1, the Industry Council on ESD Target Levels introduced the SEED concept and this document has been expanding on this concept. Developing more systematic approaches to system level ESD design has required the understanding and classification of how ESD can interact with the system and damage or interrupt system performance. The types of stress which can cause ESD issues are listed below.

- Category 1: High energy stress which can lead to physical failure
    - Category 1a: Stress delivered to IC from galvanic conduction
    - Category 1b: Stress delivered to IC without a direct galvanic path
- Category 2: Stress entering IC substrate which can cause latch-up or upset
- Category 3: Low energy pulses coupled to IOs which may cause upset

The system design techniques required to deal with each type of stress can differ. Understanding that the approaches need to be different is a first step toward defining an efficient co-design strategy. It is also important to know the limitations which various design approaches will have.

The classic SEED approach, as described in White Paper 3 Part 1, is directly applicable to the Category 1a stress. For Category 1a the stress entry point, current paths and susceptible circuits can usually be identified based on the standard circuit diagrams for a system. It is then conceptually straightforward to model the system's response to the ESD stress. This does not mean, however, that the Category 1a stress condition is totally understood and easily modeled. The major bottleneck today, as pointed out earlier in this chapter, is the lack of model files for sensitive ICs and TVS components.

Category 1b can also be addressed with the SEED approach, with the added requirement of understanding how the high energy stress enters the system. Identifying the entry path can be a challenge and is easiest to determine if the time and circumstances of failure are well known. Indirect high energy stress of a system can result from several events, such as:

- ESD stress to a metal case which then arcs to a sensitive circuit node
- ESD stress to an IO line which induces a large transient, either capacitively or inductively, to a nearby internal trace.
- ESD air discharge to a keypad or seams in the plastic case of a system
- Coupling of energy from a system housing ESD strike into internal IC bondwires

Once the entry point of the stress is identified it is possible to use the same SEED techniques as used for Category 1a stress. The additional information needed is a model for the input path. This could require a model for an arc between the ESD source and the entry point on the PCB or a capacitive or inductive model for the coupling between signal traces on the PCB.

Being able to design to protect against Category 2 stress requires an understanding of how current injected into the substrate of an IC can cause upset or latch-up. Today, there are no accepted methods for characterizing integrated circuits for their behavior when current is injected into their substrate. Development of such a test method and then how to model the ICs response is needed before Category 2 stress behavior can be predicted.

Category 3 stress levels are similar to the stress levels for EMC compatibility and advances in the robustness for EMC will also make systems more robust to lower energy strikes from low level ESD stress.

6.4   Standard Model and Analytical Tool Needs to Support SEED - Chapter 5

As has been discussed numerous times in this chapter and document, the ability to predict the behavior of systems to ESD stress requires detailed knowledge of the behavior of integrated circuits and TVS components in the voltage, current and time domain of ESD events. This component understanding must also be in a form which can be used by simulation tools used by system designers. SPICE models have been mentioned numerous times in this document. SPICE models are used for system and board design but are not the only tool. IBIS (Input/output Buffer Information Specification) models are used extensively for system and board design. IBIS models do not presently have the necessary features required for ESD system design. The necessary additions to the model need to be defined and procedures developed for extracting the relevant parameters from integrated circuits, TVS components and other components as well as the necessary PCB parameters such as coupling between traces.

The primary tools for determining component properties in the ESD range of voltage, current and time are standard and very fast TLP test methods. Standard TLP, with pulse lengths of 100 ns, is the ideal tool for obtaining I-V curves that describe component behavior during the long second peak of a system level ESD pulse, which carries the bulk of the energy of an ESD stress. VFTLP uses pulses in the range of 2 to 5 ns which have rise times similar to or faster than a system level ESD stress. VFTLP produces I-V curves which are characteristic of a component's performance during the initial current spike of a system level ESD event. VFTLP can also be used to look at the time dependence of ESD protection circuits and components within the critical first nanosecond of an ESD event.

The TLP and VFTLP tools are of no use, however, if the needed component properties are not well defined. The behavior of TVS components, which can often be analyzed as two terminal components, may be straightforward. One area that will need improvement is the analysis of the turn-on characteristics of TVS components. The requirements for integrated circuits are more complex. The important current paths within an integrated circuit may not be well defined. Power and ground pins are usually thought of as being very isolated nodes. The large power-to-ground capacitors, which are routinely placed to filter out power noise spikes in a system, will look more like a short to an ESD pulse than separate nodes. Chapter 5 showed how TLP curves for an IO vary depending on how power and ground pins are handled. The correct measurement conditions for predicting integrated circuit behavior during a system ESD event will require more experience than is available today.

6.5   Application Specific Information on System ESD Related Tests and Their Targets – Appendix A

Different product applications are used in a number of varying environments, and as such have different system ESD test standards tailored to those environments. Similar design strategies for ESD should be applicable across the application spaces, but the varying requirements and environments must be taken into consideration. Most test methods for system level ESD are based on IEC 61000-4-2. Some applications might benefit from re-evaluation of the particular ESD test method use to better match use and stress environments.

For example, mobile phones are covered by a wide variety of EMC standards, but most ESD testing is done using IEC 61000-4-2. Testing of small hand held devices based on IEC 61000-4-2 may give a different stress than actual stress in their normal environment, and might need some updating, since the IEC test standard was developed with an assumption of a table top system, rather than a portable system which is typically used when electrically floating.

The automotive industry is one which has developed a test standard, ISO 10605, which is substantially modified from IEC 61000-4-2. Automotive electronics also operate in an electrically noisy environment and are very susceptible to coupling of signals due to long wiring harnesses and a multitude of separate electronic control units (ECUs) distributed throughout the vehicle.

The avionics industry is even more susceptible to coupling between systems due to the very long lengths of wire harnesses which exist in modern aircraft. The electrical environment is also rapidly changing in the aircraft environments as airplanes are increasingly using new materials and moving more and more to electrical control of the aircraft.

The challenge is to enable by SEED a common, unified design approach, which can be used across the industry, allowing optimum use of devices and minimizing cost and time to market.

6.6   Technology Roadmap and Direction – Appendix B

Even as better methods for system level protection are understood and implemented in a more optimized manner, technologies will continue to advance at a faster rate. As an overview, these advances in IO speeds, broader system applications, and innovative developments from 3D IC packages to complex integration of multiple functions on a single board make the continued demand of IEC protection even more challenging. But at the same time there are other innovations such as optical interconnects and polymer transient suppressors that can alleviate some of these difficulties. However, the real sense of the IEC requirements needs to be understood and perhaps revised in the context of where the trade-off begins to play a role in performance versus robustness.

Looking more into details, the ever-increasing circuit complexity and scaling of silicon integrated circuit technologies is placing increasingly severe restrictions on the effectiveness of system level ESD protection components, and design methodologies employing these new circuits. Analysis of the present state-of-the-art in technology as well as a forecast in technology trends in the components which make up a system, along with the use of methods described in this document, is critical to system level ESD success in future systems.

Advances in IC technology from the present state-of-the-art at the 40 nm process node to the 22 nm process node and beyond are forecast in the next five years. The gate oxide breakdown voltages in these technologies are comparable to junction breakdown voltages, and circuit ESD protection paths must be carefully designed to maintain the ESD levels above those levels which the assembly factories can guarantee by ESD factory control. It is critical as well that IO and power supply protection continue to advance at the lower operating voltages to allow efficient system level ESD protection. Faster circuit speeds resulting from these technology advances will lead to faster IO bus protocols. USB frequencies of 15-20 Gbit/s will place increasingly severe limits on the range of system level protection that can be used. HDMI, which has the additional requirement of minimal signal swing at 3-6 GHz operation, will also see an increasingly severe environment for system ESD.

Regarding advances in automotive technology, the trend of increasing use of electronics to make automobiles safer and user-friendly will continue. The electronics in automotive systems tend to lag in technology behind their computer technology counterparts by 1-3 process generations. However, technologies of 65 nm and up will be increasingly used in automotive electronics in the next five years in applications such as digital signal processing (DSP) in audio systems, micro-electrical-mechanicals (MEMs) devices, voltage regulators and LIN / CAN automotive communication networks.

Advances in packaging technology resulting from 3D interconnects (such as through silicon vias (TSVs)) connecting multiple IC overlapping die, as well as enabling more portable system on a chip circuitry, will make system level ESD protection and failure diagnosis more difficult. More efficient scanning and diagnosis techniques such as those described in this document will go a long way towards improvement of system level ESD protection for both hard and soft failures. One of the complex issues for future technology will be the EMI effects of stacking die. In the case of two die or three die stacking, the dominant effects on EMI coming from the interposer design and the die to die coupling needs to be considered. These become more complex as multiple die are stacked and could greatly impact the design of system level ESD protection in future applications.

Although traditional FR4-based boards and surface mount technologies will continue to dominate the system board / interconnect market, advances in board and interconnect technology such as molded interconnection device (MID) for 3D board design and roll-to-roll (R2R) flexible system board technology will become more common. These new methods for integration of system level ESD protection technology in new system environments will become a development priority.

Polymer-based ESD suppression components are also finding new applications in system level ESD protection. Although their turn-on characteristics and component operating characteristics can vary with multiple ESD strikes, they are finding applications as secondary protection for existing system ESD protection strategies.

Finally, it can be said that system level ESD design will continue to ride the wave of technology developments into the future. While developments that may relieve system exposure to ESD threats will occur, the advances and complexities of system design may outweigh these design developments; with the net effect that the system level ESD design will continue to become more challenging as the system performance level increases. Thus, one might still need to consider the following: "What is the required "correct" level of system level ESD robustness for an application?"

6.7　Outlook

System level ESD will continue to be a challenge in the future, but as this white paper has shown, there are a number of techniques which can be used to improve the success rate of system level ESD design. Tools such as TLP and VFTLP should be used to understand the properties of both sensitive circuits and the components used to protect them. Scanning tools, as explained in Chapter 2, provide useful insight into the stress propagation on boards and within systems and enable detection of sensitive circuit nodes. These tools will not reach their full potential unless the data they collect can be used within the standard design flow for an electronic system. This requires that design tools can use the additional ESD data obtained from TLP measurements and scanning tools. EDA vendors must therefore enhance their tools to use the new information. Model files such as IBIS and SPICE, which describe the electrical properties of components, need to include the description of component behavior in the ESD range, which is well beyond normal operating range. Suppliers of components, from integrated circuits to ESD protection components, also need to characterize their products in the ESD range. This characterization must be compatible with the EDA tools and enhanced model files. None of this will happen without a great deal of communication between EDA tool vendors, component suppliers and system designers. This white paper, however, provides a major step towards understanding the technical challenges. Based on this, the industry can move forward to an aligned system level ESD design concept assuring a high first pass success rate at minimum system cost.

Appendix A: Application Specific Information on System ESD related Tests and their Targets

Harald Gossner, Intel Corporation
Leo G Henry, ESD-TLP Consultants
Ghery Pettit, Intel Corporation
Wolfgang Reinprecht, Austria MicroSystems
Pasi Tamminen, Nokia
Vesselin Vassilev, Novorell
Joost Willemen, Infineon

A.0   Introduction

This Appendix addresses the typical stress conditions and design solutions used in various electronic systems including mobile phones, medical devices, computers, consumer electronics, automotive components and avionics. It provides an overview of interference test standards to be applicable beyond even IEC 61000-4-2, and discusses the impact of the multiple stress requirements on the design solutions. In addition, the applicability of the previously introduced SEED concept is analyzed for the various types of systems.

A.1   System Tests and Targets of Mobile Phones

A.1.1 Overview of System ESD Related Mobile Phone Tests

ESD qualification is part of the suite of EMC tests for mobile phones. EMC/ESD immunity and emission acceptance tests are based on international and national standards. These standards specify measurement setups, frequency ranges and corresponding acceptance limits. Over 40 standards for immunity and emission qualification exist for mobile phones. A number of these commonly used standard families are listed below:

- IEC 61000-4-2 [1]
- CFR 47 §xx.xx [2]
- EN 300 xxx [3]
- EN 301 xxx [4]
- EN 55020 [5]
- EN 55024 [6]
- TS 25.xxx [7]
- TS 36.xxx [8]
- TS 51 xxx [9]
- TS 151 xxx [10]
- GBT 22450.1-2008 [11]
- YDT 1592.1 [12]
- PTCRB NAPRD.03 [13]

Mobile phones can be exposed to all possible hostile environments where people move. However, strong direct ESD discharges from the environment through the mobile phone to the earth ground are not common as mobile phones typically are not grounded during operation. There is a greater possibility to experience discharges between a phone and a person and between the phone and a cable (CDE). To guarantee operation over these environments, phone ESD acceptance tests are performed in the worst scenario environment where the phone is close to the earth and may have also high inductive ground cables in place.

Mobile phone system and device level ESD measurements are adopted from the ETSI standard, ETS 300 342-1 [14]. The standard stress for electrostatic discharge is IEC 61000-4-2, and typically level 4 requirements must be fulfilled with the end products. In the IEC 61000-4-2 test setup, the mobile phone is placed on a table in different positions and has a charger and accessory cables attached. The phone is typically in connection with a base station and operation is monitored in real time during the stress tests. Contact discharge testing to conductive surfaces of the system and to coupling planes is performed at 8 kV. Lower stress levels are also commonly tested, from 1 kV upwards. Air discharge testing to insulated or non-conducting surfaces is performed up to 15 kV. Each stress point is stressed multiple times with both polarities. Figure A1 depicts the two principle testing methods of contact and air discharge. The total ESD strike count is typically hundreds of pulses.

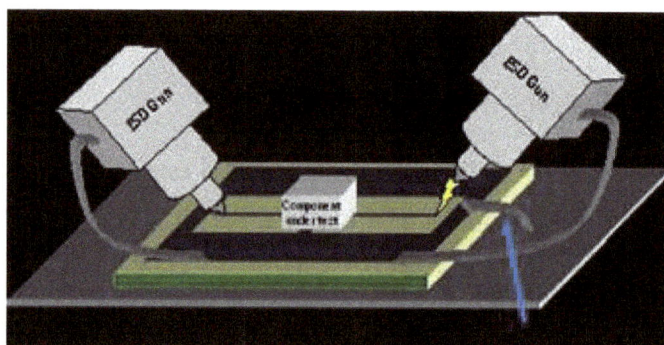

Figure A1: Principle picture of contact and air discharges with an ESD gun often used to characterize PCB designs during development

System level ESD testing is often required for those IC components themselves, within the mobile phone, which have interface pins. Interface connections with single devices are qualified with the system level ESD waveform by using case specific test boards. The test board must have all the important transmission lines; antenna, transmitter (TX), receiver (RX), control, power supply and ground connections in place. The test board can also have all those external components in place which are needed for the IC components' operational testing between the stress pulses.

A.1.2 State of the Art System Design Concept

The main challenge with mobile phones is to meet the system level EMC requirements. The product itself contains several radio systems which have to operate simultaneously without disturbing each other. There may be 4 to 6 additional radio-antenna sub-systems within the phone operating at the same time in a very small space. The main purpose of system level EMC design is to get product internal and external EMC challenges under control. Also, certain interface pins and covers may get direct ESD discharges which may disturb system operations.

ESD protection can be divided into two main areas, interface pin protection and indirect / direct ESD/EMC stresses on the cover area.

A.1.2.1 Interface Pin Protection

Interface pins, which are directly stressed in a mobile phone, have both on-board and on-chip protection for system ESD protection. On-board protection components also contribute to EMI attenuation. Interface connections include USB, HDMI, subscriber identity module (SIM) card, accessory connector, audio plug, memory card and charger plugs. These connections require case specific matching, and here simulation tools and SEED is the most beneficial way to optimize the design.

A.1.2.2 Indirect and Direct ESD/EMC Stress on the Cover Area

ESD/EMI can leak inside the phone when product covers are stressed with IEC 61000-4-2 contact and air discharges. Covers have gaps and the air discharges can leak inside through the openings. Metallic parts (covers, decoration marks, frames, etc.) may also transmit RF noise onto electronics and cause EMI problems. These challenges are typically controlled with detailed design rules for mechanics and grounding principles. Some internal and interface connections can also get extra stress when discharge energy conducts or radiates onto board traces. Another way to estimate EMC/ESD noise levels is to use 3D simulations where the mechanics and the stress waveforms are modeled in detail.

A.2 System Tests and Targets of Automotive Electronic Components

A.2.1 Overview of System ESD Related Automotive Electronics Tests

Automotive ECUs are subjected to numerous tests to ensure proper functionality in the particularly hostile vehicle environment.

The first group of tests is based on the test specs that are issued by standardization bodies. These publicly available standards define test methods and test stress criteria; however, pass/fail limits are not fixed by these documents but depend on the particular application. In addition many OEMs and Tier 1 suppliers have developed their own "in-house" spec or "consortium" specification documents. These documents often refer to the above mentioned public test specs, but also contain the required pass/fail limits for the particular application.

First, the public specs will be discussed. For system level ESD, two test standards are applicable:

1. IEC 61000-4-2 *Electromagnetic compatibility (EMC)—Part 4: Testing and measurement techniques—Section 2: Electrostatic discharge immunity test*. This document defines which ESD tests are to be applied to electronic equipment in general.
2. ISO 10605 [15] – *Road vehicles – Electrical disturbances from electrostatic discharge*. This document defines which ESD tests are applied to components for vehicles as well as vehicles.

The ESD generators used in both test specs are similar. ISO 10605 defines two different discharge networks for testing against ESD stress from persons outside or inside a vehicle. Although both test specifications define that ESD zaps should be applied to all kind of surfaces that can be touched by a user, often the same test methods are being used to stress individual pins of electronic components (i.e. ECUs, sensors).

In addition to ESD tests, various other EMC tests are required. There are differences between tests that are applied on the IC level and tests that are applied to the system level component (ECU, sensor) and the ECU-system level (ECU with peripheral). The expectation of the component level tests is that if a component is EMC robust, the system built with this component will be EMC robust as well, which is not the case.

Examples of IC level EMC tests are:

- IEC 61967 [16] – *Integrated circuits - Measurement of electromagnetic emissions, 150 kHz to 1 GHz*
- IEC 62132 [17] – *Integrated circuits - Measurement of electromagnetic immunity, 150 kHz to 1 GHz*

Examples of system level EMC tests are:

- ISO 7637 [18] – *Road vehicles – Electrical disturbances from conduction and coupling.*
- ISO 16750-2 [19] – *Road vehicles – Environmental conditions and testing for electrical and electronic equipment – Part 2: Electrical loads. Start profile/load dump, formerly part of the ISO 7637 standard (test pulse 4a, 4b, and 5).*
- CISPR 12 [20] – *Vehicles, motorboats and spark-ignited engine-driven devices – Radio disturbance characteristics — Limits and methods of measurement*
- CISPR 25 / EN 55025 [21] – *Radio disturbance characteristics for the protection of receivers used on board vehicles, boats, and on devices - Limits and methods of measurement.*
- ISO 11451 [22] – *Road vehicles – Vehicle test methods for electrical disturbances from narrowband radiated electromagnetic energy*
- ISO 11452 [23] – *Road vehicles – Component test methods for electrical disturbances from narrowband radiated electromagnetic energy*

The different parts of the Society of Automotive Engineers (SAE) J1113 standard [24] describe equivalent test methods, as in the above mentioned standards, and are commonly used by the US automotive industry. The ISO standards are referred to in the SAE J1113 documents.

The Automotive Electronics Council (AEC) has compiled (in the AEC Q100 group of standards [25]) many stress tests applicable to automotive systems and components. Included are ESD and EMC tests, however, system level ESD testing is presently not a standard within AEC Q100. The test methods refer mostly to public test specs as mentioned above. But in the case of device level ESD tests, AEC Q100 provides its own test method, which is different in detail compared to the commonly used JEDEC and ESDA standardized tests.

Numerous car system suppliers and car makers have issued their company specific ESD and EMC requirements. For example, Ford's specification is publicly available under www.fordemc.com [26, 27]. In other cases, Tier 1 or OEM consortia issue joint specification documents. For example:

- Generic IC EMC Test Specification. Bosch, Infineon, Continental (BISS)
- OEM Hardware requirements for CAN, local interconnect network (LIN) and FlexRay interface (consortium of German automotive OEMs) [28].

These documents refer mostly to public test specifications and also contain specific pass/fail requirements. For the most part, these documents are not publicly available.

A.2.2 State of the Art System Design Concept

Considerable effort in IC and ECU development is done to meet stringent system level ESD requirements included in the EMC qualification of ICs and ECUs. To discuss the commonly used strategies in the system design for EMC, it is necessary to differentiate between different interface pin categories. The following categories illustrate the main strategies for protecting different pin types in typical automotive ECUs. The applicable strategy followed depends on the functionality of the pin in question.

1. Supply lines (connected to the car battery):
   a. Local (PCB level) load dump clamps
   b. Reverse polarity (due to reversal of battery connection) protection diodes
   c. Supply line decoupling capacitors
   d. System level ESD protection clamps (TVS, varistors)

2. Power switches (connected to electromechanical actuators and indicator devices, such as lighting bulbs, motors, relays, displays, etc.):
   a. Self protecting switches: Power output switches which are large enough to divert system level ESD zaps
   b. Board level EMC capacitors: If switching speed allows usage of EMC capacitors at the switch output, they contribute to diverting system level ESD zaps
   c. System level ESD protection clamps: TVS diodes and varistors, for example, must be used if the switching device is not robust against system level ESD zaps

3. Transceiver buses (LIN, CAN, Flexray automotive communication bus systems):
   a. IC level ESD protection: Many transceiver devices or integrated transceiver modules are equipped with IC-internal system level ESD clamps
   b. System level ESD protection clamps: Are to be added if the IC-internal ESD protection is not sufficient for system level ESD zaps
   c. Common mode chokes: Applied to reduce EMC, may have influence on robustness against system level ESD zaps

4. Other IOs (for instance, sense lines):
   a. EMC capacitors
   b. TVS diodes
   c. RC networks

In addition to protection strategies for the individual pins of an ECU, the board level design (especially the layout of ground planes) has significant influence on the system level ESD and EMC properties. Ford has published a good introductory guideline document entitled "EMC Design Guide for Printed Circuit Boards" which is available from www.fordemc.com.

A.3   System Tests and Targets of Computing Devices

A.3.1   Overview of System ESD Related Computing Device Tests

Computing devices, also known as information technology equipment (ITE), are subject to the immunity requirements in CISPR 24:2010 [29]. CISPR 24 has specific requirements for, among other items, ESD testing. The test methods in IEC 61000-4-2:2008 are called out with the following specific requirements:

- Contact discharge testing to conductive surfaces of the device and to coupling planes is performed at 4 kV IEC. No lower levels are required.
- Air discharge testing to insulated or non-conducting surfaces is performed up to 8 kV IEC. Lower levels are also evaluated.
- The performance criteria for ESD testing in CISPR 24 is "Performance criterion B". The basic requirement is that "After the test, the equipment under test (EUT) shall continue to operate as intended without operator intervention." The EUT may react to the ESD event (for example, a pop heard in a speaker), but it must self-recover without help from the operator.
- Electrostatic discharges are only applied to points and surfaces of the EUT which are expected to be touched during normal operation, including user access as specified in the user manual.
- Discharge to contacts of open connectors is not required.

A.3.2   State of the Art Design Concepts

Typically the only special design features used on ITE products for ESD protection are diodes on IO ports to shunt ESD currents to ground to protect circuitry from damage.

A.4   System Tests and Targets of Medical Electronic Components

A.4.1   Overview of System ESD Related Medical Electronics Tests

A medical device is an electronic component which is used for medical purposes for patients in diagnosis, in therapy and/or in surgery. Medical devices include a wide range of products varying in complexity and application. Examples include, but are not limited to, medical thermometers, blood sugar meters, X-ray machines, pacemakers, hearing aids, and heart rate machines.

The strict requirements for medical devices are set forth by the Food and Drug Administration (FDA). Risks are related to reliability and duration of use. The standard quoted for medical devices is ISO 14971:2007 [30], which specifies processes to identify hazards and evaluate risks associated with the lifecycle of medical devices. FDA requirements for these products, record keeping and reporting, are included/contained in Title 21 Code of Federal Regulations Parts 1000-1299 (21 CFR 1000- 1299) [31]. Medical devices emitting radiation are subject to the provisions of the FD&C Act and listed in 21 CFR 1000.15 [32]. Radiation emission includes ionizing electronic radiation, particulate radiation, ultra violet (UV), visible, infra-red (IR), microwave, radio and low frequency, laser, infrasonic, sonic and ultrasonic.

Medical Device and equipment testing include product safety compliance, EMC compliance, and FDA Guidance Documents and consensus standards necessary for FDA clearance, including IEC 60601 [33], American Society for Testing and Materials (ASTM), ISO and Association for Advancement of Medical Instrumentation (AAMI) standards. Associated standards are: UL 60601-1, CSA C22.2 No. 60601-1, EN 60601-1, EN 60601-2-XX, IEC 60601-1, and IEC 60601-2.

IEC 60601 is a series of technical standards for the safety and effectiveness of medical electrical equipment, published by the IEC. First published in 1977, IEC 60601 is now a widely accepted benchmark for medical electrical equipment and compliance. As of 2011, the IEC 60601 consists of a general standard, about 10 collateral standards (IEC 60601-1-X), and about 60 particular standards (IEC 60601-2-X), and the standard includes a process for Certification.

The general standard *IEC 60601-1 - Medical equipment/medical electrical equipment - Part 1: General requirements for basic safety and essential performance* - gives general requirements of the series of standards. This IEC 60601-1 standard applies to all medical products which are not implanted inside the human body.

Collateral standards (numbered IEC 60601-1-X) define the requirements for certain aspects of safety and performance, such as Electromagnetic Compatibility (IEC 60601-1-2); or Protection for diagnostic use of X-rays (IEC 60601-1-3) or *IEC 60601-1-9* for Environmentally Conscious Design of Medical electrical Equipment. This Part 9 standard asks manufacturers of medical to consider the environmental impacts of their devices throughout the product's entire life cycle and to minimize these where possible. Medical EMC testing is typically performed in accordance with EN/IEC 60601-1-2 for most devices. IEC 60601-1-2, a collateral standard, states that the ESD requirements are +/- 2, +/- 4 and +/- 8 kV air discharge 12 (to non-conductive accessible surfaces) and +/- 2, +/- 4 and +/- 6 kV contact discharge (to 13 conductive accessible surfaces).

Particular standards (numbered 60601-2-XX) define the requirements for specific products, such as MR scanners (IEC 60601-2-33). IEC 60601-2 is Part 2 and addresses specific requirements for a particular type of medical devices. There are over 40 Part 2 devices. The XX is a number from 1 to 40+, representing the applicable Part 2 device. Additions and/or deviations to the requirements of EN/IEC 60601-1-X are published in the appropriate EN/IEC 60601-2-XX for particular devices.

In order to market medical devices within the member countries of the European Union, a product must comply with the essential requirements of the medical device directive (MDD) 93/42/EEC [34]. Compliance with harmonized standards (such as EN 60601-1-2:2001 for EMC, and EN 60601-1 for electrical safety) provides conformity to specific requirements.

Medical device testing also includes static, dynamic, fatigue, wear testing and failure analysis on medical devices and orthopedic implants according to ASTM and ISO standards.

A.4.2   State of the Art System Design Concept

EMC is a major concern and design challenge in medical electronics, as the consequences of any malfunction may be life threatening [35]. Design constraints, notably leakage current limitations for some devices, make the EMC design of medical electronic devices more difficult than their non-medical counterparts. Typical EMC design practices include shielding and filtering.

A.5   System Tests and Targets of Avionic Components

A.5.1   Overview of System ESD related Avionic Component Tests

EMC/EMI robustness is primarily considered for the on-board service electronics (avionics) in the aerospace industry. Long wiring (total wire lengths of several km are typical for the aircraft architectures) is susceptible to picking up EM noise from the environment and transferring the generated impulses to the avionic components. This can lead to a disturbance of electronic component operation, or a functional failure. The typical failure during an ESD event in aerospace (from a system point of view) originates from field induced transients. The most commonly referenced IEEE surge standards used in practice are:

- IEEE C62.41.1 [36] – *2002(IEEE Guide on the Surge Environment in Low-Voltage (1000 V and Less) AC Power Circuits)*
- IEEE C62.41.2 [37] – 2002 *(IEEE Recommended Practice on Characterization of Surges in Low-Voltage (1000 V and Less) AC Power Circuits)*
- IEEE C62.45 [38] – 2002 *(IEEE Recommended Practice on Surge Testing for Equipment Connected to Low-Voltage (1000 V and Less) AC Power Circuits)*

IEC 61000-4-2 is used as the ESD standard model at the board level. However, actual in-flight or in orbit conditions (humidity, temperature, pressure, etc.), which affect the surge waveforms compared to those on ground level, are not covered by this.

Other standards in use for EMI robustness at the component level are: MIL-STD-461 [39] (military), RTCA DO-160 [40], and EUROCAE ED-14 [41] (commercial). The aircraft level standards include US Military MIL-STD-464 [42], RTCA DO-160 and EUROCAE ED-14.

A.5.2   State of the Art System Design Concept

In the aerospace industry, for various practical reasons, a typical system design strategy is to consider the protected ICs very sensitive to ESD events.

The most common technique used at present to protect avionics is the placement of so–called "surge protection devices" (SPDs) at the power and IO lines of each sub-system. These SPDs are designed to act as filters, and can either be discrete devices or circuits, turning on at specific overvoltage or frequencies according to the overvoltage/overcurrent transient characteristics.

A particular issue to be addressed in the present and forthcoming protection/design approaches is the introduction of composite materials for the aircraft body and/or the penetration of non-conforming and/or counterfeit materials and components (which have virtually "unknown" properties related to EMC/EMI/ESD). All of these factors introduce a variety of unknowns related to the stress waveforms seen by the avionics system and impact their susceptibility and robustness to the generated overvoltage/overcurrent transients.

A.6   System Tests and Targets of Consumer Electronic Devices

A.6.1   Overview of System ESD Related Consumer Electronics Tests

Consumer electronics covers quite a wide range of devices, and the scope in this appendix will be limited to components of audio/video equipment, such as TVs, camcorders, MP3s, iPods, and mobile displays for example. In general, test methods according to IEC 61000-4-2 are used to perform contacted, air, and indirect discharge. The level of robustness depends on the environment in which the product is used and the incidence of human handling. The following test methods and specific requirements generally apply:

- IEC Contact discharge testing - to conductive surfaces of the device and to the interface pins which can be touched by humans during production and handling of the final product. 8 kV is a common level for portable products and ports often handled by humans, and 4 kV for products not in human contact.
- IEC Air discharge testing - to isolated or non-conducting surfaces and connectors is performed up to 15 kV. Discharges to displays, keyboards, etc must not cause the product to stop functioning.
- Indirect IEC discharge to metal planes (horizontal/vertical) on which the product is located to ensure functionality during operation. Levels of 8 kV contact and 15 kV air often apply. This test is done in an operating mode to check the immunity to the electromagnetic fields.
- Specific test for light emitting diode (LED) TV: CISPR 13 Class B [43]
- Handheld Audio applications also require:

  1. EN 55022
  2. EN 55024
  3. IEC 61000-4-2, -4-3, -4-8

Beside international standards, there are various examples for company specific testing:

- CDE robustness - tested by system level ESD pulses to system ports which are not specified by IEC 61000-4-2.
- Charged board event (CBE) testing - to simulate the discharge from a CBE or charged modules, caps of 100 nF – 800 nF are precharged with 25 V - 40 V and then discharged to the pin under test. This could be an exposed IC pin (Feedback) or the OUT-pin of the controlled external power.
- Stress testing of a powered IC - tested by applying MM-type pulses up to 300 V to check the sensitivity of a device through onboard coupling of an external pin to an internal pin because of the board wiring. Typically, stressed pins are control and feedback pins for external power devices or pins connected to lines wired in parallel to signals going offboard.

- Discharge to contacts of open connectors for main interface ports (Audio/Video) - tested by applying MM-type pulses up to 300 V under operation. System interruption with recovery is acceptable.

A.6.2 State of the Art Design Concepts

To protect the exposed pins of the devices, large on-chip diodes to the supply rails are commonly used. If the integration of these very large on-chip protection devices is not possible due to performance reasons, external TVS diodes are placed on the board. For TV and other display driver components, the output pins are usually protected against GND, which must also be robust against various EOS/ESD stress conditions. System level ESD protection and test concepts are widely discussed for LCD applications [44-50].

In some cases, high CBE and CDE robustness of certain exposed pins are requested to prevent damage when connecting the display with the driver boards. Typically, large primary on-chip ESD protection devices are used and an extra-large external capacitance is connected to the supplies.

Consumer electronic devices often have a very tight material cost target, which forces the use of lower cost EMC/ESD protection methods. Protection by the mechanics is the desired approach. Additional basic passive chip components may be needed for residual ESD and EMC filtering. When enhanced protection is needed, onboard varistors, ferrites and diodes can be placed at the most critical signal lines. Even then, the target is to limit this to a minimum and later cost driven redesigns will lead to a removal of expensive ESD/EMC devices.

A.7 Applicability of System-Efficient ESD Design (SEED)

Basic SEED concepts, which provide a methodology to match board protection and IO behavior for prevention of damage, are applicable to a number of interfaces which can be exposed to direct or indirect ESD stress with large stress energy. They are usually protected by a combination of PCB clamping elements like TVS devices or varistors, and current limiting serial elements such as ferrite beads. Examples for these pins are listed in Table A-1.

Table A-1: Typical interfaces exposed to IEC stress

IC pin type	Exposure	Field of Application
USB	direct	mobile phones/computing
Headset	direct	mobile phones/computing
Antenna	direct	mobile phones
Battery	direct	mobile phones
Key pads	indirect	mobile phones
LIN	direct	automotive
CAN	direct	automotive
Flexray	direct	automotive
LAN	direct	computing
DVI	direct	computing
TV driver output	direct	consumer

A.7.1 A SEED Case Study for USB

Soft fails, which represent most of the observed fails during system development are related to an indirect stress path triggered by crosstalk or noise injection. In this case, the methods described as advanced SEED will be useful to optimize ground planes, pulse mitigating discrete elements, and PCB trace layout. It should be emphasized that the optimization of the mechanical layout of the case is not in the scope of SEED, while it plays an important role in the system ESD robustness. However, mechanical layout is often predefined by product design and functional limitations.

A possible design flow applying SEED for a USB 3.0 interface with challenging high speed requirements is discussed, focusing on the USB port as an entry point of the ESD pulse. The design of a high performance interface like USB 3.0 ranging up to 5 Gbit/s requires the knowledge of the complete transfer line from transceiver to receiver. The eye diagram at the receiving port is commonly taken as figure of merit. The design of the IO circuit needs an early definition of the maximum capacitive load by on-chip / on-board ESD protection elements and of the worst case serial resistance. The application of a basic SEED approach allows selection of an optimum TVS diode with minimum capacitance while still matching the high current IO characteristic. An optimization of PCB protection elements and serial filters for a USB (3.0) following the SEED approach is demonstrated in [51]. Based on this initial choice of protection and the draft of the PCB layout, the capacitive and inductive coupling to adjacent lines can be extracted and the level of cross-coupling can be simulated for an ESD waveform. The transient latch-up robustness of IOs exposed to induced current pulses determines whether the coupling must be reduced or mitigating elements need to be added to the victim lines. Also, changing the placement of the TVS diode will influence both cross coupling to adjacent lines and stress received by the USB pin.

In the next step, the supply noise level can be assessed by these simulations, as long as the supply RLC net on the board is appropriately modeled. If the amplitude in the noise power spectrum exceeds the sensitivity level of the IC, mitigating measures might be implemented such as reduction of resistance in the supply net or use of buffer caps. Finally, the impact on signal integrity has to be estimated.

Together with functional simulation of the transmission line, this allows a simulation based co-design of the function and ESD protection of a USB subsystem, taking into account IC behavior, board components and PCB layout.

References

[1] IEC 61000-4-2, 'Electromagnetic compatibility (EMC) – Part 4-2: Testing and measurement techniques – Electrostatic discharge immunity test', Ed. 2.0, 2008. -- http://webstore.iec.ch/

[2] CFR 47 §xx.xx -- http://www.access.gpo.gov/nara/cfr/waisidx_02/47cfrv2_02.html

[3] EN 300 xxx -- http://www.etsi.org/WebSite/Standards/Standard.aspx

[4] EN 301 xxx -- http://www.etsi.org/WebSite/Standards/Standard.aspx

[5] EN 55020 -- http://www.etsi.org/WebSite/Standards/Standard.aspx

[6] EN 55024 -- http://www.etsi.org/WebSite/Standards/Standard.aspx

[7] TS 25.xxx -- http://www.etsi.org/WebSite/Standards/Standard.aspx

[8] TS 36.xxx -- http://www.etsi.org/WebSite/Standards/Standard.aspx

[9] TS 51 xxx -- http://www.etsi.org/WebSite/Standards/Standard.aspx

[10] TS 151 xxx -- http://www.etsi.org/WebSite/Standards/Standard.aspx

[11] GBT 22450.1-2008 - - http://www.ptsn.net.cn/ (Chinese only)

[12] YDT 1592.1 - http://www.ptsn.net.cn/ (Chinese only)

[13] PTCRB NAPRD.03 -- http://www.ptcrb.com/

[14] ETS 300 342-1 -- http://www.etsi.org/WebSite/Standards/Standard.aspx

[15] ISO 10605 -- http://www.iso.org/iso/iso_catalogue.htm

[16] IEC 61967 -- http://webstore.iec.ch/

[17] IEC 62132 -- http://webstore.iec.ch/

[18] ISO 7637 -- http://www.iso.org/iso/iso_catalogue.htm

[19] ISO 16750-2 -- http://www.iso.org/iso/iso_catalogue.htm

[20] CISPR 12 -- http://webstore.iec.ch/

[21] CISPR 25 / EN55025 -- http://webstore.iec.ch/

[22] ISO 11451 -- http://www.iso.org/iso/iso_catalogue.htm

[23] ISO 11452 -- http://www.iso.org/iso/iso_catalogue.htm

[24] SAE J1113 -- http://standards.sae.org/

[25] AEC Q100 -- http://www.aecouncil.com/AECDocuments.html

[26] Ford Motor Company, EMC Design Guide for Printed Circuit Boards, Engineering Specification ES-3U5T-1B257-AA, http://www.fordemc.com/docs/download/EMC%20Design%20Guide%20for%20PCB.pdf

[27] Ford Motor Company, Electromagnetic Compatibility Specification for Electrical/Electronic Components and Subsystems, General Specification Electrical/Electronic, EMC-CS-2009.1, http://www.fordemc.com/docs/download/EMC_CS_2009rev1.pdf

[28] Audi, BMW, Daimler, Porsche, and Volkswagen, Hardware Requirements for LIN, CAN and FlexRay Interfaces in Automotive Applications, version 1.1, 2009.

[29] CISPR 24:2010 -- http://webstore.iec.ch/

[30] ISO 14971:2007 -- http://www.iso.org/iso/iso_catalogue.htm

[31] 21 CFR 1000- 1299 -- http://www.accessdata.fda.gov/scripts/cdrh/cfdocs/cfCFR/CFRSearch.cfm

[32] 21 CFR 1000.15 -- http://www.accessdata.fda.gov/scripts/cdrh/cfdocs/cfcfr/CFRSearch.cfm?FR=1000.15

[33] EN/IEC 60601-1-2 -- http://webstore.iec.ch/

[34] MDD 93/42/EEC -- http://eur-lex.europa.eu/

[35] W. Kimmel and D.D. Gerke: "Exploring systems EMC in Medical Devices", http://www.eetimes.com/design/medical-design/4212205/Exploring-systems-EMC-in-medical-devices

[36] IEEE C62.41.1 – 2002 – http://ieeexplore.ieee.org/

[37] IEEE C62.41.2 – 2002 – http://ieeexplore.ieee.org/

[38] IEEE C62.45 – 2002 – http://ieeexplore.ieee.org/

[39] MIL-STD-461 -- http://assist.daps.dla.mil/quicksearch/

[40] RTCA DO-160 -- http://www.rtca.org/downloads/

[41] EUROCAE ED-14 -- http://www.eurocae.net/

[42] MIL-STD-464 -- http://assist.daps.dla.mil/quicksearch/

[43] CISPR 13 -- http://webstore.iec.ch/

[44] Tai-Ho Wang; Wen-Hao Ho; Lin-Chien Chen; "On-Chip system ESD protection design for STN LCD drivers", EOS/ESD Symposium 2005, 8-16 Sept. 2005, pp. 1 - 7.

[45] Pavey, LD.; Kerr, J.; "ESD simulation and cure: a case study", ESD and ESD Counter Measures IEE Colloquium on 28 Mar 1995, pp. 1 -4.

[46] Ming-Dou Ker;Cheng-Cheng Yen; "Investigation and Design of On-Chip Power-Rail ESD Clamp Circuits Without Suffering Latch-up-Like Failure During System-Level ESD Test" IEEE Journal Solid State Circuits, Nov. 2008, Vol.42, pp. 2533-2545

[47] Ming-Dou Ker,Wan-Yen Lin1,Cheng-Cheng Yen1,Che-Ming Yang,Tung-Yang Chen,Shih-Fan Chen, "Protection Design Against System-Level ESD Transient Disturbance on Display Panels", 2010 Asia-Pacific International Symposium on Electromagnetic Compatibility, April 12 - 16, 2010, Beijing, China, pp. 438-441

[48] Cheng-Cheng Yen;Wan-Yen Lin;Ming-Dou Ker;Che-Ming Yang;Shih-Fan Chen;Tung-Yang Chen; "New transient detection circuit to detect ESD-induced disturbance for automatic recovery design in display panels", Design & Technology of Integrated Systems in Nanoscale Era (DTIS), 2011, pp. 1-4

[49] Xiao, J.; Pommerenke, D.; Drewniak, J. L.; Shumiya, H.; Maeshima, J.; Yamada, T.; Araki, K.; "Model of Secondary ESD for a Portable Electronic Product", IEEE Trans. EMC, Oct. 2011, pp. 1-10

[50] J.Xiang, D. Pommerenke et.al. "Model for ESD LCD Upset of a Portable Product", IEEE International Symposium on EMC, 2010, pp. 354-358

[51] S. Bertonnaud, C. Duvvury; "IEC System Level ESD Design for OMAPTM System", private communication (International ESD Workshop 2011, Lake Tahoe)

Appendix B: Technology Roadmap and Direction

Charvaka Duvvury, Texas Instruments
Pasi Tamminen, Nokia
Matti Uusimaki, Nokia
Joost Willemen, Infineon Technologies

B.0   Introduction

Several important issues for ESD/EMC/EMI have been discussed in detail and efficient methods for protection against these effects have been offered. Both ESD and EMC qualification requirements for electronic parts from various fields in the industry have been outlined. While all of the requirements must be met for the present state-of-the-art systems, the continuing trend in the scaling of silicon technologies is certain to place additional constraints on how effective ESD/EMI/EMC protection can remain. Limitations can come from various directions; it is likely though that new trends may alleviate these limitations to some extent based on future development and actual applications. These issues need to be examined in the context of future road maps.

B.0.1   Scope

The purpose of this appendix is to address changing technologies from different points of view that would have an impact on the near-term (2012-2022) design capability for system level ESD protection.

We will specifically cover integrated circuit technology roadmaps and trends, with consideration given to the system level ESD performance of:

- IT, communications, and automotive electronics.
- IC Package technology development and roadmap
- Advances in board and assembly technologies
- Advances in interconnects such as optical interconnects
- Advances in new suppressive materials such as polymers for IEC protection

A final objective of this appendix is to assess the impact of rapidly advancing technologies on system level ESD requirements and designs.

B.0.2   Background

The purpose of this appendix is to address changing technologies from different points of view that would have an impact on the design capability for system level ESD protection. In this context, we plan to address only the issues that would have an impact in the immediate future. Despite consideration for system level ESD, advancements in the electronics industry will continue to include scaling of IC technologies, new IC package development, and new consumer applications. Each advance has to be examined with a proper perspective to see where the IEC protection can be included or whether there will be a significant impact to on-chip protection from these developments. We review these trends for the selected topics and offer a view of how they might affect the future of IEC system level ESD.

B.1 Roadmap for ICs: Microprocessors

B.1.1 IC Technology Development

Continued scaling of silicon technologies is necessary for achieving higher speed circuit demands and higher density chips. Both White Paper 1 [1] and White Paper 2 [2] have reviewed these technology advances and have established the impacts on component level ESD targets. 3-D transistors like FinFETs [3] can have a serious impact on ESD sensitivity, but this is most likely limited to component ESD. Products built with advanced technologies which have applications in a system can effectively have an impact on the overall system performance. Thus, the impact from scaling in most cases comes directly from the sensitivity of the interface pins. Technologies in production now are at the 40 nm process node, with the 28 nm node under broad deployment, the 22 nm currently in production and the 14nm node under development (2012).

B.1.2 New Functions

As IC technologies continue to be scaled down, newer functions continue to evolve for achieving greater speed and performance. Some of the most important functions for IOs are high speed serial links known as SERDES macros. At the 40 nm node, SERDES operating frequency has been established at 12.5 Gbit/s, and soon will be at 15 Gbit/s for 28 nm and progressing to 20 Gbit/s at the upcoming 20 nm node. Similarly, double data rate (DDR) applications also have a high demand for speed: 1.5 G at 40 nm to 2.3 G at 28 nm. Fortunately, these SERDES and DDR pins are not system interface pins, and hence are not restrictive for IEC protection designs. In contrast, high frequency and high speed pins such as USB and HDMI can have direct interfaces and thus can pose severe challenges for IEC protection design. The same challenges also apply to RF Antenna pins that would have a direct interface while having low tolerance to capacitance from external protection devices.

B.1.3 USB Requirements

The roadmap for USB evolved from 12 Mbit/s (USB1) to 48 Mbit/s (USB1.1) and to 480 Mbit/s (USB2). The extension of USB frequency in the next few years will be to 5 Gbit/s for USB3. This is shown in Figure B1.

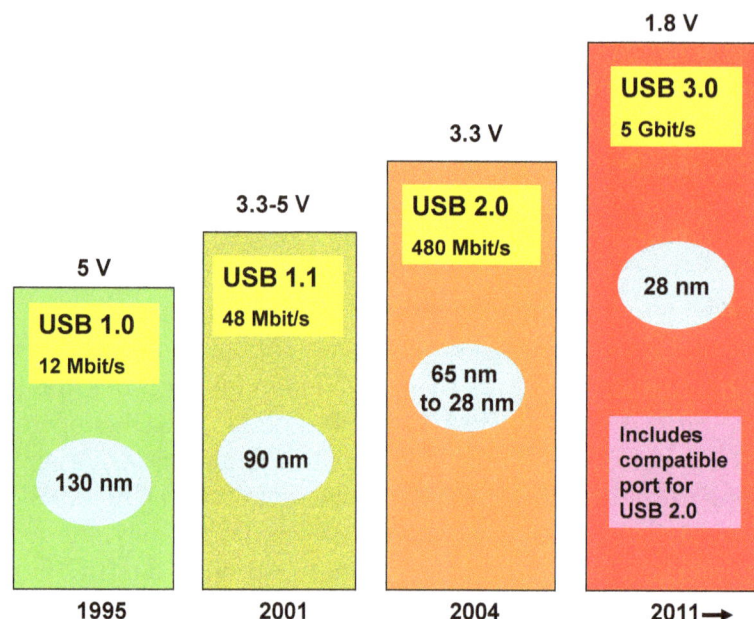

Figure B1: Trends in USB applications and data speeds. The insets represent the respective technology nodes of introduction (Courtesy of TI/Nokia)

At the present time, applications for USB 2.0 require meeting IEC protection challenges. These USB receptacles interface with IC pins that can often shunt at least 2 A of ESD current in the 100 ns regime as part of their component ESD protection. Thus, the board design to meet the IEC stress levels is not difficult as long as board components such as common mode filters (CMFs) are effectively used. There is always the question that if the USB 2.0 pins are recessed; do they really need IEC protection? Connectors for USB 3.0 would be an improvement, and thus might not face the same difficulty, but the higher speed of USB 3.0 would make it even more challenging to meet IEC protection. With USB 3.0, the interface to USB 2.0 would still be needed, and thus the same challenges would remain even if future applications switch to USB 3.0. But if customers demand IEC protection for USB 3.0, simultaneously meeting the frequency requirements of 5 Gbit/s versus designing for 8 kV IEC will require innovative approaches. In such cases, migration to polymer type protection might become necessary provided enough technical advances are achieved to make these materials practical for ESD applications.

B.1.4 HDMI Requirements

The roadmap for HDMI reflects a much more severe impact from new applications than for USB. Before HDMI, S-Video had been common but the S-Video data rates were slower. As the demand for high definition TVs has grown, HDMI data rates have increased from 1.8 Gbit/s to rates of 3.4 Gbit/s and higher. Eventually HDMI is expected to be replaced with Display Ports. These trends are indicated in Figure B2.

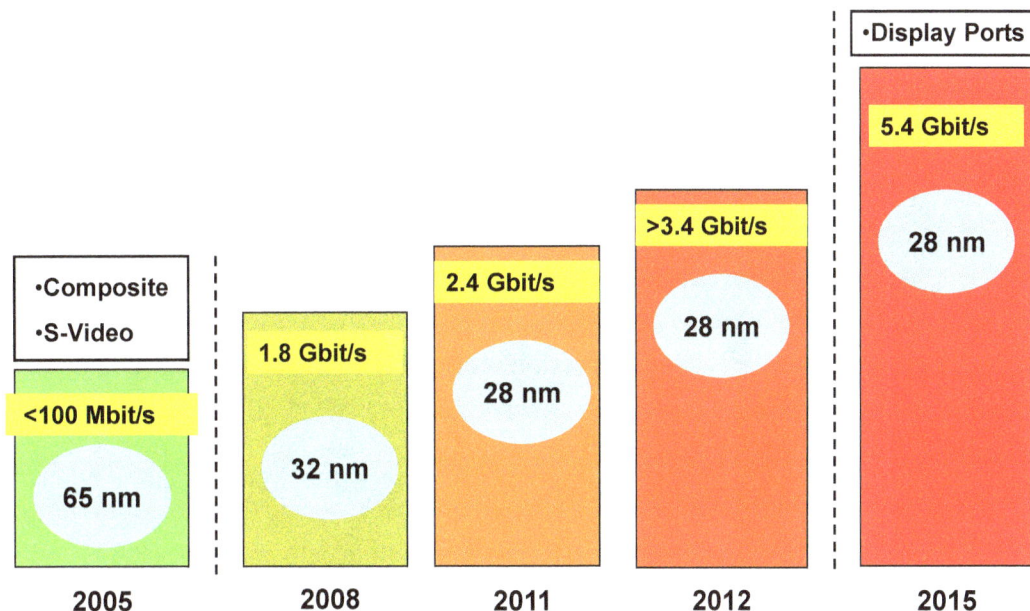

Figure B2: Trends in HDMI applications and data speeds. The insets represent the respective technology nodes of introduction (Courtesy of TI/Nokia)

It should be noted that Figure B1 and Figure B2 indicate that these speeds are achieved with 28 nm technologies whereas USB transfer rates of 5.4 Gbit/s have been achieved on 90 nm technologies and HDMI transfer rates of 3.4 Gbit/s have been achieved on 130 nm technologies.

The protection challenge for HDMI mainly comes from the intolerance of these interface pins to any value of series resistance larger than 2-3 Ohms. This is reflected in the relatively lower levels of component ESD performance for these pins (typically meeting only about 1 A TLP in the 100 ns regime). Moreover, as technology nodes shrink, the breakdown voltages at the IC pins are lower and thus the tolerable residual voltage at a particular residual current has to be smaller. This trend, combined with the fact that any board components such as a CMF or ferrite bead (FB) cannot be used due to their frequency limitations, requires the use of a very efficient TVS component (low capacitance, low breakdown voltage, very low on-resistance) to meet the IEC requirements. While this is possible to some extent today, within the next five years the problem will become much worse, and it may not be possible to meet an 8 kV requirement for HDMI at > 3 Gbit/s. Even if polymer ESD suppression components are to be used as a replacement for the TVS, the practical trigger voltages that would be available for these polymers would still make it difficult. Thus it seems that for the future, meeting expected IEC performance for HDMI could very well become a bottleneck.

B.2   Roadmap for Automotive Applications

B.2.1   Automotive Technology Trends

Automotive semiconductor products employ a variety of different IC technologies. The products made in the various automotive technologies have to be (a) robust in the harsh automotive environment (such as operating temperatures typically ranging from -40 °C to 175 °C), (b) subjected to strict reliability and safety requirements, and (c) directly or indirectly powered from a car battery supply network, which is exposed to many electrical disturbances.

Modern cars contain on the order of 50 to 100 ECUs. Typically, in an ECU we can distinguish different functional blocks:

- Voltage regulators: providing stabilized supply voltages from the car battery supply.
- Actuator control outputs: controlling electrical motors, valves, displays, lighting.
- Sense inputs: measuring physical quantities in the car, such as velocity, pressure, acceleration, temperature, etc.
- Communication: data communication networks (LIN, CAN, Flexray) are used to interconnect different ECUs in order to distribute sense and control signals inside a vehicle.
- Microcontrollers: for digital processing of sense signals and actuator control signals for monitoring and safety control functions.
- Memory: one time programmable (OTP), random access memory (RAM), electrically erasable programmable read only memory (EEPROM), Flash: for storing status, embedded software, trimming set-points, etc.

Many automotive ECUs consist of most or all of the above-mentioned functional blocks. Since the introduction of ECUs in vehicles, we see an on-going trend to miniaturization and reduction of component count per ECU. This is combined with an improvement of driver comfort, fuel efficiency, safety, and an ever-increasing expansion of features.

Reduction of component count is realized by combining functional blocks into single integrated circuits. Depending on the chosen system, partitioning different combinations of functions is possible. For instance, system base chips can combine regulators, small actuator drivers, monitoring sense inputs, and communication functions on one IC. Full system ASIC components combine even more functions in a single IC. In the future, the trend for further integration of functional blocks and enhancing features is expected to increase. The choice between different partitioning options is decided based on cost considerations. A further trend for reducing component count is the system-in-package (SiP) option. Here, multiple ICs, fabricated in different technologies, are mounted inside a single IC package.

Automotive IC technologies can be classified into different groups. For each of these groups, different technology trends can be identified.

- Low integration, such as smart switches (power devices with integrated simple diagnostic functions such as over temperature protection). The products in this group are mainly used as high current switches and are usually self-protecting regarding system level ESD. A cost reduction trend, while maintaining or improving the basic device characteristics, is expected for the coming years.

- Medium integration, such as multiple switches and power regulators. These technologies will see a cost reduction, combined with an increase in logic density. The products in this class are often used for switching large loads and thus are self-protecting to system ESD stress.
- High integration, such as System Base Chips and system ASICs. This technology class comprises bipolar-CMOS-DMOS (BCD) technologies. The present state-of-the art is in the 0.13-0.25 um feature size range. In the next 5 years it is expected that technology feature size will scale into the sub-100 nm range. The chip area required for system level ESD protection is not able to scale down at the same rate as the logic feature size, as the energy of the ESD zap defines the chip area required to divert the ESD pulse.
- Integrated sensor technologies, containing sensing elements for different physical quantities, combined with signal processing and communication functions. The expectation for the coming years is an increase of logic density and signal bandwidth.
- Microcontroller technologies. Here the present state of the art is in the 90 nm to 130 nm feature size range. In the coming years it is expected to scale down to the 65 nm and 40 nm nodes. At present, there are no specific system level ESD requirements for microcontrollers. However, it is expected that system level robustness will be required for certain pins (GPIO, analog digital converter (ADC) inputs) with external PCB-level protection measures.

B.3   Roadmap for IC Package and Applications

B.3.1   Background to Package Technologies

It is well known that rapid advances in silicon process technologies are essential to enhance IC performance limits. At the same time, newer circuit development techniques must be state-of-the-art for achieving the required high speed in the 10+ GHz range. These effects, when combined together, are already taking a major toll in the ability to meet the ESD protection design requirements. The third dimension to this negative impact for ESD design can come from large package size effects. Other features such as stacked packages are introducing further complexity to ESD design and testing.

Package advances are based on requirements of the different market segments [4]. For computer applications, these package advances boost performance. For consumers, package advances are most guided by price and robustness. For automotive and military applications, package advances are often driven by increasing temperature sensitivity and reliability. Each type of package is then designed and selected according to the application. This package evolution has progressed from the standard dual-in-line (DIP) packages to multi-chip modules (MCMs) and to flip-chips and stacked die or even stacked packages. The future will see wafer scale packages (WSPs).

Although not a primary consideration historically, during package development, attention should also be given as well to ESD impacts. The aggressive technological advances into newer types of packages might very well determine the achievable ESD performance for overall adequate reliability. Looking to the future, moving from ceramic to organic types of substrates (or even for substrates without a die) may create susceptibility to damage of the package interconnect.

B.3.2 Novel Packages

The trend towards stacked die and stacked packages, as shown in Figure B3, is already showing various effects on ESD. When two different die are interconnected within the package and the power and ground are commonly shared, the effect is minor. But when the IO pads on both die interface with each other and the chips have no common ground, the ESD current paths will involve traveling through these chip-to-chip bonds (for example from ground of one die to IO of another die) and these ESD paths must be understood through modeling and simulations.

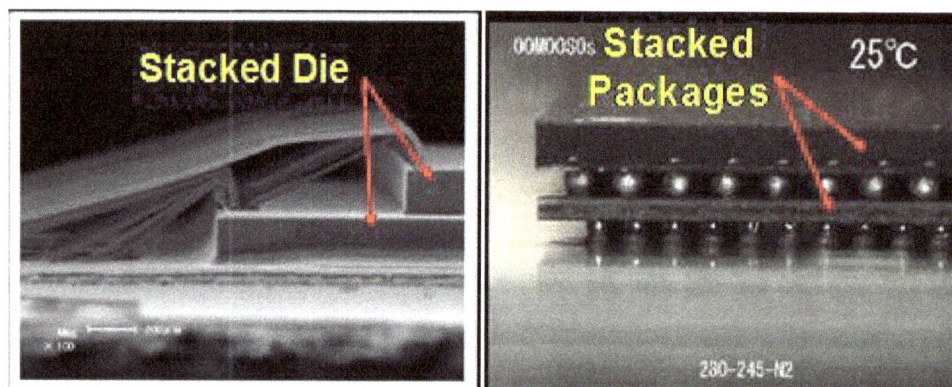

Figure B3: Stacked die versus stacked packages [4]

For stacked die packages, this can increase the complexity for ESD testing; therefore design methods must include analysis of the parasitic interactions between the die. Package development engineers are asked to develop new package materials to minimize the capacitance for large scale packages, and these new materials can have an ESD impact. Also, modeling is needed to understand the ESD current flow between die that have common bond wires and the inductive wiring effects for one package on top of the other.

B.3.3 System-on-Chip (SoC) and System-in-Package (SiP)

Novel SoC products continually require more features in a smaller space, and it is increasingly important that the power consumption, design time, and costs are optimized. SoC technologies help designers of these SoC products since they speed up system design when the core functions of the chip are well known.

SiP package developments (where multiple die stacks are packaged as a single unit) are evolving to the next phase of 3D devices involving connections using TSVs. In these cases the bottom die may have external connections while the power and ground connections are often shared among die stacks. The question then becomes what impact these new structures will have on system level ESD designs.

B.3.4 Applications with Impact on System Level ESD

A system design may contain several separate SiPs or SoCs from different sources, consequently the system designer needs to carefully match all IC and product interfaces. The matching phase includes EMC/ESD compatibility issues, as a single IC may contain for example, digital, analog and RF blocks. It may not be possible to separate noisy RF interfaces on a board from other sensitive interfaces, and as a result, the board design gets more complicated. To improve on board integration and matching, more detailed information is needed from SoC blocks, and most likely early ESD/EMC simulations are needed to optimize the system design.

B.4   Advances in Board and Assembly Technologies

B.4.1   Changes in Board Technologies

Board technologies have evolved with new board construction materials and joint technologies. There will be more technology options, especially for high end, high mix, and low volume product areas. The existing surface mount technology (SMT) with FR-4 printed circuit boards will most likely continue to be the dominant board technology due to low price, high manufacturing capacity, and existing design tools. Low cost, high volume products will use these well tested board technologies, but they will use chips in advanced technologies. Very often high volume chips are made with the latest efficient silicon processes which will provide the best operations and cost efficiency. Combining traditional board technologies with novel IC and SoC technologies may bring challenges for the system design.

High end boards will use more complex constructions with anisotropic connection technologies, optical connections and flexible printed circuit boards. The base construction with these products will most likely be the basic SMT printed circuit board. Portable products with very thin designs will have electronic components distributed around the board. The design will get more complex when these distributed sub-components will be connected to the mother board through flexible cables and optical/electrical connections. These sub-assemblies may contain more sophisticated board technologies depending on the required function. Buried ICs and components in PCB cavities may be used to increase board integration. IC components will contain a number of functions and will eventually contain multiple chips in one package.

Board 3D design with stacked components, embedded components in PCB, SiPs, system-on-packages (SoPs) and other 3D constructions will be in the mainstream within coming years. There are also other special 3D designs like MID technology which may advance. In MID technology, a laser system writes the interconnect pattern onto an injection molded component from the computer aided design (CAD) data. This technology has been available for several years, but new laser and gluing technologies (isotropic conductive adhesive, anisotropic conductive adhesive and non-conductive adhesive) may speed the technology's advance.

A completely new high volume product family will be the R2R electronics as shown in Figure B4. These applications can combine passive and active electronics, optic and optoelectronic components, different indicators, sensors, and diagnostic products which are produced on paper, plastic foil or other more novel materials, onto a flexible medium. These products can be ultra-low cost and can be integrated into disposable products like a sales item package.

Figure B4: Roll-to-Roll electronics (Courtesy of VTT)

Printable electronics is close to R2R technology and can be used both with low volume and high volume areas. Printable electronics can be used economically to construct very low volumes due to low tooling costs. At the moment, these printable electronics products typically operate in lower frequencies and have lower reliability levels than conventional electronics. However, technology advances may enable high frequency solutions for printable and R2R electronics. Future manufacturing can employ nanoscale features which will be printed with Gravure inkjet printing, Flexography printing and Nanoimprint lithography. Advances with printable electronics rely mainly on the material technology development and how well the technology fits on a mass manufacturing area.

Board level interconnections will start to use nano/composite soldering processes such as shown in Figure B5. These new materials will improve interconnection strength, enable more dense joint connections, and may also decrease component package cost due to lower soldering temperatures. There is also development ongoing with board level conformal RF shielding. This can replace metal can shields, but will also pose limitations regarding reworkability of electronics.

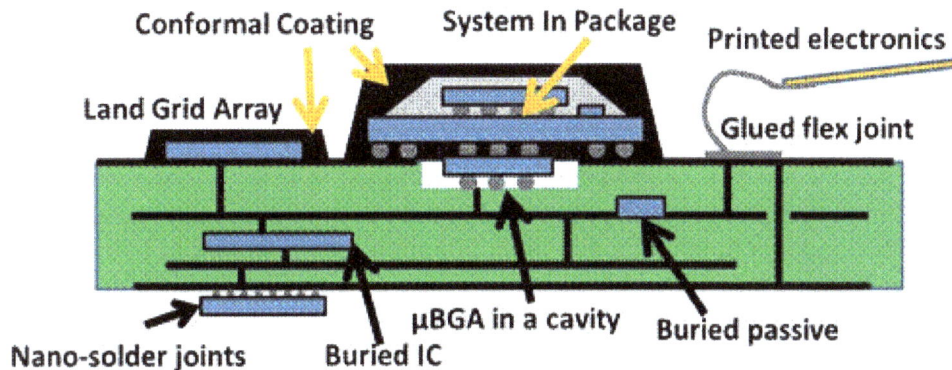

Figure B5: Board design with 3D designs, glued joints, buried electronics (Courtesy of Nokia)

All the mentioned advances in board and assembly technologies will have an impact on ESD and EMC design.

- The overall target of most of the changes is to decrease product dimensions and enable higher integration levels. From an ESD protection and EMC design point of view, smaller designs typically increase board signal coupling, crosstalk and interference challenges.
- Small physical design may limit system ESD robustness.
- On-board SiP and SoP components will have both interface and internal connections. There can be high level EMC noise in a chip when interface pins are stressed.
- Printable and R2R electronics will require both new design tools and also possible new EMC/ESD protection components which can be integrated into the new manufacturing processes.
- 3D designs require board designers to use simulation and modeling tools to optimize supply planes and signal routings due to a higher level of complexity.
- Components with very dense joint spacing require complex board designs where signal integrity and EMC issues will both need to be optimized.

Ultra low cost products will have additional challenges with ESD/EMC on new system designs. These products can also contain high-end technologies such as advanced SoC and SiP packages and high frequency connections. At the same time, the whole design must be made with the lowest cost. This requires using PCBs with only two to four layers, where common supply or ground planes may not be achievable. The amount of EMC and ESD filtering or matching components may also be highly limited to keep material costs down. Product physical designs need to be small, with covers and other mechanics designed for lowest cost. At the same time, the product may have the same design and acceptance targets from an ESD/EMC point of view as with high-end products.

B.4.2   Changes in System Boards

Technology advances will bring more functions to electronic devices in all product ranges. Advanced sensors (haptic control), high speed display technologies (3D displays), > 3 GHz data transmission, and optoelectronics will most likely increase system complexity. At the same time, product size, power issues, design time, and overall cost of design have very tight limits. These new systems on board will also be made with technologies which are not re-workable anymore (for example most glued joints cannot be repaired). This requires high quality design to reduce or eliminate risk of a re-design to fix problems in these areas.

B.4.3   3D ESD and EMI Effects

For stacked die packages, the complexity for ESD testing also increases and therefore the design methods must include analysis of the parasitic interactions between the die. Some basic structure analysis with simulations [5] indicates the importance of interposer design and the metal layer designs when stacking a couple of die together. When one starts to extend to several die on top of each other, the importance of bond wires and bond wire resistance/impedance becomes a dominant effect. The basic rule is to guide ESD and EMI energy back out of the component without traveling inside the component too deeply.

One of the complex issues for future technologies will be the EMI effects from stacking die [5]. As described in this reference, in the case of one or two die stacking, the dominant effect on EMI is the interposer design and die to die coupling. When there is more stacking, the bond wires become more and more important. That is, basically EMI effects depend on how the wires are implemented. In stacked structures, the proper grounding arrangement is the key. For example, if there is an EMC filtering die, it should be the bottom die of the die stack. Then, to minimize the ground impedance, grounding should be done with short bond wires. However, if the filtering die is on the top, there is extra ground impedance which creates unwanted ground bounce. Die to die coupling is very important. A basic EMC design rule is to guide the ESD/EMI out of the component as soon as possible. Going into future applications of die stacking, these effects could dominate the design of system level ESD.

B.5   Optical Interconnects

B.5.1   Introduction

Traditional metal interconnects within a die are rapidly becoming road blocks for RF speed and performance. To help resolve this issue, local RF wireless optical communications are being considered to communicate between chips on a circuit board, but these are typically only for special applications. It is expected that advances in optical interconnects will outpace those in conventional metal interconnects, and there will come a time when optical interconnects will be seen in all but the shortest and slowest on-circuit links [6]. Meanwhile, short distance wire connections will still dominate. As clock speeds increase, power efficient interconnection through metal wires becomes increasingly difficult. Thus, electrical signal interconnects will soon limit on-board communication between chips.

B.5.2   Impact on System to IC Interface

Once chip to chip optical interconnect becomes mainstream, it will also offer benefits for system level ESD. This is due to the fact there is no electrical coupling between ICs through signal lines, so no ESD protection is required to the signal lines. This can be highly beneficial from a chip area perspective. However, at this stage it seems these technologies are still more than 5 years away from commercial use.

B.6   Polymer Applications

B.6.1   Need for Alternate Materials

While on-chip protection components can be designed to meet system level ESD requirements, they are definitely not practical due to their on-chip area requirements, and often may not consistently meet the specification requirements. These issues have been discussed in White Paper 3, Part I [5]. The design of on-chip IEC protection with compatibility to the circuit function is very difficult. For example, some external interface pins that require this system level protection may use SCR components, but these need to be free of latch-up issues during the stress. If an SCR component (which is very efficient) is used, its holding voltage at < 2 V it would not be compatible (without causing latch-up issues) for 3.3 V applications. Even if such large protection components are safely designed, they would require very wide (but impractical) metal widths to meet the current density requirements. Therefore, to effectively meet this stress challenge, an external protection method might remain as the only feasible approach.

The external protection components consisting of TVS components can be effective if the residual pulse entering the IC pin is compatible to its breakdown characteristic under transient conditions as outlined in the SEED approach. But with the USB roadmap, as described in Section B.1.3, the TVS component will start to have limitations. However, can we replace the TVS with polymeric materials? These polymers will be examined for their practicality.

B.6.2   Polymers

Polymer materials essentially transition from a nearly insulative state to a rapid conduction state with a response time of < 4 ps [7]. Their behavior can be understood to a first order with characterization using TLP. This is somewhat similar to the behavior of spark gaps but with two noticeable differences. In the case of a spark gap, when a high voltage pulse such as 1000 V is applied, the spark gap will allow the pulse to pass through without any clamping until it breaks down. But for a polymer, the voltage is clamped to between 200-400 V even before the polymer component breaks down. The second important difference between the two is the time needed to break down (statistical time lag). A spark gap can breakdown in as little as a few nanoseconds, but the polymer will first turn on in the picosecond to nanosecond range. Thus, the polymer will only allow a pulse of less than 1 ns and passing through 200-400 V while the spark gap will let a pulse of at least a few nanoseconds at 1000 V pass through. After breakdown both components have approximately the same clamping voltage of 25-50 V.

Systems will require a polymer designed/rated for the highest voltage used in the system, or at least used on a given PCB, and consequently may not be suitable for protecting low voltage pin applications (sharing the same PCB) that likely need most of the operational benefits that polymers offer (such as allowing for less capacitive loading ESD clamps on chip). Whereas engineering the polymer for the low voltage applications does not guard against the high voltage analog nodes from "triggering" the polymer's breakdown voltage during fast switching. If everything operated on the same (relatively) voltage, then this is not a problem, but such a scenario is not necessarily common in analog IC applications.

B.6.3 Polymer Impact on System to IC Interfaces

While it might appear that polymers are not suitable for low voltage applications, where the IC component breakdown values are typically in the 5-7 V range, the clue to its effective integration would depend on the manner in which the polymer goes into rapid succession of increasingly conductive states, thereby shunting the current away from the IC pin. At the same time as the IEC stress current is increasing, the on-resistance of the internal component may allow even more current to go through the polymer under breakdown. This type of efficiency, with these desirable qualities, would still require more research into making these materials while achieving good control, and allowing matching to the particular application.

Meanwhile, for higher voltage applications, polymers can offer immediate applications. As an example, for GaAs components, the breakdown voltages are in the 200-300 V range and thus a polymer in parallel can effectively protect them for both component HBM and the system level IEC stress. Other applications include micro-electrical-mechanicals (MEMs) with similar breakdown voltages.

How would implementation be done in the above applications? Integration onto the system board can be possible with a surface mount connector. Another approach uses an interposer board that will have the polymer matching the ball pattern of the IC. A third approach is to build custom PCBs with embedded polymer. Beyond that, the package integration of the polymer is under exploration.

B.6.4 Expediency for Board ESD Applications

ESD protection components have a number of characteristics (on-resistance, triggering and clamping voltage, capacitance, turn-on speed, leakage current, linearity, price, size, availability, etc.) and any requirement for ESD protection typically involves some tradeoffs to get an optimal solution. For example, polymer ESD suppression components typically have low capacitance, which is a desired feature for high speed signal line protection. However, polymer components also have the limitations of higher leakage currents after ESD pulses, and some of the polymer components also have relatively high on-resistance during ESD events. Polymer technology can provide reasonable ESD protection in many applications, but most likely polymers will be used along with other on-board protection methods (such as varistors, diodes, ferrites, and passive discrete components) depending on the required protection and signal integrity level.

B.7 Future Compatibility to IEC Protection Requirements

Part of system level ESD/EMC validation is the IEC 61000-4-2 stress tests. Here a system is stressed with up to 8 kV contact discharge and 15 kV air discharge. The same requirement is also increasingly being requested by suppliers for component qualification. It has been common to request up to 8 kV for those IO pins that may be stressed in a system. ESDA WG5.6 has created the HMM standard practice document to specify a corresponding system level test method at the component level. One option for this test is based on the IEC pulse waveform applied with an IEC "gun" (150 pF / 330 ohm circuit). In addition, a TLP-based pulse source can also be used for this component testing.

There are several open questions related to component level testing with IEC waveforms.

- Is the IEC current pulse a realistic stress waveform for a component pin which is on a board inside a system?
- Since the IEC pulse has been shown to have a wide variation in current and energy levels according to the standard, what is a "correct" stress level?
- Is there variation between the different ESD generators, both in the low frequency content and in the static behavior – how do we deal with this?
- How can possible component EMC/EMI related challenges be evaluated on a test board which differs from the final product?
- How much of the ESD pulse energy should dissipate within the IC, as system mechanics and board design will always have an effect on the final stress levels?
- New SiPs will typically contain interface connections. These components can be made with advanced wafer processes, and building up a high on-chip ESD protection level in these processes can be both challenging and costly.

Electrical products also vary in size from a small coin size battery operated electronic component up to a car size system. Small products can have a very high capacitance when tested on a test bench during IEC qualification while some products with a large plastic housing can be more or less electrically floating on a test bench. In the floating case, corresponding component ESD stress levels in the system can be totally different between different designs with the same initial voltage stress level. This can lead to both under- and over-estimations of the real component stress levels.

This raises the issue of whether it is beneficial to try to validate electrical components with similar setups and pulse waveforms as those delivered in a system qualification. Instead of trying to estimate the component behavior with IEC pulses, it could be more efficient to try to measure component operation and robustness during different stress situations. This data could be used by the system designer to build up required shielding and filtering for the component to withstand ESD/EMC stress in a system.

B.7.1 Impact of the Microprocessor Roadmap

Advances in microprocessor technologies will continue to pose a tremendous challenge for system level IEC robustness if the required and expected protection levels remain the same. The challenge of system ESD robustness comes at the expense of meeting the speed performance requirements for USB and HDMI. This raises the question of whether there should be new development for connector techniques, such that the ports are exposed for IEC zap tests. As these applications cannot rely on on-chip solutions, the use of external clamping needs to be more effective and relatively more efficient than what is commonly practiced today. One example of a solution could be the application of polymers. As long as the need persists where applications of these pins requires testing of the exposed entry points, innovation will continue.

B.7.2 Impact from the Automotive Roadmap

In the automotive industry typical maximum requirements for system level ESD robustness involve packaging and handling tests. System level testing for automotive applications is done with components connected to the cable harnesses with an 8 kV IEC requirement (contact discharge) for automotive electronic control units. However, there is no universally required ECU system ESD level test, and the requirements differ from customer to customer.

The 8 kV target value is historically defined. In some cases, it is empirically determined in the laboratory by increasing the zap level until failures that appear in production could be reproduced. This has led, for instance, to the 6 kV requirements for transceiver pins from German OEMs [8]. At the moment, no trend to change from 8 kV contact discharge as a typical requirement is to be expected.

One method of system level testing is to test the system in the unpowered state. Minimum pass/fail testing involves monitoring of post-stress leakage currents at the stressed pins to detect IC damage. Improved pass/fail assessments are done with functional retesting after system level ESD zaps.

The importance of system level testing in the powered state will increase in the future. In this case, transient latch-up (TLU) events due to ESD zaps can also be detected. ESD zapping in the full functional state, as is possible at the completed ECU level, is not feasible on individual components. This, however, is required to detect possible ESD induced system upset.

At the moment, automotive OEMs are not aligned to support the recommendations of lowering HBM and CDM levels to 1 kV and 250 V, respectively. Their concern is that lowering HBM and CDM on local pins negatively affects the system level performance due to increased exposure to crosstalk between system level stressed PCB lines and neighboring board-internal traces, which they contend charge internal components and sensitize ESD current paths which can be tested by HBM and CDM.

The main challenge for system level ESD in future automotive products is expected to be in the high-integration ICs (refer to Section B.2). Several aspects have to be considered here, for more details see reference [9].

- The crosstalk issue, mentioned above, must be resolved. This should happen both at the ECU board level as well as at the IC level. If a real crosstalk threat is identified, design guidelines and simulation methods should be developed to identify weaknesses early in the design phase.
- If the occurrence of harmful board-level crosstalk can be excluded on physical grounds, the HBM/CDM levels for ICs that have only local, ECU-internal, connections can be lowered.
- If on-chip crosstalk between high current carrying system level pins and other on-chip nets turns out to be a real threat, the present trend of making certain IC pins system level robust will change and move off-chip, PCB level protection measures will be introduced. This will help avoid the high current peaks flowing into a possibly sensitive IC.
- Shrinking the technology feature size leads to higher packing density and higher clock rates for logic circuit blocks. The on-chip system level ESD protection components will not shrink at the same rate since the required ESD component areas are correlated to the amount of energy to be dissipated, which stays basically the same.

- With advancing automotive technology designed into smaller areas, logic and analog circuit blocks will become more sensitive to ESD if they are indirectly connected to exposed system level pins (such as low voltage devices cascoded with high voltage devices). Here, greater design effort must be invested to provide secondary on-chip ESD protection measures. This poses a risk to increasing area requirements for ESD protection.
- Scaling technology goes hand in hand with a decrease in supply voltage and signal levels. Lower signal levels increase the susceptibility to EMC types of disturbances and therefore EMC caused by system level ESD zaps. Again, this is a possible scenario that contradicts the present trend to have more on-chip system level ESD protection and drives towards more board-level protection.

New system level ESD test methods may become more prevalent in the future. For instance, with the LIN transceiver tests, defined by Volkswagen, a new indirect zap test has been introduced. In this test method, an IEC 61000-4-2 ESD pulse is applied to the shield of a coax cable whose inner lead is connected to a LIN terminal. This setup leads to an indirect coupled pulse on the LIN pin. The test is a novel way to emulate LIN bus line ESD response inside a vehicle chassis, which was identified as a source for destructive ESD coupling in some cases.

It is expected that comparable test methods may be defined in the future, in order to assess the ECU functionality on a test bench during qualification, before testing the ESD robustness of the completed vehicle.

In any case, the effectiveness of the new ESD test methods must be carefully assessed in order to optimize, and not to unnecessarily increase, the number of test methods. Careful studies of reproducibility, ESD generator dependence, occurring waveforms, calibration methods, sensitivity to changes in the setup, and other factors must be carried out before introduction of new test setups.

B.7.3 Conclusions from Technologies and IC Systems

There is strong competition in the system market. Here, the speed of component, system, and software design plays a major role in the electronics business. ESD and EMC design is one small part of the big picture, but it is critical to avoid non-optimal, unrealistic target levels or test methods which hamper technological advances. Arbitrary fixed stress levels, qualification standards, and targets will only create unnecessary conflicts between the component and system manufacturers. To avoid these conflicts, we need to verify that each component or system qualification and validation method provide useful information for the next technology advancement.

One future scenario could be that system manufacturers would leave behind all component ESD requirements, but request information on the component behavior and tolerances during different stress situations. Component suppliers would include this information and simulation models in the datasheets. In this case, the system designer could select the component based on parameters which best fulfill the need, and use SEED simulations to optimize the design and verify the system robustness with final stress tests.

B.7.4 Cost of System Level ESD Design

At present, there is confusion regarding whether the system designer (customer) or component designer (supplier) should bear the cost of ESD design for the system boards. Combined with this confusion is the pressure of time to market to deliver the product for applications. One OEM comments that time to market is a primary factor for system designers with low cost category products. For this category of products, the price constraints do not allow assignment of numerous EMC/ESD designers to achieve safe and effective solutions. Therefore, OEMs ideally want delivery of an IC with built-in IEC protection, ready for board implementation to improve confidence in the board level IEC protection.

This scenario means the cost shifts to the IC manufacturer. However, IC development costs could be very high if trying to meet the speed requirements with the extra load (resulting from requirements of system ESD immunity) in a large on-chip protection device. In the case of critical pins such as high speed USB / HDMI interfaces, this is not practical – that is, the on-chip solution approach is not acceptable. This means either more effective, shielded connector pins have to be developed or novel on-board solutions have to be created. Polymeric materials are one example.

What about the use of more "commercial packages"? System designers may be buying certain IC and layout designs for a specific purpose, such as the assembly of a simple mobile phone in just a few days. For these quick systems there would not be much room to squeeze in the EMC/ESD design components. So these challenges would continue into the future. All of the above point to an urgent need for constant dialogue between the IC supplier and the system board designer for each product / application.

References

[1]   Industry Council on ESD Target Levels, "White Paper 1: A Case for Lowering Component Level HBM/MM ESD Specifications and Requirements," August 2007, at www.esda.org or JEDEC publication JEP155, "Recommended ESD Target Levels for HBM/MM Qualification", www.jedec.org

[2]   Industry Council on ESD Target Levels. "White Paper 2: A Case for Lowering Component Level CDM ESD Specifications and Requirements," Revision 2, April 2010, at www.esda.org or JEDEC publication JEP157, "Recommended ESD-CDM Target Levels", www.jedec.org

[3]   H. Gossner, C. Russ, J. Schneider, K. Schruefer, T. Schulz, C. Duvvury, R. Cleavelin, W. Xiong, "Unique ESD Failure Mechanism in a MuGFET Technology," IEDM Tech. Digest, pp. 101-106, 2006.

[4]   D. Edwards, "High Performance IC Package Design and Electrical Reliability," EOS/ESD Sympoisum Proceedings 2003, pp. 1-7

[5]   Industry Council on ESD Target Levels, "White Paper 3 System Level ESD Part I: Common Misconceptions and Recommended Basic Approaches," December 2010, at www.esda.org or JEDEC publication JEP161, "System Level ESD Part I: Common Misconceptions and Recommended Basic Approaches", www.jedec.org

[6]   David Miller, "Rationale and Challenges for Optical Interconnect to Electronic Chips," Proceedings of the IEEE, Vol. 88, No. 6, June 2000, pp. 728-749

[7]   M. Uusimaki and A. Renko, "A systematic Approach and comparison of different 3-D chip structures for electromagnetic compatibility" IEEE 0-7803-8443-1/04, 2004 IEEE International Symposium on EMC, pp. 734-739

[8]   Based on emperical evidence coupled to the German OEM requirements and through personal communication with M. Hell, Infineon Technologies

[9]   http://www.esdforum.de/Files/Impact%20of%20Lowering%20the%20ICHBM%20Robustness.pdf

Appendix C: Fast and Slow ESD Stress (High and Low Frequency Spectrum)

Matti Uusimaki, Nokia

C.0:   Discussion

In ESD design at the product level, one should understand that there are several different phenomena involved. The first thing to understand is that there are two parts to an IEC 61000-4-2 ESD pulse; fast and slow ESD. The fast ESD stress refers to the high frequency content initial current pulse and the slow ESD stress refers to the low frequency content current after the initial current pulse. The fast and slow currents tend to use different current paths to ground. Current will always flow preferentially through the lowest impedance path. The impedance of any current path is, however, frequency dependent. Current paths dominated by capacitive elements will have their lowest impedance at high frequencies, while current paths dominated by inductance will have their lowest impedance at low frequencies. It is therefore not surprising that the initial current spike in the ESD waveform with its high frequency components may travel through a different path than the later parts of the ESD stress which is dominated by lower frequencies. See Figure C1 for more details.

As an example, in Figure C1 there is a large metal sheet which is grounded only at one end and floating throughout the rest of the area. Also shown in the figure is a simplified parasitic circuit element drawing for the mechanical structure (in dotted lines). Clearly visible is a capacitive and inductive path for the ESD current. In Figure C1, there is a fast stress capacitive path through the small board which can either overstress the components on the small board or could generate unwanted operation effects. The slow stress flows through the inductive path to board ground.

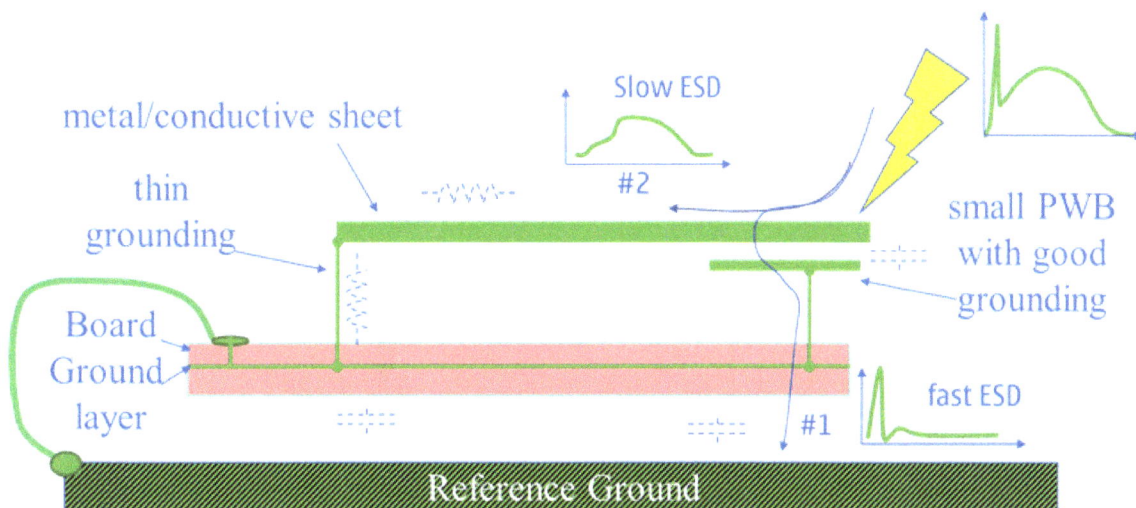

Figure C1: Fast and slow current ESD paths

For corrective actions, the ESD current flow needs to be redirected. In Figure C2, some options are shown:

C2a: Adding better grounding to all the metal corners can redirect the ESD current in a different path. This will reduce the high frequency impedance of the inductive path creating less fast capacitive ESD current coupling to the small board.

C2b: Add an additional metal shield over the small board. This will guide the fast ESD current to the main board and ground without disturbing the small board.

C2c: Improve the pig-tail grounding (i.e. reduce the inductive impedance) and additionally try to increase the connection impedance of the small board to the main board by adding coils or ferrites. However, in this solution, note that the impedance should be added to all lines. As a consequence the small board would "float" from an RF perspective, making the component more sensitive to RF fields. This solution needs careful tuning and should not be the primary fix.

C2d: Create some additional mechanical structures to introduce other current paths for the ESD pulse. Again, this is not as good a solution as proper shielding, but may work in some cases with proper fine-tuning.

Figure C2a: Good overall grounding

Figure C2b: Small board's own shield

Figure C2c: Better grounding, while increasing the impedance of the small board grounding. Note that this solution makes the small board more sensitive to other RF noises.

Figure C2d: Create an additional mechanical "ESD trap" to guide ESD for an alternative route.

When completing the ESD design, it is good to remember that ESD has a rather high momentary current. Let's look, for example, at Figure C3 in which the upper metal sheet is actually a printed circuit board and the thin pig-tail grounding is still in use. Assume, for example, that the high frequency impedance (inductive reactance) of the pig-tail is equal to 1 Ohm. If an ESD pulse creates an 8 A fast ESD peak current which sees the impedance of inductive reactance this will create a momentary 8 V (up to) ground bounce on the upper board, meaning the whole upper board voltage level will rise relative to the lower board reference level. Any sensitive signal connected between the boards with these pig-tails will have a momentary glitch in the signal between the boards. The example values given in these calculations are just relaying the idea of the phenomenon, not describing it in exact detail where the time effect should be addressed in more detail.

In this case, the best option is to have a differential signal transmission between the boards which has higher immunity to such ground bounces. However, the differential signal needs to have high enough swing with good drivers and receivers to minimize the noise effect. It is also a good idea that the logic connected in such interfaces is not asynchronous nor edge triggered. Level triggered (synchronous logic) is more immune for momentary glitches. That should be valid for all signals, including a reset signal for example. Reset signals are often edge triggered so small glitch can cause unwanted resets. There are use cases where an immediate reset is important or necessary for application, but if possible, some minor glitch filtering can be a good idea. This glitch filtering would of course need to balance the need between speed and sensitivity to small noise glitches.

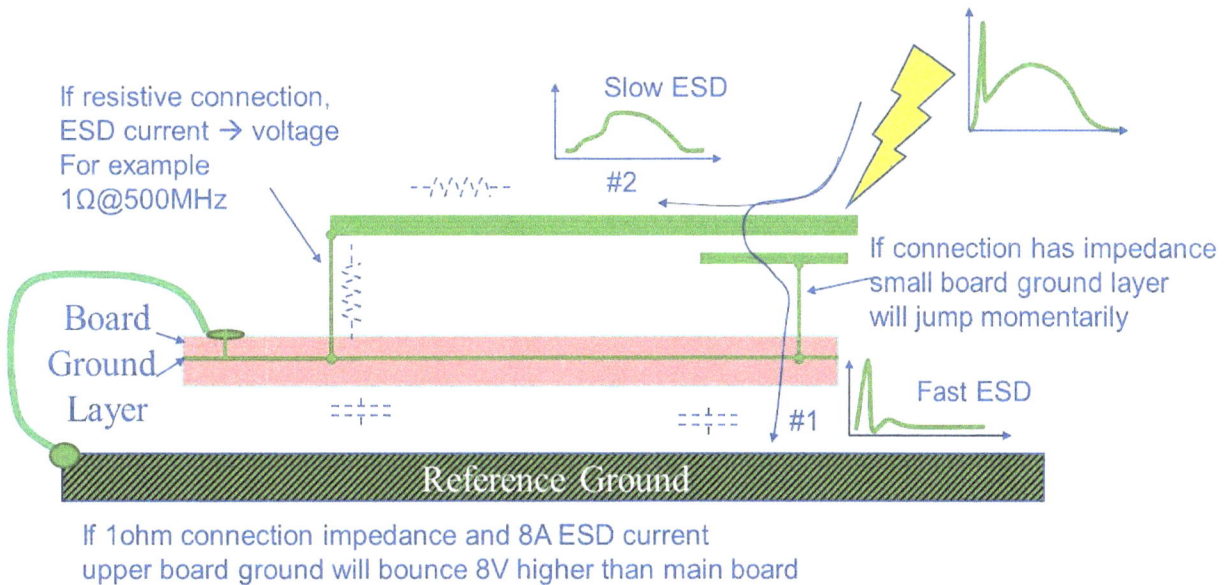

Figure C3: ESD current causes ground bounce

C.1: Some Words on ESD Debugging

In thinking through the ESD effect, one should remember that there are both ESD current and ESD voltage involved. The majority of the problems come from ESD current, but there are cases where ESD voltage (through the associated electric field) can cause a problem also. For example, a floating trace with weak pull-up or pull-down can cause edge sensitive logic to react to ESD noise induced voltage on the gate input. In the worst case scenario (Figure C4) the logic will react before the ESD event, just due to the charged ESD gun. In such cases shielding of traces and improving the pull-ups or pull-downs will help.

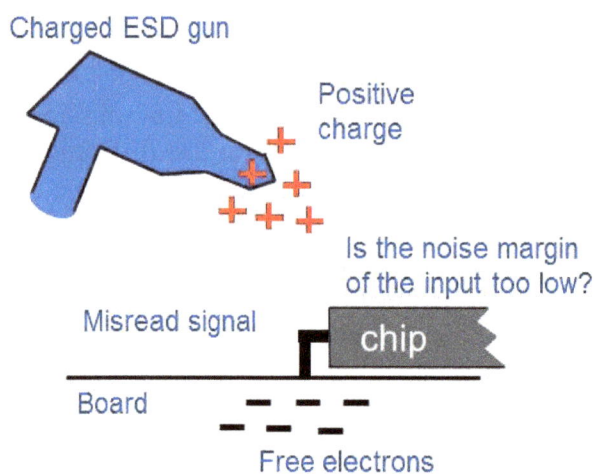

Figure C4: ESD induced noise on signal

A charged ESD gun (or some other charged material) can affect the board and could cause unexpected voltage levels on floating traces. All traces should be pulled to some known status. Also, all the unused IC IO pins should be set to an output mode to avoid unnecessary interrupts due to floating IO inputs. These unexpected interrupts can cause problems in software.

Understanding the ESD current path will help dramatically during ESD analysis. For example rearranging the grounding points of components or board subassemblies during ESD testing can help debug ESD induced malfunctions by isolating areas where the effect is reproducible. Root causes are more easily identified when the susceptibility is localized to a small area of the system. The root cause and the associated local area or sensitive IC can be identified using susceptibility scanning in which strong, but local transient fields are produced by driving electric or magnetic field probes using a transmission line pulse generator or equivalent disturbance generators.

As an example of isolating an area, one could begin inducing ESD into the board grounded on one end as shown in Figure C3, and then move away from this point (increasing voltage levels and alternating polarity in multiple passes) until failures begin to appear. Then one could move the ground to the other end of the board and identify the robustness of the area around that point until a susceptible area is found. Ideally this procedure will identify a localized area (or set of areas) in the system where the susceptibility is obvious. At that point the root cause analysis of individual components and signals in the area can commence. Of course this procedure is mostly relevant to soft failures and upsets, since permanent damage during hard failures is readily apparent and isolated to a particular component.

Appendix D: Review of White Paper 3 Part I

Robert Ashton, ON Semiconductor
Charvaka Duvvury, Texas Instruments

D.1   Part I Chapter 1: Purpose and Introduction

Part I Chapter 1 introduced not only the motivations for the document, as discussed above but also some of the common terminology to provide a basis for the paper. The most important was the distinction between *internal* and *external* pins in a system as illustrated in Figure D1. Pins that connect between chips on a board will not need the same level of system ESD protection as pins that attach to a bus which connects between circuit boards within a system. Pins which go to a connector, such as USB port, would require proper external clamp design to protect against direct system level ESD stress events. By the same token, some purely inter-chip pins may require more protection or design care due to crosstalk with signal lines connected to external connectors or long wiring harnesses.

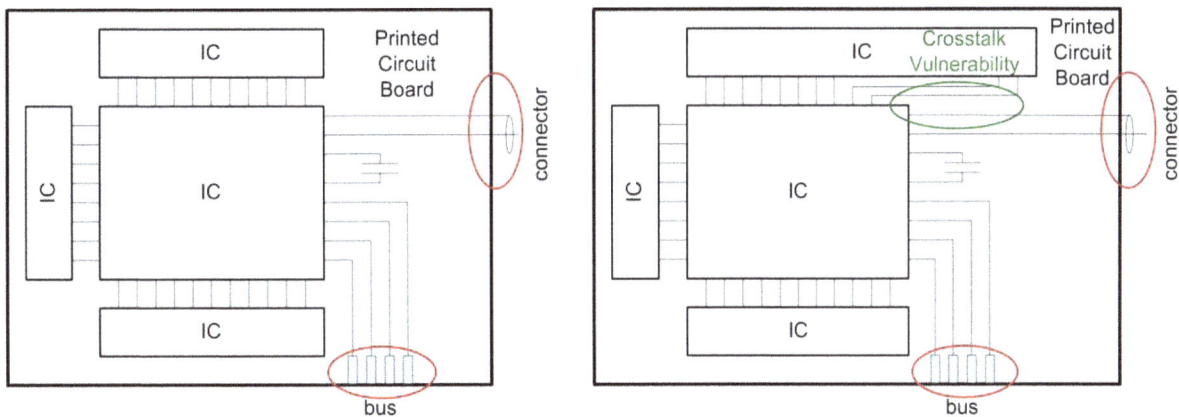

Figure D1: Classes of pins for specific system level ESD considerations including external and inter-chip coupling.

D.2   Part I Chapter 2: Test Methods and Their Field of Application

Part I Chapter 2 described how systems are tested for ESD with a focus on the most widely used standard, IEC 61000-4-2. This standard defines the stress waveform, the test environment and defines how a test plan should be developed. The IEC 61000-4-2 test bench is shown in Figure D2, showing direct stress to the unit under test as well as stressing to horizontal and vertical coupling planes which can induce errors due to coupling. The response to a system level ESD stress is not a simple pass/fail but has 4 levels of possible response as defined by IEC 61000-4-2:

   a) Pass: Normal performance within limits specified by the manufacturer
   b) Conditional Pass: Temporary loss of function or degradation of performance which ceases after the disturbance ceases. Equipment under test recovers its normal performance without operator intervention

c) Conditional Fail: Temporary loss of function or degradation of performance. Recovery requires operator intervention

d) Fail: Loss of function or degradation of performance which is not recoverable, owing to damage to hardware or software, or loss of data

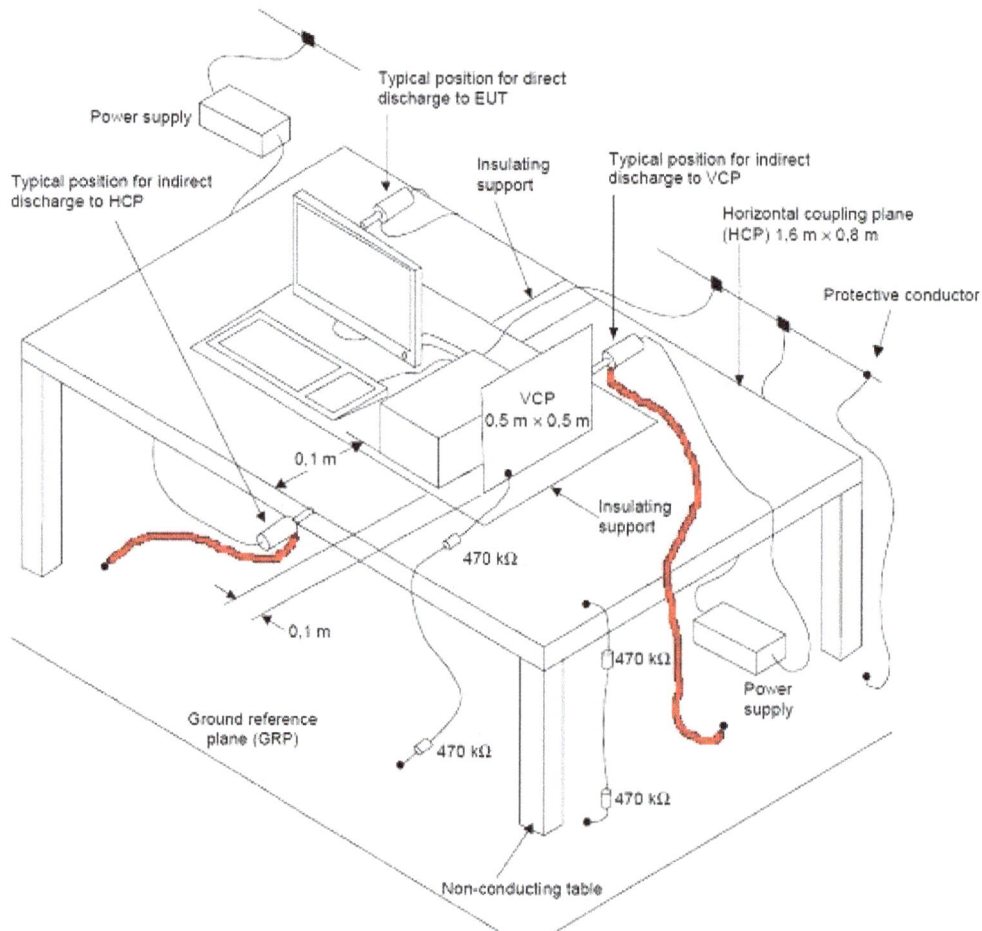

Figure D2: Test bench for IEC 61000-4-2 ESD testing [1]

Part I Chapter 2 also described several other system level standards that are based on IEC 61000-4-2, such as the automotive standard ISO 10605 and avionics DO-160 standard. The automotive and avionics test methods differ in detail from the more general IEC 61000-4-2, but the basic premises of the documents are similar and the failure modes are a combination of both recoverable failures and non-recoverable or hard failures. Chapter 2 also discussed two test methods in which system level ESD waveforms are applied to components. These include IEC 62228 for CAN bus transceivers and the more broadly applicable American National Standards Institute (ANSI)/ESD SP5.6-2009, which is known as the HMM. Both of these component ESD tests focus on physical failure and not system upset. It should be noted that IEC 62228 includes a suite of EMC tests other than ESD which do deal with system upset, however the ESD test does not deal with system upset.

D.3 Part I Chapter 3: Proven System Level Fails

Part I Chapter 3 presented analysis of 58 system level case studies provided by Council member companies. The type of failures and when the failures occurred are shown in Figure D3. The case studies cover the full range from system qualification, to field installation and field failure after installation. The types of failures include both physical and soft failures.

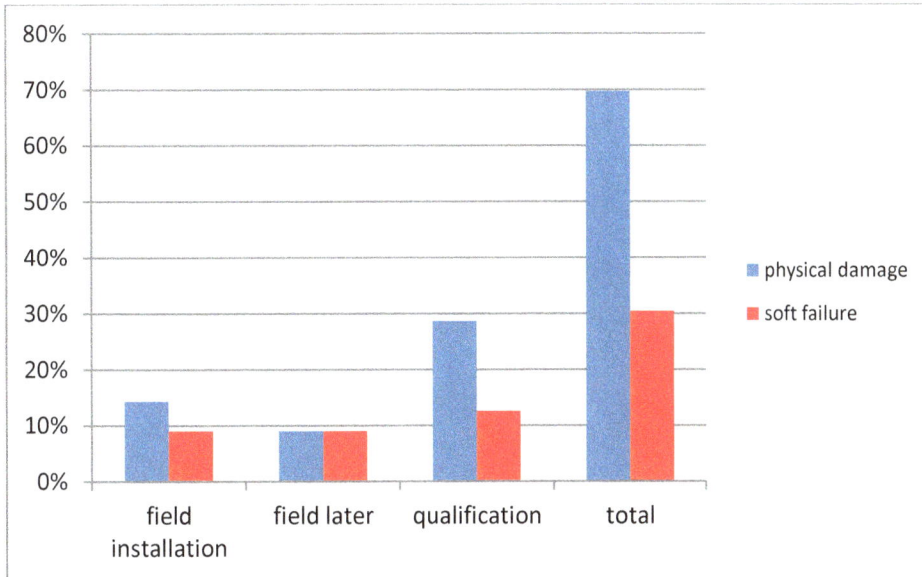

Figure D3: Types of failures and when failure occurred

How these problems were solved is shown in Figure D4. IC redesign was found to be the solution in only a small percent of the cases and that was only in cases in which the damaged pin was directly connected to a system IO. In one case, improving the HBM level of the IC actually resulted in a decrease in system level ESD robustness. The most interesting outcome of the case studies was that no correlation could be found between component ESD (HBM and CDM) and the system ESD failure levels of the products. Chapter 5 of Part I examined this in more detail in Section 5.1.5.

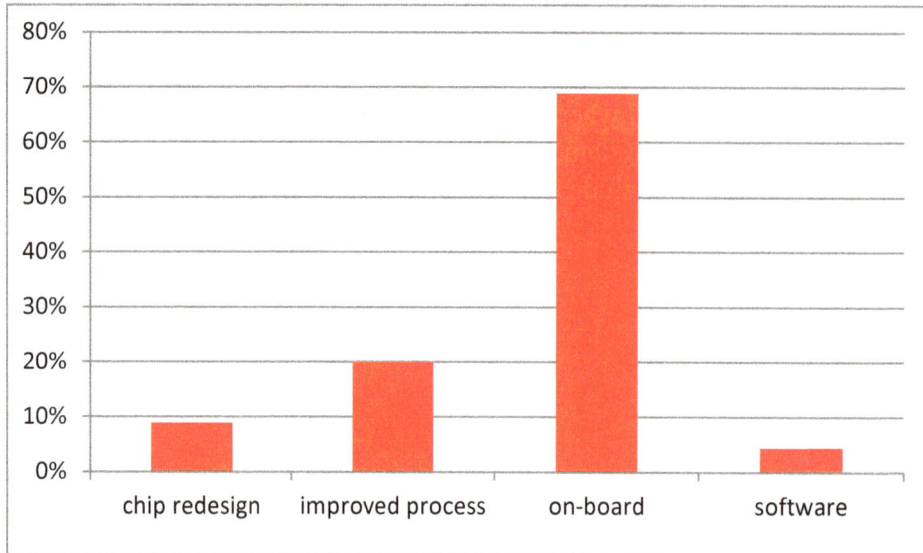

Figure D4: Solutions of system level failures

D.4   Part I Chapter 4: OEM System Level ESD Needs and Expectations

Part I Chapter 4 described OEM system level ESD needs and expectations. This chapter described three potential paths to designing ESD robust systems.

1. The components, including integrated circuits, chosen for the system are all inherently robust to system level ESD and the OEM does not need to think about system level ESD.

2. Not all components, including integrated circuits, chosen for the system are inherently robust to system level ESD, but component suppliers provide clear rules and procedures for using a set of system level ESD robust and non-system level ESD robust components that will produce an ESD-robust system.

3. Not all components chosen for the system are inherently robust to system level ESD, and the OEM has to find a solution on their own to design an ESD robust system.

The chapter then showed that Path 1, while appearing desirable, is not realistic for a variety of technical, practical and economic reasons. Path 3 is obviously undesirable since it is little more than a trial and error approach. Path 2 becomes the path of choice in which the properties of both ESD robust and non ESD robust components are understood and a design methodology is available to design a system with high probability of success. This path has been called system-efficient ESD design or SEED. This approach was described in the summary of Part I Chapter 6 and is a major topic of this white paper to be analyzed in more detail.

D.5   Part I Chapter 5: Lack of Correlation between HBM/CDM and IEC 61000-4-2

Part I Chapter 5 took on the question of correlation between HBM/CDM and IEC 61000-4-2 in detail. For the 58 case studies from Part I Chapter 3, only 9 were found to have both well documented HBM levels and system level data. The limited data in Figure D5 shows no evidence of a correlation between HBM level and system level ESD. The remainder of the chapter focused on reasons why a correlation between HBM levels and system level ESD should not be expected.

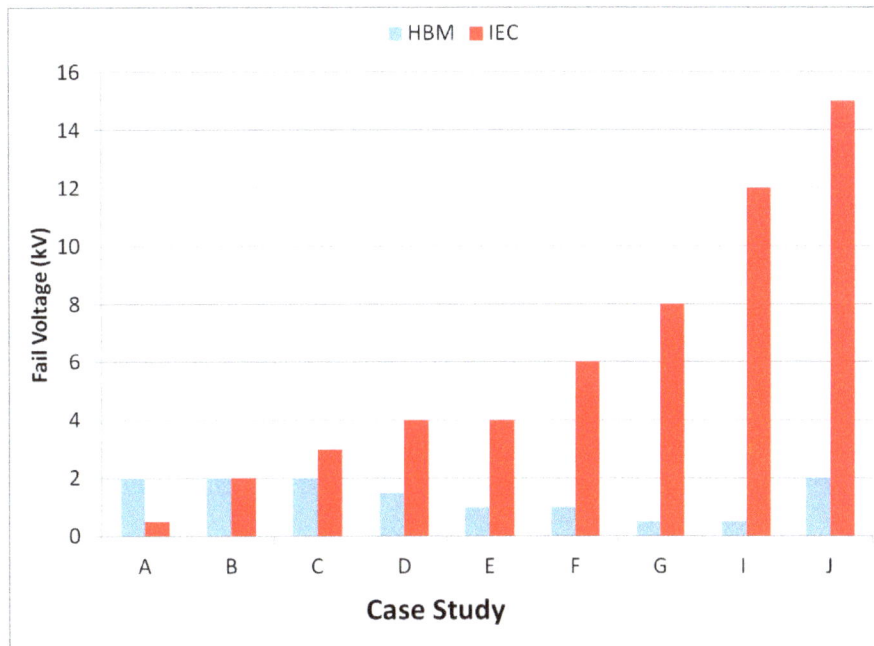

Figure D5: Component level HBM and IEC 61000-4-2 system level results compared

Component level HBM and IEC 61000-4-2 stress simulation circuit diagrams have a superficial similarity, but the waveforms are substantially different. HBM has a classic double exponential wave shape with a 2 to 10 ns rise time and 150 ns decay time constant, while the IEC waveform has an initial current spike with a sub-ns rise time followed by a second broader peak with a 50 ns decay constant. The applications of the two stresses are also totally different, as outlined Table D-1. With the major differences between the two stresses and the conditions under which they are applied, it was concluded that a lack of correlation in some cases was to be expected.

Table D-1: Comparison of IC level HBM (ANSI ESDA/JEDEC JS-001-2010) and System Level ESD (such as IEC 61000-4-2, ISO 10605)

	IC level ESD test	System level ESD Test
Stressed pin group	Multitude of pin combinations	Few special pins
Supply	Non-powered	Powered & non-powered
Test methodology for 'HBM'	Standardized	Application specific using various discharge models
Test set-up	Commercial tester & sockets	Application specific board
Typical qualification goal	1 ...2 kV HBM	8 ...15 kV
Corresponding peak current	0.65 ... 1.3 A	> 20 A
Failure signature	Destructive	Functional or destructive

Chapter 5 also looked for correlations in other system level tests such as CDE, transient latch-up (TLU), human metal model, extended length TLP, and CBE. While some correlations were found, such as between CDE and extended length TLP, no evidence could be found for higher levels of HBM or CDM leading to more robust system level ESD.

D.6 Part I Chapter 6: Relationship between IC Protection Design and System Robustness

Part I Chapter 6 dealt with the relationship between IC protection design and system robustness for physical, hard, failures by first introducing the concept of SEED. Subsequently this new concept was discussed in more detail and with specific examples. This involved detailed descriptions of a number of different considerations. IC ESD protection design methodologies were reviewed and it was shown that when the IC is included within a system, these methodologies are often not relevant to system level robustness. The chapter also examined the tradeoffs between adopting system level ESD protection within an IC (on-chip design) versus using board components (on-board design) for protection. On-chip system level ESD protection conceptually simplifies system design, but it significantly increases chip area and introduces on-chip protection techniques that are not always well suited for system level protection.

Full system ESD design was then introduced as shown in Figure D6. ESD stress can enter into a system by several paths, 1) stress to the system's cover creating ground currents, 2) stress entering through seams or vent holes, 3) stress entering through IO connections and 4) stress entering by arcs from nearby hardware such as screws in the system case. Of these paths, 1, 2, and 4 are best fixed by changes in the physical case to eliminate the stress. Only stress through a connector or to input/output devices, such as key pads, need to be protected by on-board or on-chip protection elements, or a combination of both. This is illustrated by the proper order of ESD solutions, a) cover redesign, b) air gap redesign, c) shielding and grounding, d) on-board protection and finally e) on-chip protection.

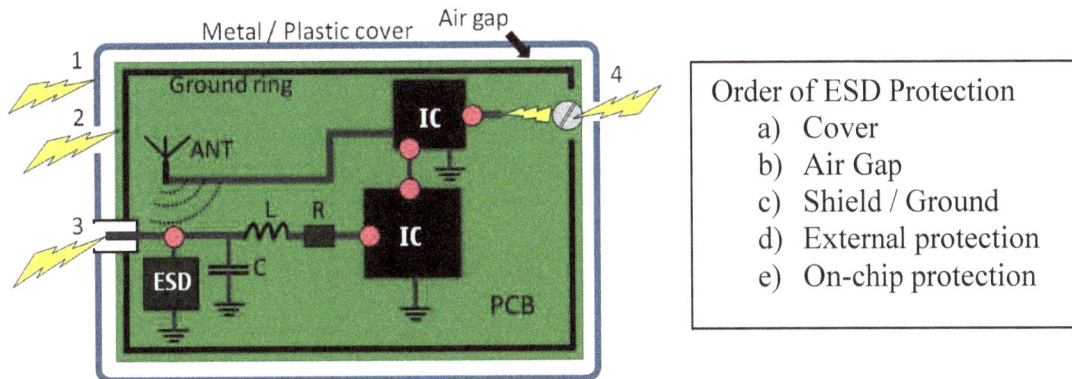

Figure D6: System ESD protection depends on product physical protection (covers), shields and groundings, on-chip and external signal protection and signal integrity targets.

There are several considerations for protection at the printed circuit board (PCB) level. On-board protection products including diodes, varistors and polymer components were introduced and the advantages and disadvantages of these products were compared. For protection of the most sensitive ICs, diodes have an advantage in terms of the balance of protection capability and minimal degradation in circuit performance. The chapter then introduced an overview of the SEED approach and its applications.

SEED primarily requires the understanding of the properties of the IC, on-board protection elements and the properties of the board and passive components on the board. This is illustrated in Figure D7. The properties of the IC and external transient voltage suppression (TVS) component are determined using TLP measurements. A standard 100 ns TLP system to mimic the long pulse duration (after the initial current spike) of the IEC stress pulse can be used. Knowing the properties of the TVS component and its response to the ESD stress, usually the IEC ESD waveform, it is possible to determine the residual voltage at the output of the TVS component. Applying this residual voltage in simulation to the IC, as filtered by on board components (such as resistors, capacitors, trace impedance, common mode filters, and chip ferrite beads), it is possible to determine the stress voltage and current which the interface IC pin actually experiences. If the calculated voltage and current are within the IC pin's safe operating area (SOA), the protection design can be considered acceptable. If, however, the voltage and current are outside of the SOA, either due to too high a current or too high a voltage, the protection design needs to be modified. This can be done in a variety of ways. A different TVS component can be chosen with lower turn on or lower dynamic resistance or the on-board components can be modified to reduce the stress on the IC. In a more detailed analysis, as described in the chapter, the first spike of the IEC pulse may also need to be considered using very-fast TLP (VFTLP).

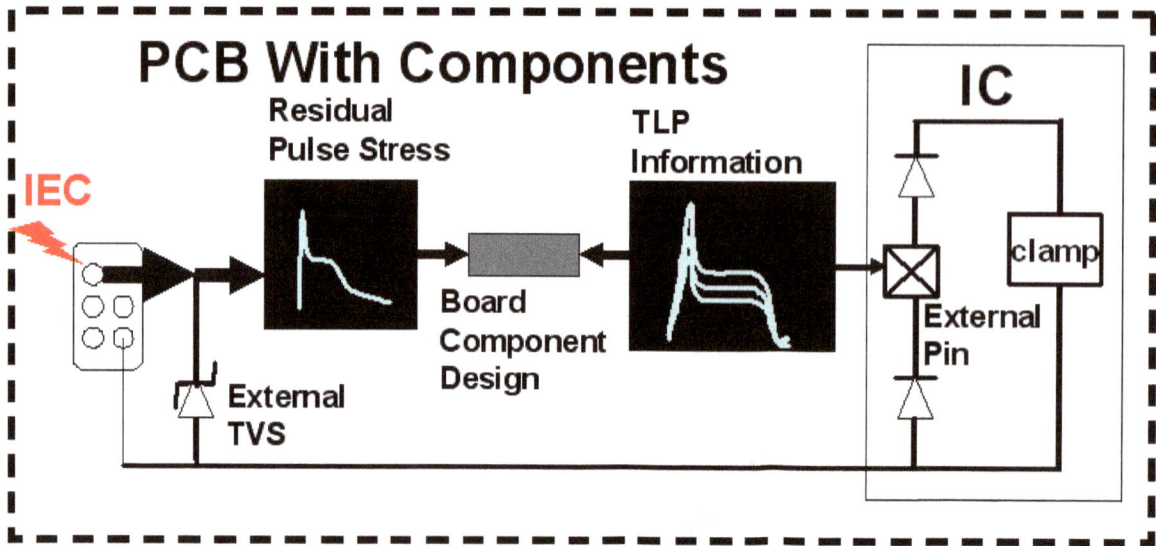

Figure D7: System-Efficient ESD Design (SEED) Methodology

D.7   Part I Chapter 7: Summary, Conclusions and Outlook

After reviewing the misconceptions about system level ESD and introducing a more realistic approach to efficient system level protection, the final chapter summarized the key takeaways from Part I. The summary of the method for producing successful system level ESD designs has the following key features:

- ESD test specification requirements of system providers must be clearly understood as a separate domain from IC level ESD specifications. IC level ESD specifications should not be used as a basis for system level requirements.
- Understanding ESD failure and upset mechanisms is critical to recognizing their relevance for robust protection design and for correlating them to the IC specifications.
- Most importantly, for achieving proper system level ESD protection, the responsibility must be shared between system designers and IC providers. This can happen only when a communication path is established for dealing with system level ESD design.

The document finally illustrated some product examples where, without this communication taking place, robust system level protection design could involve time to market delays and inefficient solutions. From the background information conveyed in Part I, the present document (Part II) examines more of these issues in greater detail and includes more details on soft failures for a more comprehensive understanding of system level ESD.

References

[1] IEC 61000-4-2, 'Electromagnetic compatibility (EMC) – Part 4-2: Testing and measurement techniques – Electrostatic discharge immunity test', Ed. 2.0, 2008.

Appendix E: Frequently Asked Questions

Q1: Why is the Industry Council addressing Non-Correlation issues between Component Level and System Level testing?

Answer: Some OEMs have been under the impression that higher levels of system robustness can be achieved by designing and measuring greater than necessary IC ESD levels. Our focus is to show that the system ESD measurement is relevant only when the IC is placed on the printed circuit board (PCB) and that stress data obtained at the IC level does not often correlate to system ESD capability when running the IEC 61000-4-2 test procedure.

Q2: Is there a correlation between component failure thresholds and real world system level failures?

Answer: There is rarely correlation between component (IC level) failure thresholds and real world system level failure <u>in the field</u>. Component failure thresholds are based on a simulated ESD voltage and current injected directly into the component (IC) with the component in a powered down condition. Real world system level failures <u>in the field</u> occur in many different conditions, most of which are powered. In addition, there is no clear definition of soft failure robustness for ICs, and many real world errors are soft failures. First one needs to establish reliable methods for soft failure evaluation of ICs before one can attempt to compare IC level and real world failures. In addition, system level ESD tests for hard and soft failures under operating conditions, while IC level mainly tests for damage under un-powered conditions.

Q3: Why wouldn't you expect to see correlation between IC level and system level testing?

Answer: Since the tests are done in different environments (unpowered versus powered or stand-alone versus on board) along with the different stress current wave shapes for the two tests, it is not surprising that they would be uncorrelated. In some instances external IC pins with higher IC level ESD robustness may result in less of a load on the board ESD protection components. However, there are many examples where an improved HBM level for an IC resulted in lower system level ESD as discussed in Chapter 5 and Chapter 6 of White Paper 3, Part I. The approach of relying on IC level ESD for system level ESD protection is not only impractical and unpredictable, it also detracts from the need for an efficient system ESD design in which the on board and on chip protection work together.

Q4: Can components really be designed to withstand real world system level events?

Answer: It is certainly possible to design ICs that can withstand system level ESD stress, but it is a complex and often unwise path. It is hard to know the exact details of the stress that will reach an IC on a board due to circuit board parasitics, making the design difficult and prone to overdesign. Additionally, IC protection for IEC stress consumes considerable area and is likely not to be the most economical path.

Q5: Do all pins on a component need to be tested using system level events?

Answer: <u>External pins</u> (such as USB data lines, Vbus line, identification (ID) and other control lines; codec and battery pins, etc.) need to be tested if the IC is not to be protected with on board components. But if the pin is to be protected by on board components, TLP characterization of the pin is more useful. <u>Some internal ESD sensitive pins</u> (such as control pins, reset pins, and high speed data lines, etc.) can be inductively coupled during a discharge to the case and/or to an adjacent trace of an external pin undergoing system testing. These pins need to be identified and may need to be tested using system level events.

Q6: If system level ESD testing at the IC level does not guarantee system level ESD performance, aren't higher target levels of IC HBM ESD better than nothing?
Answer: This would only give a false sense of security and could result in extensive cost of analysis, customer delays and a circuit performance impact. (Remember, higher HBM ICs may be harder to protect in a system!). System ESD protection depends on the pin application and therefore requires a different strategy. System level ESD is clearly important, but targeting excessive IC level requirements could pull resources away from addressing and designing better system level ESD.

Q7: Since ICs are now designed for lower IC ESD levels, why would this not be reflected by a sudden change in the overall health of a system for ESD capability?
Answer: The overall health of a system is dependent on a comprehensive approach to the protection methodology that includes a number of factors including on board protection components, optimized board signal routing, component packaging, mechanics (covers), and, as a last line of defense, the IC level protection.

Q8: How will you reach all the different system designers for their inputs?
Answer: The present document has been reviewed by a number of OEM representatives and they are in agreement with the conclusions of the document. We expect that the publication of this document will result in further input from the system design community.

Q9: If the system designers who are not involved in this document do not agree that it is a shared responsibility then what is next?
Answer: The system designers need to be educated in terms of system ESD versus IC protection design. Education with regard to these issues is a major focus of the Industry Council and we are convinced that as the benefits of the shared ESD responsibilities become evident more system designers will become convinced of their shared responsibility.

Q10: What is the purpose of IEC 61000-4-2?
Answer: The purpose of the IEC 61000-4-2 test is to determine the immunity of systems to ESD events during operation. The document states that it relates to equipment, systems, subsystems and peripherals, without further defining them. Its scope and description clearly indicate the purpose: to test electrical and electronic equipment that may be subjected to ESD from operators directly to the system under test or from indirect discharges from personnel to adjacent objects. See Section 2.0.1 of Chapter 2 in White Paper 3, Part I.

Q11: What is HMM?
Answer: HMM stands for human metal model. It is a method to assess the robustness of external IC pins against a system level ESD pulse. See Section 2.2.2 of Chapter 2 of White Paper 3, Part I for details.

Q12: Do all system level ESD standards use the same waveform?
Answer: In short: No. However, most use the waveform as defined in the IEC 61000-4-2, which is determined by a 330 Ω, 150 pF resistor/capacitor (RC) network. An example of a standard that uses a different waveform is ISO 10605. This standard uses the same type of ESD gun but the RC network is modified for some of the tests, using a 2 kΩ resistor and/or 330 pF capacitor instead of the values used in IEC 61000-4-2. See Section 2.1.1 of Chapter 2 in White Paper 3, Part I.

Q13: Is SEED considered to reproduce real, physical behavior of board and IC?
Answer: SEED is a concept to limit damaging current pulses reaching the internal IC pin. So in this sense it represents what the physical effect would be on the IC pin coming from an IEC stress at the external port of the PCB. What it represents for the board depends on how well the scenario is represented during the SEED analysis.

Q14: How can system/board designers get the required information about the IC IO behavior?
Answer: First, both the OEM and the IC supplier must define the 'external pins'. Following this, the IC supplier provides the TLP curve of the pin under interest with either bias applied or without bias which would depend on the pin application in the overall system board. The measured TLP at the pin will not only represent the pin's internal ESD clamp behavior but it will also include the IO design behavior to the transient pulse analysis.

Q15: What is the required degree of accuracy of the simulation models?
Answer: The simulation models can only be as accurate as the measured waveforms at the external clamp under IEC pulses along with the variations, and the internal IC clamp under the TLP conditions. Experience will teach us what level of accuracy is needed. Even if early attempts at simulation do not have the level of accuracy we may desire, the simulations will still provide insight into the ESD properties of a design.

Q16: Can the SEED concept ensure system robustness against any EMC and ESD related fail?
Answer: SEED is a concept to develop a robust system based on characterization of its components when subjected to IEC ESD stress conditions. There is a wide range of EMC type interferences and resulting fails. Established best practice EMC design approaches and quantitative analysis by simulation are not replaced by the SEED concept. Moreover, the robustness of the system against high energy pulses like an ESD discharge is built on the assumption that low amplitude interference can be handled by the system through appropriate EMC design.

Q17: Can I also use SEED for systems not listed in this White Paper?
Answer: Definitely. SEED is a basic concept which can be applied to any system. The main constraint is that all relevant components of the system have been characterized accordingly and modeling information is provided to the system designer.

Q18: OEM customers require specific system level ESD tests like (MM-like) bipolar testing. Do component suppliers need to satisfy both SEED characterization and additional company specific test requirements?
Answer: SEED is built on the knowledge of the failure mechanisms known by the large group of experts contributing to White Paper 3. It is the expectation that the SEED concept also addresses design optimization measures which will improve robustness against bipolar pulses. The requested tests are a matter of supplier-customer relation. Industry Council recommends the efficient test and design flow which is described within SEED.

Q19: White Paper 3, Part I introduced the concept of system-efficient ESD design (SEED) approach to system / component ESD co-design. Are there more details in Part II extending this SEED approach to other than "hard" system failures?

Answer: Yes. White Paper 3 Part I introduced the term system-efficient ESD design (SEED) by giving the basic example of a hard fail at an external pin. This has been expanded upon in White paper 3 Part II. SEED includes characterization and design optimization for hard fails at external pins, transient latch-up at internal and external pins and soft fails due to noise injection.

Q20: Can I apply the SEED concept right away for my products?

Answer: In principle yes, as long as the models and tools are available for the specific system. However, in practice there is still a ways to go. For example, both transient latch-up and noise characterization methods need to be defined in more detail and finally standardized. White Paper 3's main intention is to guide the industry towards developing the ecosystem to enable successful use of SEED.

Q21: In addition to the "hard" system failures described in White Paper 3, Part I, what additional failure types are described in Part II?

Answer: Beside irreversible hard failures, which exhibit physical damage of the component, so-called soft failures are also discussed. Soft failures are reversible by definition. These failures might require intervention of the user or the system could also exhibit a short interruption of normal operation and regain full functionality without intervention. Detailed definitions for hard and soft failures are given in Chapter 2. The document also provides a novel categorization of hard and soft failure types linking them to the underlying physical stress mechanisms and the applications of SEED (Chapter 4).

Q22: What are the most common ways external events happen and couple into a system?

Answer: External electrical events include ESD (Electrostatic Discharge), EFT (Electrical Fast Transients) from switching of inductive loads and running motors and nearby lightning (Surges) which are all transient effects and can also include RF (Radio Frequency) being radiated or conducted into a product. In this document we are only dealing with ESD. However, EFT is in many ways similar to ESD and can be coupled into a system in a similar manner. Further details on how ESD is coupled into a system are given in Chapter 2. Another way to categorize the way external events happen is by what the external root cause is: 1) a charged person discharges into a product; 2) a charged product discharges when it is plugged into a system; 3) a charge cable discharges a product and 4) a charged product discharges into another product when they are joined together in use.

Q23: Are there ways ESD / EMI can enter a system that is not well understood and modeled at this time? Does Part II describe these and provide guidance or direction for understanding?

Answer: While the physics of direct or indirect injection of charge into a system are well understood, it is hardly ever possible to model the complete entry path of ESD due to the complexity of real systems if both enclosure and PCB details are to be included in the same model. For example, it is very difficult to model the upsets of LCD displays including their control circuits and the direct coupling of ESD transient fields into those integrated circuits. However, the spread of ESD currents into a system can be modeled using EM simulators, and the propagation and absorption or reflection at ESD protection components and IC input can be modeled using SPICE simulators. Use of past experience, system component susceptibility and 3D case simulation where practical are all encouraged during system design. However, it is important to note that poor realization of a good design can break all design efforts to shield a system.

Q24: If I am developing a synthesized IC (with external IP) for system applications, what would Part II propose regarding how to investigate its system ESD susceptibility? Do simulation tools exist (or can they be developed) to investigate this?

Answer: Simulations tools (SPICE, VHDL-AMS, VERILOG-AMS, etc.) are capable of nodal analysis that can identify susceptible pins and/or internal circuits. However, these tools are only as good as the netlists and the models that they are provided as input. Often the netlists provided by external IP providers are focused on functionality and do not include the parasitic elements (for example, lateral or vertical bipolars in an IC's IO) nor do they model the high current regions of operation that are necessary for system ESD modeling.

Q25: How would troubleshooting a "soft" internal system failure differ from troubleshooting a "hard" failure? Does Part II give guidance on this?

Answer: Soft failures can be understood and fixed by three methods. The most common is trial and error in diverting the ESD current, shielding using ferrites or filtering of lines. A more in-depth method is to identify sensitive nets or ICs using susceptibility near field scanning which locally injects ESD derived noise into traces or ICs to determine at which level of local injection the same soft failure is observed as in system level testing. The third method uses software and firmware to identify which process or trace has been disturbed.

Q26: Regarding common shielding / isolation methods for improving system ESD performance, which method works best for soft failures? Does it depend on the type of soft failure?

Answer: At first one should determine which IC is affected, and in which traces or cable connection the noise couples. This is best done by susceptibility scanning the device for sensitive areas. After the sensitive nets, connections and ICs have been determine countermeasures such as improved field confinement of wanted signals, guiding of ESD currents, filtering of wanted signals, shielding or software changes can be implemented. Still, there is no simple solution for improving system ESD performance to reduce soft failures. In one case, shielding can be effective, but with another design isolation of a sensitive circuit might be effective. Use of shielded cables can certainly be effective but this is not always cost effective; isolating sensitive circuits may not be possible when the system consists of a single PCB with interfaces to the outside world. The type of soft failure depends on the system functions and which may be critical to operation. Chapter 3 contains information regarding the use of shielding to prevent ESD upset.

Q27: I have read that system ESD / EOS failures can result from isolation from / reconnection to ground. I know the IEC 61000-4-2 test provides for system testing of systems not having a path to ground. But this test passes on my system.

Answer: Regarding the ungrounded product which passes the IEC test: If during the test the product becomes charged an ESD event could certainly occur if some metallic part on the product suddenly becomes grounded -- by connecting a grounded cable or dropping it on a metal surface for example. If this is a concern, then an IEC test should be expanded to include discharging the product after a test.

Q28: Does EOS cause different system failures than system ESD failures? What is the difference?

Answer: This depends on the interpretation of the vaguely defined term electrical overstress (EOS). In the broad definition ESD is a subset of EOS. But often one uses the term more narrowly to refer to failures and damage due to transients of high power. Usually, the hard failures caused by a system level ESD will show less severity than an EOS due to a power transient. While ESD applies a limited pulse of stored energy to a component which may damage it during the short discharge pulse, EOS usually results from a condition where a component is subjected to a relatively large source of energy. As a result, when a failure occurs, EOS damage tends to produce extensive melting and charring at the damage location while ESD stress tends to leave burned junctions, blown oxides and melt stringers in its wake. The boundaries between ESD and EOS physical failure are not firm and some ESD, such as charged board events and cable discharge events have sometimes been initially mistaken for power-transient induced EOS. In some cases an ESD can trigger a subsequent stress such as latch-up, which may then appear more like an EOS.

Q29: Are there "non-invasive" methods of system board analysis for system ESD immunity?

Answer: Non-invasive methods for system level analysis are mainly software based. In many cases it is possible by reading register values, by checking bus errors, by looking for terminated processes to identify the root cause of a soft failure. This might require that the firmware is written with ESD soft failures in mind to provide sufficient error recording and register read out.

Q30: What board-specific design considerations in Part II will be useful to help reduce the incidence of "soft" system ESD failure? For example, placement of components, type of component used, trace isolation, shielding, etc.?

Answer: Primarily, all nets in a board design need to be grouped into nets which are connected to IO ports on the board, (nets connected to other boards or to system IO ports: HDMI, USB, etc.), nets that stay on the same board but have a high likelihood of coupling, and low risk (short, well shielded) nets. If available from IC soft error qualification, the sensitivity of each IC pin for all ICs on the board with respect to ESD-induced soft failures should be known. Those nets which connect to sensitive IC pins and that are connected by IO ports or have a high probability of coupling need to be filtered or rerouted if possible. Further, in software and firmware design, possible bit-error detection strategies should be considered to allow for transparent error recovery. The enclosure design is a primary methodology for guiding the ESD current and the associated fields such that the coupling to the interior of the enclosure and to PCBs and board to board connections is minimized. Connector shells are typical entry points. These should be connected to the enclosure using a 360 degree connection. Even an inductance of 2 nH between a connector shell and an enclosure will allow a 5 kV ESD to create a pulse of about 35 V between the connector shell and the enclosure. This pulse will drive a current onto the PCB increasing the likelihood of soft failures. The second aspect of the enclosure design is to avoid any non-grounded metal as these can cause secondary ESD.

Q31: From this white paper, are there test methods other than IEC 61000-4-2 that are relevant to understanding system level ESD failures?

Answer: The most common ESD scenario relevant to system level are the human metal discharge (IEC 61000-4-2), the discharge of cables which are plugged into connectors and the discharge of furniture, such as metallic lab carts. The discharge of cables is certainly of growing importance as these ESDs often lead to damage because the ESD currents can be injected directly into the pins of high speed interfaces. At present no tests are conducted for the discharge of furniture, however in the 1980's a "crossed vane" simulator was used. Experience showed that discharges from the IEC 61000-4-2 ESD generator at about 2-3 times the charge level provide the same failures as the crossed vane simulator. There is a plethora of application specific test methods that relate to system level ESD failures as described in Chapter 6. However, many of them refer back to IEC 61000-4-2 as specification of the stress waveform. Special ESD tests which are conducted for satellites are not covered by this white paper. Tests have also been developed, but not yet standardized for Charge-Board Events (CBE) and Cable Discharge Events (CDE). The ESDA is working on a CDE test method.

Revision History

Revision	Changes	Date of Release
1.0	Initial Release	September 2012
2.0	Added references to IEC 61433-6 and ESDA WG26 (system level ESD modeling) work in the Executive Summary and Section 5.1.1.1	March 2019

White Paper 3
System Level ESD
Part III: Review of IEC 61000-4-2 ESD Testing and Impact on System-Efficient ESD Design (SEED)

Industry Council on ESD Target Levels

October 2021

Revision 1.2

The Industry Council on ESD
http://www.esdindustrycouncil.org/ic/en/

The Electrostatic Discharge Association
http://www.esda.org/

JEDEC – Under Publication JEP164
http://www.jedec.org/

Abstract

This document is the third of the Industry Council white papers dealing with System Level Electrostatic Discharge (ESD). The previous two white papers addressing system level ESD are the Industry Council's WP3 Part I and WP3 Part II.

In WP3 Part I, the misconceptions common in the understanding of system level ESD between supplier and original equipment manufacturer (OEM) were identified, and a novel ESD component / system co-design approach called system efficient ESD design (SEED) was described. The SEED approach is a comprehensive ESD design strategy for system interfaces to prevent hard (permanent) failures. In WP3 Part II this comprehensive analysis of system ESD understanding to categorize all known system ESD failure types was expanded, and described new detection techniques, models, and improvements in design for system robustness. WP 3 Part II also expands this SEED co-design approach to include additional hard / soft failure cases internal to the system.

This third document on system level ESD takes this further while focusing on system testing of ports and the shortcomings of air discharge testing.

Part A of the document highlights the need of a consistent and standardized specification of IO Port contact discharge which is widely used in industry. This is supported by real world discharge scenarios like cable discharge. A well specified testing procedure and the related target levels for IO port direct pin injection are the base for a SEED simulation and co-design approach which can commonly be executed by IC suppliers and system customers.

Part B addresses air discharge testing, which is most relevant for field fails, while its specification in IEC 61000-4-2 and its practical application suffers from both missing repeatability and reproducibility. In the first part, the arguments of maintaining air discharge as a relevant, mandatory test method are explained. Various scenarios leading to real world discharge events which correlate to IEC 61000-4-2 testing are analyzed in more detail. They can lead to soft and hard fails reproducible by system ESD testing. Secondly, in this document new considerations to better calibrate the air discharge test and reliably perform a repeatable air discharge test are given.

About the Industry Council on ESD Target Levels

The Council was initiated in 2006 after several major U.S., European, and Asian semiconductor companies joined to determine and recommend ESD target levels. The Council now consists of representatives from active full member companies and numerous associate members from various support companies. The total membership represents IC suppliers, contract manufacturers (CMs), electronic system manufacturers, OEMs, ESD tester manufacturers, ESD consultants and ESD IP companies.

Core Contributing Members	Core Contributing Members
Robert Ashton, Minotaur Labs	Guido Notermans, Nexperia
Andrea Boronio, ST	Motostugu Okushima, Renesas
Ashok Alagappan, Ansys	Nate Peachey, Qorvo
Brett Carn, Intel Corporation	David Pommereneke, Graz University of Technology
Ann Concannon, Texas Instruments	Alan Righter, Analog Devices
Jeff Dunnihoo, Pragma Design	Theo Smedes, NXP Semiconductors
Charvaka Duvvury, Retired (Chairman) c-duvvury@gmail.com	Andrew Spray, Synaptics
Harald Gossner, Intel Corporation (Chairman) harald.gossner@intel.com	Pasi Tamminen, EDRMedeso
	Scott Ward, TI
Reinhold Gärtner, Infineon Technologies	Terry Welsher, Dangelmayer Associates
Robert Gauthier, GlobalFoundries	Joost Willemen, Infineon Technologies
Steffen Holland, Nexperia	Heinrich Wolf, Fhg
Masamitsu Honda, Impulse Physics Laboratory	Mike Wu, TSMC
Stevan Hunter, ON Semiconductor	Guido Notermans, Retired
Hiroyasu Ishizuka, Synaptics	**Advisory Board***
Nathan Jack, Intel Corporation	Stephan Frei, Technical University of Dortmund
Chanhee.Jeon, Samsung	Mart Coenen, EMCMCC
John Kinnear, IBM	Michael Laube, Renesas
David Klein, PSemi	Henning Lohmeyer, Bosch
Peter Koeppens, ESD Unlimited	
Timothy Maloney, CAI	

Associate Members	Associate Members
Arnold Steinman, Simco Ion	KH Lin, Amazing IC
Bernard Chin, Qorvo	Kitae Lee, Samsung
Brian Langley, Oracle	Larry Johnson, Avagotech
Chang Kim, Samsung	Marcus Koh, Everfeed
Che Hao, Amazing IC	Marty Johnson, Retired
C Hillman, Ansys	Mike Chaine, Micron
Christian Russ, Infineon	Ming-Dou Ker, NCTU
CJ Chao, Richwave	Melissa Jolliff, Aero
Efraim Aharon, Towersemi	Markus Mergens, QPX
Fred Bahrenburg, Dell	M Lee, Semtech
Frederic Lafon, Valeo	Morphy Gao, Hisilicon
Gery Pettit, Retired	MyoungJu.Yun, Amkor
Gaurav Singh, Dialog Semiconductors	Natalie Hernandez, Ansys
Graver Chang, ma-tek	Nobuyuki Wakai, Toshiba
Greg O Sullivan, Micron	Peter de Jong, Synopsys
Hang Kim, Samsung	Philip Baltar, Renesas
Mike Heaney, Amazon	Ramon Del Carmen, Amkor
Henning Lohmeyer, Bosch	Reza Jalilizeinali, Qualcomm

Acknowledgments:

The Industry Council would like to thank all the authors, reviewers, and specialists who shared a great deal of their expertise, time, and dedication to complete this document.

Editors:
Brett Carn, Intel Corporation

Authors:
Graz University of Technology and MS&T EMC laboratory:

- David Pommerenke
- Sen Yang
- Jianchi Zhou
- Omid Hoseini Izadi
- Ahmad Hosseinbeig
- Yingjie Gan
- Wei Zhang
- Giorgi Maghlakelidze
- Marathe Shubhankar

Robert Ashton, Minotaur Labs
Jeff Dunnihoo, Pragma Design
Charvaka Duvvury, iT2 Technologies
Harald Gossner, Intel Corporation
Nate Peachey, Qorvo

Mission Statement

The Industry Council on ESD Target Levels was founded on its original mission to review the ESD robustness requirements of modern IC products to allow safe handling and mounting in an ESD protected area. While accommodating both the capability of the manufacturing sites and the constraints posed by downscaled process technologies on practical protection designs, the Council provides a consolidated recommendation for future ESD target levels. The Council Members and Associates promote these recommended targets for adoption as company goals. Being an independent institution, the Council presents the results and supportive data to all interested standardization bodies.

In response to the growing prevalence of system level ESD issues, the Council has now expanded its mission to directly address one of the most critical underlying problems: insufficient communication and coordination between system designers (OEMs) and their IC providers. A key goal is to demonstrate and widely communicate that future success in building ESD robust systems will depend on adopting a consolidated approach to system level ESD design. To ensure a broad range of perspectives the Council has expanded its roster of Members and Associates to include OEMs as well as experts in system level ESD design and test.

Preface

This white paper presents the recent knowledge of system ESD field events and air discharge testing methods. Testing experience with the IEC 61000-4-2 (2008) and the ISO 10605 ESD standards has shown a range of differing interpretations of the test method and its scope. This often results in misapplication of the test method and a high test result uncertainty. This white paper aims to explain the problems observed and to suggest improvements to the ESD test standard and to enable a correlation with a SEED IC/PCB co-design methodology. This white paper is divided into a part on direct pin stressing (Part A) and IEC 61000-4-2 testing of chassis and display which is typically a type of air discharge (Part B). Part A discusses the necessary industry alignments and standardization to support a unified SEED design procedure. Part B provides the input to improve the calibration and test reliability of IEC 61000-4-2 in the air discharge mode.

Disclaimers

The Industry Council on ESD Target Levels is not affiliated with any standardization body and is not a working group associated with JEDEC, ESDA, JEITA, IEC, or AEC.

This document was compiled by recognized ESD experts from numerous semiconductor supplier companies, contract manufacturers and OEMs. The data represents information collected for the specific analysis presented here; no specific components or systems are identified.

The Industry Council, while providing this information, does not assume any liability or obligations for parties who do not follow proper ESD control measures.

Table of Contents

Glossary of Terms

AC	alternating current
ADS	Advanced Design System
ASIC	application specific integrated circuit
CAN	controller area network
CAT	category
CDE	cable discharge event
CDM	charged device model
CST	Computer Simulation Technology
DAC	direct attach copper
DC	direct current
DNA	deoxyribonucleic acid
DUP	device under protection
DUT	device under test
DVI	digital visual interface
EOS	electrical overstress
ESD	electrostatic discharge
EUT	equipment under test
FDTD	finite-difference time-domain
FIT	finite integration technique
GND	negative voltage supply in digital logic, neutral voltage supply in analog logic
HBM	human body model
HCP	horizontal coupling plane
HDMI	high definition multimedia interface
H-M	human-metal
IBIS	input/output buffer information specification
IC	integrated circuit
IEC	International Electrotechnical Commission
IEEE	Institute of Electrical and Electronics Engineers
IO	input/output
ISO	International Organization of Standards
IV	current-voltage
LAN	local area network
LCD	liquid crystal display
PC	personal computer
PCB	printed circuit board
PHY	physical layer interface
PIFA	planar inverted F antenna
PoE	power over Ethernet
RC	resistor capacitor network
RF	radio frequency
RH	relative humidity
RLC	resistor inductor capacitor network
SEED	system efficient ESD design
SFP	small form-factor pluggable
SPICE	simulation program with integrated circuit emphasis

TLM	transmission line matrix
TVS	transient voltage suppression
USB	universal serial bus
UTP	unshielded twisted pair
UV	ultraviolet
WB	wet bulb
VCP	vertical coupling plane

Definitions

<u>EMC:</u> electromagnetic compatibility – The condition which prevails when telecommunications (communication-electronic) equipment is collectively performing its individual designed functions in a common electromagnetic environment without causing or suffering unacceptable degradation due to electromagnetic interference to or from other equipment/systems in the same environment (MIL-STD-463A).

<u>SEED:</u> system-efficient ESD design - Co-design methodology of on-board and on-chip ESD protection to achieve system level ESD robustness.

<u>Soft Failure:</u> Failure of a system, not due to physical damage, in which the system can be returned to a functional state without the repair or replacement of a component. Return to a functional state may or may not require operator intervention. Operator intervention may include rebooting or power cycling. Soft Failures can involve software issues and software fixes but in the context of this document they are primarily due to ESD events injecting unwanted signals into the system which place the system into a state in which it does not function as intended.

Executive Summary

The Industry Council previously published two white paper documents describing the nature of component level electrostatic discharge (ESD) versus system level ESD, and the methods to ensure efficient system level ESD robustness. WP3 Part 1, also known as JEP 161, established that there is no correlation between component and system level robustness and introduced the concept of *system-efficient ESD design*, known as SEED, while WP3 Part 2 (JEP 162) expands on the concept of SEED, which utilizes simulations to achieve desired system ESD robustness with minimal impact on system performance.

We present here White Paper 3 Part 3 that addresses real-world system level ESD testing and suggests improvements to the existing International Electrotechnical Commission (IEC) 61000-4-2 standard widely used to test for ESD reliability of electronic systems. Before presenting a summary of this new white paper, highlights of the two previous white papers are reviewed to place the challenges of system level ESD in a full perspective.

Review of WP3 Part 1 document highlights:
- The misconceptions common in the understanding of system level ESD between supplier and original equipment manufacturer (OEM) were identified.
- A novel ESD component / system co-design approach called system efficient ESD design (SEED) was described.

Review of WP3 Part 2 document highlights:
- Comprehensive analysis of system ESD understanding to categorize all known system ESD failure types and describe new detection techniques, models, and improvements in design for system robustness.
- Discussed system ESD stress application methods and introduces new system diagnosis methods to detect weak ESD failure areas leading to hard or soft failures.
- Outlined present-day state-of-the-art electromagnetic compatibility (EMC)/electromagnetic interference (EMI) design prevention methods that have been developed to prevent system level ESD failure.

Introductory Note for WP3 Part 3:
While the two previous documents have set the stage for approaching system level ESD in a more realistic manner, the nature of testing for the well-established IEC 61000-4-2 test method has not adequately addressed all the challenges of testing from an increasing variety of modern electronic systems. This third document on system level ESD is intended to be a comprehensive review of more effective test procedures and recommends them for consideration as part of an overall improved IEC 61000-4-2 test method.

Overview of the Document:
As an important note, compilation of this document involved contributions from university and industry experts and includes a survey of the most salient findings from published academic research papers. The main goal is to derive the most important information and to leverage this to formulate recommendations to an overall improved system level ESD test method.

What is the real motivation here?
All electronic systems, from mobile phones to large computers, need to be tested for robustness from damage or upset from electrostatic discharge. System ESD testing is done using an ESD pulse generator, often called an ESD gun. The vast majority of system level ESD testing follows the test procedures and ESD pulse generator specifications defined in IEC 61000-4-2. Unfortunately, even when carefully following the procedures in IEC 61000-4-2, there is a high level of uncertainty in the test results, and successfully passing the test requirements is often not a complete prediction of ESD robustness in the field.

The uncertainty in system level ESD test results is often attributed to the large differences in the calibration current waveforms which can exist between ESD guns from different manufacturers, even when all the ESD guns meet the specifications in IEC 61000-4-2. But this is only a part of the problem.

The reason the Industry Council has addressed this topic is regarding the need for a more relevant and reproducible system ESD test procedure to better correlate integrated circuit (IC) and board co-design approaches as discussed in WP3 Part 1 and Part 2. The Industry Council provides recommendations based on technical data in this document which can be used for further discussion in standardization committees which are chartered with system level ESD test methods.

This White Paper explores the reasons for the high level of uncertainty in the test results, proposing test procedure modifications to improve repeatability, reproducibility, and predictability in field performance. Modeling methods are also discussed which can give insight into system level ESD testing and the design of ESD robust systems.

Organization of the Document:
The document is divided into two main parts: Part A describes direct pin discharge, and Part B considers discharge to chassis and display, also known as air discharge. Specifically, Part A presents real world scenarios such as cable discharge events (CDE) to introduce an extension to the IEC 61000-4-2 standard on direct port discharging. To further facilitate this, reference circuit models are described so that SEED, as introduced in White Paper 3 Part 1, can be used more effectively. Whereas Part B describes the shortcomings of the currently practiced air discharge method and recommends improvements for calibration and test measurements.

Synopsis of Part A:
For shielded connectors, the IEC 61000-4-2 standard does not specify direct pin stress, but this is important for system qualification involving uncertain results, variations in individual set ups, and repeatability of the results. This section promotes an industry wide alignment for the IEC 61000-4-2 standard to include direct pin discharge. For printed circuit board (PCB) designs to address this issue, standard models for SEED are described.

In Part A of this document, the approach is first to describe the real-world events (Chapter 1) that must be understood, then detailing the direct-injection method (Chapter 2) and concluding with information on the important target levels for these tests (Chapter 3). For these issues on direct pin testing to be viewed in an analytical perspective, the ESD gun model details (Chapter 4) and the techniques of SEED simulations (Chapter 5) are also presented. An alternate characterization method using 50-ns TLP is also introduced that can be directly applied to the IC as a way of characterizing the external pin high current I-V response. One very important outcome of the investigations is that for universal serial bus (USB) cables meeting standard shielding requirements a 2 kilovolts stress is enough.

Synopsis of Part B
After studying the air discharge test method, it is concluded that both a lack of precise test methods and specifications led to extreme variability and uncertainty in qualification results. Removing this method altogether is not an option since in the real-world events discharges to chassis do occur. Details of this inadequacy and the subsequent improved recommendations are given.

In Part B, the document focuses on improved test methods for displays using the air discharge method. This includes a review of the challenges associated with the IEC 61000-4-2 test methods and the test results' uncertainties (Chapter 6). Improvements to the calibration of the ESD simulators for both contact-mode discharge and air discharge are suggested (Chapter 7). With the extensive background of the test methods and the description of corresponding various opportunities for improved applications of these methods, the document recommendations useful test improvements (Chapter 8). Supporting technical details of the work presented in this document are given in the appendices at the end.

Nuances of System Level ESD Testing:
The most important points of this document can be understood only after the real issues of system level ESD robustness are clearly stated and defined. This will also give a better understanding of why the test method improvements recommended here are important to the industry.

System level ESD testing is performed using a combination of stresses to each system while in a powered, functional state. Contact discharge is applied to conductive surfaces by placing a pointed tip of the ESD gun against the system and initiating an ESD pulse. Air discharge is performed on insulative surfaces, with emphasis on locations frequently touched by the user, such as keyboards and touch screen displays, and locations where ESD can enter the system, such as seams, vent holes, etc. In this test a rounded tip on the ESD gun is charged to a high voltage and the gun is moved toward the system under test until an air discharge occurs or the surface of the system is touched. In the next test discharges are made to horizontal and vertical coupling planes near the device under test to see if the fields generated by these stresses can upset the system.

It is important to note that test results from system level ESD testing can have a variety of results:

- System suffers no ill effects from the testing
- Soft failure
    - System is upset but returns to normal operation without user intervention
    - System is upset but returns to normal operation after user intervention which may require a reboot of the system
- Hard failure in which there is permanent physical damage requiring repair.

The causes of testing uncertainty and failure to predict field reliability covered in this white paper are numerous and will be summarized here in bullet form.
- Important for both contact and air discharge

- IEC 61000-4-2 compliant ESD guns have substantial variability in current waveforms and power distribution in the RF spectrum between models and manufacturers.
- Number of stresses per discharge point (10) is insufficient to adequately predict soft failures, which can depend on precise timing between the stress and system timing.
- Test results can depend on a previous stress if charge is not removed before the next stress - this phenomenon can lead to *secondary discharge*.
- Transient electromagnetic fields from ESD guns are not specified or limited. This leads to an uncertainty of soft failure testing with different guns. Pulse to pulse variation of the same gun can be eliminated by multi-zap testing.

- Contact discharge
  - Contact discharge is generally more repeatable than air discharge but does not represent the real world which is typically air discharge
  - Discharge current waveform is only verified for a low resistance load:
- Air discharge
  - There is no calibration procedure for ESD guns in air discharge
  - Test severity depends on arc length which is sensitive to a number of factors
    - Humidity
    - Approach speed
    - Geometry
    - Materials on each side of arc
- Discharge to coupling planes
  - Geometries specified in IEC 61000-4-2 are sometimes unrealistic for modern form factors
- Expanding the scope of IEC 61000-4-2
  - Cable discharge event (CDE) is not included
    - There are a number of forms of cable discharge event (CDE)
    - Cable quality can affect CDE
  - Tests conditions are not realistic for wearable devices, which could not have been realistically anticipated in the earlier IEC 61000-4-2 specification creation. These include items such as smart watches or mobile phones held in a case attached to a belt.

Important Recommendations from This Work:
Some important recommendations are summarized here for both contact discharge and air discharge tests.

Direct Discharge:
- Establish industry wide alignment for direct discharge to ports
- Consider alternate 50-ns TLP test with 2 ampere goal applied to IC pins
- Consider 2 kilovolts contact discharge as a standard spec for USB pins meeting cable shielding requirements

Air Discharge:
Consider recommendations to improve the testing at a chosen discharge point that include:
- Maintain air discharge as a field relevant stress method, but improve its reproducibility by:
    o Improved gun calibration for air discharges
    o Controlled arc length using
        ▪ A well-controlled humidity environment
        ▪ Introduction of ionization
    o Controlled approach conditions by use of robotics
    o Increased number of discharges per test point
    o Controlling secondary discharge
    o Measurement of radiated gun emission
- Address cable discharge as a relevant system level ESD event

Test Reproducibility Recommendations:
The white paper presents data to support an understanding of state-of-the-art system ESD testing and makes suggestions for improving the repeatability and reproducibility of system level ESD testing. The following bullet list summarizes the improved understanding and suggestions for improving future testing.

- ESD gun calibration
    o Contact discharge waveform verification with loads other than a low impedance resistive load
    o Introduction of a method for ESD gun calibration in air discharge mode without the uncertainties of air discharge
    o Measurement of radiated emissions, including the RF power spectrum
- Methods to improve air discharge arc repeatability
    o Ionization to stabilize arc length
    o Different air discharge tip materials
    o Improvements in temperature and humidity control
    o Charge removal techniques
    o Robotic testing to improve control of approach speeds
- Improvements in test procedures
    o Recommendations on increasing the number of stresses at each test location to improve capture of weaknesses due to soft failures
    o Improved understanding of ESD discharge voltage as a function of humidity
    o Robotics to improve test repeatability
    o Improvements in documentation of test procedures used, including video recording

Finally, the white paper also presents modeling methods which can both improve the understanding of system level ESD testing and help in the design of ESD robust systems.
- Simulation program with integrated circuit emphasis (SPICE) level models of ESD guns
- System Efficient ESD Design (SEED) for simulating systems when ESD stressed
- 3D models of ESD guns and systems

Part A: Direct Pin Discharge

Chapter 1: Real World Stress to Port Pins

Dr. David Pommerenke, Graz University of Technology
Marathe Shubhankar, MS&T EMC laboratory
Yingjie Gan, MS&T EMC laboratory
Nate Peachey, Qorvo

1.0 Introduction

The IEC 61000-4-2 test uses the discharge of a human via a handheld metal object as its reference ESD event. However, a variety of other ESD events occur which are not within the scope of the test standard. Many attempts have been made to use the IEC 61000-4-2 test generator to perform testing which will ensure robustness against other ESD events.

1.1 Human-Metal Discharge

The IEC 61000-4-2 standard uses the discharge of a human via a small metal part as its reference event [Chu2004]. The waveform shown in the IEC 61000-4-2 standard is similar to the discharge waveform of a human via a small handheld metal part at about 5 kilovolts at an arc length of 0.8 mm. Due to the nonlinear behavior of the arc, human-metal ESD discharges at shorter arc length will have faster rise times and longer arc lengths will have slower rise times. The reference event covers only one situation. Its current derivative is in the range of 4 A/ns, which is at the lower end of the current derivatives observed for human-metal ESD, over a large range of voltages and arc lengths.

1.2 Human Skin ESD

A human skin ESD has much longer rise times (in the range of 1-10 ns for < 8 kilovolts) and lower peak current values compared to a human-metal discharge. However, in the case of multiple discharge pulses, the subsequent pulses will have much faster rise times even if the amplitudes are lower.

1.3 Cables and Cable Discharge Events

In system level testing, discharges to connector pins should be included if the connector geometry allows for such discharges. This requires understanding of the threat levels. Directly applying IEC 61000-4-2 testing at system level voltage settings (e.g., 8 kilovolts) can lead to over testing and unnecessary additional costs. The situations which lead to connector currents need to be identified and the stress levels, as seen by the sensitive data wires, need to be determined and used as guidance for setting test levels.

If a charged cable is inserted into a device under test (DUT), a current will flow onto the connector. Typically, events which cause connector currents are:

- Discharge to a DUT which is connected to ground in a system from a charged source such as a USB cable

- Discharge to another DUT, when both DUTs are connected to each other by a cable.

- Attaching a charged DUT to a grounded cable or a docking station.

The present system level IEC 61000-4-2 standard does not consider a cable discharge event, often referred to in industry as cable discharge event (CDE). However, the IEC 61000-4-2 standard does give some guidance as to which part of a connector should be exposed to a direct ESD discharge.

Protection design and an appropriate testing of system input/outputs (IOs) is certainly a critical topic. Multiple companies have reported field failures on ports such as USB or local area network (LAN). The assumption that a discharge will be safe because the cable is properly shielded, and a ground contact will be made first due to connector design, is often not true. The analysis of cables bought from a wide range of cable suppliers has shown that about 30% of the USB2 cables did not have any shield connection. Missing shield connections have also been found on USB 3.1 cables [Mar2017c, Stad2017].

Due to faster IO speeds, and influenced by reduced human body model (HBM) and charged device model (CDM) robustness levels, more emphasis must be placed on the trade-off between PCB and IC level protection. This is covered by the SEED concept. To enable a SEED based design there must be an understanding of the amplitude and the waveform of the cable discharge. Real world conditions for cable discharges of LAN and USB cables are discussed in the following.

For Ethernet, one needs to distinguish between Ethernet ports that use a transformer, often referred to simply as "magnetics" at the port, and those which have a direct connection to the circuit. The transformer provides a high voltage isolation barrier of at least 1500 volts, and for short term, nanosecond long pulses probably > 5000 volts. It also provides a common mode return path via the Bob Smith termination circuit, and via ferrites for power over Ethernet (PoE) devices. Further, the transformer will saturate at rather low currents. This saturation will reduce the coupling between the primary and secondary side; thus, it reduces the coupling of the physical layer interface (PHY) IC. This section analyzes LAN interfaces, such as unshielded twisted pair (UTP), which uses a transformer.

In contrast, small form-factor pluggable (SFP) + direct attach copper (DAC) cables directly provide an entry path, without magnetics to the physical layer interface IC. These cables are shielded, have a shielded connector that mates at the shield first. However, direct discharges to the pins are, although not likely, possible. The ESD performance of such cables has not been investigated yet.

Cables have been found which allow discharges to the signal wires of a USB, as they contain ungrounded decorative metal parts above the unshielded signal wires and connector pins. The cable shown in Figure 1 had a decorative metal that is not connected to the shield or the connector. The decorative metal is above the signal wires which connect to a PCB which is within the connector. The signal wires are not shielded underneath the metal part, as the foil and braid are only connected to an unshielded PCB on one side. This allows sparking between the unconnected, decorative metal and the signal wires of the cable and the PCB.

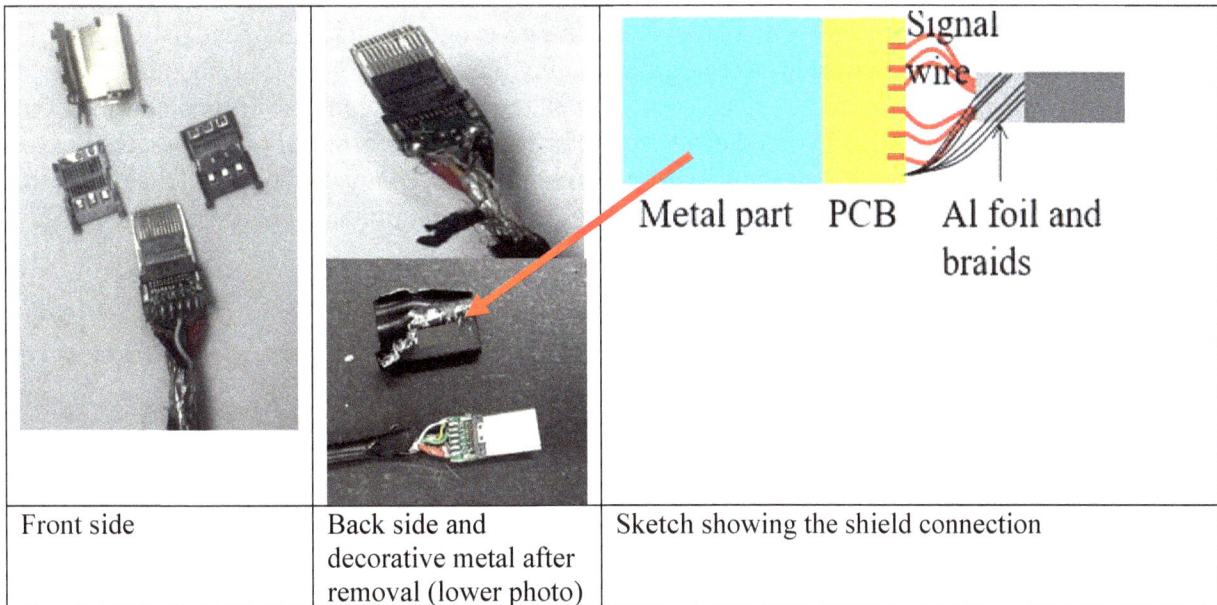

Front side	Back side and decorative metal after removal (lower photo)	Sketch showing the shield connection

Figure 1: Photo and drawing explaining the shield structure of a badly shielded USB 3.1 cable.

In general, CDE discharges are a rather complex process. The effect of CDE depends on:

- The charge voltage of the cable

- The charge mode (most cables will charge in common mode: All wires and the shield are at the same potential). A cable may also be polarized by a charge on the jacket. This charge will cause polarization of the metal structure, but no net charge on the metal structure. No matter whether the cable carries a net charge or if it is polarized, sparking can occur if the cable is inserted into a grounded connector.

- The type of shield structure: No shield, such as UTP; shield present but not connected; foil shield; foil + braid shield; connection of the shield to the connector from pigtail to $360°$ shield.

- The contact sequencing will vary from an undefined sequence, as every pin has the same chance to mate first, to a typical sequence observed in a good USB cable: Shell first, then GND or 5V, then data.

- Exposure of pins: Some connectors will not allow an ESD to a pin as the ESD will be diverted to the shell. However, other cables may not have a shell, or may not connect the shell and may allow direct ESD to pins.

- Probability of ESD: Connections like USB are inserted rather often, other connections, such as a video interface will be inserted into most devices less often.

- Form factor and chassis material: For example, some USB memory sticks often have no shield, no shell and may allow discharges via the plastic into the circuit.

To highlight the essential aspects, a brief analysis is provided for two interfaces; LAN and UTP. The details can be found in the cited literature and in Section 1.4 (for LAN).

1.3.1 Charging and Discharging of USB Cables

Electronic systems must tolerate electrostatic discharges in a typical user environment. The existence of badly shielded cables and poor contact between the shield and the connector requires studying real world waveforms for USB cable discharge events [Mar2017c, Stad2017]. Charged cable events have been investigated by charging a cable to a known voltage and then inserting the cable into a connector to measure the currents. However, environmental conditions and materials can change the charge level and related electrostatic charge voltage. But, for many practical cases, the voltages that a cable can reach via triboelectric charging are not well known. To address this concern, a set of experiments were conducted where the total accumulated charge was measured.

High charging voltages have been observed for:

- People sitting up from a chair
- People removing a garment, such as a fleece jacket
- Handling of plastic sheet material

The voltages reached by walking on a carpet are usually much lower than the voltages reached by the situations described above.

An experiment was conducted measuring the voltages on cables that were rubbed against various clothing materials. The aim of this study by Rezaei et al. was to enhance the understanding of cable charging behavior under different humidity conditions, with a variety of rubbing materials [Rez2017]. The voltage levels and various dependencies for such cable charging were investigated. The charge levels translate into a voltage if the capacitance of the cable to ground is considered. The insertion of a charged cable into a USB connector may lead to a CDE. Throughout this section the act of plugging in a cable is called a "plug-in event". The USB cable connected devices are potential sources of a CDE. A CDE occurs because of the USB cable being charged and then discharging to the USB connector of the connected device. Typical examples include discharge scenarios such as an ESD to USB drive with no shield, a USB cable connected to cellphone inserted into a grounded device such as a personal computer (PC), a USB HUB connecting multiple electronic devices together via USB cables, etc. The goal of the work by Mardiguian et al. and Stadler et al. was to understand the range of stress levels seen by the USB 2.0 A/B (type A and type B connectors) cable connected devices due to various USB cable discharges as a function of the charging scenario, cable shield quality, and cable connections [Mar2017c, Stad2017].

1.3.2 Overview of Charging Scenario

Investigation on the effect of relative humidity and materials on triboelectric charging of USB cables was performed experimentally. The measurement setup involved the use of different sweater materials to rub on the outer jacket of various cables. This action of rubbing of different sweaters on cables led to the charging of cables by a triboelectric charging process. Figure 2 depicts one of the sweaters and one of the cables used in the study.

Cable

Figure 2: Rubbing of the sweater on the cable leads to a charge developed on the cable.

1.3.3 Overview of Discharge Scenarios

Various USB discharge scenarios are possible. For example, the direct ESD discharge to a USB drive, or a charged human holding a cable and plugging it into a USB connector. The focuses of the study were the discharge into a USB connected device and a discharge into a USB connector during a plug-in event. They are illustrated in Figures 3 and 4, respectively.

Figure 3: Discharge into a USB connected device scenario.

In Figure 3 a cellphone is connected to a grounded device such as a personal computer (PC) by a USB 2.0 cable. A charged human can discharge into the cellphone and the ESD discharge source point is at the cellphone. The current flows via the USB cable to the grounded PC. In Figure 4, a charged cable is inserted into the PC. The other end of the cable is connected to a USB device (a cell phone is depicted) which leads to a plug-in event.

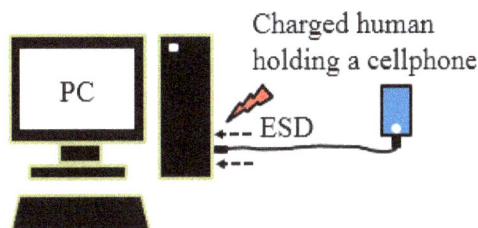

Figure 4: Discharge into a USB connector during a plug-in event.

1.3.4 Classification of USB 2.0 Cables Based on Shield Quality

In this study, randomly purchased USB 2.0 A/B cables (USB 2.0 type B to type A) are classified into three different types based on their cable shield quality. A resistor inductor capacitor network (RLC) meter was used to measure the USB connector shell to shell DC resistance of the cable. This gives a first impression on the shield's properties. If the resistance is within several hundred milliohm and does not change much when the cable is twisted, the cable is classified as a well-shielded cable. If the resistance is in the range of 1-80 ohms and it changes during twisting, the cable is classified as a badly shielded cable. A few of the USB 2.0 cables purchased had no shields. In addition, the shields of a few USB cables were cut for experimental purpose. Cables with a cut shield or no shield are classified as an unshielded cable.

1.3.5 Discharge into a USB Connected Device

A block diagram for this series of tests is shown in Figure 5. A 2 kilovolts contact mode ESD was injected into the conductive enclosure of the DUT box, resembling a situation in which an ESD occurs to a USB connected device (such as a USB drive) connected to a PC. For such a situation, data traffic can be active, while there cannot be data traffic during a plug-in event.

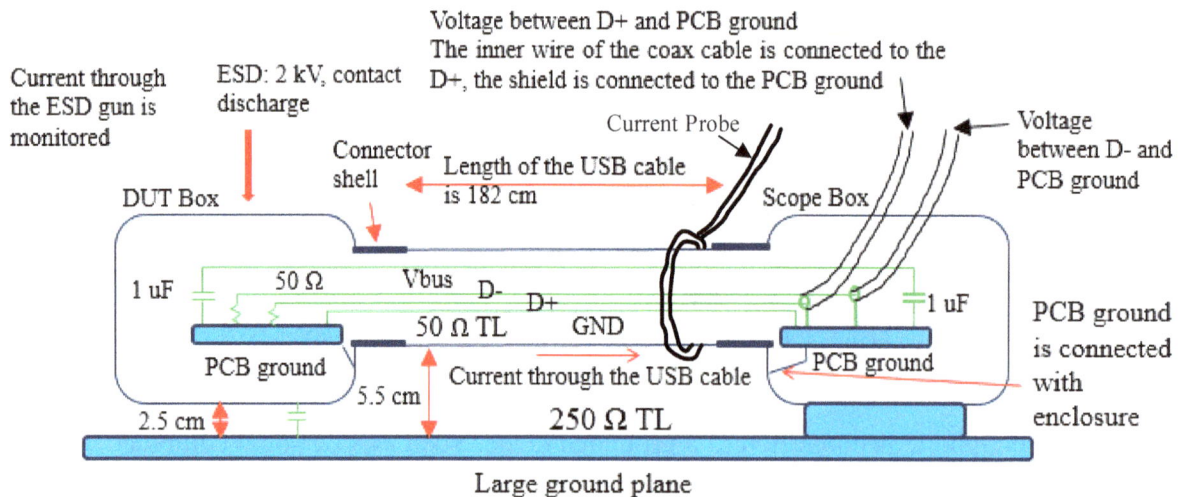

Figure 5: Discharge into a USB connected device.
Note: D+/D- lines terminated into 50-ohm impedance scope

The following figures show the common-mode current on the USB cable as captured by the F-65 monitoring probe clamped close to the scope box and the voltage induced between D+ and ground on the scope box side. The measured waveforms for the well-shielded, badly shielded, and unshielded USB cable types are shown in Figure 6.

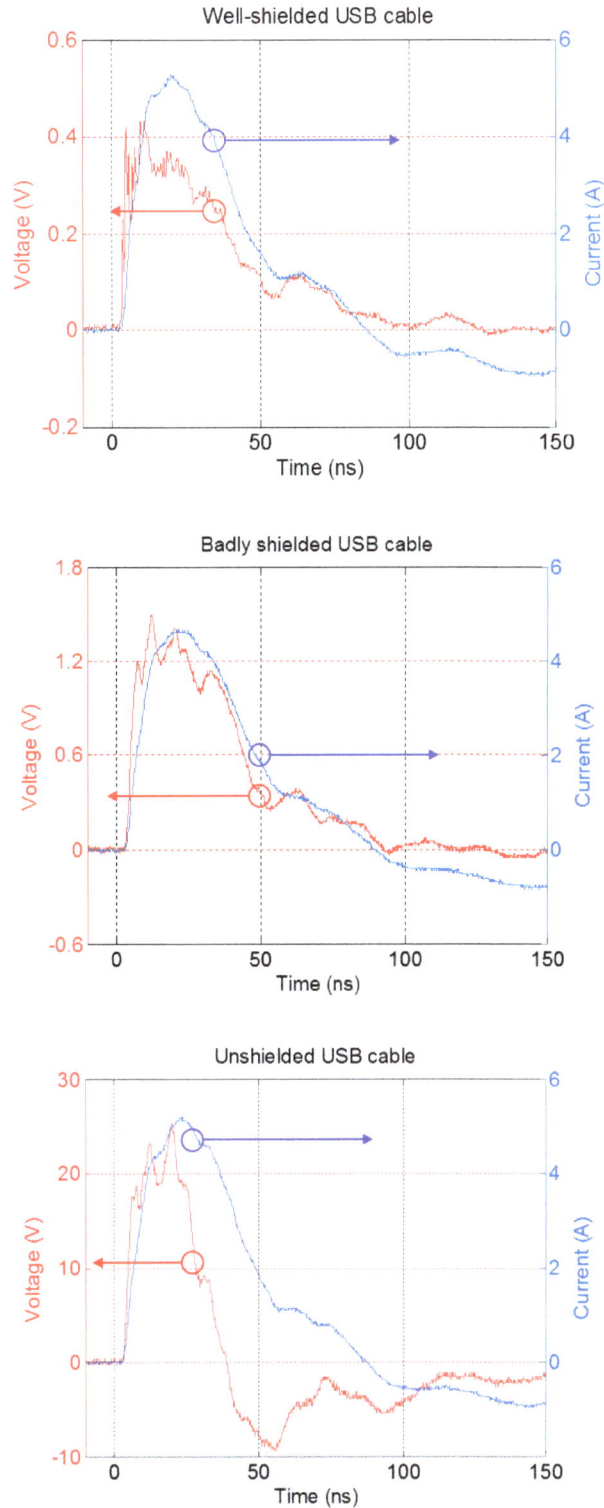

Figure 6: Common-mode current and voltage induced between D+ and ground for a discharge into a USB connected device scenario. Data is shown for well-shielded, badly shielded, and unshielded type USB cables. The ESD gun is discharged at 2 kV in contact mode. Note: the negative current readings are a measurement artifact of the F-65 current probe.

At a 2 kilovolts contact mode discharge, a noise voltage of 0.4 volts is measured for the well-shielded cable, 1.2 volts is measured for the badly shielded cable, and approximately 20 volts are measured for the unshielded cable (without taking the reaction of ESD protection into account). In contact mode the ESD gun discharge current has a linear relationship to the ESD gun voltage setting. If the same measurement is performed at a higher ESD gun voltage, the amplitude of the measured waveform is expected to increase linearly. At 2 kilovolts for an unshielded cable the peak current occurs at 20 ns and the amplitude is approximately 5 amperes as shown in Figure 6. The initial peak current would be 2 x 3.75 amperes, but it returns via the DUT box to ground capacitance and is therefore not visible in the F-65 current clamp data. The designer is often interested in a realistic worst-case specification. Although possible in some rare situations, direct discharges into pins were disregarded and the discharge into a USB connected device for the case of an unshielded USB cable was considered the realistic worst case.

If the ESD gun had been set at 8 kilovolts, a current amplitude of 20 amperes would be expected. It should be noted that the current measured using the F-65 current clamp is the common-mode current and it is not a function of the type of USB cable used. In the case of a badly shielded or unshielded cable it is reasonable that the common-mode current splits equally over the four wires of the USB 2.0 cable. Approximately a 5-amperes current will flow on each of the four wires, GND, Vbus, D+, and D-. This assumption is based on the fact that all wires are connected via low impedances to ground. The 5 volts (Vbus) is usually connected via a large value capacitor, and D+ and D- have ESD protection diodes either on the board or in the IC. A current of 5 amperes with pulse widths of 10's of nanoseconds on the IO lines (D+ and D-) can cause hard failures. Here the pulse width is not determined by the cable length, but by the source.

1.3.6   Discharge into a USB Connector during a Plug-in Event

When a charged person holds a cellphone connected to a USB cable, the charge will be distributed on the USB cable via polarization and/or charge migration. When the person inserts the USB cable into a grounded PC, ESD occurs. Figure 7 illustrates the test set-up used to investigate plug-in events.

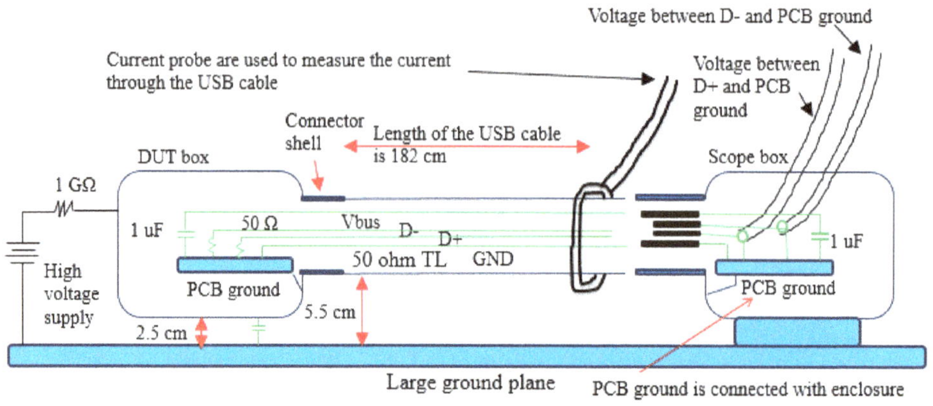

Figure 7: Discharge into a USB connector during a plug-in event.

Because of the design of the USB connectors, the shell will contact first, then the GND/Vbus pin follows, and finally D+/D- will contact. Even if the cable is unshielded, GND/Vbus will contact first.

1.3.6.1   Plug-in event sequencing

During the plug-in measurements the cable was charged via 1 GΩ. The unconnected end of the USB cable was held by hand using insulated gloves, then the operator inserted the cable into the scope box. Multiple pulses can occur during the insertion process. To be able to capture pulses that occur with microsecond delay times, a fast re-trigger mode was selected. Another application of the fast retrigger mode or sequence mode acquisition is shown in [Mar2017b], and [Mar2018], which explains the use of the fast re-trigger mode in capturing multiple pulses. The measurement was performed using ultra-segmentation (Rohde & Schwarz) or sequence mode acquisition (Teledyne LeCroy) activated on the oscilloscope. The oscilloscope can be triggered on the channel connected to the F-65 current clamp or the D+ signal.

The USB cable connector may not always enter the scope box in parallel to the connector. It may hit the scope box at an angle to be aligned for entering the scope box connector. This action leads to multiple triggers for a single plug-in event measurement. A possible sequence is illustrated in Figure 8.

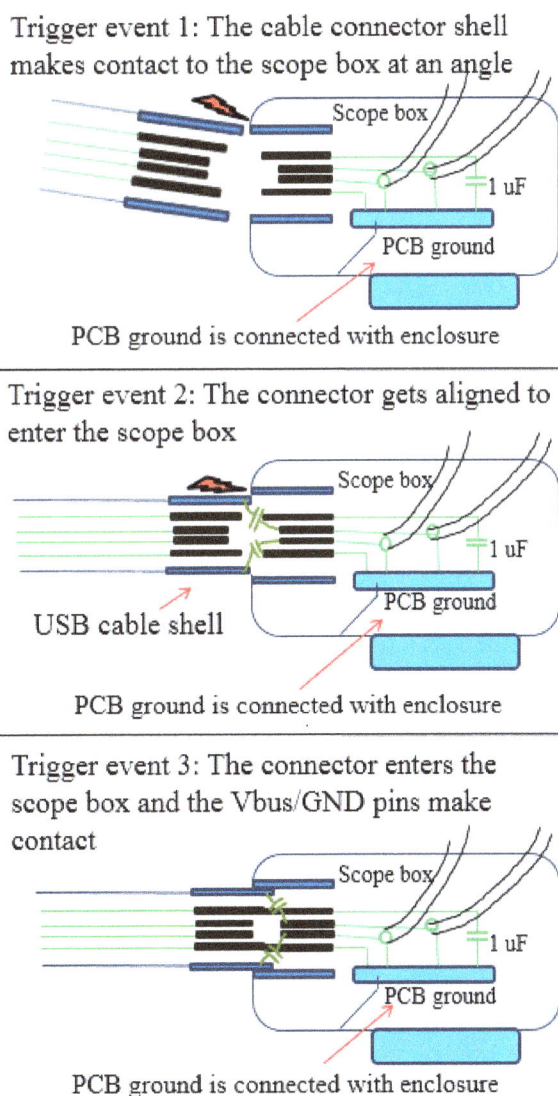

Figure 8: Plug-in event sequence is explained using three triggered events. It should be noted that multiple triggered events can occur in a plug-in measurement.

1.3.6.2   *Plug-in event measurement results*

A set of measured typical waveforms during a plug-in event is shown in Figure 9. They are based on charging the DUT box to 8 kilovolts and using a well-shielded cable. In this set-up the USB cable behaves as a transmission line (wire above the ground plane) having an impedance approximately equal to 266 Ω (1 kilovolt/3.75 amperes). In this measurement set-up, the USB cable GND wire was directly connected to the scope box PCB ground. The USB cable Vbus wire was terminated with a 1 uF capacitor on the PCB inside the scope box. Neither of these wires were monitored during the plug-in event measurements. However, emphasis was placed on the IO lines (D+ and D-) of the USB 2.0 cable. The waveforms induced on the D+ and the D- IO signal lines due to the plug-in event were comparable.

The F-65 current clamp, D+, and D- channels can be used to trigger the oscilloscope. In this case the oscilloscope was triggered on the D+ channel and the ultra-segmentation/sequence mode was enabled. It should be noted that the oscilloscope is triggered on the D+ channel to observe the waveforms induced on the IO signal lines due to the plug-in event. This leads to a few of the initial current waveforms through the USB cable not being acquired by the oscilloscope.

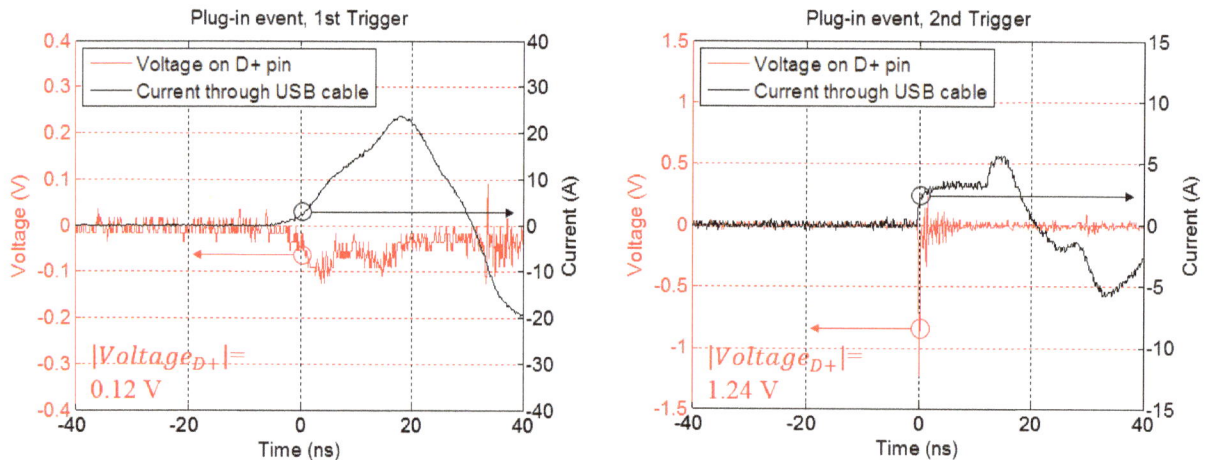

Figure 9: Discharge into a USB connector during a plug-in event. Measured waveforms for the first and second triggered event are shown.

During the plug-in sequencing, the first event caused lower ESD-induced voltage (0.12 V) on the D+ pin because most of the current flowed between the USB shell and the scope box. The second plug-in triggered event had a higher ESD-induced voltage on the D+ pin (1.24 volts).

1.3.7   Summary of USB Risks

A triboelectric charging scenario was investigated for various types of cables [Rez2017]. Two USB 2.0 cable discharge scenarios were investigated [Mar2017c, Stad2017]. The first scenario investigated ESD on a USB cable connected system. The second scenario investigated a plug-in event of a connected USB cable.

In the measured waveforms for the first discharge scenario, the discharge into a USB connected device shows pulse widths on D+ (IO signal) of 20-40 ns and 2-10 ns rise times at 2 kilovolts contact mode ESD to the USB connected device. If 8 kilovolts is set as the expected reliability level for the user environment, then the ESD protection needs to be able to handle 5 amperes for tens of nanoseconds on the D+ and D- at the same time. Here, damage, as well as latch-up or soft failures may need to be considered. The plug-in of an already device-connected cable has a lower chance of causing damage. The voltages induced in D+ by a discharge between a USB cable and a grounded connector during the insertion show less than 5 volts at 8 kilovolts for the well-shielded and the badly shielded cables. For the case in which the USB cable has no shield, a voltage of approximately 12 volts has been induced at 8 kilovolts. This translates to a current of 0.24 amperes (oscilloscope channel impedance of 50 ohms) on the D+ IO line. It should be noted that even though this voltage amplitude may look significant, the duration of the pulse width is less than 10 ns as shown in Figure 9. Soft failures may still be possible if a cable is inserted into a USB HUB servicing multiple USB connections. Here, other active connections could be interrupted by the insertion of a charged USB cable even if the voltage is rather small.

The system designer must account for the fact that a large portion of the USB 2.0 cables on the market have no shield. About 1/3 of the thirty USB 2.0 cables randomly selected for this study had no shield or no shield to USB connector shell connection. ESD protection circuits should be designed to protect electronic device from CDE stress levels up to as much as 5 amperes for tens of nanoseconds on the D+/D- IO lines for an 8 kilovolts device stress level.

1.4 Cable Discharge of LAN Cables

1.4.1 Introduction

One needs to distinguish between Ethernet ports that use a transformer, often called magnetics at the port, and those which have a direct connection to the circuit. The transformer provides a high voltage isolation barrier of at least 1500 volts, and in the short term, nanosecond long pulses probably > 5000 volts. It also provides a common mode return path via the Bob Smith circuit, and via ferrites for PoE devices. Further, the transformer will saturate at rather low currents. This saturation will reduce the coupling between the primary and secondary side, thus, reducing the coupling of the physical layer interface IC. This section analyzes LAN interfaces, such as UTP, which use a transformer.

In contrast SFP + DAC cables (small form-factor pluggable + direct attach copper) directly provide an entry path, without magnetics, to the physical layer interface IC. These cables are shielded, having a shielded connector that mates at the shield first. However, direct discharges to the pins are, although not likely, possible. The ESD performance of such cables has not been investigated yet.

The protection strategies for an electronic system against ESD are mainly developed based on the type of ESD source, the discharge mechanism, and the likelihood of its occurrence. An IC located internal to an electronics system is at a lesser risk to an ESD event when compared to an IC having an interface that is exposed to the external world. For example, Ethernet physical layer (PHY) transceivers may be subjected to CDE, which is not the case for on board interfaces between ICs such as DDR interfaces.

When a charged LAN cable is plugged into an Ethernet connector [Gan2016], a cable discharge event will occur. The discharge voltage and subsequent current can damage the transformer, the common mode chokes, and especially the Ethernet PHY IC transceivers.

Over the past few years, researchers have understood that CDE needs to be treated separately from other types of ESD tests due to vast differences between the effects caused by CDE. LAN cable discharge pulses can be quite long and have the potential to stress the IO more than those discharges described by an HBM or IEC 61000-4-2 discharge model. Additionally, the LAN cable discharges lead to shorter rise times compared to the 0.8 ns rise time of the IEC 61000-4-2 standard. These shorter rise times may require faster turn on performance of the electrostatic discharge protection circuit. Some of the important differences are listed in the Table 1.

Table 1: Comparison between Different ESD Models

Parameter	HBM	CDM	IEC 61000-4-2	CDE
Rise time	2 to 10 ns	100 to 500 ps	~ 0.8 ns	100 to 300 ps
Pulse duration	150 ns,	~ 2 ns	~ 80 ns	up to several milliseconds (depends on the cable length)
Peak current	0.7A/kV	1- 16A (depends on the size of IC)	3.75 A/kV	Peak: ~5A/kV Plateau: ~3A/kV

The effects of LAN cable discharges leading to malfunction or even damage is presently not covered by any of the ESD immunity test methods. IEC 61000-4-2 is used to verify the ESD withstand capabilities, i.e. immunity of equipment, directly at its enclosure port and the metal shells of connector ports (excluding direct discharges to the pins). Complying with the system level IEC 61000-4-2 standard may not ensure enough robustness of the Ethernet system against LAN cable discharge. Several semiconductor manufacturers dealing with these communication interfaces have developed their own in-house CDE testers to qualify their ICs. However, the need for standardizing the test procedure is ever growing.

1.4.2 LAN Cable Charging Scenario

When a LAN cable is pulled over a surface, the cable may get charged due to a triboelectric effect. A LAN CDE can also occur if a person, who carries a laptop and is charged, inserts a LAN cable which is at ground potential. Other LAN CDE sources may be a charged device close to a cable which polarizes the cable.

The voltages of a LAN cable can appear between: (1) All twisted pairs and ground (common mode); (2) Wires of a single pair (differential mode); (3) Different pairs in the cable (mixed mode). The LAN cables can be charged in common mode and, although with a smaller likelihood, in differential mode. However, during the discharge sequence (pin sequencing) a cable which is charged in common mode can become charged in differential mode.

In [Gre2009], the average surface charge density on a short cable was experimentally measured by pulling it through an inline tribo-electrification charger consisting of a sleeve fabricated from numerous different plastic materials, including PTFE and aluminum, to provide friction against the cable jacket. The surface charge was measured using an electrometer. Examples of the surface charge measured densities are:

(1) For a PTFE charging sleeve: $\sigma = 0.1 *10^{-8}$ C.cm^{-2};
(2) For aluminum charging sleeve: $\sigma = -0.15 *10^{-8}$ C.cm^{-2}.

In another study [Wan2013a], a similar experiment has been performed where a spooled category (CAT) 6 cable is unwound over flooring materials like a high-pressure laminate, a typical surface material for a raised floor system. An example of the measured voltage (1.1 kilovolts) is shown in Figure 10 and Figure 11.

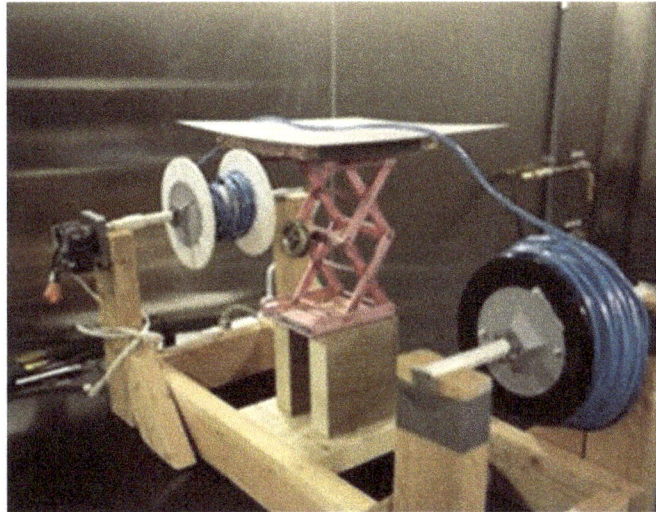

Figure 10: CAT 6 Cable Unwind Test Set up

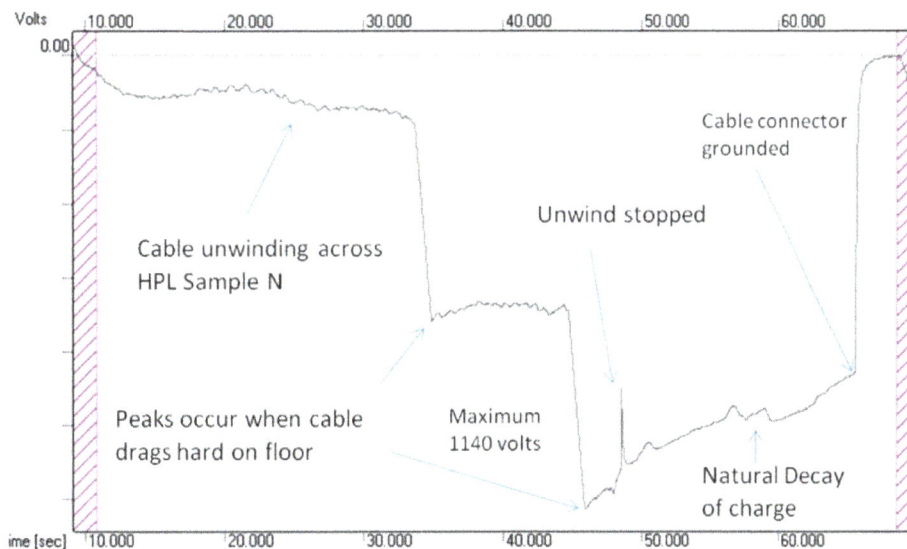

Figure 11: Voltage on the inner conductor of a CAT 6 cable during unwinding across a high-pressure laminate floor

1.4.3   LAN Cable Discharging Scenario

As shown in Figure 12, a typical LAN cable discharge situation includes: (1) Charged cable; (2) transformer possibly having common mode chokes included in an Ethernet connector; (3) PHY IC transceiver interface pins. The primary-side center tap of the magnetics is usually connected to a

Bob Smith termination circuit which consists of a resistor and a high voltage capacitor. During a LAN CDE, the discharge sequence, the Bob Smith circuit, the magnetic group and the PHY IC pin dynamic impedance will all affect the discharging scenario.

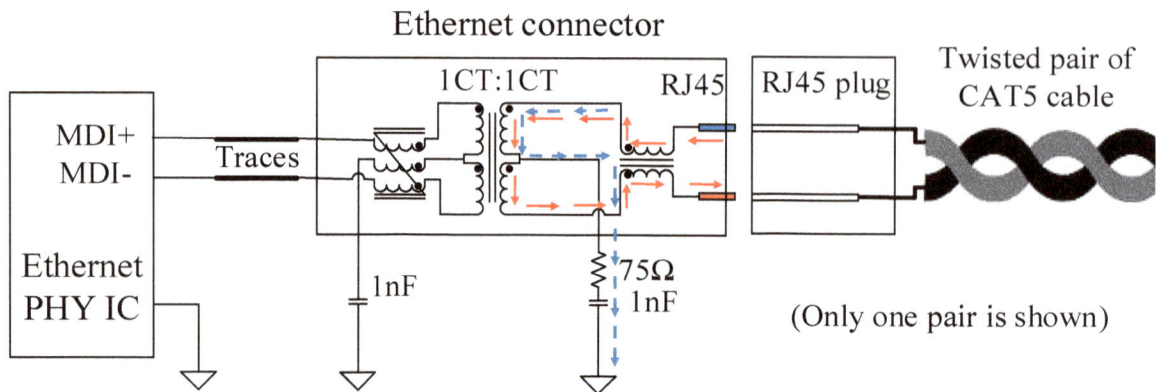

Figure 12: Schematic diagram of the LAN cable discharge situation. The system contains four pairs, here, only one pair is shown

When a charged LAN cable is plugged into an Ethernet LAN connector, not all pins make contact at the same time. One of the pins will spark over first. The first pin that contacts initiates a current flow between that wire and the LAN connector. Then, at least two different modes can be initiated by the second pin that mates. The second pin can either be from the same twisted pair, or from another twisted pair.

The assumption is that the LAN cable is charged in common mode. In other words, all 8 wires of the cable have the same potential with respect to a reference plane. When the first pin contacts, the discharge current flows via the common mode choke, and one half of the transformer primary winding to the Bob Smith termination circuit. The first pin discharge current path is illustrated by the blue arrows in Figure 12. This is referred to as a differential mode discharge, and the current magnitude mainly depends on the Bob Smith termination type, the common mode impedance the cable forms with respect to reference ground, and the charge voltage. Typical values for the resistor and the capacitor in the Bob Smith circuit are 75 Ω and 1 nF respectively, but other realizations are possible. For example, if power over Ethernet is provided, the capacitor value may be much larger. Furthermore, many center taps from different pairs or even cables may connect together and share the same termination circuit. This makes the first pin discharge current path become more complicated as multiple PHYs may be affected by one discharge.

If the second pin that mates is from the same twisted pair, the second pin discharge current path is illustrated by the red arrows in Figure 12. In this case, the current will flow through the transformer primary winding, and the first pin will be the main current return path. This is also referred to as a differential mode current.

As the discharge currents are in differential mode, the common mode chokes in the magnetic group cannot suppress the CDE currents. Differential mode discharge situations can occur under two circumstances:

- First contact causes a current flow from a pin, via the transformer, to the Bob Smith circuit

- If the first discharge caused a differential voltage on a pair (the analysis of this requires knowing the capacitance distribution between the wires, pairs and to ground) and the second pin that contacts is from the same pair, then a differential current will flow.

Most discharge currents will transfer to the PHY IC pins through the magnetics. The failure level of the PHY IC will depend on the magnetic module which are manufactured in several different configurations. Due to the large currents during an ESD event, nonlinear effects such as core saturation must be considered. The saturation of the magnetic cores can limit the amount of energy which is coupled into the transceiver circuit. In [Gan2016], the LAN CDE second pin discharge simulation result shows that the long pulses or high current levels will saturate the transformer, and this effectively reduces the PHY IC current, protecting the IC (see Figure 13).

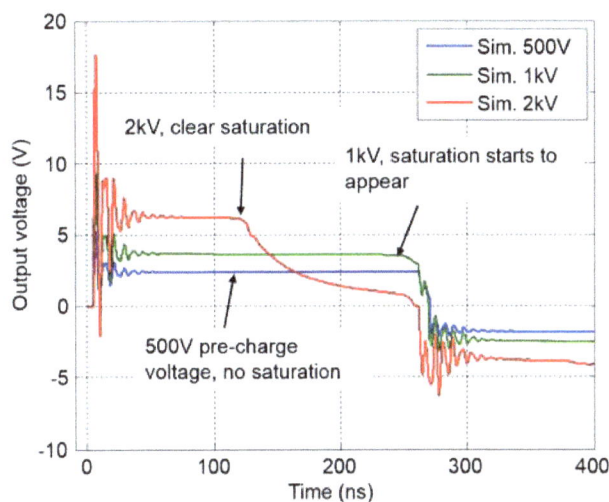

Figure 13: Voltage at the PHY IC pin side for a 25m long cable at different charge voltages [Gan2016].

1.5   Protecting Radio Frequency (RF) Antennas

With the increase of mobile and wireless applications, protection of the RF antenna ports from ESD becomes both critical and challenging. Antenna reception and signal integrity are important figures of merit for these products and virtually anything that is added to the circuit to protect it from ESD will impact its performance. Thus, considerable effort must be focused on either making the antenna self-protecting or minimizing the impact of the ESD protection elements.

One of the most common antennas in use for wireless applications is the planar inverted F antenna (PIFA). This antenna has a ground leg and thus can be inherently self-protecting. An example of a tunable PIFA antenna is shown in Figure 14.

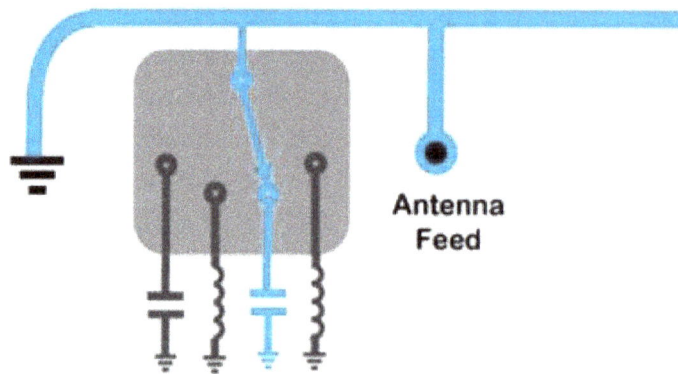

Figure 14: Tunable PIFA antennae

While the PIFA antenna is indeed grounded, care must be taken in designing the actual ground network. An antenna connected to a noisy analog ground will impact performance and can dramatically lower the antenna's performance. Thus, there is a tendency to design a DC ground that also provides some RF isolation. This is typically done by adding inductance in series with the ground connection. But this can defeat the purpose of an effective ESD ground. The Fourier transform of the initial peak in the IEC 61000-4-2 contact discharge pulse contains significant contribution from frequencies as a large as 1 GHz. Furthermore, there are additional high frequency components in the typical ESD gun pulse. Any inductance to ground in the PIFA antenna ground net will degrade the grounding of the high frequency components in the system level ESD pulse. Effective grounding of the antenna will require designing a ground net that provides some level of RF isolation as well as an effective ESD ground.

The second type of antenna in common use is the monopole antenna. This antenna tends to be higher performance than the PIFA antenna, but it is much more difficult to protect. An example of a tunable monopole antenna is shown in Figure 15. Since the antenna has no ground of its own, ESD protection design becomes critical.

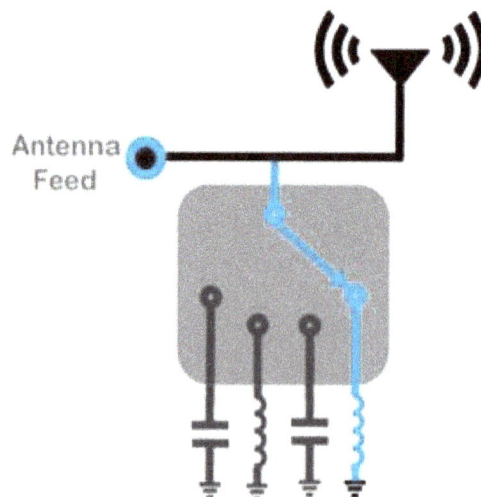

Figure 15: Tunable Monopole antennae

There is a tendency by companies that find the ESD protection of the antenna system challenging to push the requirements for ESD protection back on their component suppliers. This will lead to inefficient ESD designs since all the components connected to the antenna such as antenna switches, tuners, multiplexers, or other components must have their own protection. This not only duplicates ESD protection in the individual components, but it limits the ability of the system engineers from developing the most effective ESD protection that also minimally impacts performance. Ideally, the ESD protection of the antenna circuitry should be done at the system level so that there can be an efficient co-design of the ESD protection and RF performance circuit elements. With the development of ESD simulations at the system level through such approaches as SEED, the design of the RF antenna protection can be done much more intelligently.

Antenna ports that are external to the electronic component must, of course, be tested using the system level IEC 61000-4-2 test. For most wireless applications, including mobile phones, the device must pass Level 4 testing. Consequently, if the antenna is exposed, it will need to withstand 8 kilovolts contact discharge testing. For antennas that are not exposed, they need only be protected sufficiently to withstand the 15 kilovolts air discharge testing that is applied to the non-conductive exterior of the electronic device.

Chapter 2: Direct Pin Injection

Harald Gossner – Intel Corporation

2.1 Purpose of Direct Pin Injection of IEC 61000-4-2 Pulse

In real world stress situations IC pins are usually well shielded from ESD discharges. This includes mechanical design measures like pre-mate contacts of cable connectors, grounded shield metal of the connector and the chassis and electrical protection by filtering components on the PCB which mitigate the transient pulse on the PCB line towards the IC pin. Nevertheless, a partial stress, typically referred to as a residual pulse on the PCB line, can reach the IC component and cause damage or a malfunction. Also, an electromagnetic coupling can occur between the direct discharge path and nearby sensitive PCB paths.

Currently the direct discharge of an IEC 61000-4-2 pulse to a connector pin is excluded from IEC 61000-4-2 if a grounded shield is provided by the connector. However, as system design needs to consider the aspect of a residual pulse, it is industry practice that IEC 61000-4-2 pulses are applied to the connector pins even though IEC 61000-4-2 excludes this test. Since there is no specification of this test, performing this stress and the target values are largely differing. Primarily IEC 61000-4-2 ESD gun simulators are used due to the wide availability of the test equipment. Even if the waveform of the IEC 61000-4-2 spec does not match the waveform of the cable discharge, the practical experience of many generations of systems designs produced in high volume has shown that it provides a relevant test of the robustness in the field. This can be explained in that typically IEC 61000-4-2 failure mechanisms are dependent on the power or energy of the stress pulse and not on the details of the waveform. With increasing speed and downscaling of technologies a well-defined IC/PCB co-design is important to satisfy performance as well as robustness requirements. The essential base is the common understanding of the type of stress and the targeted stress levels by the IC and system design. **This requires a clear specification of the direct pin discharge to the connector pins.** The purpose of the chapter is to recommend a relevant and practical direct pin injection test at the system level.

2.2 Applying IEC 61000-4-2 Gun Pulses to IO Lines

Applying an IEC 61000-4-2 pulse to a connector pin using the commercially available ESD simulators poses various difficulties. To ensure a reproducible and well-defined discharge, a contact discharge to the pin under investigation should be performed. Due to the geometry of the connector pins with small distances and the large tip of the gun it can be difficult to impossible to contact a single pin (see Figure 16). In addition, the pins are often situated deeply inside the connector. A viable solution is a fan-out board with test pads of each line under test where the gun tip can reliably be contacted (see Figure 17). The board needs to have an appropriate jack to the connector and the board parasitics should be minimized.

Figure 16: Comparison of a USB-C connector Size and the Tip of an IEC 61000-4-2 gun

Figure 17: Example Fan-out Board used for Direct Pin Injection

As the PCB trace between the device pin and transient voltage suppression (TVS) diode acts as an inductor with efficient high frequency filtering in case of a low impedance state of the TVS diode, a 2-3 cm length of the PCB line between the branch point of the TVS diode and the IC pin landing pad is sufficient to block any overshoots related to fast transients in the first 5-10 ns of the residual pulse. Longer traces help to divert more current into the TVS path over the full pulse length. An example is given in Table 2. For the same IC component protected by the same TVS diode the direct pulse injection robustness can improve from < 1 kilovolt to 10 kilovolts when the trace length increases from 0 to 8 cm. While this can be a design optimization parameter for the form factor design, it needs to be considered for reference designs in order to avoid a too optimistic conclusion for a later form factor design with shorter trace lengths.

Table 2: Comparison between Different ESD Models

Trace length between TVS and IC pin	Current of first peak into IC pin	Current of second peak into IC pin
0 cm	11 A	6.5 A
2 cm	6.5 A	6 A
4 cm	No peak visible	6 A
8 cm	No peak visible	6 A

A direct pin stress of a fully assembled PCB, including all protection devices of the pin under test, enable applying the test procedures following IEC 61000-4-2. In the case of a partially dissembled board without a TVS diode the filtering of high frequency content of the IEC 61000-4-2 pulse by the PCB line is much less efficient. This can trigger failure mechanisms which are not apparent in a design using a TVS diode. To correlate the robustness against residual pulses a filtering of the 1st peak of the IEC 61000-4-2 pulse is recommended. This can be done using ferrites on the fan-out board (see Figure 18). In the resulting waveform the first peak of about 5 ns has vanished (see Figure 19).

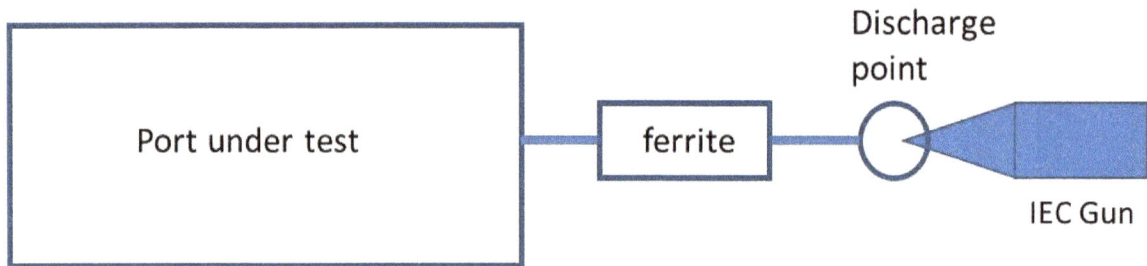

Figure 18: Set-up for IEC 61000-4-2 Direct Pin Discharge using a Ferrite for Fast Transient Suppression

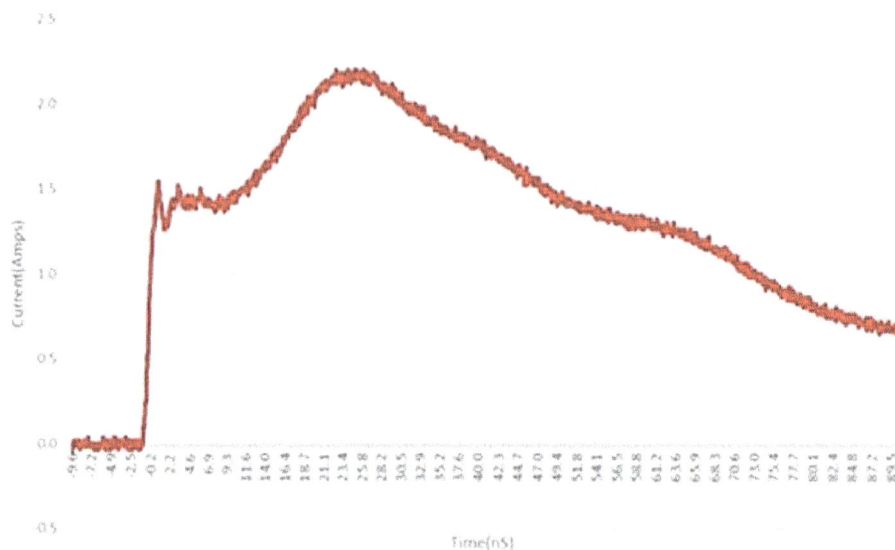

Figure 19: Discharge waveform into 2 Ohm of an IEC 61000-4-2 gun when a ferrite is mount near the discharge point. The first peak in the initial 1.2 ns has been attenuated

2.3 Transmission Line Pulsing

As an alternative waveform to an IEC 61000-4-2 ESD pulse, a trapezoidal transmission line pulse (TLP) can be used to characterize the ESD robustness. While this method has widely been used for characterizing IC pins, it has only been reported for a few applications on the system level. An obvious advantage is the similarity to the cable discharge waveform. In contrast to the IEC 61000-4-2 waveform, the TLP pulse enables varying the rise time and the pulse length over a wide range to explore the robustness of the pin or a TVS diode. TLP pulsing also provides a detailed analysis of the transient response of a semiconductor device to a fast transient, high current pulse. Additionally, the clamping behavior of the TVS devices can be extracted.

TLP testing will deliver the IV characteristic and the failure level by the same TLP testing procedure. It can be applied to an IC pin, TVS and discrete components as well as to the system port. An appropriate choice of the TLP pulse conditions is important to correlate to an IC pin and TVS diode behavior under IEC 61000-4-2 pulse conditions. Various TLP pulse length have been discussed in literature ranging from 30 ns to 100 ns. The longer pulse length leads to self-heating effects and to an increased differential resistance seen in the IV characteristics. With increasing pulse length, the power-to-fail level will drop as described by the Dwyer model [Dwy1990]. **A pulse length of 50 ns is recommended which leads to a good correlation of TLP failure current It2 to the hard failure threshold associated with the IEC 61000-4-2 second peak pulse current.** If the system port is stressed to emulate a cable discharge, the pulse length must be chosen following the discharge time of the selected cable. This can be estimated by the length of the cable and the speed of light which results in a value for the chosen pulse length of 14 ns/m as a rule of thumb. A recommended typical rise time is 1 ns, but in the case of extracting modelling parameters for transient turn-on of TVS diodes, various rise times should be used [John2011].

The use of needle probes for the purpose of TLP testing does have one advantage. Probes enable selecting the point of injection on the board in a flexible way, e.g. the probe can touch the soldering pins of the connector on the board avoiding the contact problem associated with the port pin inside a connector with a complex 3D geometry (see Figure 20).

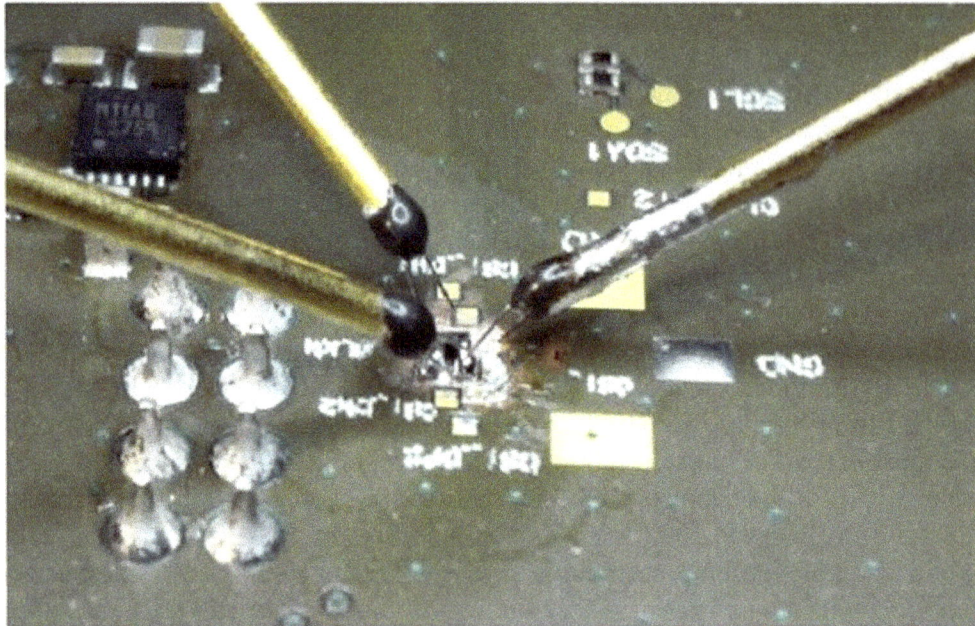

Figure 20: Use of needle probes to inject TLP pulses directly to the traces on the board. Voltage probes are added to sense the voltage drop.

2.4   Detection of Failure

A direct pin injection can result in various types of failure signatures. A fundamental distinction can be made between hard fails due to irreversible, physical damage and soft fails due to signal noise, power noise or latch-up, which are reversible. The latter leads to various categories of malfunctions described in IEC 61000-4-2. On the system level more severe soft fails can lead to a shutdown, a system hang, or battery drain and overheating due to latch-up.

In respect to the presented discussion, only hard fails are considered for the determination of the failure threshold of the system pin under test. This can be evaluated by leakage or impedance test of the pin and a subsequent functional test of the port interface. The functional test of the interface under test requires the removal of the stress test board or probe contacts and a connection to a client or host system or a loop-back test. Before the functional test a power cycling of the system is recommended to remove any malfunction of the system due to soft failures.

There are limitations in system level testing to detect minor physical damage in the IO. In particular, differential interfaces have low ohmic paths to ground making it difficult to detect small (in the microamp range) leakage paths. The interface might still be operating within specifications in the case of smaller damage levels. However, this can lead to a lifetime problem with an early fail of the system.

Chapter 3: Target Levels for Direct Pin Injection

Harald Gossner – Intel Corporation

3.1   Purpose

The definition of a target level for direct pin stress on IC pins supports the IC/PCB co-design approach for system efficient ESD design (SEED) protection. It opens a design window for the optimization of the system board protection. Knowledge of the target level for IC pin direct pin discharge and the target level of the system direct pin discharge allows the TVS vendors to develop appropriate TVS diode solutions.

3.2   Injection into IC Pin

The design window of the IC is determined by a destructive current level It2 and a destructive voltage level Vt2. In most designs, the required system pin robustness exceeds It2 of the IC and additional PCB protection circuit elements must be implemented matching the IC's protection design window. Vt2 provides an initial design window without considering the voltage drop on the PCB trace to the IC pin. Taking It2 into account increases the effective voltage design window for the TVS diode by adding resistance and impedance to the trace between TVS branch point and IC pin landing pad. With this knowledge a co-optimization with functional performance criteria can be performed.

The following recommendations address the **robustness of an IC pin (connected to an external port)**:

- A minimum It2 level of 2 amperes is recommended. This is extracted for a recommended TLP pulse length of 50 ns.
- The Vt2 window is set in relation to the maximum operational voltage VIOmax. The difference between Vt2 and VIOmax provides the useful design window. The recommendation is to maintain a minimum voltage delta Vt2-VIOmax > 4 V. This enables physically feasible TVS protection diodes with 0.3 Ohm on-resistance for high protection level at the port up to 8 kilovolts.
- To have a standardized and system relevant condition for the buffering cap between VCC and VSS it is recommended to attach a 1 µF cap with low serial impedance to the rails.
- To ensure a high robustness at all system power states, It2 and Vt2 should be extracted both for an unpowered domain and a powered domain at maximum VDD.

3.3   Injection into System Port

A direct pin discharge to the system pin of a reference design serves the purpose of proving feasibility of a protection concept. The testing of the final form factor design should provide the confidence to deliver a robust design to the market.

The following recommendations address the **robustness of a system pin**:

- The target level is ultimately a topic of the supplier-customer negotiation. There is system specific knowledge with most of the system designers.
- The following guidance can be used: a reasonable maximum requirement to pins of a cable connector is the required IEC 61000-4-2 level to the metallic parts of a system. A typical value is 8 kilovolts contact discharge. However, more recent investigations of USB cable discharges indicate that a level of 2 kilovolts would be sufficiently robust for USB cables fulfilling the standard shielding requirements [Stad2017]. Avoiding excessive target levels helps to balance performance, cost, and area.
- A reliable system pin test requires a fan-out board to guarantee a reproducible contact discharge.
- All relevant power states should be investigated.

Chapter 4: ESD Generator Modelling

Sen Yang, MS&T EMC laboratory

4.0   Introduction

The purpose of this chapter is to compare the different ESD generator models used for system level ESD simulations. The average measured value of peak current and rise time for each load impedance from Chapter 7, Section 7.1.2 were used as reference values in this chapter. For SEED simulations of direct pin-related contact discharge, SPICE based models are most suitable, but there is a wide range of SPICE based models to choose from. In this chapter, differences between the models and their impact on discharge scenarios into a load are discussed.

4.1   Current Source Models

Several authors [Wu2013, Kat2010, Yua2006, Son2003 and Fot2006] have introduced mathematical descriptions of current source models for an IEC 61000-4-2 gun discharge. As these models do not include any output impedance of the ESD generator, they might fall short in the prediction of currents for non-shorted circuit loads. Others have also proposed models for International Organization of Standards (ISO) 10605 pulses [Mer2012].

The main current source model for system level ESD generators is given in the IEC 61000-4-2 standard. The waveform describes an idealized contact mode discharge waveform. The waveform is somewhat related to observed human-metal discharges. However, most human-metal discharge measurements do not report such a clearly distinguished second peak after the initial peak. Here the IEC 61000-4-2 waveform deviates from most measurements of human metal discharge and the clearly double peak structure of the IEC 61000-4-2 waveform must be better understood from a legacy perspective. Changing the waveform to a waveform that closely resembles the measured human-metal discharge waveforms would require changes to all existing ESD generators.

A current source model into a short provides misleading results if the model is applied to situations in which the DUT is small and non-grounded. Such DUTs (for example, a cell phone) form a capacitance to ground or to the horizontal coupling plane (HCP). If a current is forced which equals the IEC 61000-4-2 waveform, then the small capacitance of such a device may be charged to a voltage much higher than the initial charging voltage. Forced current source models must be handled with care for all situations except a discharge through a short.

On the other hand, authors like Wang et al. [Wan2013b] and Zhou et al. [Zhou2014] proposed equivalent circuit models that are derived from the mathematical equations of the IEC 61000-4-2 standard waveform or from measured contact mode currents. Figures 21 and 22 are the equivalent circuits for the Zhou et al. model and Wang et al. model, respectively. These models can be more relevant for a broader range of loads.

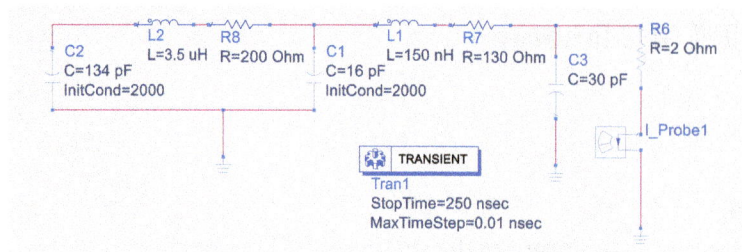

Figure 21: Zhou et al. model [Zhou2014].

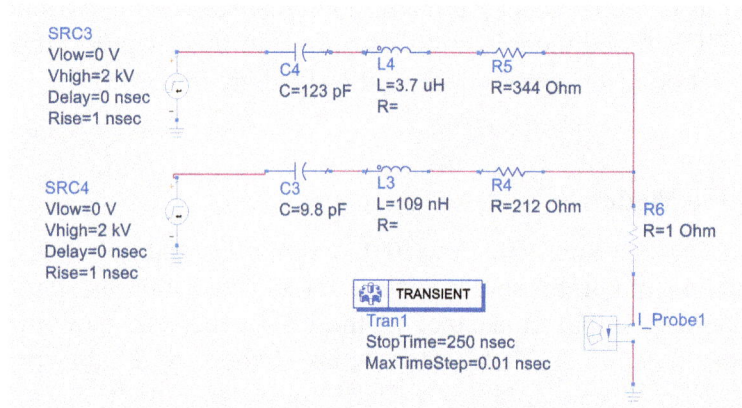

Figure 22: Wang et al. model [Wan2013b].

4.1.1 Physical-based Models

A physical-based model uses components to describe actual structures or components of the ESD generator. For example, the main resistor capacitor network (RC) constant is obviously expressed as the RC component block, but the ground strap's inductance can be explained as an inductance (Figure 24, Figure 27, and Figure 28) or a transmission line (Figures 25 and 26). Further, the excitation of the ESD generators can be expressed by low pass filtering a voltage source which rises in < 500 ps (which is closer to describing the voltage collapse in the relay) or by a slowly (e.g., 1 ns) rising voltage source (Figure 25); i.e., avoiding any representation of the relay and associated low pass networks. The coupling from the body of the ESD generator to the surrounding is often expressed as an RLC component block (Figure 24 to Figure 28). The following figures (Figures 23 to 28) represent different ESD generator circuit models.

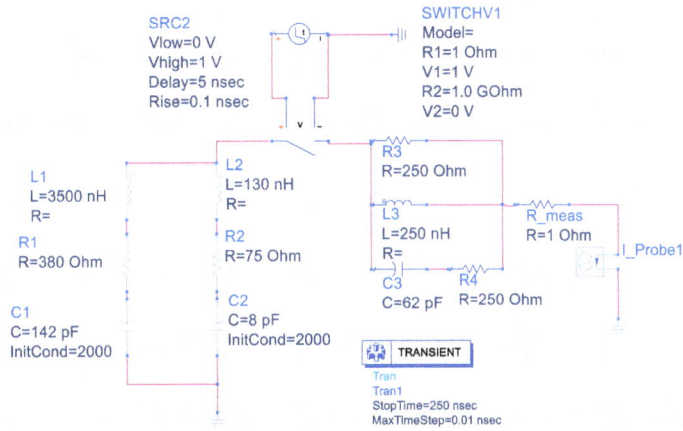

Figure 23: Tamminen model [Tam2016].

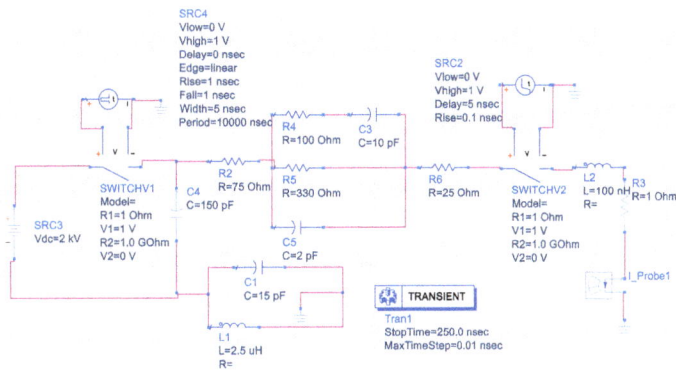

Figure 24: Notermans model [Not2016].

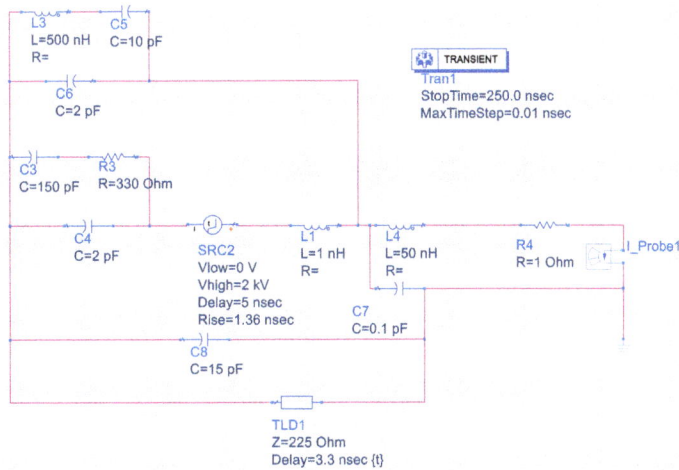

Figure 25: Sekine model [Sek2013].

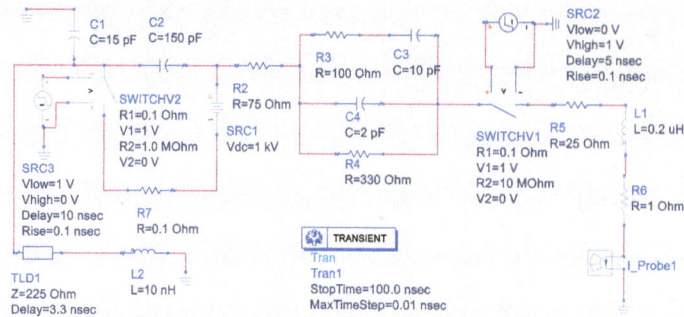

Figure 26: Caniggia model [Can2006a].

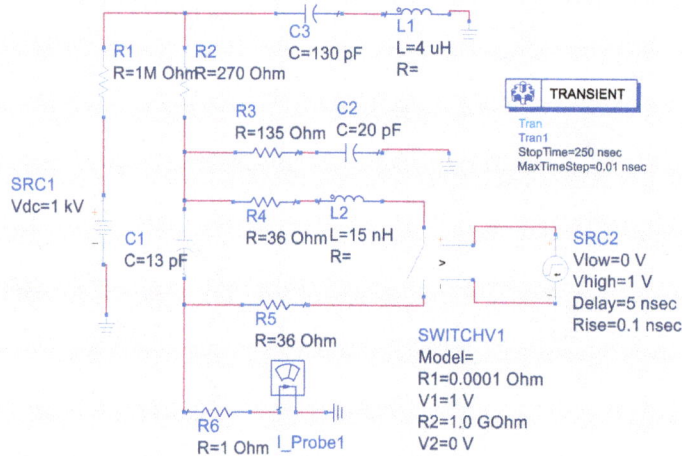

Figure 27: MST model [Wan2003][Li2015].

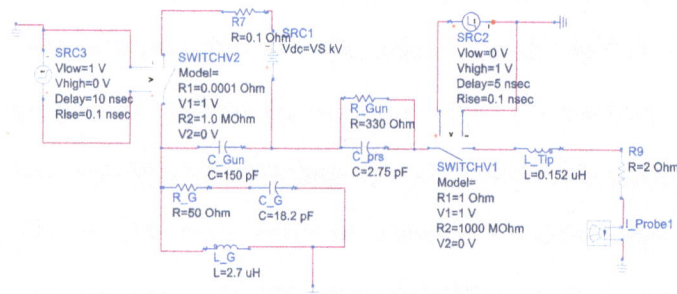

Figure 28: Yousef et al. model [You2017].

4.1.2 Simulation Results

As a first step, the current waveforms for discharge into a short are simulated using PathWave Advanced Design System (ADS) and shown over the entire time frame and over the first 7 ns (Figure 29). As shown in Figure 29, the Zhou et al. model has a peak value of 9.62 amperes, which is 28.3% larger than the IEC 61000-4-2 specification of 7.5 amperes. Furthermore, the simulation of different load impedances also shows that the Zhou et al. model deviates strongly from the measurement results and other models. Thus, it was excluded in the analysis of the effect of load impedances. The simulation results from the models are summarized in Table 2 with respect to the specifications of the IEC 61000-4-2 standard.

Table 2: Comparison of 2 kV peak values and rise times, data that violates the IEC 61000-4-2 specification is shown in red.

Model	1st peak current [A]	10%-90% rise time [ns]	30 ns current [A]	60 ns current [A]
Caniggia	7.04	0.94	2.49	1.81
Notermans	8.26	0.56	4.04	1.76
Wang et al.	7.23	0.94	4.00	1.97
MST	7.17	0.81	4.14	2.37
Tamminen	7.66	0.75	3.86	2.24
Sekine	7.46	0.91	2.32	2.52
Zhou et al.	9.62	1.04	4.40	2.36
Yousaf et al.	7.57	0.67	4.93	2.31

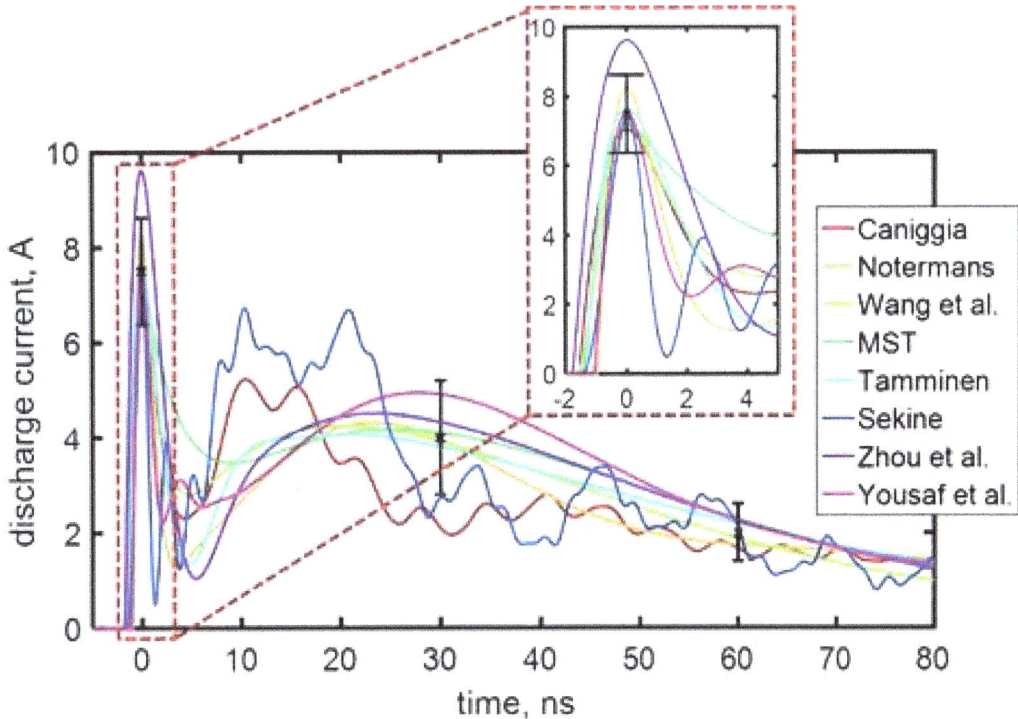

Figure 29: Simulated currents for discharges into a short at 2 kV. Waveforms are time shifted to arrange the maxima at 0 ns. Error bars indicate the IEC 61000-4-2 limits.

Two models, the Caniggia model (Figure 26) and the Sekine model (Figure 25) introduced transmission lines instead of inductances for the ground strap. Ringing in the waveforms of actual ESD generators can sometimes be attributed to the length of the ground strap. If the ground strap is parallel to a conducting surface [Can2006b], it forms a low loss transmission line. The transmission line causes ringing within the tail of the waveform, which may cause the current values for 30 ns and 60 ns to fall out of the allowed tolerance of the IEC 61000-4-2 specifications. However, after replacing the transmission line with an inductance of about 2 μH, the current waveform tail is within the tolerance. In the case where the ground strap forms a triangular shape, as the calibration set-up describes [Wang2003], it will radiate above 100 MHz, leading to little ringing.

The next step is to compare the models for different load impedances. Here, a set of loads is selected that matches the ESD target loads used in the ESD generator measurement in the previous section.

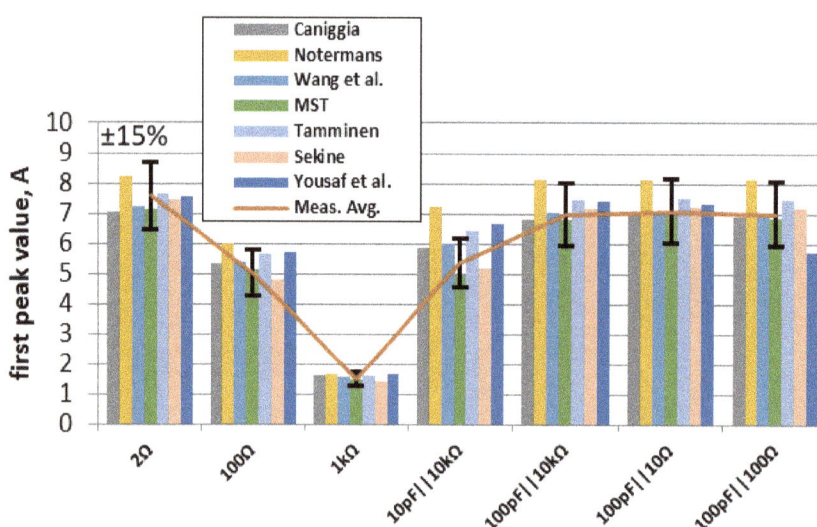

Figure 30: First peak values for different ESD generator models. The error bars indicate ±15% variation of average measured values.

The peak values are shown in Figure 30. For all models, the peak value for a 2 Ω load is very close to the average measured peak value which again is in good agreement with the IEC 61000-4-2 standard's current specification (3.75 A/kV). All modeled and measured waveforms show a similar tendency for discharges into other loads. This indicates that the "output impedance" of these models is rather similar and not far away from the measured data.

The rise times are shown in Figure 31. Significant differences are observed for the 1 kΩ load. Only the MST model and Wang et al. model's rise times are within the range of measured rise times for all load impedances (variation ± 25 % of average measured values).

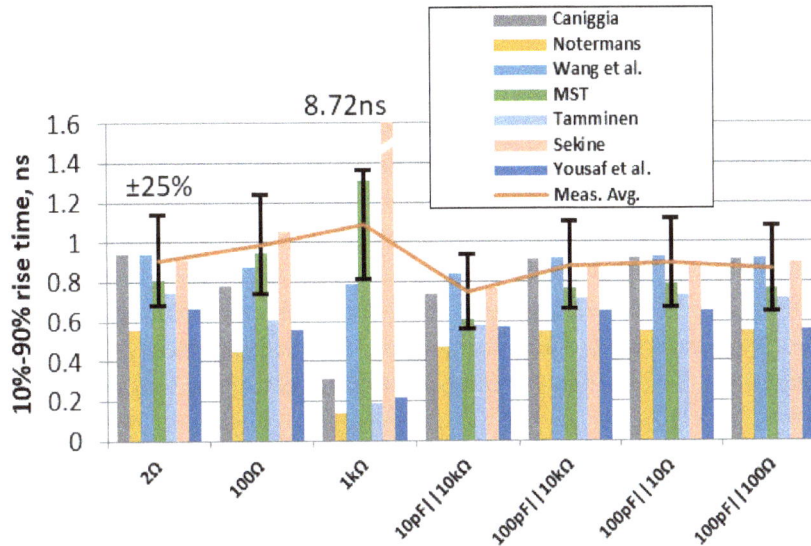

Figure 31: Bar diagram of 10%-90% rise time of the first pulse (Unit: ns). Note: the y-axis is limited to 1.6 ns for better comparison, all though the Sekine model rise time is 8.72 ns. The error bars indicate ±25% variation of average measured values. The IEC 61000-4-2 standard (2008) allows for +/- 25% variation around 800 ps for discharges into a 2 Ohm target in contact mode.

4.1.3 Discussion

As a SEED design is based on SPICE models, a similar question can be raised for simulation models: do these models predict the behavior observed in measurements for different load impedances? Or are there any models which strongly deviate from the behavior observed in experiments? Most ESD generator SPICE models have the first peak values close to the measurement of a real ESD generator. However, a pronounced difference is observed for the rise time in the case of a 1 kΩ load. Only two models, the MST model and the Wang et al. model yield the rise time which is within the range of 0.7-1.2 ns for all load impedance conditions simulated. The MST model is a physical model which means that it represents some details of a real ESD gun. On the other hand, the Wang et al. model is a waveform-based model, which is derived from mathematical approximation of the IEC 61000-4-2 standard waveform. Both models performed well when loaded by impedances other than a short.

4.1.4 Conclusion

The simulation result of the ESD models considered here (except the Zhou et al. model) agree well with the IEC 61000-4-2 standard (2 kilovolts, 2 Ω load), for both peak values and rise times. The peak values for all models are close to the measurement results in each load impedance scenario. On the other hand, pronounced differences are observed in the rise times in various load impedance conditions compared to the real ESD generator results. Further, some models such as Noterman and Yousuaf et al don't meet the tolerance of 15%. All the other models can provide rise times around 1 ns for all load impedances except for the special case of 1 kΩ where only the MST and Wang et al. models comply with the measurements.

Chapter 5: Impact on PCB Protection Design

Jeff Dunnihoo, Pragma Design

5.0   Introduction

Notwithstanding explicit directives and reasonable opinions on both sides regarding the merits of striking pins directly on external system connectors, the problem exists for the designer as to what should be done when the IEC 61000-4-2 qualification tests fail for that port.

Hard errors may create physical damage which clearly indicates the beginning point for the investigation of improved robustness. However, the failing device is not necessarily the problem to be addressed, nor is it always an option.

For example, when a high definition multimedia interface (HDMI) application specific integrated circuit (ASIC) fails during system ESD testing but the transient voltage suppression (TVS) protection remains undamaged, the designer needs to (quickly) design an improved protection strategy to protect the HDMI ASIC. The designer probably has no immediate pin-compatible device available to replace the HDMI ASIC, therefore a SEED analysis may be necessary to determine a better TVS option and/or other shielding, layout, or other mitigation strategies.

The case might be made that the test condition or level is too severe, according to the old adage about the patient complaining to the doctor, "Doctor, it hurts when I do this." To which the doctor replies, "Then don't do that."

But what of the condition where other systems (or worse, competing systems) have passed this onerous and "unfair" test condition? In this case the designer really has little choice but to try to improve the system robustness which has been a tedious job of trial and error in the qualification lab but has recently been alleviated by advances in simulation techniques.

5.1   SEED Concept and Impact of Interface Specific IEC 61000-4-2 Targets

SPICE-based simulations typically assume that the circuit will remain "functional" during a transient analysis. The nature of ESD and electrical overstress (EOS) suggests that normal operating conditions of the devices will be exceeded during the analysis, and this means that the dynamic response of devices is not only modulated by currents, voltages and time, but that the devices change their response based on recent abuse. What happens when the devices are taken far beyond their operating boundaries for short periods of time is a critical consideration. Sometimes it is enough to know that one or more devices have failed, and that is that. In some cases, it is interesting to know if a device will "fail open" or "fail short" and though damaged and more vulnerable, if the interface may still operate after the destruction.

First order protection circuit design analysis is often based on datasheet parameters of TVS devices, such as ESD rating (target IEC 61000-4-2 robustness rating, etc.), and clamping voltage, etc. These parameters are usually tested under the one condition which they will never see in a circuit; by themselves!

Since TVS devices are always included in a circuit to divert strike current away from the device under protection (DUP), the actual clamping voltages at the two devices are distinct and the voltage at the protected device during a strike is not zero. The current diverted by the TVS is not 100 %, and the residual current into the protected device is also not 0 %.

Second order modeling of this interaction comprehends the Kirchhoff current division between these two dynamic devices, and the current, voltage, power and energy maximum limits which can cause latent or permanent damage in either the TVS or DUP (or even the PCB traces themselves if the pulses are sufficiently energetic.) Most device input/output buffer information specification (IBIS) or SPICE models available today provide information about "clamping devices" in device IOs, but these elements were intended to model signal integrity issues like overshoot and ringing within 5-10% above and below VDD and VSS. ESD/EOS strikes inject levels 1000x or more than what are contemplated in those models, and while simulators will happily extrapolate those models out to +/- 50 amperes peak for a 4 mA clamp, there is no information on when the device will fail and how it will behave on the way there and beyond. In the framework of SEED this led to the introduction of high current models for behavior of IO pins as well as TVS diodes [InCWP3PI]. SEED models are applicable to short pulse currents up to several tens of amperes. They need to account for package as well as IO circuit behavior and should flexibly be used for various simulators like Spectre, ADS, and SPICE simulators [ESDATR26].

Given meaningful device models in the ESD/EOS regime, this level of approximation provides superior estimations of the system level robustness for a given conducted pulse applied to a given node for the specific devices. However, it does not typically address soft errors, system upsets, secondary discharges or coupled pulses into adjacent conductors and devices.

Third order modeling attempts to virtualize the entire 3D system assembly and solve the aggressor E- and B-field interactions predicted by Maxwell's equations. Given the exorbitant amount of accurate physical and electrical model input required, this can theoretically provide the most complete and accurate representation of an ESD/EOS strike on a system. It is also extremely difficult and time consuming. While elegant and expensive, 3D field-solvers are commercially available and extremely powerful, given the dearth of accurate ESD-regime electrical models for devices, they can also produce prodigious amounts of "Garbage-In, Garbage-Out" results.

For most quantitative "compare and contrast" analysis, though, the second order modeling as discussed above, with validated models, can provide excellent results for "better or worse" decisions.

For example, the simulation in Figure 32 (red line) analyzes a system ASIC failing with a "zero Ohm jumper" resistor between the TVS and its GPIO input. Selecting a 2 Ohm resistor instead that also doesn't affect the I/O operation reduces the residual ESD current by almost half. Further testing is thus warranted to ascertain if the system can then pass the target ESD qualification level in the lab with potentially only a BOM change and no additional cost.

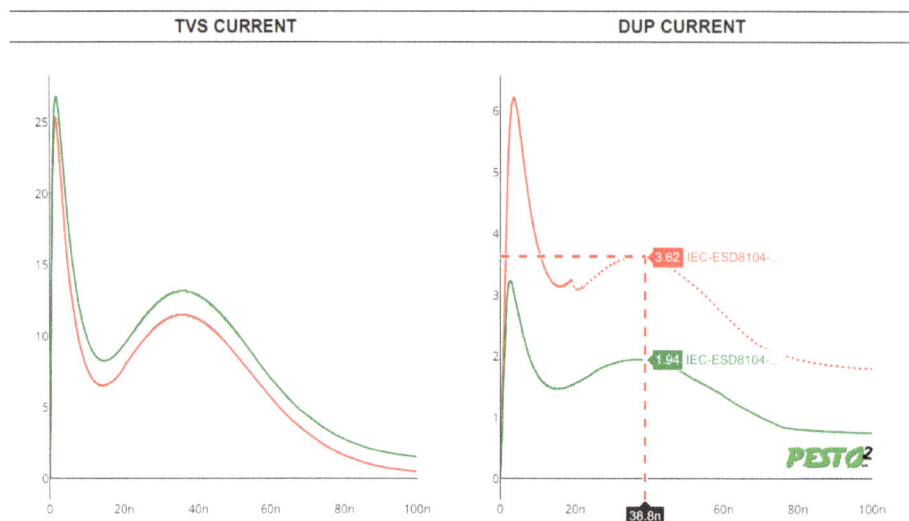

Figure 32: TVS clamp current at 8 kV IEC 61000-4-2 left, ASIC GPIO residual current right, "zero Ohm jumper" in red, 2 Ohm resistor in green.

5.2   Limits of Simulation

Of course, no simulation under any circumstance should be assumed to answer all questions, nor be extrapolated outside its limited sphere of valid inputs. The pass/fail criteria of a system are defined at the system level. For example, one TVS device may clamp to a lower voltage and faster than another device. This additional shunt current may inject undesirable currents and rise times into power rails or ground, causing secondary upsets on other devices.

5.3   Direct Pin Currents

5.3.1   Reasoning for Pin Specific Stress

Generally, it is assumed that connector shielding precludes direct strikes, and therefore pins should not be subjected to ESD testing for characterization. Below are several cases where direct and indirect stress can be applied to an "external" line exposed to the outside world.

1) When a line is exposed at the opposite end of an attached cable. Even when a peripheral such as a mouse or thumb drive is attached without a connection related event, a device may expose a path via plastic housing joints or button/LED entrance, etc.
2) Sequenced connection related (CDE) events
3) GND/shield pulses inducing current into adjacent signal lines.
4) Actual zaps to connector pins are possible, even when most discharges prefer shields (as seen in this multi-zap overlay at 20 kilovolts in Figure 33).
5) In specific applications where there is no shielding at all, such as an automotive or ethernet environment.

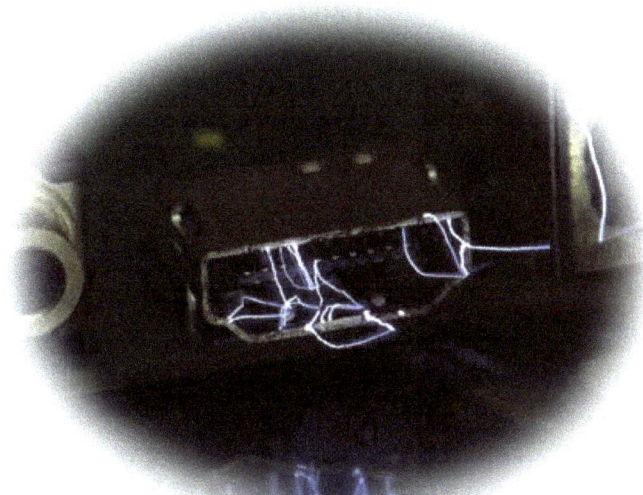

Figure 33: Multi-zap Overlay at 20 kV

5.3.2 Interface Types (USB, HDMI, Ethernet, audio, antenna, etc.)

There is an extended set of application specific interfaces which are exposed to system ESD discharge energy and need to be protected by TVS devices, see Table 3.

Table 3: Examples of interface type exposed to IEC 61000-4-2 ESD stress pulse energy for various applications.

Automotive	Wireless	Wireline
Controller area network (CAN) Bus	Antenna Port	A/B Line Driver
LIN Bus	USB Interface	Ethernet
Camera Interface	Camera Interface	HDMI
Display Interface	Touchscreen Interface	Display Port
USB Interface	Headset	eSata
Sim Card Interface	Sim Card Interface	

5.4   SEED Pass Criteria

Given that the residual pulse that a protected device sees from an air- or contact-discharge will be substantially modified, attenuated and morphed from the reference IEC 61000-4-2 calibration waveform, the failure mechanism may be dramatically affected for a given device, depending on what kind of protection is placed in front of it.

It is not possible to generalize a single static failure level in one parameter (such as peak current) that will properly indicate damage (or potential soft failure in some cases) under all stressors. However, physics-based models for current-and-time (fusing) and power-and-time or energy may often come close to approximating more than one particular failure mode for more generalized stressors.

In practical terms of guiding a useful design decision for a matching PCB design which can successfully protect the connected pin, model parameters such as peak failure currents (and others as mentioned above) for various pulse durations can be defined. These failure criteria characterize the robustness of the specific pin under consideration from the chip-level perspective [ESDATR26] However, that failure criteria may be completely unrelated or uncorrelated to the more general system level testing failure criteria. Pin level failure models may be defined by a specification limit guard banded to guarantee, for example, a defined leakage limit caused by ESD damage across process and temperature. ESD currents injected in this example that exceed the model's limit may cause a slight increase in leakage, but this might not actually affect operation in any detectable way during IEC 61000-4-2 testing.

In most cases, the actual measured robustness for a limited test sample will perform better than the worst-case, guard-banded specification guarantees. Designers at the chip and system level must both fully communicate and consider the likely divergence in robustness testing and simulation due to the selection, type and guard-banding of system and chip-level failure criteria.

5.5   Soft Fail SEED

Even while using an IEC 61000-4-2 ESD gun that is fully compliant very different pass and fail levels can be found depending on the brand of the gun, the set-up of the stress experiment and even the operational mode and conditions of the system.

In most cases a difference is seen in soft fails, but also a variation of the hard fail threshold can be observed. Reasons for uncertainties in the test results are discussed in previous chapters and improvement measures are proposed. But a soft failure, where an ESD direct pin injection, or EMI-coupled noise event causes a system signal or data corruption state that is detectable as an ESD-induced failure criteria is exceptionally difficult to simulate due to the complexity of the software and hardware functionality to be simulated.
It is possible, however, to simplify the failure criteria to a worst-case window and evaluate protection options again with the SEED approach.

For example, suppose a -500-mV glitch on an exposed signal line of at least 1ns in width is known or observed to cause a system upset. This might be due to a corrupted state machine or triggering on chip protection in a non-damaging way. This pulse might require rare alignment and timing to make the upset occur (thus the multiple strikes replicating real-world pulses described in Chapter 8 are also important to help catch a system in just the right timing to cause an upset.) Assuming that a large number of zaps allow this condition to be observed in the real world, albeit intermittently, we can assume that the prudent design strategy is to assume that "if this vulnerability can happen, it will happen and at the worst possible time." (Murphy's Law applies.)

Thus the SEED failure criteria may utilize the same direct pin injection simulation as for hard-failures, but instead of grading the pass/fail result by a damaging It2 level or other energy related failure, the simulation may rely on a signal integrity limit signature such as the glitch example above. If that envelope is exceeded during a simulation, it may be useful in predicting a possible failure in IEC 61000-4-2 gun testing.

With so many variabilities affecting test results, it may not be possible to encapsulate all the situations in a complete test suite. However, certainly when comparing multiple ESD mitigation or protection component options in the same circuit, a relative susceptibility to a soft-error may be weighed reasonably with this approach. Here again, SEED simulation can provide advantages for the designer even with soft-failures.

5.6 Conclusions and Correlations

When a system fails IEC 61000-4-2 qualification testing, the designer is tasked with identifying the unexpected susceptibilities and vulnerabilities that were exposed via gun testing. Many simulation and analysis tools are now available, and the designer can extract information about the system design and components. However, part of the overall qualification system as defined by IEC 61000-4-2 is not necessarily as easily represented in even the most elementary testing situations due to anomalies detailed elsewhere in this paper. The more reliably and rigorously refined the IEC 61000-4-2 test is, and the more repeatable gun testing results are, the more readily predictions and "first time right" design decisions will lead to robust, cost-effective and *passing* systems.

Part B: Discharge to Chassis and Display

Part A of this WP focused on discharges to pins. Part B takes the complete chassis into account and derives suggestions for improved IEC 61000-4-2 testing. Here the goal is to improve the repeatability of test results and to provide guidance for meaningful testing. This is partially done by the analysis of existing problems in the testing, and partially directly formulated for implementation in test laboratories.

Chapter 6: Historical Problems Observed when using the IEC 61000-4-2 Standard

Dr. David Pommerenke, Graz University of Technology

6.0 Motivation: High Test Result Uncertainty of System Level ESD Testing

ESD testing in accordance to IEC 61000-4-2 or derived standards has a high-test result uncertainty. Soft failures are especially hard to reproduce. Multiple reasons contribute to this difficult situation.

The present method of calibrating ESD generators covers the voltage and the discharge current in contact mode into a large ground plane (with the ESD generator held perpendicular to the plane) . From this, it is possible to determine the calibration uncertainty. The calibration uncertainty [Leu2001, Sro2003, Jan2010, Bar2008, Bar2010 & Mor2011] analyzes the effect of oscilloscope bandwidth, ESD target used and cables on the contact mode short circuit discharge current calibration.

These papers point at small effects of the calibration chain on the current or the voltage displayed by the oscilloscope. However, the differences caused by calibration test setups are insufficient to explain the test result variations observed in multiple test laboratories. Variation of soft-failure test results in which different manufacturer's ESD generators are used, often reach 1:2 (expressed in fail voltage levels) and may reach 1:3 [Koo2008]. Neither the uncertainty of the current waveform test results, nor the variations of the waveforms (there is no specification on the waveform, but only a specification on 4 parameters) can explain the large variations in the observed test results.

Consequently, improving the calibration of the contact mode discharge current cannot reduce these test result variations.

The test result variation is caused by a wide range of effects, not all of them related to the ESD generator itself. The dominating reason for soft-failure level variations, caused by changing the ESD generator brand, is due to the differing transient fields of the various ESD generator brands. This is an unintended consequence of using an ESD generator when running ESD testing. However, there are also many other contributing factors.

Parameters that contribute to the large test result uncertainty are:
- The transient field of ESD generators varies strongly from brand to brand

- No air discharge specifications, consequently, there are larger differences between generators in air discharge compared to contact mode
- Repeatability of air discharges due to natural variations in the spark length
- Pre- and Post-pulses
- Micro-amp pre currents due to field emission currents in the relay prior to the discharge
- No contact mode current specification into loads other than a short, this is especially important for discharges into small, not directly grounded DUTs
- Insufficient number of ESD pulses applied to each test point to have a statistical meaningful result considering time varying sensitivities of DUTs
- Testing at a level, such as 4 kilovolts, but not determining the failure threshold. To illustrate this, consider a case in which the actual failure level would be at 4.1 kilovolts. It would pass 4.0 kilovolts. But even small uncertainties would show the DUT passing in some labs and failing in other labs. Testing should be done to levels significantly above the limit such as to at least 6-8 kilovolts.
- Test point selection
- Not recognizing that secondary ESD has occurred (secondary ESD is when one part of a system under test becomes charged and then discharges to another part of the system)

6.1 Test Result Uncertainties

6.1.1 Importance of Air Discharge as a Mandatory Test

The majority of real world ESD events are air discharges. Most of the approaches of charged humans or objects will be toward plastic enclosures, screens, and connectors. These approaches may lead to (a) a discharge with sparking, (b) a corona surface discharge which has no visible sparking, but still may disturb or damage, e.g. a display, or (c) no relevant effect. Air discharge is the largest threat in the field. Testing according to the IEC 61000-4-2 standard should address this. In the maintenance cycle of the IEC 61000-4-2 standard the maintenance group MT-12 has suggested to remove air discharge from the mandatory testing requirements. However, the Industry Council on ESD Target Levels considers air discharge as a relevant and necessary system ESD test and proposes approaches to improve repeatability of the test.

By removing air discharge testing from the mandatory testing requirements, discharges to systems with plastic enclosures and displays could become an optional testing routine leading to coverage gaps in the system qualification. As most portable enclosures are made from plastic, and since those enclosures are especially prone to receiving ESD events, a large increase in devices susceptible to field failures would be expected.

This white paper describes several crucial steps to improve test result uncertainty, including the documentation of discharge currents during the ESD testing and a better calibration method for air discharge.

6.1.2 Air Discharge – Variation due to Arc Length and Approach Speed

A challenge in today's IEC 61000-4-2 testing is that air discharge current waveforms are not very repeatable. This is mainly due to the physics of the spark formation which strongly depends on ambient conditions as well as the test equipment and its operation. Even if the same ESD generator is discharged, applying the same approach speed and voltage, the waveform may vary significantly.

The underlying reason is the interplay of statistical time lag and approach speed (see references for details [Pom1995, Pom1996, Chu2004]). For the cases of either a rising voltage in a fixed gap, or a closing gap at a constant voltage the time lag is defined as such: The time lag is the time between the moment the field strength in the spark gap reaches the value at which a breakdown is possible, and the moment when the breakdown actually happens.

Currently there is no industry standard method defined to obtain repeatable waveforms for air discharge. This can be technically addressed by application of a very short time lag (the time between when a discharge becomes possible and the time it occurs) to achieve the maximum arc gap distance. In the case of homogeneous fields, the maximum distance is given by Paschen's law. The values for more complex electrode arrangements is discussed in [Zho2017]. However, this approach leads to the lowest peak currents and di/dt values. Thus, the severity of the test may be significantly less compared to real world events.

As the arc length variations and the resulting waveform variations couldn't be avoided for air discharge testing in the past, in the 90's the decision was made to introduce a contact mode ESD testing procedure in IEC 61000-4-2 and to require the contact mode test whenever the discharge can occur to a conducting surface.

Within the normal variation of the arc length, which is determined by the approach speed, surface properties, voltage and humidity, there are discharges which have short arc lengths. For example, at 50% of the Paschen length the discharge current rise time can be very short (<300 ps) and can carry exceptionally high frequency spectral components and large time derivative values, leading to soft failures. Recently, it has been shown that the probability of those discharges can be reduced by using an ionizer during air discharge testing [Zhou2019]. The ionizer provides charge carriers that can initiate the breakdown once the field strength in the gap reaches a value that allows discharges to occur. An additional advantage of using an ionizer is the ability to readily discharge charged plastic surfaces. This white paper explains the initial results of a study that aims at reducing the variability of air discharge and indicates a possible path for an improved ESD test standard with reduced test result uncertainty during air discharge testing.

6.1.3 Air Discharge – ESD Generator Calibration

Many discharges that occur at the customer side are air discharges. Contact mode discharges are not recommended to plastic or glass surfaces, while air discharge may lead to sparking through gaps into metal parts, or to surface charging by corona. Thus, air discharge must be included in system level ESD testing.

Presently, there is no calibration for air discharge in the IEC 61000-4-2 standard. To address this gap, this white paper suggests a methodology to measure the step response of an ESD generator in air discharge mode. This can become a base for the calibration. It avoids the fundamental problem of all previous air discharge calibration methods as it does not involve an arc discharge [Yan2017]. Instead, it directly measures the properties of the ESD generator with high repeatability.

6.1.4 Number of Discharges per Test Point

It is well known that the ESD sensitivity of DUTs depends on the transient state of the system. This is usually caused by temporal changes in the software running on the DUT. As the time variation is not known during system level ESD testing, one cannot guarantee that the set number of pulses and pulse rate will detect the most sensitive phase. While a rigorous statistical approach is described in [Ren1993, Wen1999, Mar2016, Rit1992, Har2016], this white paper further details test point selection, number of points and voltage levels in Chapter 8.

A relevant question could also be: Does this need to be done? If the sensitive phase is short and rare, then the likelihood of a real ESD event hitting the phase is low.

If the consequence of this ESD is not endangering safety, then the appropriate product-level cost-benefit of adding protection in such a case needs to be evaluated both by the vendor and the OEM. Such a weakness needs to be discovered before this trade-off can take place. However, the present IEC 61000-4-2 standard requires only 10 pulses per test point. Practical experience has shown that this is often not enough for achieving reliable test results. If, for example, 20 pulses per second are used, and each contact mode test point receives a few seconds of pulses, then easily 100 pulses can be applied without impacting the overall test length. This will certainly increase the likelihood of detecting sensitive phases of the DUT (note that secondary discharges must be avoided by removal of charges after each pulse as discussed in Section 6.1.11). Using such a pulse rate is common practice in many ESD test laboratories. In some cases, a test point is re-tested at 1 pulse per second if a failure was detected at 20 pulses per second. The rationale is that the DUT may not have fully recovered within 50 ms. If a failure occurs in the re-testing, also applying 100 pulses, then it is considered a failure of the test at this test point. This method seems to be a step in the right direction. For air discharge it is more difficult to apply a large number of pulses if this is performed by hand. Still, a number larger than 10 pulses per test point is needed to provide a larger coverage against software-induced sensitivity variations. Further, the arc length variation, and the consequential variation of the rise time and peak current, lead to further uncertainty in air discharge testing. Here the best method is to use automatic ESD testing with a robotic system, and to capture the discharge current at the ESD generator tip. This will identify the current waveform which caused a failure and also identify any secondary ESD. A more detailed method of selecting discharge points, voltage levels and dissipation methods is part of this white paper.

6.1.5 Transient Fields

Transient fields are not specified, and the difference in the transient fields between different brands of ESD generators is rather large: a factor of 3x at a given frequency is quite common. If a DUT has a more narrowband susceptibility, e.g., due to a resonance, then the test result may vary up to 3x by just changing the brand of ESD generator. This behavior and its root cause have been reported in [Koo2008, Koo2008b]. The present IEC 61000-4-2 standard (2008) provides information on the fields in the informative annex, but transient field calibration is not required. Further, one needs to consider that the fields of the generator depend on the orientation of the generator. Only the magnetic field within a few cm of the tip will have no angle dependence, as it is determined by the current in the tip.

6.1.6 Position of the ESD Generator

The current of the ESD generator will depend on the angle between the ESD generator and the DUT. The generator is supposed to be held perpendicular to the surface. If the ESD generator is at a different angle, the injected current will increase significantly.

6.1.7 Insufficient Contact to the Metal Part in Contact Mode

If the tip of the ESD generator in contact mode is not contacting well to the metallic surface, e.g., because of a thin paint layer, then a spark gap is created between the tip and the metal. Upon closure of the relay in contact mode, this gap will breakdown. In most cases the rise time of the current will be much less than the 850 ps of the main discharge. This causes strong RF components and may disturb the system. Due to the uncontrolled nature of the gap this will lead to difficult to repeat results. The standard should emphasis on the importance of having good contact between the tip and the meal part.

6.1.8 Generator Calibration

Often accreditation bodies confuse calibration uncertainty with test result uncertainty. The user cares about test result uncertainty and the calibration uncertainty is often just a small contributor to the test result uncertainty. Many papers describe the calibration of ESD generators and the analysis of different influencing factors, such as the ground strap routing, the effect of the ESD current target, or the oscilloscope (+cable) bandwidth.

It is believed the contact discharge current calibration uncertainty is low, and its influence on the test result uncertainty is not strong. Therefore, to improve the repeatability of ESD test results, there is no need to improve the contact discharge current calibration. An air discharge calibration can be introduced following the method explained in [Yan2017].

6.1.9 Approach Methods for Air Discharge

The approach speed of the ESD generator in air discharge affects the development of sparks in the air discharge. The physics is well understood [Pom1995, Pom1996, Pom1998]. On average, faster approach speeds leads to shorter spark lengths, faster rise times and higher peak currents. The effect is rather strong: Reducing the spark length from 2.7 mm at 10 kilovolts (this is the Paschen value) to 2 mm at 10 kilovolts will increase the peak current time derivative from a few A/ns to > 1000 A/ns. Even if it is not possible to obtain the same rise time and peak value from each discharge, controlling the approach speed, such as is done by robotic testing, is certainly the right direction to improve test result repeatability. As mentioned above, this white paper explains initial results of the effect of using an ionizer during air discharge testing. The ionizer will help to avoid very fast rising ESD currents during air discharge testing, thus, it opens a path for reducing the test result uncertainty during air discharge testing [Zhou2019].

6.1.10 Insufficient Test Setup Specification

Presently, the IEC TC77B working group MT-12, which is responsible for of updating and maintaining the IEC 61000-4-2 standard is discussing a variety of possible changes.

6.1.10.1   *ESD discharge current calibration in contact mode*

Different aspects are discussed, for example, the uncertainty of the ESD current measurement and the effect of oscilloscope input reflections on the captured data. Here the Industry Council on ESD Target Levels is of the opinion that the present calibration method for contact mode is not the main reason for test result uncertainty. The calibration could be improved by testing ESD generator discharges into impedances higher than the present 2 Ohm target. The uncertainty of the captured current data is small. Any improvement of the calibration for contact mode will not help to reduce the test result variations observed by many laboratories.

Large test result uncertainties are caused by the difference in electromagnetic radiation between different brands of ESD generators. These fields are created by the voltage collapse inside the relay which occurs in < 100 ps, while the current rise time is 850 ps. Thus, the spectral content of the current is lower than the spectral content of the voltage collapse at the relay.

So far this has been only addressed in the informative annex of the 2008 version of the IEC 61000-4-2 standard. The Council suggest that all manufacturers at least need to measure the fields around the ESD generator e.g., at a distance of 20 cm from the ESD generator at 0, 90, 180, and 270 degrees (in a plane parallel to the ground plane being discharged into) for contact mode at 5 kilovolts and publish this as specifications using a bandwidth of ≥ 3 GHz.

6.1.10.2   Humidity specifications

An on-going discussion topic is to narrow the range of temperature and humidity that is allowed during ESD testing. There can be various effects humidity has on ESD testing:

A) In Contact mode: The actual discharge is confined to a relay; thus, humidity does not affect the current waveform of the ESD generator. But charge decay is affected. If an ungrounded DUT is subjected to ESD, without external grounding or ionization the charge will remain on the DUT a long time. In high humidity, the charge decay times are reduced. However, this is not relevant for IEC 61000-4-2 testing, as the remaining charge must be removed before a new ESD pulse is applied as per IEC 61000-4-2 Section 7.2.4.1. A third effect relates to secondary ESD. Here, one may consider that in high humidity the statistical time lag of a secondary gap is reduced. However, most secondary gaps do not have a homogeneous E-field, as they are e.g., formed by sharp corners on PCBs. In this case the time lag is rather small. Another argument for a small-time lag is that the charge up times of secondary gaps are often very short, on the order of tens of nanoseconds. It is determined by the 330-ohm output impedance of the ESD generator (approximated) and the capacitance of the device that is not grounded, typically a few pF. Thus, a voltage much higher than the static breakdown voltage of the gap occurs. Discharges across such gaps will have very small statistical time lags.

In general, one can conclude for the contact mode: Humidity has no relevant effects in contact mode testing.

B) Air discharge: In air discharge, humidity has a drastic effect on the statistical time lag. Thus, approaching electrodes will show (on average) much higher peak currents and faster rise times in dry air. Here, one may feel a need to narrow the allowed range of humidity. Ishida [Ish2017] provided evidence. Here, air discharge testing was used to discharge between the air discharge tip and the standard current target. Both sides of the gap are rather smooth and from stainless steel. These factors lead to somewhat longer statistical time lags. The data from Ishida did not measure the spark length, but an increased spark length (probably approaching the Paschen value) is visible especially at the high absolute humidity corner of the IEC 61000-4-2 allowed specification.

6.1.10.3   Removal of the vertical coupling plane (VCP) testing

Discharges to the VCP are performed to expose a DUT to a rapidly rising electric field and to the magnetic field of the current spreading on the VCP. In addition, the DUT may be exposed to the fields from the relay structure of the ESD generator. However, the test geometry will keep the ESD generator at a distance from the DUT, such that the field exposure from the relay structure may be small.

Practical testing has shown that VCP very rarely leads to a failure that is not detected in other parts of the test. From that point of view, VCP testing may not be considered as a relevant, mandatory test.

6.1.10.4 *Horizontal coupling plane (HCP) capacitance*

The HCP to ground capacitance depends on:

- Distance to grounded structures, such as walls

- Size and shape

- Distance to the operator

- Equipment placed on it

It has been noted that the capacitance may vary by a factor of 1 to 2 or larger if the HCP is close to a metal wall in a shielded room (this may not be the suggested test setup, but such setups have been observed by the author many times). The capacitance of the HCP is mainly relevant after about 10 ns of a direct discharge into the HCP (in indirect ESD testing). For the first 10 ns the waves bounce on the surface of the plate and couple into the DUT. The fact that the HCP acts as a capacitor is not yet visible. Only after the reflections of the waves have ceased will the plate be considered a capacitor. It will reach a voltage that can be determined by a capacitive divider ratio from the ESD generator capacitance and the HCP-to-GND capacitance. The DUT will be exposed to a more or less static field (decay time: C_HCP * 1 megohm). Only very few DUTs are sensitive to such slow decaying fields. Keyboards, and high impedance turn on/off circuits are examples of such circuits.

6.1.10.5 *Insulating layer on top of the HCP*

Especially for testing of displays on mobile devices, the specification of the thickness of the insulator on top of the HCP is critical. The current standard specifies the insulator as a 0.5 mm thick plastic sheet. Two problems have been observed:

- The insulator is often not flat. Here, using a polycarbonate insulator can provide a long-term flat material

- In testing of tablets and cell phones, one test configuration is to place the display towards the HCP and to discharge to the phone. In this case, a large capacitance is formed between the DUT and the HCP. The value depends on the size of the DUT, geometry, flatness of the insulator etc. Values from 100-300 pF are typical for cell phones and tablets. Due to the discharge to the DUT all injected current will flow as displacement current through the insulator. Local variations in the flatness will lead to local variations in the displacement current density. Many soft and hard failures have been observed in the up-side down testing. A discussion is needed regarding changing to a thicker insulating layer, e.g., 5 mm. This would still lead to a large displacement current through the display, and display weaknesses could be discovered by an air discharge to the display. It would also remove the repeatability problem caused by the present test requirement.

6.1.11 Avoiding Effects of a Previous ESD Event on the Next ESD Event: Software Recovery and Charge Removal

The basic idea of the IEC 61000-4-2 testing is that each discharge is independent of any previous discharges. This has two consequences to be aware of:

1) **Software:** Any error correction or recovery triggered by the previous discharge should have completed before the next ESD pulse is applied. Now the test engineer may ask: How do I know if this is the case? There is no clear answer. For that reason, many will often apply 20 pulses/sec in contact mode and, if a failure occurs at a test point, the test may be repeated at 1 pulse/sec. Now applying 1 pulse a second for approximately 100 pulses (a large number of pulses are needed to achieve repeatability, see [Ren1993, Wen1999, Mar2016, Rit1992, Har2016]) will take time. But this is only needed at the few points at which a problem was detected.

2) **Electrostatics:** A previous ESD can charge the DUT. In this case the solution is easy, connect the DUT to ground via some high impedance path. People have used carbon fiber wire, wires with 470 kilohm resistor at each end etc. All of those will (see the exception below) not really impact the testing as they are rather invisible for RF, but they drain the charge after the ESD. The exception is secondary ESD. Consider a cell phone connected to AC power via a 2-wire charger. The 2-wire charger has no connection to ground. So, if a discharge is applied to the phone, the phone and the DC side of the charger will charge up, this may lead to a secondary ESD event inside the charger. This secondary event is known to often destroy the charger (and sometimes the phone due to over voltage from the charger). The other aspect are charges on the glass or on the plastic surface. An air discharge to an insulating surface will lead to surface charges. Although no spark is visible, current levels of up to 10 amperes can be reached as shown in Figure 34.

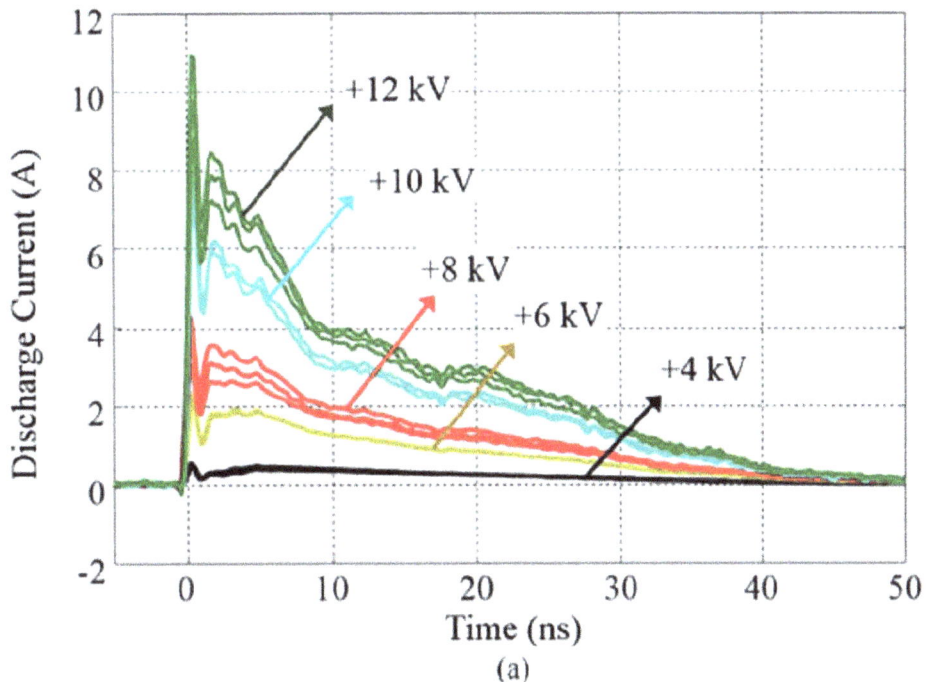

Figure 34: Discharge current for spark less discharges to a display glass [Gan2017]

The surface charge deposition can be visualized by blowing laser toner dust onto the charged surfaces. The toner powder will be attracted to the charges on the surface. The figures created this way are called Lichtenberg dust figures (see Figure 35). The method was initially published in 1776.

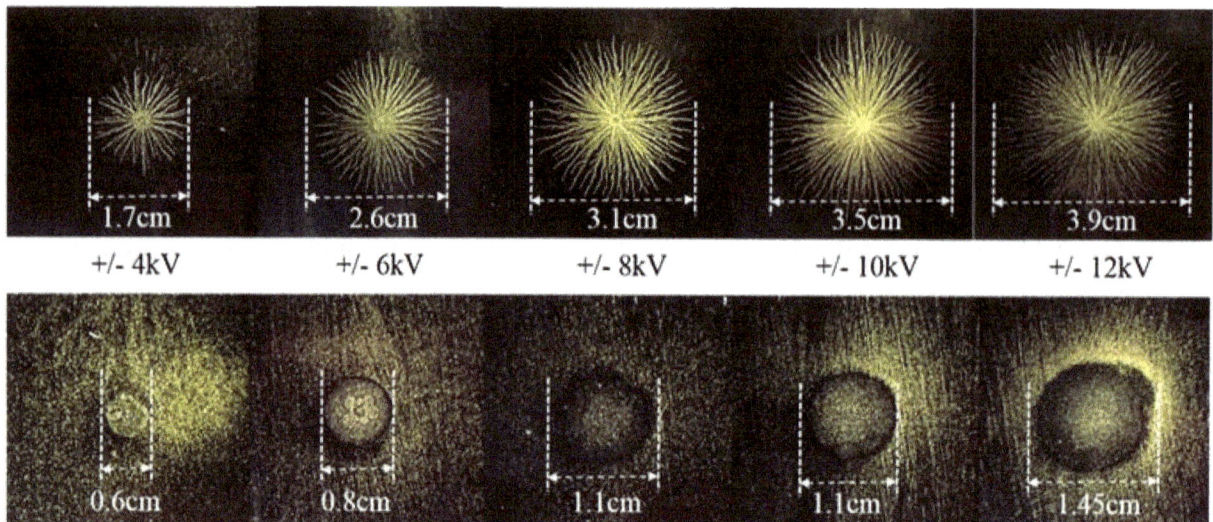

Figure 35: Lichtenberg dust figures obtained for different discharge voltage levels and polarities on a display glass surface. ESD simulator approaching speed: 0.3 m/s. Relative humidity (RH): 40% [Gan2017]. Note: top figure is positive charge voltage, bottom figure is negative charge voltage.

These charges should be removed from the glass or plastic surface. Here a brush seems to be a good method. While an ionizer will also remove the charges quickly, there is the potential risk that an ionizer will change the characteristics of the air discharge. This change can be positive for the test result repeatability. The stream of ions during air discharge testing will reduce the likelihood of very fast rising air discharge currents at higher voltages.

To allow independent testing without impact of the previous discharge the following ground rules are given:

- Use 20 pulses/sec for contact mode testing, 100 pulses at each test point and voltage level seems to be a good value (at least much better than the 10 pulses as suggested in the present standard).

- If a soft failure occurs at 20 pulses per second, retest at 1 pulse per second. If the failure does not re-occur, assume it was a result of the fast pulse rate.

- After each ESD pulse the charges must be removed to avoid secondary discharge situations.

  o For contact mode testing to parts that connect to the body of the phone a ground wire with >1 megohm is appropriate.

  o If a 2 wire AC/DC converter is used the time constant should be about 1 ms – 10 ms to GND (global GND) capacitance.

- Surface charges on glass or plastic surfaces from spark-less air discharge should be brushed off using a carbon brush or an ionizer. Note that the carbon brush should be in good shape (no worn out bristles for example) and ensure that appropriate locations be brushed (i.e. locations of ingress, floating conductors, and insulators).

6.1.12 Secondary ESD
When an ESD event reaches a non-grounded metallic part within a product, the voltage of this metal part with respect to ground will increase. If the isolation to ground is insufficient, a secondary ESD

event can occur. The discharge occurs across a spark gap between the floating and the grounded metal. The spark gap is often referred to as highly "over-voltaged." Over-voltaged means that the voltage is larger than the static air breakdown voltage of the gap. This can occur if the voltage on a spark gap rises quickly. Even if the voltage surpasses the static breakdown voltage, the breakdown may not occur due to the lack of initial electrons. This delays the initiation of the breakdown. If the voltage rise is fast, nanoseconds, the gap may reach double or triple its static breakdown voltage. Such highly over-voltaged spark gaps lead to very fast current rise once the discharge is initiated [Wan2014, Wol2015, Xia2011, Xia2012, Whi2012, Mar2017, Mar2017b, Mar2018]. The voltage across the spark gap leads to the breakdown of the spark gap and the initiation of the secondary ESD current. Secondary ESD events are especially harmful to electronic products for multiple reasons. First, the peak discharge current within the secondary spark gap can be more than five times larger than the current of the primary electrostatic discharge (ESD) event from the ESD gun [Mar2017a, Rez2017, Mar2017b, Mar2018]. Next, the rise time can be much faster than the discharge from the ESD gun. This is a consequence of discharging a highly over-voltaged gap. Third, the secondary ESD is within the product, thus, it can couple more strongly to the electronics. This can lead to soft and hard errors. From a testing point of view, another problem results from the repeatability of secondary ESD. The secondary discharge varies much more than the primary discharge due to the variability of the statistical time lag [Wan2014].

Secondary ESD is often found for:

- Non-connected metal parts in a product. These are often metal parts placed for decorative reasons
- Two wire connected AC/DC power supplies

It is important to monitor secondary ESD in a test setup. This can be done by attaching a current clamp at the tip of the ESD generator or on the ground strap of the ESD generator. The discharge currents, as captured by a current clamp can then be monitored by an oscilloscope to determine if secondary ESD occurs. If it occurs, it is essential to note this in the test report. It is suggested that monitoring for secondary ESD in ESD testing be required. It needs to be addressed when planning the experimental setup.

The secondary ESD event can be detected using software-assisted measurement techniques. The ESD gun discharge current is monitored using an F-65 current clamp at the tip of the ESD gun. The acquired current clamp waveform is further analyzed for waveform parameters such as the vertical threshold of the rising edge, the di/dt of the current waveform, and total charge delivered, which enable automatic detection of secondary ESD [Mar2018].

6.1.12.1　Timing sequence of secondary ESD

The secondary ESD event follows the primary charging event by a variable time delay (statistical time lag), ranging from nanoseconds to milliseconds. Most modern oscilloscopes offer the capability to collect separate events as a sequence of individual captures. These methods allow the scope to re-trigger very quickly (<50 ns) after an initial acquisition. The data is collected without processing to increase the re-trigger ability. Only after a sequence of trigger events is captured will the data be processed and displayed. This allows to capture sequences of pulses having a very low chance to miss a pulse. A sequential acquisition is ideal for capturing many rapidly occurring ESD events, or for capturing intermittent ESD events separated by long time gaps. The concept of capturing a sequence of events during secondary ESD is illustrated in Figure 37. Figure 36 shows the geometry and the test setup.

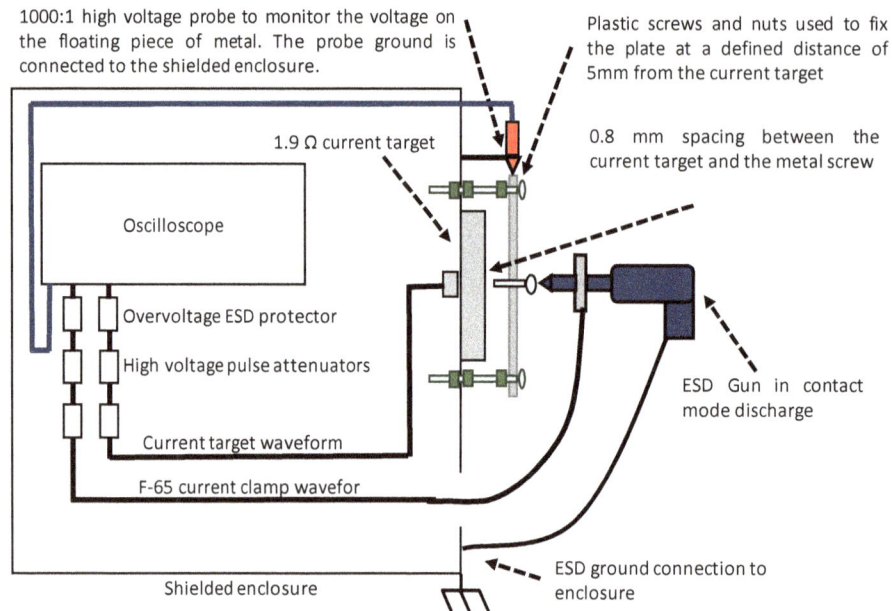

Figure 36: Geometry and test setup used for the measurement of secondary ESD of decorative metal (here a 3 mm thick Al-plate is used). The timing of the voltage charge up, primary and secondary ESD is explained in Figure 37.

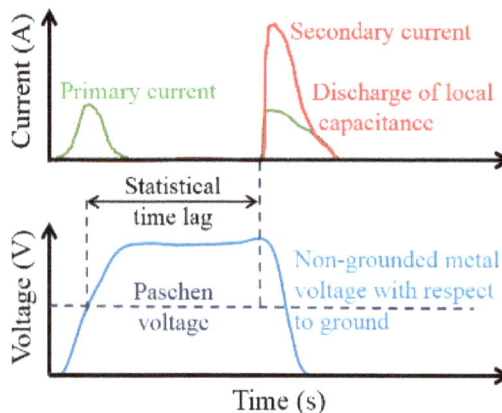

Figure 37: Graphical explanation of a typical ESD event followed by a secondary ESD waveform. The collapse of plate voltage is an indication of the occurrence of a secondary ESD event.

In general, the rise time of the secondary ESD currents is faster than that of the primary charging ESD current from the ESD gun. In real products, measuring the secondary discharge at the source location would be difficult to access and would require the use of external measurement equipment such as the wire loop antenna, F-65 current clamp, or monitoring the floating metal voltage using a high voltage probe to detect the occurrence of the secondary ESD event. In some cases, it may not be possible to access the source location of the secondary ESD event inside a real product, which will lead to a bandwidth limitation of the rise time measurements performed using the external equipment. Figure 36 illustrates a controlled setup to measure a secondary ESD event that can be used for modeling in full-wave simulation software. The simulation model assists in predicting secondary ESD induced current levels as a suggestive guideline for worst-case rise time and peak secondary ESD discharge current.

6.1.12.2 *Insufficient test documentation*

Engineers are often faced with the dilemma that a product failed ESD testing. However, it is often unclear how the testing was done and reproducing the result is often impossible. Here, the variability of the test results due to software status, configuration, wire routing, ESD generator model used, etc. cannot be easily overcome. However, many test details can be documented very well using video recording (if allowed).

It is suggested that test houses record a video of the testing while the discharges are applied. This has been implemented using a foot-controlled paddle that initiates the recording of video such that it captures the ESD generator's position and approach, as much of the DUT response as possible (like screen flicker), and it starts the recording of the current by an oscilloscope. This way one can exactly associate the current and testing to an individual failure. Of course, if no failure occurs the video and current data may be disregarded.

Chapter 7: ESD Simulator Calibration Improvement

Sen Yang, MS&T EMC laboratory
Jianchi Zhou, MS&T EMC laboratory
Dr. David Pommerenke, Graz University of Technology

7.0   Introduction

The IEC 61000-4-2 ESD generator calibration standard only specifies the discharge-to-ground scenario. In real measurements however, the ESD generators often discharges to different load conditions. The behaviors of the ESD generator under different load conditions are unknown to the engineers. Thus, a measurement setup which is based on an extension to the IEC 61000-4-2 standard calibration setup is proposed here to help the engineers understand the ESD generator behavior under different impedance load conditions.

Additionally, an air-discharge mode calibration method is introduced to determine the step response of ESD generators for air discharge mode.

7.1   Contact Mode Discharge into Impedances other than a Short

7.1.1   Modified ESD Target and Impedance Loads

The proposed method only slightly modifies the existing ESD generator calibration method, greatly simplifying its implementation and not requiring a different set-up. Different impedance loads are created by adding lumped components (resistor, capacitor or their combination) to the front of the ESD target. The impedance load will be referred to as the ESD target load in the later sections. The following loads are proposed (‖ indicates components in parallel):

- 2 Ω (present ESD target load)
- 100 Ω
- 1 kΩ
- 10 pF ‖ 10 kΩ (10 kΩ is needed as some ESD generator models do not discharge in contact mode without a resistive load.)
- 100 pF ‖ 10 kΩ
- 100 pF ‖ 100 Ω
- 100 pF ‖ 10 Ω

These values are selected to represent the load impedance of real measurements. For example, the 100-pF load could be a discharge into a non-grounded small cell phone. The 10-pF load discharge could represent the case of discharging into a small decorative metal which may cause secondary ESD or a discharge into a car key fob. The 100 Ω, 100 pF ‖ 100 Ω and 100 pF ‖ 10 Ω are the values recommended by other researchers. A high resistance load investigation was reported by Nieden in [Nie2010, Nie2009], thus, a high resistance load case (1 kΩ) will be included in these measurements.

The center discharge pad of the ESD target is removed (Figure 38) and the ESD target load is then screwed into the center of the ESD target. This offers a flexible method to change the load impedance seen by the ESD generator and to measure the discharge current with high accuracy using the same set-up. A plastic tube filled with epoxy holds the resistor or the capacitor (Figure 39) to avoid high voltage breakdown and ensure mechanical stability.

Figure 38: Load installed on an IEC 61000-4-2 ESD target. Shown is a 100 Ω resistive load

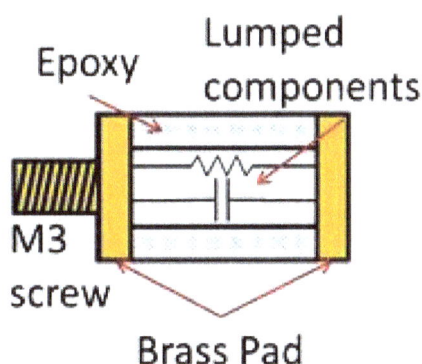

Figure 39: Structure of the ESD target load. The resistor and capacitor are embedded in epoxy.

The ESD generator discharges directly to the metal pad on the top of the ESD target load.

7.1.2 ESD Gun Calibration with a New Target Load Measurement

There are three major objectives: 1) to study the behavior of ESD generators discharging with different load impedances; 2) to offer a calibration test method for different load impedances if the experiments show that it is necessary; 3) to provide reference data for the study of the ESD generator SPICE models with different load impedances as discussed in Chapter 4.

The response to different load impedances of seven ESD generators was analyzed. These ESD generators will be referred to as "ESDGUN1 to 7" (includes 4 different commercially available models from 3 different suppliers) in the later section. Before performing the ESD target load experiments, all the ESD generators were tested using the standard IEC 61000-4-2 ESD generator calibration set-up (Figure 40, note that the use of a 20 dB attenuator may result in oscilloscope damage for higher voltage levels, use of higher attenuation values, such as 70 dB, may be necessary). The discharge current after the first several nanoseconds is sensitive to the ground strap position. Although the strap position was as recommended for ESDGUN2, its discharge current violates the requirements (Figure 41). Because of the limited high voltage tolerance of the resistors and capacitors used in the ESD target loads, only ±1 kilovolt and ±2 kilovolts discharge levels were measured. Three discharge events were recorded at each voltage level. As shown in Figure 42 and Figure 43, the discharge current of the ESD target load shows good linearity.

If a nonlinear behavior were to be observed, the following reasons should be considered:
1) Inaccurate voltage is displayed by the generator.
2) A small voltage drop occurs across the relay after the internal spark is initiated. This drop is not a function of the current (or only a weak function) and is often in the range of 25-40 volts. For low voltage settings, the drop may lead to an observable nonlinear current increase with charge voltage.
3) Resistors may show a voltage dependent resistance value. The voltage coefficient is usually negative (resistance drops with voltage).
4) Voltage coefficient of the capacitors, especially for ceramic capacitors.

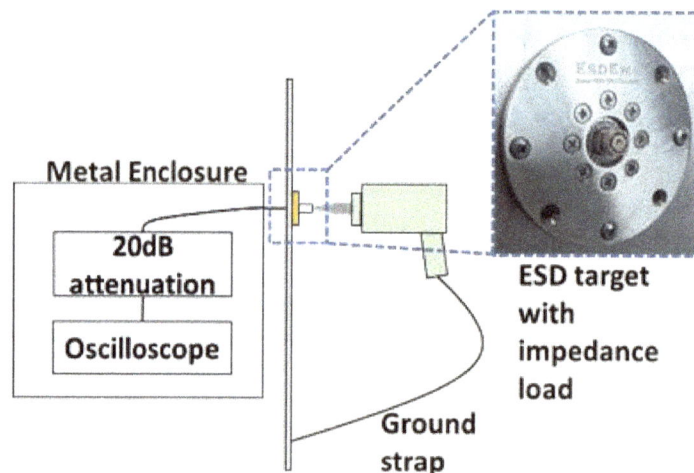

Figure 40: Measurement setup for ESD generator discharging into different load impedances, 10 GS/s, 2 GHz bandwidth. ESD target courtesy of ESDEMC Technology LLC.

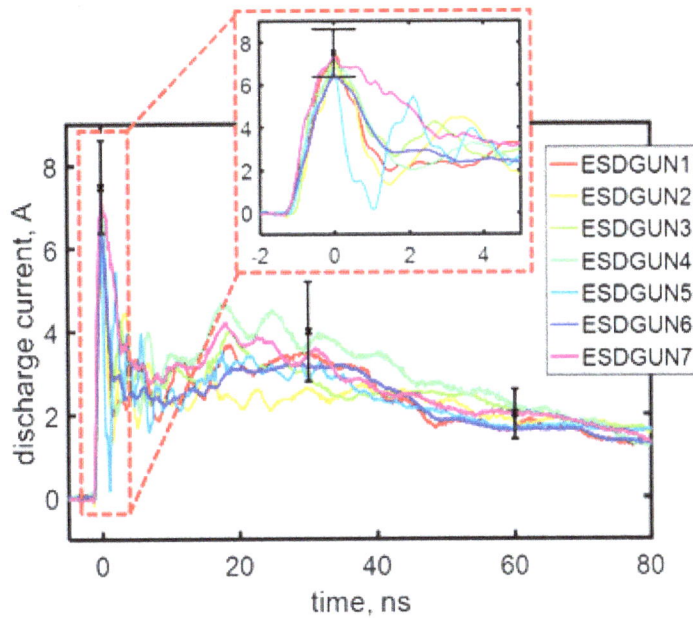

Figure 41: 2kV discharge current into a 2 Ω current target as specified by IEC 61000-4-2 (2008 release). Error bars represent the IEC 61000-4-2 limits. The error bars indicate that all but ESDGUN2 pass the requirements.

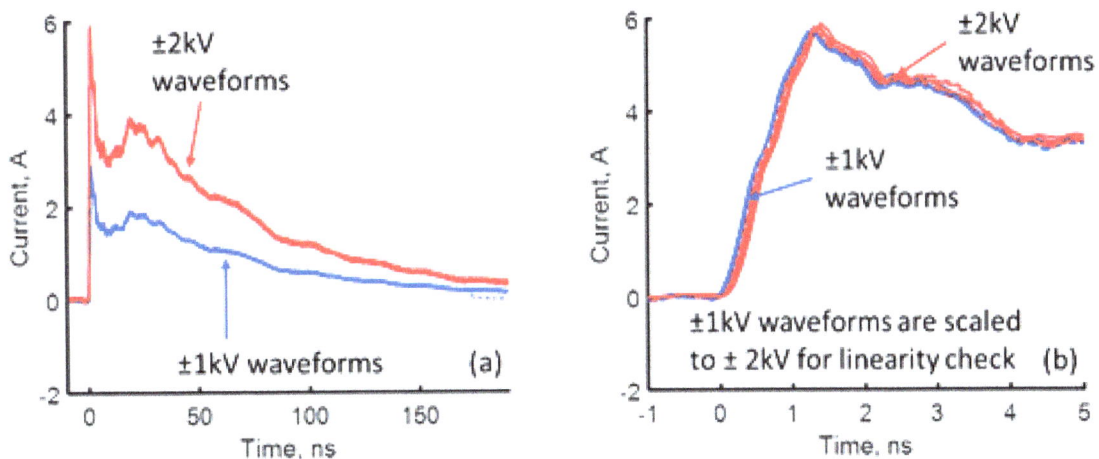

Figure 42: (a) Partially overlapping discharge current waveforms of ESDGUN7, 100 Ω load impedance. Three waveforms are shown for each discharge level. (b) Zoom-in to first peak. ±1 kV waveforms are scaled to ±2 kV; the results indicate good linearity and repeatability. Note: all negative discharge waveforms are inverted in the plot for better comparison.

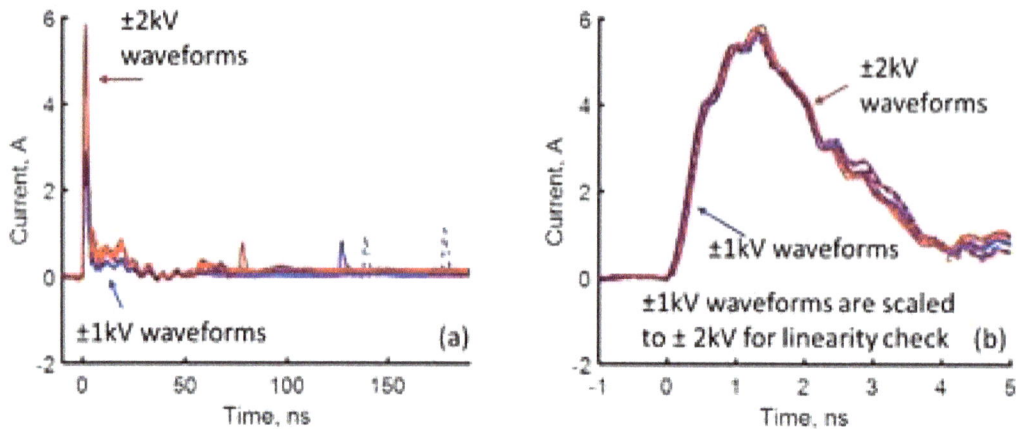

Figure 43: (a) Partially overlapping discharge current waveforms of ESDGUN7, 10 pF load impedance. Three waveforms are shown for each discharge level. (b) Zoom-in to first peak. Note: The discharge voltage levels are ±1 kV and ±2 kV. ±1 kV waveforms are scaled to ±2 kV; the results indicate good linearity and repeatability. All negative discharge waveforms are inverted in the plot for better comparison.

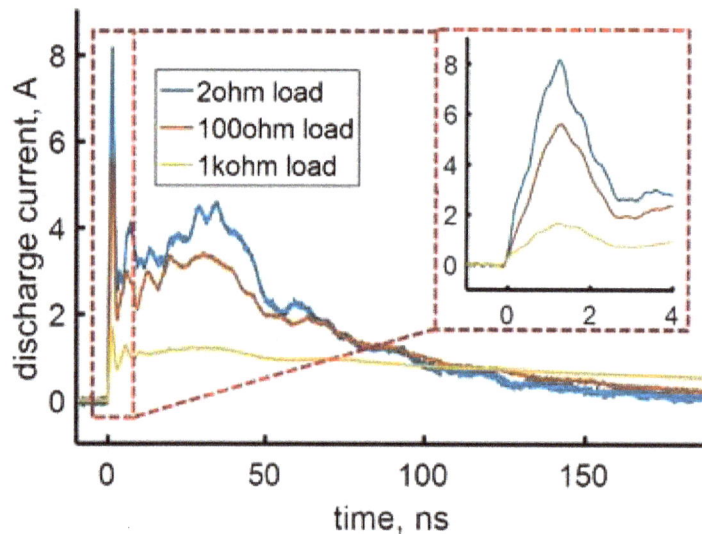

Figure 44: Discharge current waveforms for resistive loads of ESDGUN1. The discharge level is +2 kV.

To illustrate the dominating effects, the waveforms of one ESD generator are discussed in detail (Figure 44 and Figure 45). A similar behavior is observed for the other ESD generators (Figures 46 and 46). If an output impedance of a contact mode ESD generator is defined by the peak current requirement of 3.75 A/kV, a value of 266 Ω is obtained. This indicates that a generator should reduce the current from 7.5 amperes to 5.4 amperes if a 100 Ω load is used at 2 kilovolts and to 1.57 amperes if a 1 kΩ load is used. The measured average values are 5 amperes and 1.52 amperes which indicates that the simple output impedance calculation is suitable to predict the current for resistive loads. For a 10-pF capacitive load the second peak disappears while the first peak is not strongly affected (Figure 45).

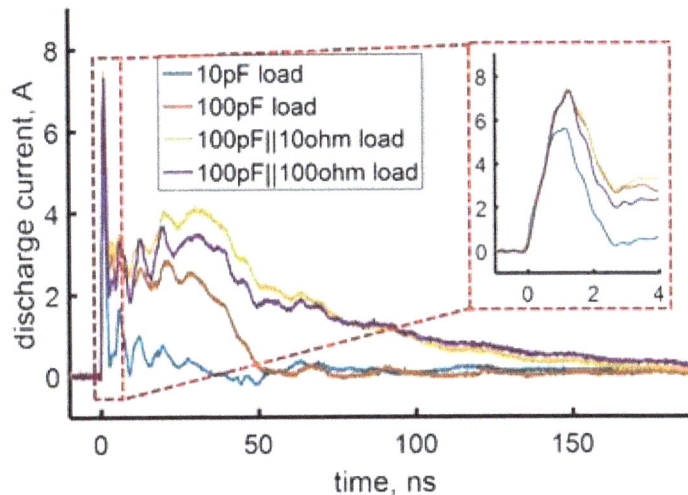

Figure 45: Discharge current waveforms for capacitive loads of ESDGUN1.

The first peak and the rise time for different ESD generators under different load impedances are summarized in Figure 46, Figure 47 and Tables 4 & 5. A load impedance of 2 Ω indicates a discharge directly into the ESD target without additional ESD target loads. Figure 46 and Figure 47 offer an easier comparison between different ESD generators. To judge if the variations between different generators at non-shorted load impedances are a concern or not, one can use the IEC 61000-4-2 peak current limits. This allows ±15% deviation from 3.75 A/kV. Figure 46 indicates that the variations for other load impedances are also within ±15% limit relative to the average values at each load impedance. Furthermore, the ESD generators show similar tendency over different load impedances; i.e., if the ESD generator has lower peak values in 2 Ω load impedance (such as ESDGUN4 vs ESDGUN7 in Figure 46), it is very likely to have lower peak values in other load impedances.

As shown in Figure 47, most ESD generators' rise times are within the range of 0.7-1.2 ns for all load impedances tested, which is within ±25% (as specified by the IEC 61000-4-2 standard for rise time variation) of the average measured values. The largest variations are observed in the high impedance load impedances (1 kΩ and 10 pF||10 kΩ). The internal structure of ESDGUN3 leads to a reduction of the rise time to only 0.3 ns rise time at a load impedance of 10 pF||10 kΩ. The average measured value of peak current and rise time for each load impedance were used as reference values in Chapter 4 which focused on ESD generator SPICE model comparison.

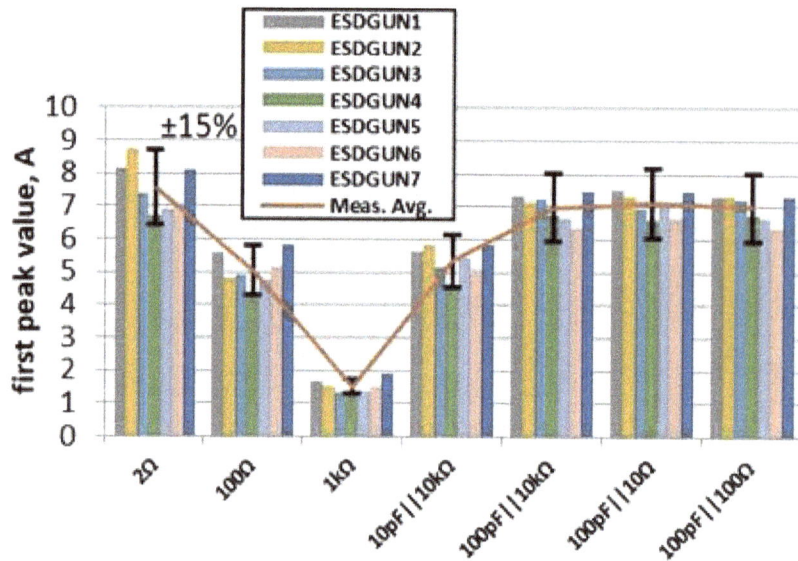

Figure 46: Value of the first peak for different load impedances. The error bars indicate ±15% variation of average measured values.

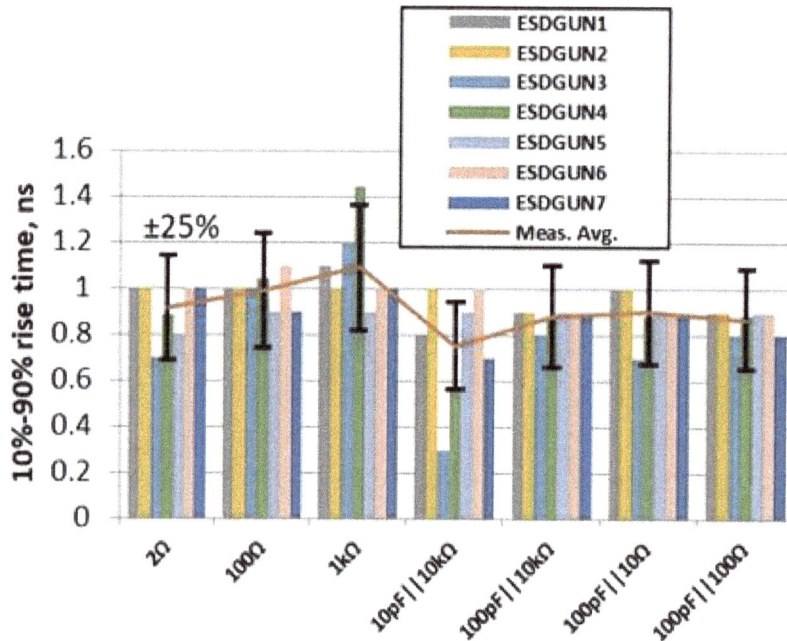

Figure 47: Rise (10%-90%) for different load impedances. The error bars indicate ±25% variation of average measured values.

Table 4: +2kV contact discharge of 7 ESD generators, first peak value. Unit: Ampere

Load impedance	ESDGUN1	ESDGUN2	ESDGUN3	ESDGUN4	ESDGUN5	ESDGUN6	ESDGUN7
2 Ω	8.18	8.74	7.39	6.72	6.92	6.99	8.11
100 Ω	5.60	4.81	4.91	4.24	4.77	5.17	5.81
1 kΩ	1.68	1.50	1.32	1.31	1.40	1.52	1.92
10 pF‖10 kΩ	5.63	5.85	5.14	4.52	5.45	5.09	5.84
100 pF‖10 kΩ	7.31	7.16	7.22	6.76	6.64	6.36	7.47
100 pF‖10 Ω	7.51	7.35	6.95	6.60	7.15	6.64	7.47
100 pF‖100 Ω	7.31	7.35	7.26	6.76	6.64	6.37	7.35

Table 5: +2kV contact discharge of 7 ESD generators, 10%-90% rise time. Unit: ns

Load impedance	ESDGUN1	ESDGUN2	ESDGUN3	ESDGUN4	ESDGUN5	ESDGUN6	ESDGUN7
2 Ω	1.0	1.0	0.7	0.9	0.8	1.0	1.0
100 Ω	1.0	1.0	1.0	1.0	0.9	1.1	0.9
1 kΩ	1.1	1.0	1.2	1.4	0.9	1.0	1.0
10 pF‖10 kΩ	0.8	1.0	0.3	0.6	0.9	1.0	0.7
100 pF‖10 kΩ	0.9	0.9	0.8	0.9	0.9	0.9	0.9
100 pF‖10 Ω	1.0	1.0	0.7	0.9	0.9	0.9	0.9
100 pF‖100 Ω	0.9	0.9	0.8	0.9	0.9	0.9	0.8

7.2 Pre- and Post-Pulses and Leakage Current Caused by ESD Generators

Besides the main current pulse, ESD generators will inject other currents into the DUT. In most cases, this is not relevant, however, if a DUT reacts to them, it can be very difficult to identify the reason. This chapter briefly treats the pre and post pulses caused by ESD generators.

The authors are aware of the following secondary currents:

- Current due to main capacitor voltage variation. Some ESD generators charge up the main capacitor and do not monitor its voltage. Now the voltage may decrease due to corona discharge. These corona currents are mainly relevant at > 10 kilovolts charge voltages. One may argue that a charged human would have the same properties: Corona may reduce the voltage. From a testing point of view one may want to have a known voltage, thus, some ESD generators have a voltage regulation circuit which keeps the voltage constant. However, this leads to another set of problems. The ESD generator may charge up a high impedance device by its recharge current.
- Leakage current. In contact mode the relay is at first open and the main capacitor is charged. However, there is a leakage current through the relay. This current, although small can charge up a high impedance device. The leakage current strongly depends on the charge

voltage, and the age of the relay. It is time dependent, initially, after charging the capacitor it may reach 1uA.

- Pre-pulses. Depending on the circuit of the ESD generator (e.g., usage of one or two relays) pre pulses can occur from the charge up process of the capacitor.
- Post pulses. After the initial pulse, additional pulses can occur due to re-ignition of the arc, and due to charge up of the main capacitor.

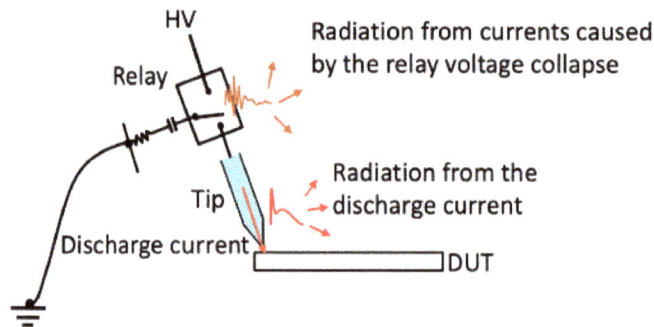

Figure 48: Illustration of the radiation sources during ESD testing

The radiation from the discharge current (see Figure 48) has been well studied [Koo2008] to correlate the failure level with the characteristics of the discharge current, e.g. the rise time, the maximum current derivative or the spectrum. However, the radiation from the currents caused by the relay voltage collapse is an often-neglected factor during ESD testing.

7.2.1 Multiple Pulses in Air Discharge

In an air discharge, the main pulse may not completely discharge the main capacitance of the ESD generator. The spark may quench leaving a residual charge in the main capacitor. Upon further approach, subsequent discharges occur which will deplete the remaining charge in the main capacitor. These secondary ESD pulses have lower peak amplitude, but much faster rise time. Here it needs to be considered that a human-metal (H-M) pulse in air discharge from a discharging person may also have the same features. The total transferred charge (transferred in multiple pulses) will not surpass the total charge stored in the main capacitor and local stray capacitance of the ESD generator.

The second situation, which may cause pulses, has been observed on ESD generators which have a cable connection to a high voltage supply. The capacitance of the cable and possible capacitances inside the base unit will recharge the main capacitance of the ESD generator. The speed of the recharge depends on the charging resistor value and having a second relay in the hand-held unit will prevent this type of recharge. However, if it occurs, it can lead to subsequent ESD pulses after the main pulse, finally (sum of the charge of all pulses) surpassing the charge value stored in the main capacitance, as additional capacitances (e.g., cable capacitance) are discharged.

7.2.2 Pre/Post Pulses

Pre/post pulses, also seen in the transient fields if the tip is not connected to anything, are usually caused by the charge-up phase of the ESD generator. The ESD generator may close a relay, which suddenly charges structures inside the ESD generator. This rapid charging can cause transient field pulses having rise times < 300 ps. A similar charging process can also occur when the relay moves back to the charging position depending on the structure of the generators.

Figure 49 shows an ESD generator model including the parasitic parameters. When the relay closes on the A side to charge the main capacitance C_1 the local ground will move in potential, and contact A will drop in potential. This will lead to a current injection at the tip via C_{p4} and C_{p7}.

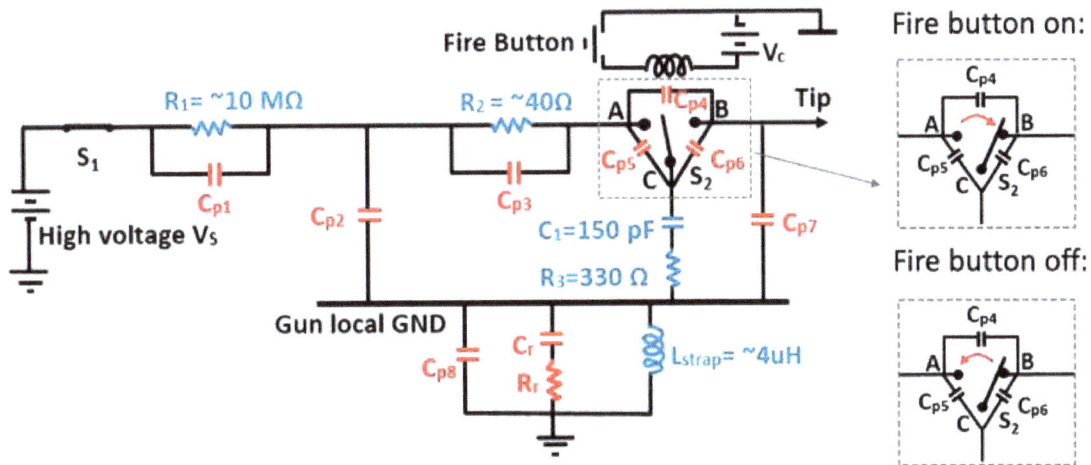

Figure 49: ESD generator model including some parasitic parameters

As the ESD-induced soft failure is mainly caused by the disturbance of the field, the induced voltage measured by a 0.5 cm^2 loop probe is used as an indicator for the transient magnetic field. Two types of post pulses were observed. The time sequence of such an example is shown in Figure 50.

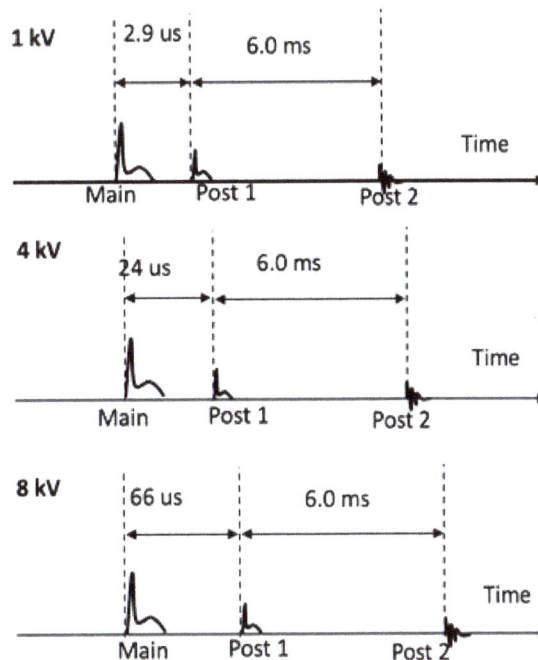

Figure 50: Time sequence of the pulses during an ESD discharge

The first post pulse is caused by the relay re-ignition of the spark within the relay. This is identified by the fact that the current waveform has similar waveform shape with the main pulse, and the time difference depends on the voltage. The second post pulse is caused by the relay. The 6.0 ms delay represents the relay mechanical movement. The timing is not affected by the charge voltage. The pulse is caused by the charge up of the structure, with emphasis on the high frequency components. The waveform at 8 kilovolts is shown in Figure 51.

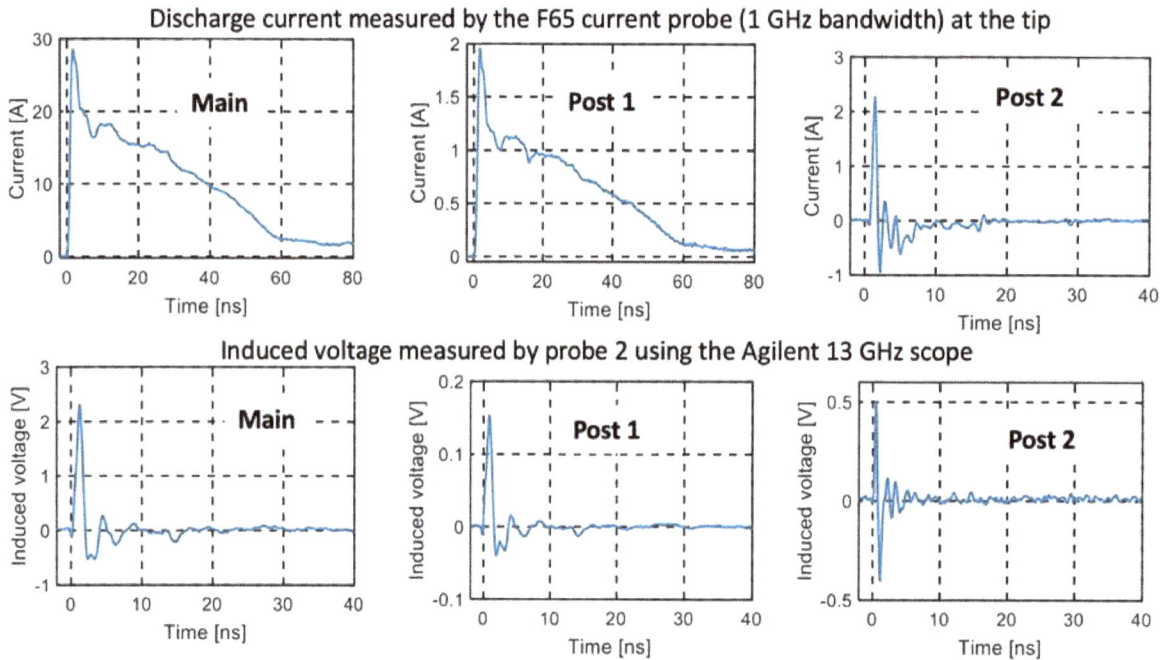

Figure 51: Discharge currents measured at the gun tip and the induced voltage measure by a 0.5 cm² loop probe at 8 kV. The probe is placed at 20 cm, oriented to capture the main field component.

As Wang etc. [Wan2004] has shown that the devices subjected to the ESD-induced fields can be sensitive to a certain range of the spectrum, it would be challenging to reproduce the test results when the post pulse is stronger than the main pulse for certain generators. An example is given for a generator that has been tested when the loop-measured induced voltage is 10 dB higher for the post pulse than the main pulse in the frequency range from 2.4 GHz to 2.8 GHz, see Figure 52.

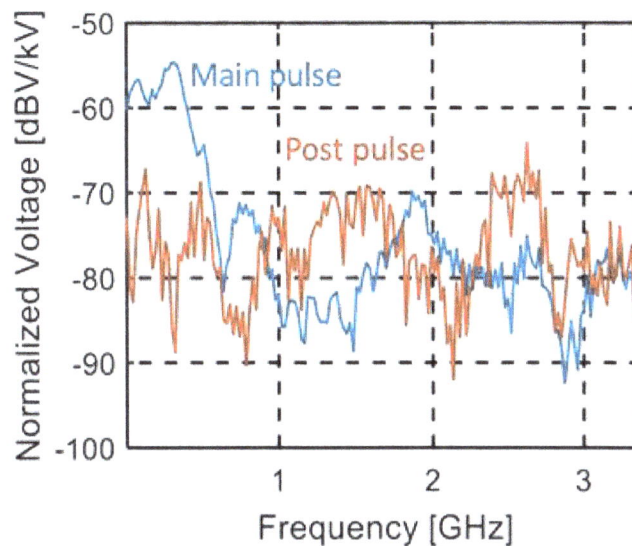

Figure 52: Comparing the spectrum for the main pulse and post pulse at 8 kV.

The Industry Council on ESD target levels suggests adding guidance to the IEC 61000-4-2 standard which:

- Informs the user and ESD generator manufacturer about pre and post pulses
- Ask manufacturers to quantify the leakage currents
- Ensures that these pre and post pulses are significantly weaker than the main pulse, both for current and fields.

7.3   Improved Air Discharge Calibration

Many electronic products must comply with IEC 61000-4-2 [IEC2001] and/or ISO [ISO2001] ESD immunity standards before entering the market. Contact discharge and air discharge are considered in these standards and have been analyzed in [She2014, She2015 and Zho2016]. In the contact discharge measurement, the ESD generator (also known as "ESD gun") tip contacts directly with the DUT. The discharge occurs when the internal relay of ESD generator closes. In the air discharge measurement, the internal relay is kept closed as the charged ESD generator approaches the DUT. The discharge in the air gap between the ESD generator tip and DUT can occur when the distance reaches a certain length. The current carrying charge carriers within the spark can either originate from surface processes or a result of gas discharge processes [Sve2002].

For contact discharge, the discharge current waveforms are highly repeatable because the high voltage spark only occurs inside the ESD generator's internal relay, which is filled with inert gases such as SF-6 and N_2. Thus, it is possible to specify a standard waveform for a contact discharge measurement. On the other hand, for an air discharge measurement, it is well known that the current waveforms have poor repeatability due to the variations of the spark resistance which results from the variation of the spark length. The variations of the arc length are a result of the approach speed and the statistical time lag. The statistical time lag is affected by humidity, surface conditions, voltage, etc. [Pom1993]. Thus, it is difficult to define a reference current waveform for air discharge calibration.

To produce better repeatability in the discharge current in an air discharge mode, two approaches exist:

1) Discharge at a spark length given by the Paschen equation [Pom1993, Rit2015, Zho2017, Yua2010 and Bor2008]. This can be realized by slowly approaching the ESD generator, possibly combined with methods that reduce the statistical time lag. Here, strong ultraviolet (UV) light, high humidity, and graphite layers on the electrodes can be used. If this method is selected, the discharge waveforms will repeat well, especially above 5 kilovolts. However, the current rise time will be rather slow (e.g., 3 ns at 10 kilovolts) as the long arc length leads to a slow drop in the arc resistance. The rate of change of the current for voltages above 3 kilovolts will be in the range of a few A/ns. Thus, the arc is stabilized at a low threat level. The step response of the ESD generator and high-frequency components of the current waveform and fields will be suppressed by the slow drop of the arc resistance. Thus, the test is more about determining the selection of the main RC components of the ESD generator. In addition, Ishida et al. proposed an air discharge calibration method based on a fixed gap discharge [Ish2016]. The idea is to use a fixed gap to replace the varying distance between ESD gun tip and the ESD target of the typical air discharge measurement. The process used spark gaps from Paschen length, down to 1/3 of Paschen length. The shorter, strongly over-voltage gaps lead to fast rise times and high peak currents. As the spark length is fixed, the waveforms are repeatable. However, they are still influenced by the time varying spark resistance. Another condition for this method to work is that the voltage rise, which is initiated by closing the internal relay is much faster than the statistical time lag. Otherwise, it is possible that the discharge may already occur while the voltage is rising.

2) If a low voltage is used and the ESD generator approaches the ESD target quickly, the arc resistance may approach an ideal step function. One would hope that its resistance changes from infinite to nearly zero in picoseconds. Achieving this would allow capturing the step response of the ESD generator. The problem is that the spark gap formed between the ESD target and the ESD generator tip does not act as an ideal switch even at fast approach speeds. The method suggested in this white paper (see Section 7.4) improves this concept by using a Mercury wetted relay that approaches an ideal switch much better.

The calibration approach presented in this white paper avoids testing ESD generators in actual air discharge mode that have a spark at the tip. The proposed method measures the step response of the ESD generator in air discharge mode using a Mercury-wetted relay. The details of the structure and measurement set-up are explained in Section 7.4. Human body discharge step response was also measured using the same Mercury relay set-up. The measured human body discharge waveform can serve as a reference for the air discharge mode waveform for air discharge calibration.

7.4 Step Response Method for Air Discharge Calibration

7.4.1 Step Response Method using Mercury-wetted Relay

To measure a good approximation of the step response of the ESD generator a Mercury-wetted relay is mounted between the tip and the ESD current target. The additional structure has a length of 16.4 mm. It substitutes the actual spark (Figure 53). ESD current targets have a discharge pad at their center [IEC2001]. In this measurement, the discharge pad is replaced with the Mercury relay which is enclosed into an epoxy filled tube to ensure mechanical stability. The relay is activated once a permanent magnet is brought into its proximity.

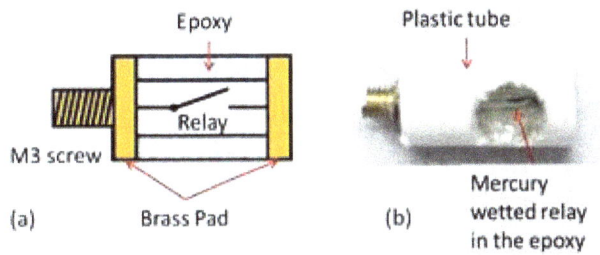

Figure 53: The diagram (a) and photo (b) of the Mercury relay tube

7.4.2 Step Response Measurement for the ESD Gun

The Mercury relay tube is screwed onto the center of the ESD target (Figure 54). The selected relay cannot withstand voltage higher than 2 kilovolts, so measurements were performed at 1 kilovolt. An Agilent DSO81304A oscilloscope was used in the measurement (40 GS/s, 12 GHz bandwidth). Three ESD generators from different manufacturers were tested. They will be labeled as "ESDGUN1", "ESDGUN2" and "ESDGUN3" in the measurement results section.

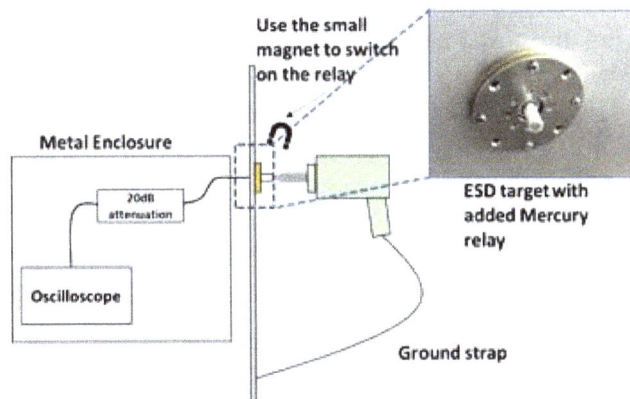

Figure 54: Mercury wetted relay attachment screwed into the center conductor of the ESD current target.

7.4.3 Step Response Measurement for the Human-metal Discharge

The event of a human discharging via a hand-held metal forms the reference event for the IEC 61000-4-2 standard, and it can be tested as to how similar the step response of the air discharge mode generators is to the human-metal discharge event. As shown in Figure 55, the person is standing on a piece of Styrofoam for insulation. A high voltage supply ensures the correct charge level. The tester holds the air discharge tip against the Mercury relay when the step response is measured. The discharge current and transient field of two testers were measured, the text refers to them as "Person1" and "Person2".

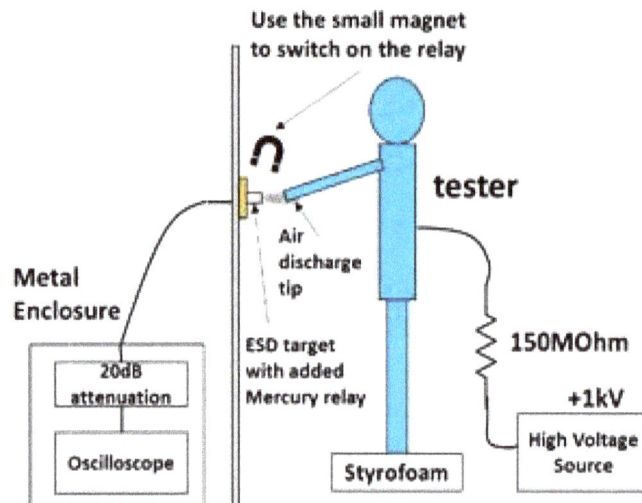

Figure 55: Illustration of human-metal discharge step response measurement.

7.4.4 Transient Field Measurement

Previous research has shown that the transient field of ESD generators in contact mode differ strongly, especially in the higher frequency region [Koo2008]. The variation in the transient field often causes different equipment under test (EUT) failure levels when using different ESD generators. Thus, it is necessary to investigate the transient fields of the ESD generator in air discharge mode. Following the methodology outlined above, the transient fields during a step response excitation were captured. Figure 56 shows the set-up of the transient field measurement. A shielded loop probe having a loop diameter of 1 cm is used for H-field measurement. The E-field sensor which was shown in [Chu2004] is used in the measurement. The E-field sensor has a flat response from 2 MHz to 2 GHz. On the other hand, waveform deconvolution [Yan2017] is needed for the H-field data as the sensitivity of the loop probe drops at lower frequencies by 20 dB/dec. The H-field probe is good up to 2 GHz. The transient fields at a 10 cm and 40 cm distance from the ESD target center were measured. The discharge current, E-field, and H-field of the same discharge event can be recorded simultaneously. It should be noted that the transient field of the ESD generator discharge event is NOT rotationally symmetric [Koo2008]. The E/H fields in this set-up were measured at different locations (Figure 56).

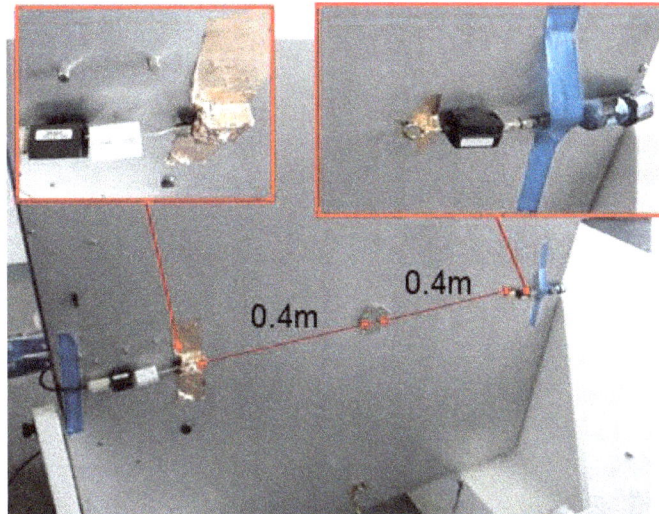

Figure 56: E- and H-field measurement set-up, based on the IEC 61000-4-2 ESD generator calibration standard. The frequency response of the E-field probe is flat from about 2 MHz to 2 GHz. The frequency response of the H-field probe has been deconvolved mathematically.

7.4.5 Repeatability

Achieving repeatability is the main challenge for every air discharge calibration method. The Mercury-wetted relay is the best possible approximation of an ideal switch. Thus, this relay has achieved excellent repeatability in both human metal and ESD generator discharges.

7.4.6 Discharge Current

The data shown in Figure 57 compares ESD generators and discharges from people holding the air discharge tip in their hand. The step response and the initial rise of each waveform is similar. The rise time is determined by the "ideal switch" formed by the Mercury relay, and it is also limited by the bandwidth of the oscilloscope. Since the ESD target shows ideal impedance up to 5 GHz, the measured discharge current data will be filtered by a 5 GHz first-order low pass filter. The discharge current of ESDGUN1 in contact discharge mode (black dotted line) is also shown in Figure 57.

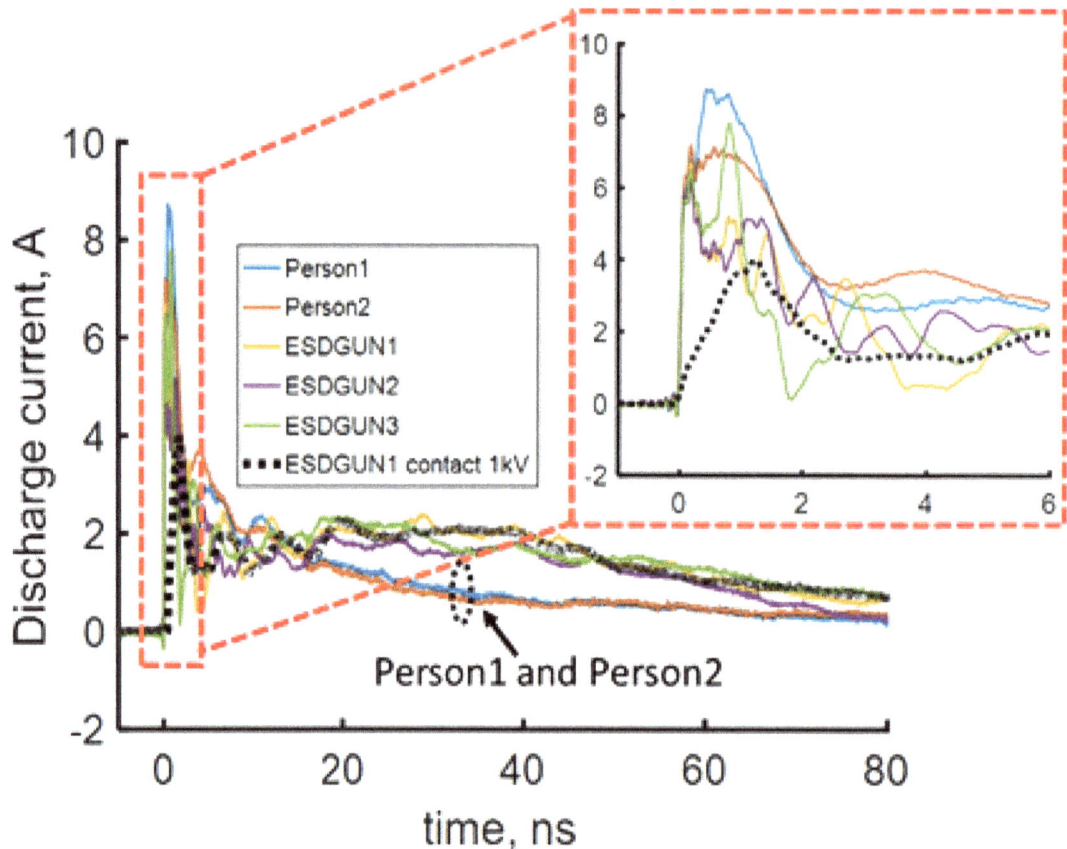

Figure 57: Step response currents for three ESD generators and two people.

Core findings from the discharge current comparison in Figure 57 are:

1) The measurement method gives repeatable step response information that allows characterization of the ESD generators in air discharge mode without having any effect of an arc (repeatability not shown in the Figure 57).

2) The peak values of these three generators varied between 6.4 amperes and 7.8 amperes, which are well within ±15 % of the average measured ESD generator peak values for a charge voltage of 1 kilovolt (a larger sample size of ESD generators may show higher variations between ESD generators).

3) Comparing the 1 kilovolt ESDGUN1 air discharge step response (yellow solid line) to the contact mode discharge (black dotted line) reveals a 2.6 amperes larger peak current value which is partially explained by the difference in rise time. There is also significantly more charge in the initial peak of the step response due the charge on the stainless steel tip. The later parts of waveforms (after 10 ns) almost overlap. This results from having the same RC network for contact mode and the step response.

4) The human metal ESD event ("Person1" and "Person2") showed larger current values of 7 amperes to 8.2 amperes. This is caused by the local capacitance of the hand that is close to the grounded wall. This structure is bulkier than the tip region of most ESD generators leading to a higher current in the step response (Figure 55). The total charge of the human metal ESD was less than the total charge of the ESD generators. This is to be expected as in most cases the capacitance of a human to ground is less than the 150 pF as specified in the IEC 61000-4-2 standard. Human to ground capacitance can be as low as 70 pF in a wood frame house

[Tal2016], but the capacitance can be double that standing (insulated) on a conductive floor. For that reason a person charged with the same amount of tribo generated charge can have twice the voltage in a wood frame house as they would have on a conductive floor.

5) All air discharge ESD generators showed ringing in a different frequency. The human metal ESD does not show the double peak structure which is (for historical reasons) part of the IEC 61000-4-2 standard's reference waveform.

6) It is known and has been reported many times that the human metal ESD only rarely shows the clear double peak structure [Pom1996].

7.4.7 Transient Field Results

As the current flows through the ESD generator, the transient field must be part of the discharge. However, there are fields that are caused by the relay. Thus, a real human-metal ESD would not have such fields. On the other hand, a human metal ESD may have much faster rise times, thus causing strong EM fields. The setup of the field measurement is shown in Figure 56.

The following conclusions can be drawn from the electric field results in Figure 58 at a 10 cm distance:

1) The peak field strength at 10 cm is between 4.5 and 5.5 kV/m at a 1 kilovolt charge voltage. In a real air discharge situation, it is not expected that the field strength increases linearly with voltage, as the rise time would typically increase with voltage.

2) The human metal ESD event shows a much larger electric field in the later time of the waveform as a result of having a charged body. In contrast the ESD generators store the energy in a discrete capacitor. Thus, these fields are not visible outside the ESD generator.

3) The rise time is determined by the relay and the field sensors' bandwidth (about 2 GHz); thus, it cannot be attributed to properties of the ESD generator.

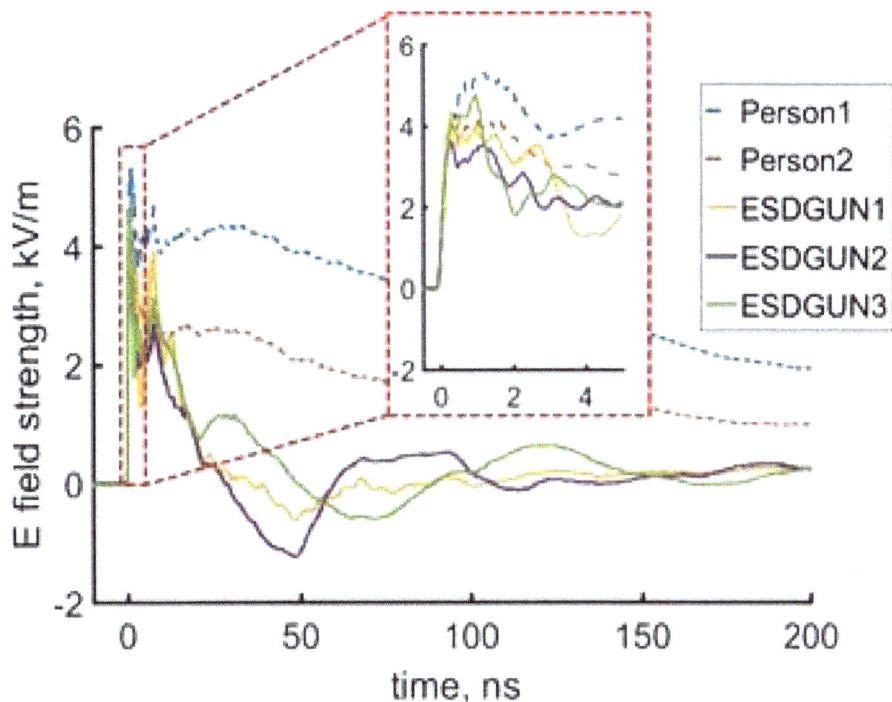

Figure 58: E-field of the step response measurement result at 10 cm at +1 kV for three ESD generators and two human metal discharges, the measurement bandwidth is limited to 2 GHz (E field sensor) at a 1 kV charge voltage. Discharge is performed via Hg relay.

The magnetic field data in Figure 59 shows that

1) The peak values are in the range of 9 A/m-11 A/m for the cases investigated at 1 kilovolt.
2) The rising edge is determined by the Mercury relay, not by the ESD generators.
3) The H-field waveform shapes are similar to the corresponding discharge current waveforms at this distance.

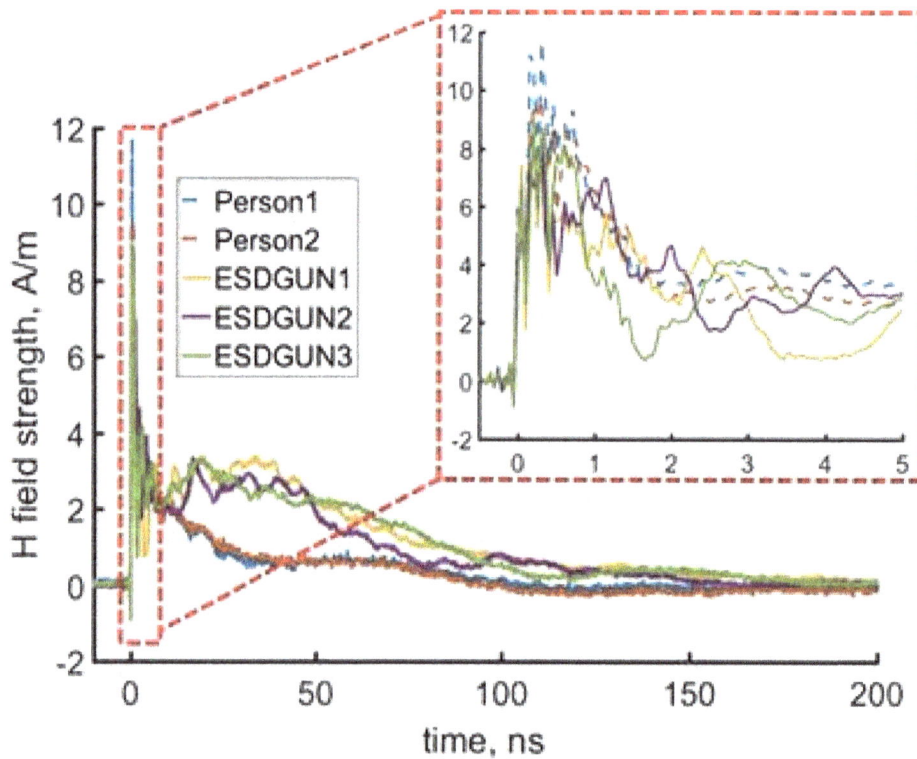

Figure 59: Deconvoluted H-field of the current step response measurement at 10 cm at 1 kV for three ESD generators and two human metal discharges. Discharge is performed via Hg relay.

The E-field sensor and H-field sensor were placed 40 cm away from the ESD target center. The setup of the field measurement is shown in Figure 56. The discharge current, E-field and H-field waveforms were recorded at the same time by the oscilloscope as shown in Figures 60 and 61.

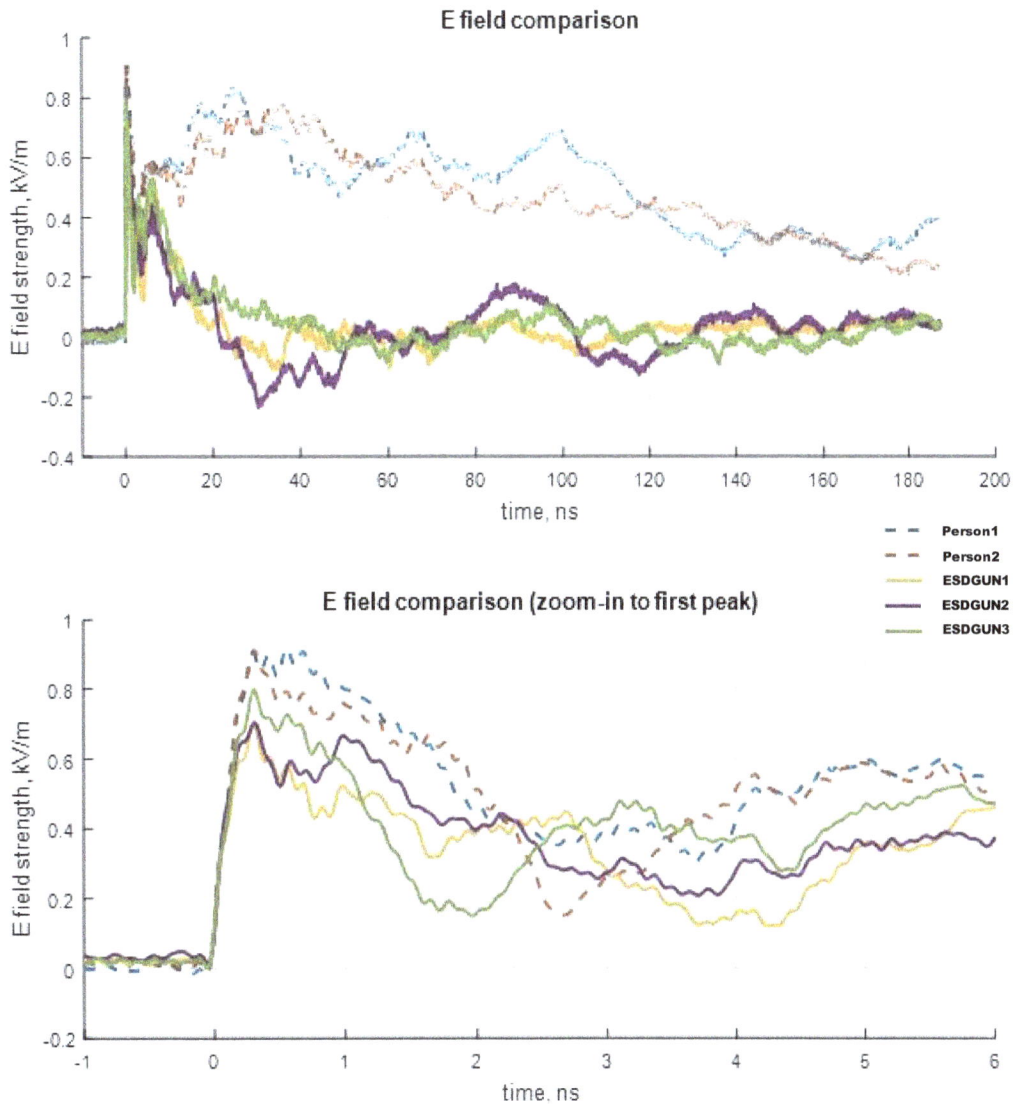

Figure 60: E-field measurement result at 40 cm at 1 kV for three ESD generators and two human metal discharges, the measurement bandwidth is limited to 2 GHz. Gun are discharged in contact mode. Persons discharged via air.

The following conclusions can be drawn:
- The field strength at 1 kilovolt and 40 cm is between 0.7 – 1 kV/m. In a real air discharge situation one could not expect that the field strength increases linearly, as the rise time would increase with voltage.
- The human metal ESD shows a much larger electric field in the later time of the discharge. This is a result of having a charged body. The ESD generator stores the energy of the charged body in a discrete capacitor, thus, these fields are not visible outside the ESD generator.
- The rising edge is determined by the relay and the scope bandwidth, thus, it has no relationship to the ESD generator.

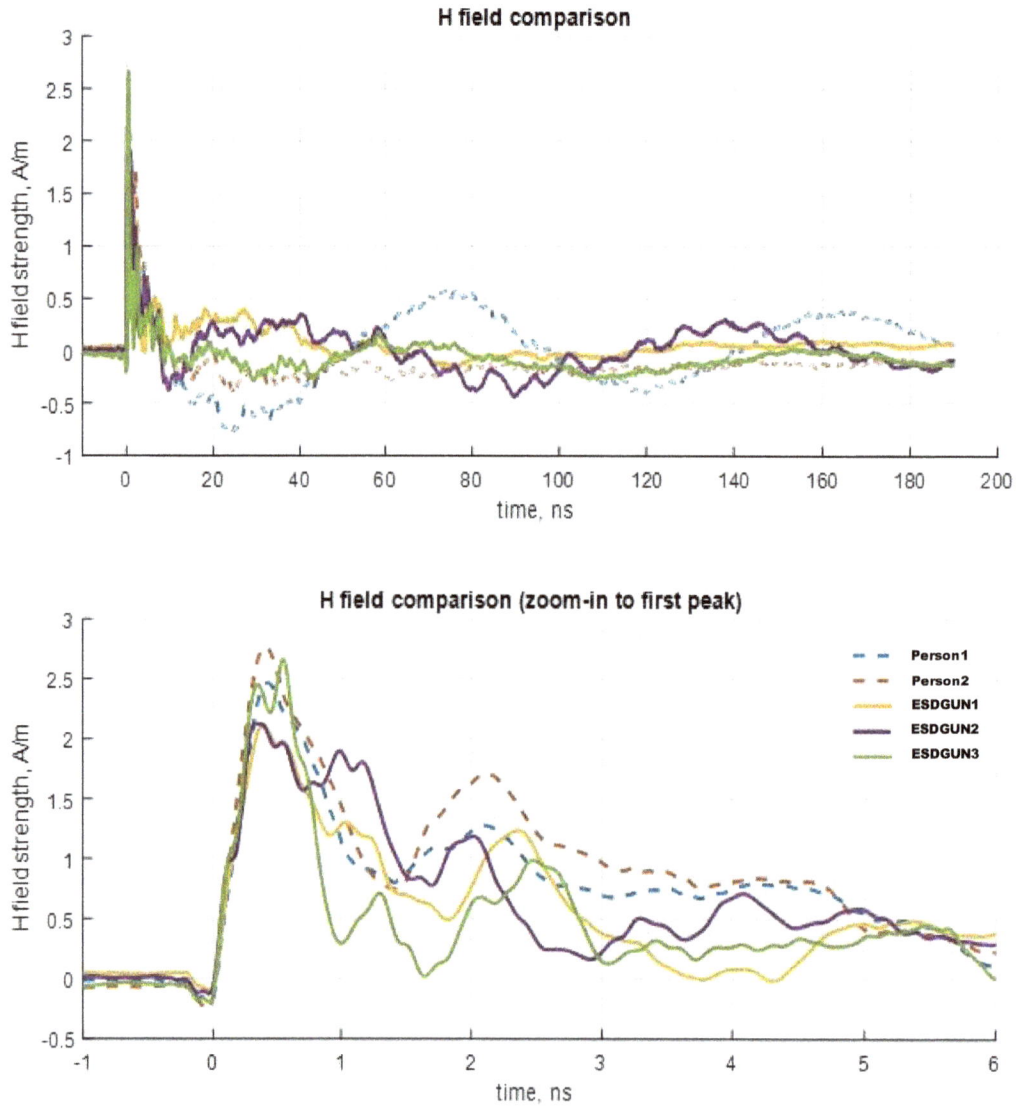

Figure 61: H-field measurement result at 40 cm at 1 kV for three ESD generators and two human metal discharges. Gun are discharged in contact mode. Persons discharged via air.

The magnetic field data shows:
- The peak values are in the range 2.1 A/m – 2.8 A/m for the cases investigated at 1 kilovolt
- The rising edge is determined by the mercury relay, not by the ESD generators
- All ESD generators and the human show ringing, the belief is that some of the ringing may be caused by the test setup (e.g., the 90 ns ringing seen in every waveform).

The Mercury relay measurements are highly repeatable. Thus, this method has the potential to be a calibration method of an ESD generator in air discharge mode.

7.4.8 Discussion

The air discharge is well known for its poor repeatability due to the variation of the spark length for approaching electrodes. Several attempts have been made in the hope of defining a calibration method for ESD generators in air discharge mode. Greatly improved repeatability can be achieved either by a fixed gap [Ish2016] or by only considering discharges at spark lengths defined by Paschen's law. Although carefully controlling the experimental parameters such as approaching speed, humidity, and air pressure can improve repeatability, it is not possible to achieve the repeatability of contact mode if a spark is part of the testing. The proposed method overcomes this by avoiding the arc and capturing the step response of the linear ESD generator. The measured currents and fields are well repeatable in using the Hg relay, repeatable enough that differences between different brands of ESD generators become clearly visible. The data also indicates that the peak current variation between different brands of ESD generators (sample size of only three) is in the same range (within ±15%) as accepted for contact mode. As the spark is substituted by a relay, no useful rise time measurement can be performed. This is not considered as a disadvantage, as the rise time in air discharge is determined by the drop of the arc resistance, and the arc physics is independent of the specific model of ESD generator used. Another possible limitation which was not investigated further is the nonlinear effects, such as the usage of a ferrite in the ESD generator which may lead to nonlinear effects at higher charge voltages. However, as the same ferrite would be used in contact mode, such nonlinear effects could be captured during the contact mode calibration. Overall, the proposed method combines the advantages of only modifying the test setup slightly, with directly measuring the step response, such that it may enable creation of a practical air discharge calibration method.

7.4.9 Conclusion

A calibration method is proposed in this section for an ESD generator air discharge measurement. The method is based on the step response which is realized by using a Mercury-wetted relay. The experiment has shown very good repeatability for discharge current and field measurement. The Mercury relay measurements are highly repeatable, and it excludes any arc effects. Thus, the data shows the effect of design choices within the ESD generator contact mode specification. This method has the potential to be a calibration method of an ESD generator in air discharge mode.

Chapter 8: Improved Test Practices

Michael Heaney, Amazon Lab126
Dr. David Pommerenke, Graz University of Technology

The IEC 61000-4-2 standard is a compliance test: it describes the minimum ESD immunity many consumer electronics products must pass to comply with legal requirements for sale in a country. However, the IEC 61000-4-2 standard says nothing about the field reliability of products that pass this compliance test: the test cannot predict the ESD reliability of the products in the hands of customers! Companies that make and sell consumer electronics products need to predict the field reliability of their products, to satisfy customer expectations, meet warranty requirements, and preserve company brand reputations. This chapter explains a new technique to test products and analyze the test results that enables accurate prediction of the ESD reliability of these products in the hands of customers. This new technique is not intended to replace the IEC 61000-4-2 standard but may prove useful guidance for future updates of IEC 61000-4-2.

8.0   Voltage Levels

People commonly associate ESD with charges accumulated by walking on carpet and touching a doorknob. However, it has been shown that the voltage levels acquired from walking on carpet are often much lower than voltages obtained from:

- Standing up from sitting in a chair

- Removing a garment, such as a fleece jacket

- Handling of plastic materials

For the removal of a sweater, voltages of > 20 kilovolts have been observed in dry air, and while performing experiments which tried to maximize the voltage, voltages of > 40 kilovolts have been measured while handling Nylon clothing materials. These extreme values do not indicate that each device should be tested to 25 or 40 kilovolts, but they are a reminder that voltages of > 25 kilovolts can occur in rather ordinary circumstances (removing a sweater on a dry winter day). In setting a test level, the manufacturer should consider:

- The likelihood of electrostatic discharges at a certain voltage level, which will depend on the surroundings, such as: consumers may use devices in a wooden house where the capacitance to ground is low (leading to higher voltages), in dry air during winter, in an automobile, or in an office environment.

- The consequences of the ESD (such as soft failure or damage).

- The usage of the device (safety-related, medical equipment, automotive, consumer electronics).

The test voltage levels in IEC 61000-4-2 (2008) are based on the synthetic fabrics line of Figure 62.

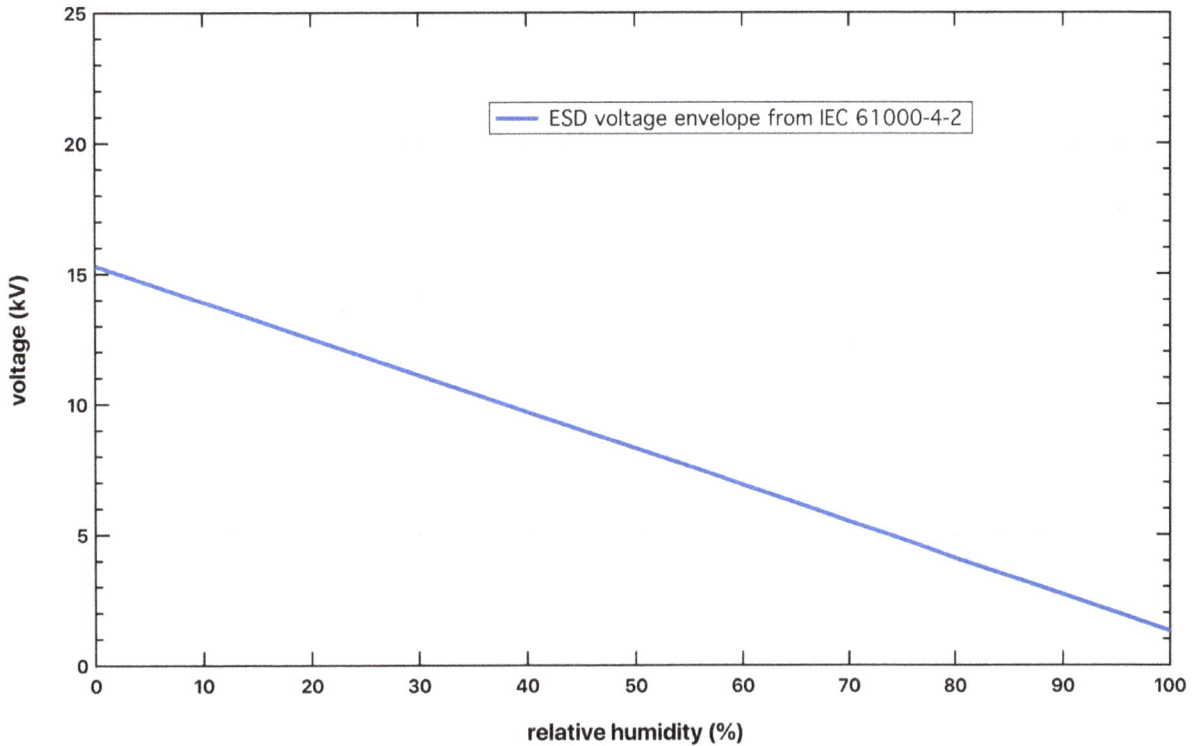

Figure 62: The maximum value of electrostatic voltage to which a person may be charged while in contact with synthetic fabrics (according to Figure A.1 of IEC 61000-4-2).

Note that there are no data points on this graph, nor references to sources, nor experimental information, nor any explanation of how this graph was made. So, an attempt was made to try and reproduce the experiment. Figure 63 shows the data from that experiment (red circles), the maximum voltage envelope of the data (green line), and the synthetic fabrics line of Figure 62 (blue line).

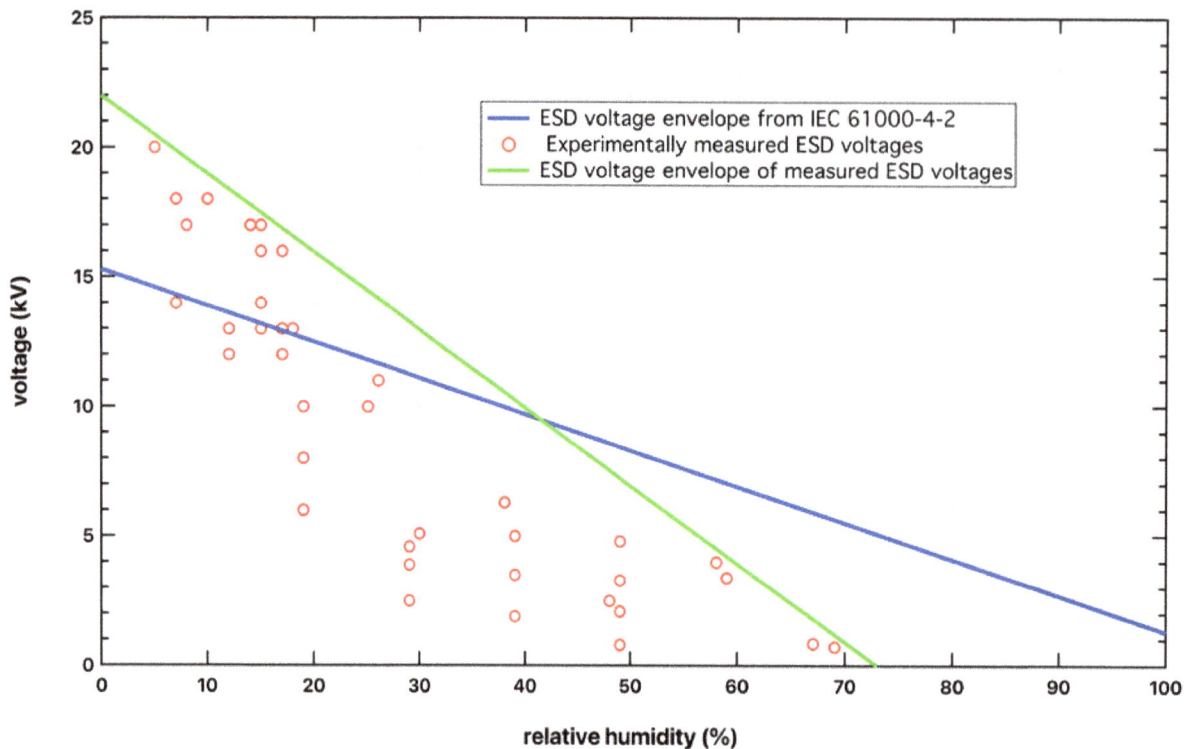

Figure 63: The measured values of electrostatic voltage to which a person may be charged after contact with synthetic fabrics (red circles), the linear envelope of these data points (green line), and the IEC 61000-4-2 maximum value of electrostatic voltage (blue line).

This experiment was done in an environmental chamber at 30 °C. The person charged themselves by putting on a synthetic fleece jacket, rubbing the jacket, then removing the jacket. The voltage on the person was measured with a Trek 341B non-contact electrostatic voltmeter with a range of 0 to 20 kilovolts and an accuracy of better than 20 volts. The lowest voltage data points at each humidity correspond to weak rubbing of the jacket, while the highest data points at each humidity correspond to aggressive rubbing of the jacket.

Note the significant differences between the results of this experiment (Figure 63) and Figure 62 of IEC 61000-4-2 (2008). In this experiment, at 70 % RH, no amount of rubbing would produce a body voltage greater than 1 kilovolt. According to Figure 62, at 70 % RH it is possible to produce a body voltage greater than 5 kilovolts. In this experiment, at 5% RH, it was easy to produce a body voltage greater than 20 kilovolts. According to Figure 62, at 5 % RH it is impossible to produce a body voltage greater than 15 kilovolts.

In summary, these experimental results suggest the test voltage levels in IEC 61000-4-2 (2008) are not realistic. It was decided to determine the voltages levels and frequencies that real handheld consumer electronics see in the hands of customers. A survey was conducted of smartphones, e-readers, and tablet computer users in the USA. The survey was planned and carried out with the assistance of a professional survey company. A total of 41,906 devices were surveyed, covering every state in the USA. Users were asked to estimate the length and frequency of sparks to their devices over the past year and describe if any type of device failure had occurred as a result. The ESD breakdown voltages were estimated from the reported spark lengths using the interpolated curve in Figure 64.

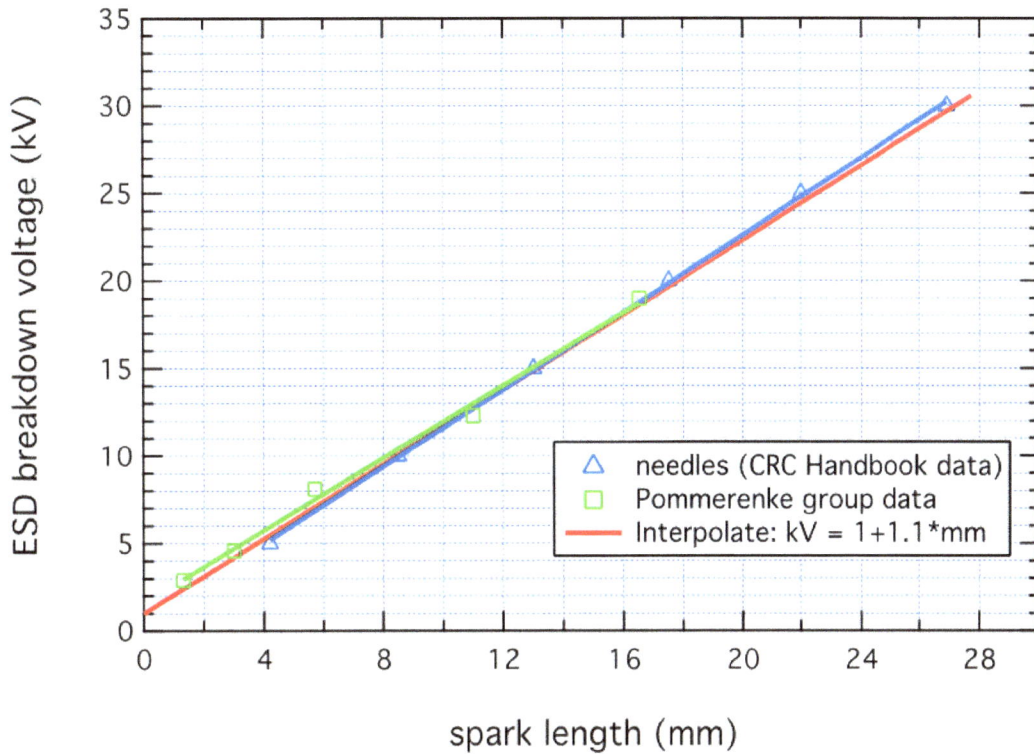

Figure 64: ESD breakdown voltage vs spark length.

The voltage and frequency data give the field ESD cumulative distribution shown in Figure 65.

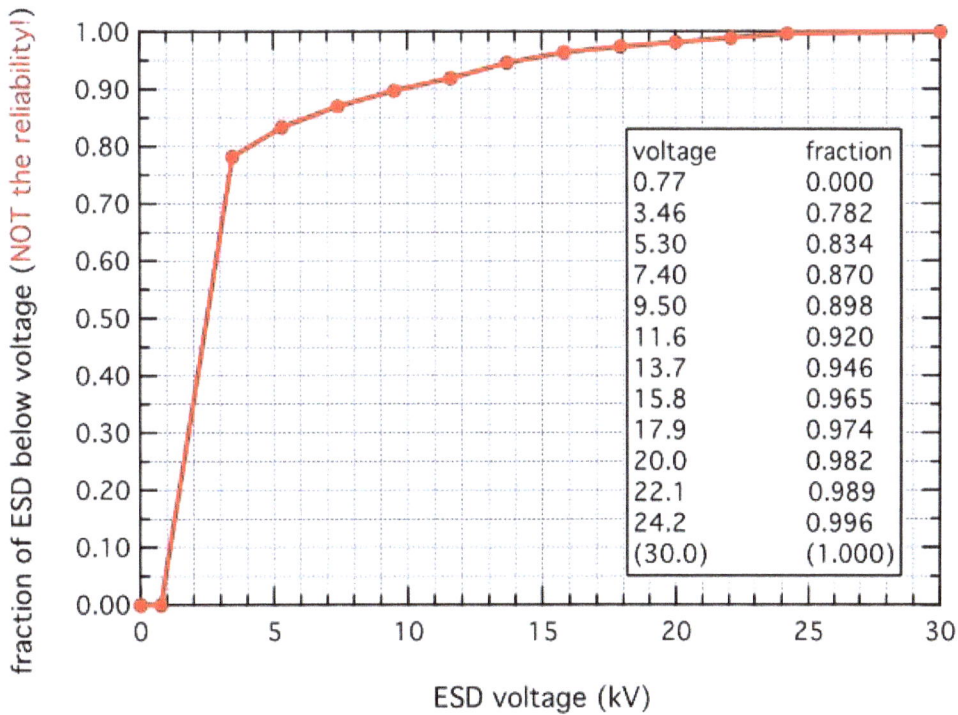

voltage	fraction
0.77	0.000
3.46	0.782
5.30	0.834
7.40	0.870
9.50	0.898
11.6	0.920
13.7	0.946
15.8	0.965
17.9	0.974
20.0	0.982
22.1	0.989
24.2	0.996
(30.0)	(1.000)

Figure 65: The cumulative distribution of ESD voltages in the field, based on a consumer survey of smartphones, e-readers, and tablet computers in the USA. The point at 30 kV is added as an estimate of the maximum field ESD.

These data show that ESD voltages greater than 20 kilovolts do occur in the field, and about 2 % of the field ESD is above 20 kilovolts. This is consistent with the experimental results shown in Figure 63. These data also show that about 5 % of field ESD voltages are greater than 15 kilovolts. These results show that testing to a maximum voltage of 15 kilovolts is neither realistic nor representative of field ESD. It is proposed to change the recommended air discharge test voltages to 4, 8 and 15 kilovolts, and, if covering rare extreme events is part of the quality goal, to test at 24 kilovolts.

8.1 Number of Test Points

The IEC 61000-4-2 standard (2008) specifies that each equipment under test (EUT) should receive at least 10 discharges at each test voltage level. No reasons are given for this specification. The field ESD cumulative distribution in Figure 65 shows that the probability of an ESD discharge to a EUT in the field decreases significantly with voltage. For example, the probability of an EUT in the field seeing a discharge between 0 and 5 kilovolts is about 16 times greater than the probability of the same EUT seeing a discharge between 10 and 15 kilovolts. These results show that doing 10 test discharges at every voltage is neither realistic nor representative of field ESD. It is proposed to do more test zaps at lower voltages and fewer test zaps at higher voltages. The optimum ratios were worked out by Renninger [Ren1992]:

$$\frac{N_i}{N} = \frac{\sqrt{f_i}}{\sum_{j=1}^{L} \sqrt{f_j}}$$

where:

i = *the test voltage number subscripts: V_1 = 4 kV, V_2 = 8 kV, etc.*
N_i = *the optimum number of test zaps at test voltage V_i*
N = *the total number of test zaps at all test voltages*
F_i = *the fraction of field zaps $\in$ the voltage range $V_{i-1}V_i$*
L = *the maximum test voltage number, e.g. L = 4*

The IEC 61000-4-2 standard (2008) also implies that each test location of the EUT should receive 10 single discharges. This does not make sense. In the field, some locations on the EUT are touched much more frequently than other locations. Figure 66 depicts the data from the same consumer survey of smartphones, e-readers, and tablet computers in the USA.

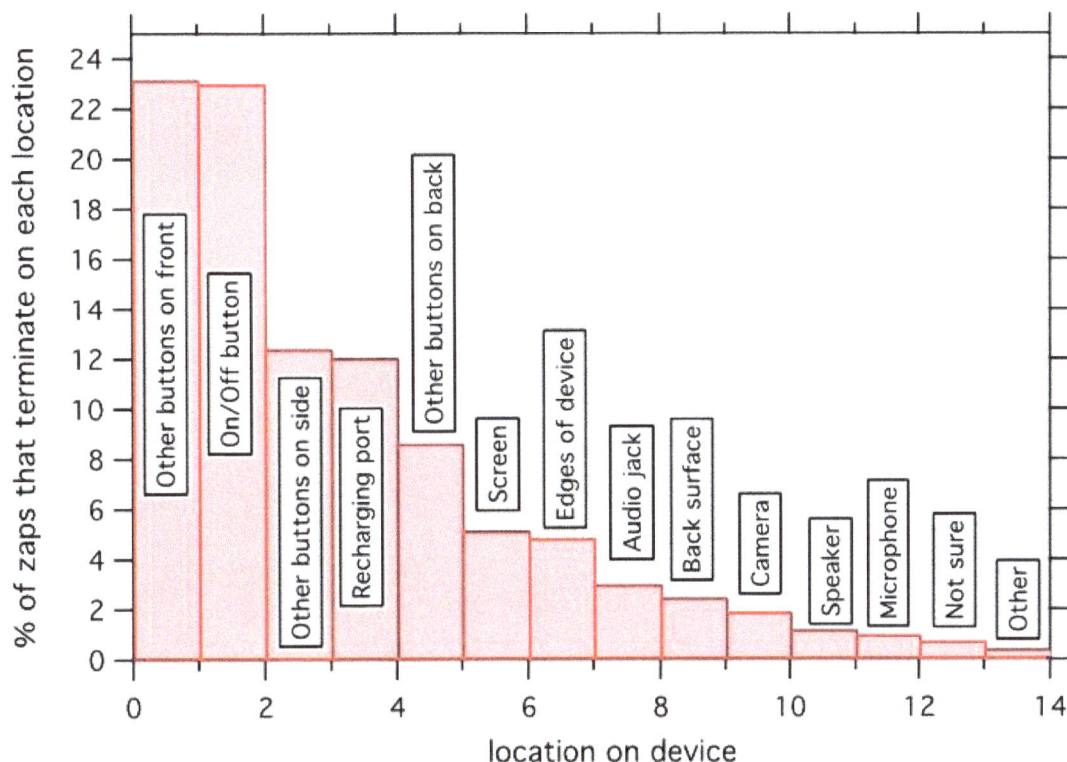

Figure 66: The frequency of ESD discharges onto different parts of handheld consumer electronics products in the field, based on consumer surveys of smartphones, e-readers, and tablet computers in the USA.

These results show that doing 10 discharges at every test location on the EUT is neither realistic nor representative of field ESD. Every test engineer will intuitively select ESD test points. Connectors and user interface devices, such as screens, knobs etc. will intuitively be given priority. This methodology substitutes the intuitive test point selection by a systematic approach to enhance the correlation between field ESD locations and test ESD locations. If the information on locations that are subjected to ESD by the user is available, it is proposed to do more test zaps at test locations that see more field zaps and fewer test zaps at test locations that see fewer field zaps. To simplify and generalize, consider four groups of locations as a useful approach:

- Group 1 (82 % of all zaps): Openings that customers will touch frequently: buttons, USB jacks, audio jacks, charging ports, etc.
- Group 2 (7 % of all zaps): Openings or seams that customers will touch infrequently: speaker holes, microphone openings, housing/mating seams, etc.
- Group 3 (5 % of all zaps): The screen (both touch and non-touch screens)
- Group 4 (6 % of all zaps): Surfaces: the back of the device, etc.

Here it needs to be reiterated that this is a concept that can be used as a quality tool in developing ESD test requirements and predicting field failure rates within a company. This is not intended to become mandatory nor to be included in the IEC 61000-4-2 test standard.

The IEC 61000-4-2 standard (2008) defines a test for the performance of electronic equipment when exposed to predefined electrostatic discharges in a specified laboratory environment but says nothing about how these test results correlate with the performance of the same electronic equipment when exposed to real electrostatic discharges in the field. Electronic equipment manufacturers and users need quantitative estimates of the reliability of electronic equipment when exposed to real electrostatic discharges in the field, before this equipment is deployed in the field. Renninger [Ren1993] developed a technique for doing this. Applying his work to the ESD survey data gives:

$$R_{lf} = 1 - 0.1765[(\sum_{i=1}^{L} f_i\, p_{ui}) + f_{max} p_{umax}]$$

where:

-R_{lf} = *the predicted lower limit on the first-year field reliability*
-0.1765 = *the average number of zaps per smartphone/ereader/tablet per year ∈ the field (the survey data)*
-i = *the test voltage number subscripts:V_1 = 4kV, V_2 = 8kV, etc.*
-L = *the maximum test voltage number, e.g. L = 4*
-f_i = *the fraction of zaps ∈ the voltage range $V_{i-1}V_i$ ∈ the field (survey data)*
-f_{max} = *the fraction of zaps above V_L ∈ the field (the survey data)*
-p_{umax} = *the upper limit on the probability of device failure at $V>V_{L,}$∧*
-p_{ui} = *the upper limit on the probability of device failure at V_i, obtained by solving:*

$$\sum_{k=n_i+1}^{N_i} \frac{N_i!}{k!\,(N_i-k)!} p_{ui}^k (1-p_{ui})^{N_i-k} = C,$$

where:

-n_i = *the number of failures at voltage Vi*
-N_i = *the number of discharges at voltage $V_{i,}$∧*
-C = *the one – sided confidence level*

An ESD Calculator was written that helps plan an ESD test, analyzes the test results, and predicts the field reliability at a chosen confidence level. To obtain and use this calculator:

1. Download and install the free CDF Player:
www.wolfram.com/cdf-player/
2. Download the free ESD Calculator:
https://drive.google.com/open?id=1XNc7GlOnGvc-6785Plpx8klk0lSY7pE5
3. Start up the CDF Player,
4. From within the CDF Player, open the ESD Calculator, and
5. Follow the directions in the ESD Calculator.

8.2   Reducing the Variation of Discharge Currents during Air Discharge Testing

In the interest of improving the reproducibility of air discharge during IEC 61000-4-2 ESD testing, different methods are investigated that reduce the variation of the spark current during air discharge. As the variation is caused by the interplay of the statistical time lag and the approach speed, methods are investigated that provide initial charge carriers for the avalanche initiation at the beginning of sparking. These methods include ionizers, cold plasma, humidity, exposure of the electrodes to UV light, different electrode materials, and surface shapes. While most of these methods reduce the variations of the spark currents, only some can be implemented during testing. The strongest effect was achieved by using cold plasma and by one surface material found in a gasket.

8.2.1 Background

The IEC 61000-4-2 ESD standard describes both air discharge and contact mode testing. Contact mode does not reflect ESD outside the test lab, but its reproducibility is better. Further, the IEC 61000-4-2 standard does not contain a calibration method for air discharge. This, and the low reproducibility of air discharge, led to suggestions to remove air discharge from the standard's mandatory requirements. However, if this step would be taken, many DUTs would not be tested for air discharge. As most critical test points are non-conducting (knobs, switches, displays, gaps in plastic enclosures, wires) the ESD standard would lose its ability to hold ESD induced field failures to a low level.

Introducing a calibration for air discharge of ESD generators has been investigated by multiple authors [Ish2016, Yan2018]. The methods either try to stabilize the spark, or, avoid having a spark. The method proposed in [Yan2018] and as discussed in Section 7.4 measures the step response of the ESD generator in air discharge mode. As this fully characterizes the ESD generator, it provides a suitable calibration method of ESD generators in air discharge mode. Practical testing requires further considerations. The waveform is not only determined by the linear step response of the ESD generator, but also by the time dependent spark resistance.

It is known that ESD air discharge currents vary strongly from ESD to ESD even if the approach speed, electrode and voltage etc. are kept constant. This is caused by the interplay of the statistical time lag and the approach speed. This phenomenon is well understood [Pom1995, Pom1993].

Every spark gap has its static breakdown voltage. This is the minimal voltage at which a breakdown can occur. For homogeneous fields the static breakdown voltage can be calculated from Paschen's law. This leads to 2.8 mm at 10 kilovolts and 1.1 mm at 5 kilovolts. If sharp edges or sparks gliding on plastic surfaces are involved the static breakdown voltage cannot be determined by Paschen's law [Zhou2018]. The core aspect of the static breakdown voltage is that a breakdown can occur, however, it does not necessarily happen right away. To initiate the spark, initial electrons are needed [Mor1953, Mee1978]. If there are no initial electrons and the electrodes are approaching, then the gap can close further until a breakdown occurs. This is illustrated in Figure 67. If the gap closes, e.g., at 10 kilovolts from 2.8 mm to 2 mm before a breakdown occurs, the current will rise faster, and reach higher peak values once the breakdown occurs. This is caused by the increased field strength within the gap. Examples of such measurements are shown in Figure 68.

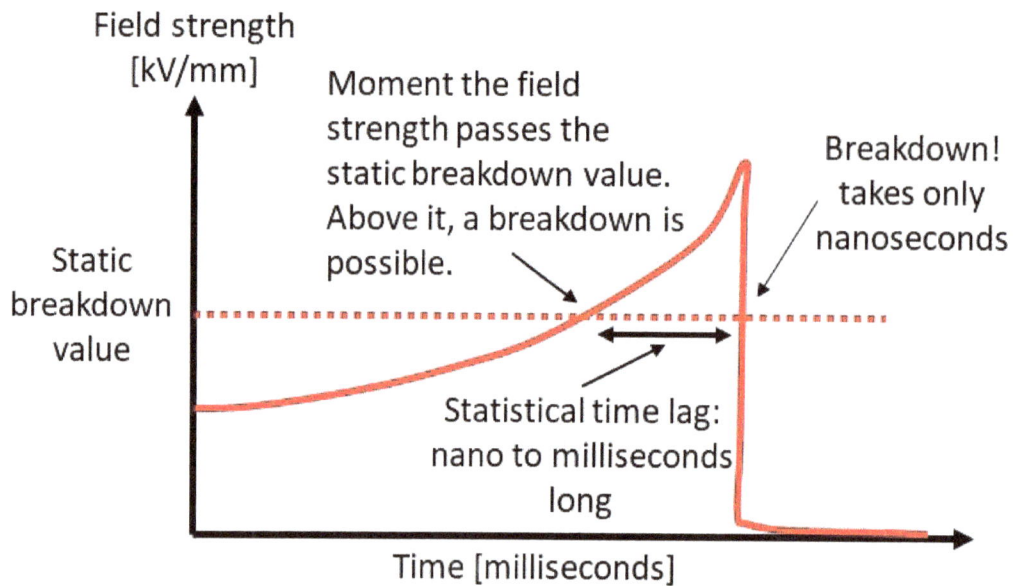

Figure 67: Graphical explanation of the statistical time lag and its influence on the breakdown.

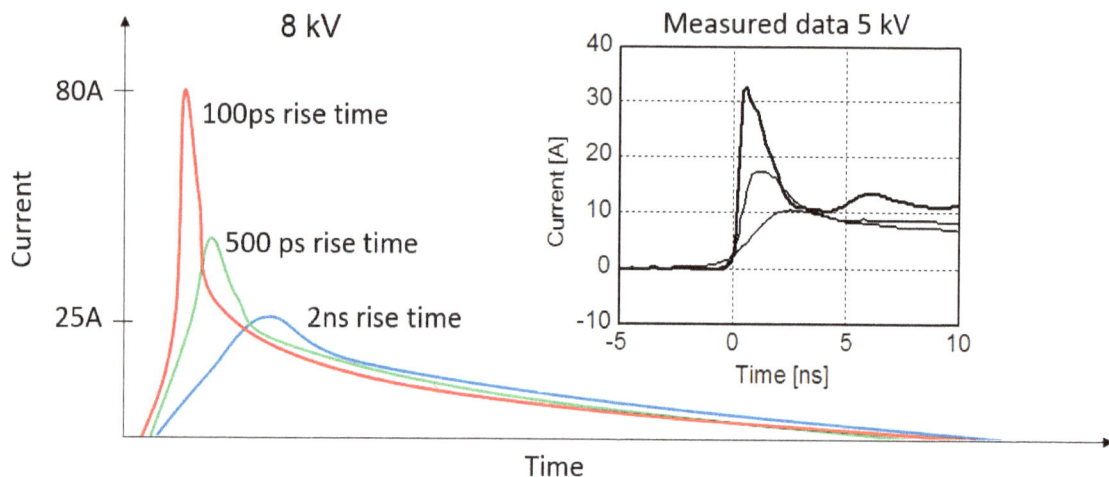

Figure 68: Human discharge currents from a person holding a metal piece for different spark lengths. Left: Illustration, Right: Measured currents.

In an environment in which initial electrons are widely available, the gap will breakdown once the field strength reaches the static breakdown field strength. If the operator repeats an air discharge at typical approach speeds of 0.01~1 m/s the interplay of the statistical time lag and the approach speed can lead to large variations of the discharge currents. This is caused by variations of the spark length. If this effect is observed or not depends on the voltage, and the availability of initial charge carriers that can initiate the avalanche breakdown of the spark.

If clean metal electrodes are used, e.g., a stainless-steel air discharge tip of an ESD generator and the ESD current target, the variability is strong for voltages larger than 2 kilovolts and diminishes above 15 kilovolts [Fri1999]. The smaller variations observed at the low end are probably caused by the fact that discharges at lower voltages are not gas discharges, but discharges that take charge carriers from the metal surfaces, and at higher voltages the edges at the current target will provide initial charge carriers.

Initial charge carriers can also be provided by a variety of mechanisms that are discussed in greater detail in this white paper. Examples are humidity, sharp edges, UV light etc.

While a proposed method that has the potential to be a calibration method for ESD generators in air discharge exists [Yan2018], it is still desirable to reduce the discharge to discharge variations during testing. One approach is to create an environment that has a large density of initial charge carriers. These will initiate the spark once the field strength reaches the static breakdown value of the gap. However, this may be impractical during testing, and a lesser goal can improve the testing. Any modification of the test method that reduces the likelihood of those electrostatic discharges that have extremely short sparks will improve the test result uncertainty.

8.2.2 Experimental Setup

In this white paper, a variety of methods are investigated that can improve the repeatability of the ESD discharge. All of the methods provide initial charge carriers to increase the likelihood that an ESD occurs at a spark gap distance that is at, or not much shorter than, the static breakdown distance. The experimental investigation captures discharges from an ESD generator to a current target, or from a modified current target having a sharp point. The current is analyzed by its peak value, variability, and maximal current derivative. Then, different methods are used to influence the spark development, such as shining UV light onto the gap to create initial charge carriers via the photoelectric effect.

This white paper compares the different methods and identifies solutions that will improve the repeatability of air discharge.

The variability of the spark current is caused by the interplay of the statistical time lag and the approach speed. An obvious choice would be to reduce the approach speed, such that the spark will occur at the distance which is given by the static breakdown value. However, this would require very slow speeds, often less than 1 mm/sec [Fri1999]. The alternative is to create a sufficient number of initial charge carriers such that the breakdown occurs at the static breakdown distance, or at distances that are not much shorter.

There are a variety of methods to create initial charge carriers. These can be externally provided, or be delivered by the electrode itself.

Methods that provide ions from the outside are listed below:
- Ionizer: An ionizer provides a stream of ions. These can be either charge balanced, negative, or positive ions that dominate the ion flow
- Cold plasma gun: A cold plasma is a non-equilibrium plasma. Its electron temperature is much larger than the ion temperature. Thus, it is not physically hot, and can be touched. Cold plasma [Plasma] is often used to disinfect surfaces or to activate plastic surfaces before gluing them. The cold plasma gun is similar to an ionizer, but the ion density is much larger.
- Radioactive materials: An alpha or beta decay material will create initial charge carriers. This is often used in smoke detectors and in low jitter (Krypton filled) spark gaps. Due to the complexity of handling radioactive material, this was not investigated in this study.

- UV light or laser: The photoelectric effect will create charge carriers. Here, either a low pressure Hg lamp or a UV laser can be used. Both emit photons in the range of 240 nm. However, a laser can focus a much larger photon density on the gap and can influence the spark development strongly.

- High humidity: It is known that water molecules attract charge carriers as the water molecule is highly polar. In a rather low field strength, the charge carriers can detach and initiate a spark. This effect causes the humidity to have a very strong impact on the statistical time lag. Thus, applying moist air may reduce the time lag, and thus stabilize the spark development [Gos1985].

Methods that provide charge carriers from the electrodes are listed next:

- Graphite ESD generator tip: It is known that graphite has a strong electron emission even at rather low field strength. It has been used as fast reacting overvoltage protection spark gap since the early days of telegraphy [Lev1982, Sta1998].

- Gaskets: It has been observed that different gasket materials, such as fabric over foam gaskets, help to initiate a spark at lower field strength. The gasket material is glued, using a conductive glue, to the ESD generator tip. Besides such gaskets, other surfaces such as steel wool, copper wool etc., have been investigated.

8.2.3 Measurement Setups

The experimental setup captures the discharge current. At first, the data is analyzed by visually inspecting the variability of the waveforms. Numerical analysis quantifies the variation of the maximal time current derivative (max (di/dt)) of the discharge current. This measure was selected as it is closely related to soft failures and the rise time of the pulse. Three different electrode setups were used, see Figure 69. The first uses a standard ESD current target, the second adds a pointed tip to the target, and the third partially covers the target with Mylar tape. The second setup was selected as most DUTs will have sharp edges, e.g., at the PCB. The last setup was selected because practical air discharge testing nearly always approaches plastic or glass surfaces to allow a spark to glide on the surface of the insulator and possibly reach a grounded metal part. It was selected to confirm that such discharges lead to slower current derivatives compared to discharges in air.

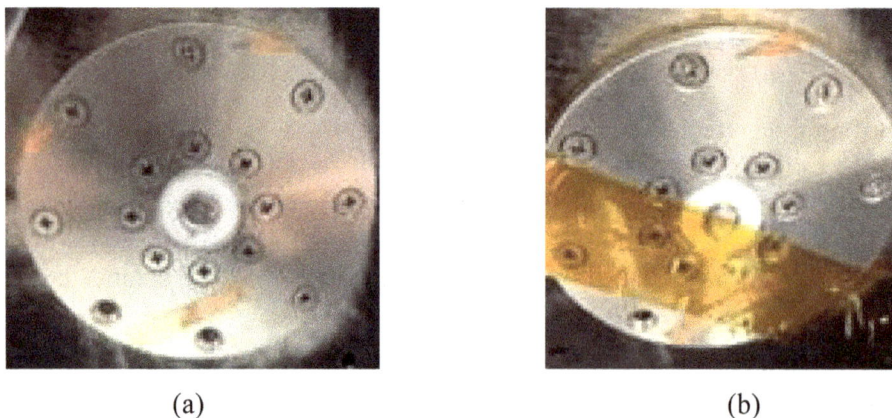

| (a) | (b) |

Figure 69: Measurement setup: (a) Current target mounted into a shielding enclosure. (b) Mylar tape partially covering the current target to mimic structures having the spark travel along a plastic surface.

The discharge currents were measured by the current target and an oscilloscope inside a shielded enclosure. The bandwidth of the measurement system is mainly limited by the oscilloscope bandwidth. A 2 GHz oscilloscope was used to capture most waveforms. Select data, used for the statistical analysis in Section 8.2.4 was captured using a 13 GHz oscilloscope.

8.2.4 Measurement Results - Current Waveforms

The fastest rising currents are obtained if the statistical time lag is long at higher approach speeds. The approach speed was manually kept approximately constant around 10 cm/sec.

A 13 GHz oscilloscope was used to obtain the reference data when the ESD generator discharged to the flat surface of the current target (see Figure 70). The test was repeated 100 times. The test environment was at 20 % relative humidity and 23 °C. A low humidity was selected as this aggravates the problem of repeatability of the air discharge. Any effect observed by a method to improve the spark repeatability would be even more efficient at higher humidity.

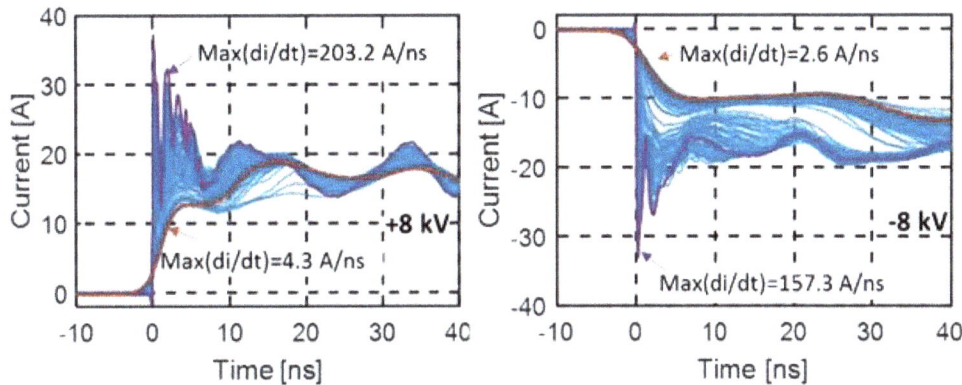

Figure 70: Current waveforms without using any methods to stabilize the waveform at +8 kV and −8 kV. 100 waveforms have been captured and the maximal current derivative extracted.

The maximal current derivative is used for the analysis as the maximal current derivative often correlates to the soft failure threshold for electronic systems. The figure also shows the range of the maximal current derivative (di/dt) for each discharge. They are obtained from either 100 or from 30 repeated measurements. The initial investigation using 30 discharges provided preliminary judgements on the effectiveness of the methods.

As it is not possible to present all the data, a selection was chosen such that the dominating effects are illustrated. Figure 71 shows current waveforms obtained while exposing the spark gap to cold plasma. The maximal current derivative is limited in a small range from ~2 A/ns to 4 A/ns.

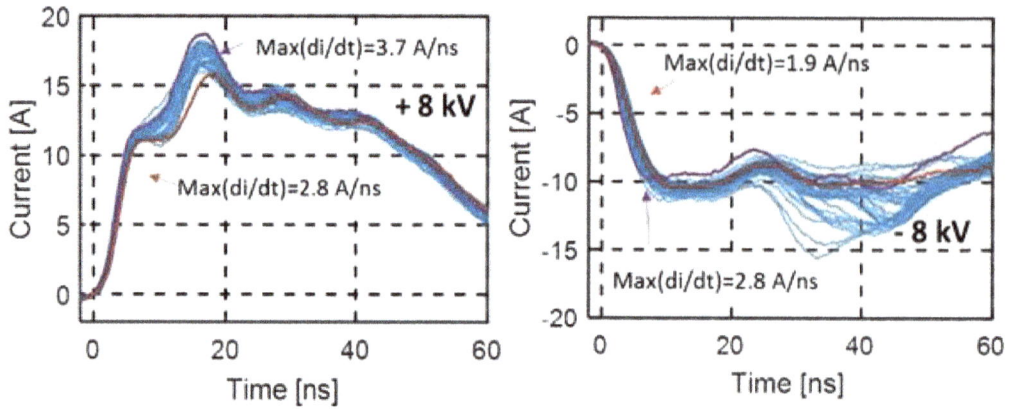

Figure 71: Current waveforms using cold plasma to provide external initial charge carriers at +8 kV and -8 kV.

Commercial ESD generator tips are usually made from stainless steel. Polished stainless steel has a low electron emission for a given field strength. In contrast, graphite is a strong emitter [Lev1982, Sta1998]. A graphite tip was machined to replace the stainless-steel tip. The current waveforms using the graphite tip are shown in Figure 72. The graphite tip only helped to stabilize the waveform at negative polarity. This can be explained by graphite being an electron emitter.

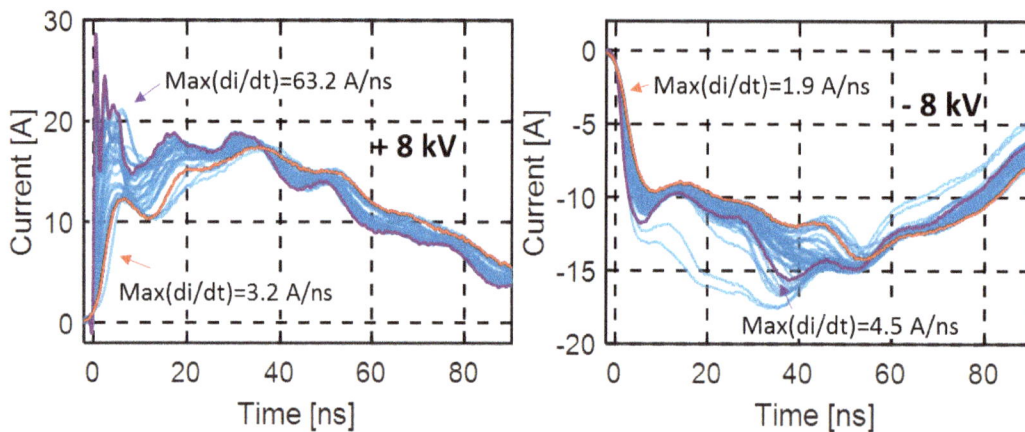

Figure 72: Current waveforms using the graphite tip at +8 kV and -8 kV.

Another data set presented here was obtained using a gasket, shown in Figure 73 with a picture of the gasket material used shown in Figure 74. This fabric over foam gasket covered the ESD generator tip during the discharges. It greatly stabilized the spark initiation. With respect to the test standard, it would be an easy change to modify the ESD generator air discharge tip by covering it with a gasket. The exact mechanism and surface properties of this gasket have not been clarified yet.

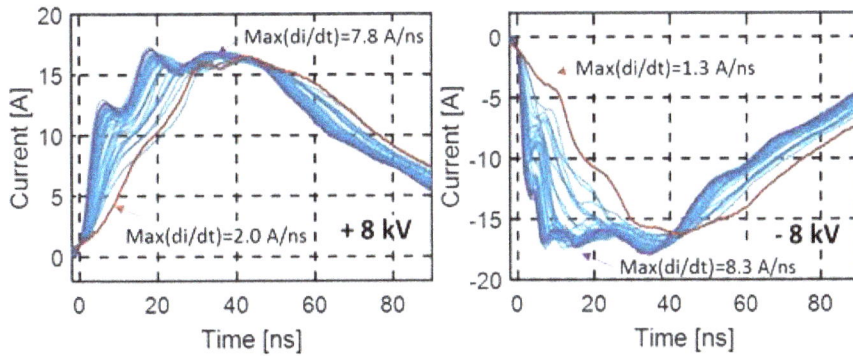

Figure 73: Current waveform obtained using a gasket to cover the tip at +8 kV and -8 kV.

Figure 74: Gasket Material Larid cf100

8.2.5 Maximal Current Derivative Distributions

The data shown in the Figures 70 to 73 give a visual impression of the variations and quantifies the max (di/dt). A statistical analysis of the distribution of the current derivative values gives further insight into the effectiveness of the methods investigated.

It had been shown that spark discharges having spark lengths close to the static breakdown gap distance have current derivatives of a few A/ns for voltages relevant to system level ESD testing in the range from 2 to 15 kilovolts [Pom1995, Pom1993]. In air discharge testing of an ESD gun against an ESD target, the current derivative may reach hundreds of A/ns. During product testing, the range of variation will depend on the shape of the surface and other parameters. However, if the air discharge occasionally leads to a very large max (di/dt) value, failures may occur that are hard to reproduce. Thus, it improves the test result uncertainty if the statistical distribution of the current derivatives does not contain very fast rising events. The methods used to stabilize the spark initiation have been statistically analyzed by observing the distribution of max (di/dt) with special attention to very fast rising electrostatic discharges.

It is not possible to show all distribution functions (multiple voltage levels and both polarities). Thus, only those have been selected which indicate successful stabilization of the spark initiation. The data shown in Figures 75 and 76 are based on 100 discharges. The green bar shows the distribution using the stainless-steel air discharge tip. The current derivatives are broadly distributed and include discharges having > 100 A/ns max (di/dt). This data can be considered as the reference, as the data is captured using the method outlined in the IEC 61000-4-2 standard. A graphite tip was machined having the same dimensions as the stainless-steel tip. Graphite was selected as it is known to have a strong electron emission under rather low surface field strengths [Lev1982, Sta1998]. For positive voltages, the graphite tip (blue bar) removes most of the very fast rising discharges, but the distribution is still rather broad. The effect is much more pronounced for negative charge voltages. Here, the initiation is clearly driven by electrons from the tip. Graphite being a good electron emitter led to discharges having low max (di/dt) values. For negative charge voltages, using a graphite tip was nearly as effective as using cold plasma (not shown in the figure). Another method selected for Figures 75 and 76 was the usage of a gasket. This foam over fabric gasket (2 mm thick) was used to cover the air discharge tip. The surface materials and microscopic structure of the gasket material have not been investigated. The gasket material stabilized the spark initiation strongly, as is shown by the red bars in Figures 75 and 76.

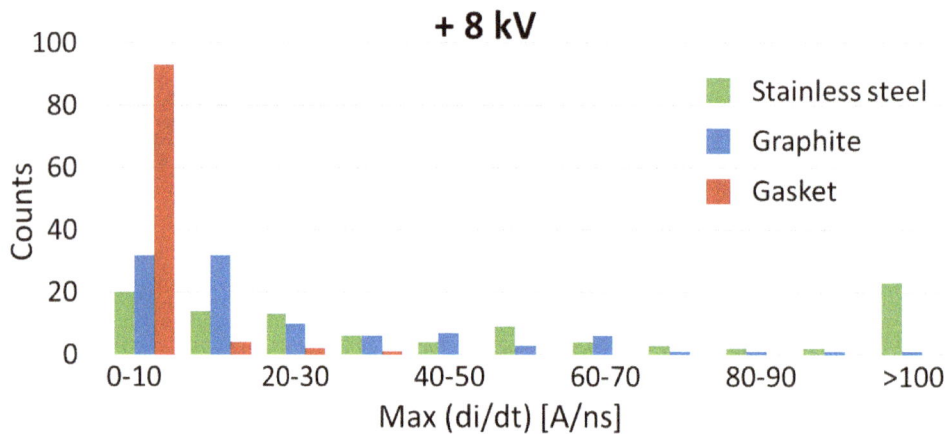

Figure 75: Maximal current derivative distributions for different tip materials at + 8 kV.

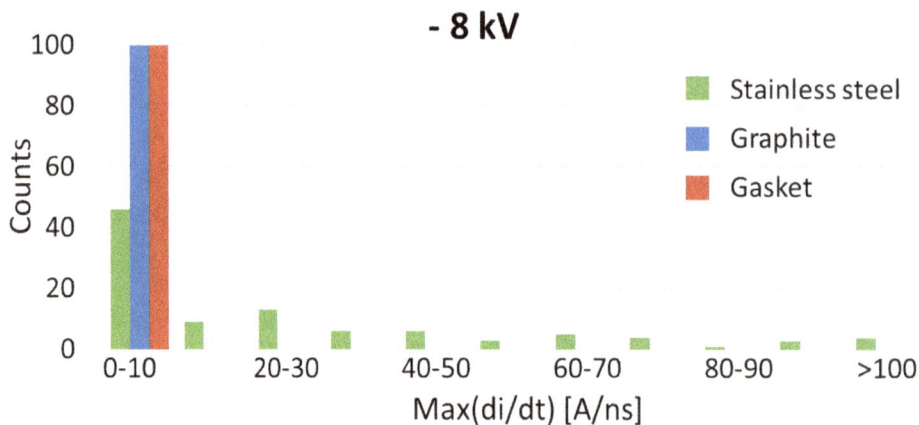

Figure 76: Maximal current derivative distributions for different tip materials at - 8 kV.

8.2.6   Comparison of the Methods

This study forms an initial investigation into methods that might be used to improve the repeatability of an air discharge in the IEC 61000-4-2 standard. The data was obtained on a limited number of discharges in just one humidity condition. However, the data already indicates possible paths to improve the repeatability. The data is summarized in Table 6.

The maximal current derivative ranges are shown, and the effectiveness is compared for each method at positive/negative 8 kilovolts. All methods, except using graphite for positive voltages, show some effect on the rise time stabilization. However, the effectiveness varies, and the complexity of implementation varies greatly. According to this limited data set, the methods that provide external charge carriers can limit the max (di/dt) to below 10 A/ns. But those methods may not be practical for engineering tests. For example, water mist helps to stabilize the waveforms, most likely due to the electron affinity of water molecules. However, one cannot expose a DUT to water mist. The graphite only helps to stabilize the waveform for the negative voltage.

The strongest effect was achieved by exposing the gap to cold plasma during the discharge. However, due to the creation of NOx gases, it would require good ventilation. The cold plasma may also interact with circuits, because it is created by a high power ultrasonic piezo device.

The UV laser used in this experiment provided 1 mW of optical power. However, the wavelength of 240 nm can cause deoxyribonucleic acid (DNA) damage if human skin is not protected. Similar arguments limit the use of low-pressure mercury lamps. Here, a 150 W lamp (AC power consumption) was used.

Ionizers may provide a practical method, as well as covering the ESD generator air discharge tip with a fabric over foam gasket. Further investigations are needed to determine if these are good options.

Table 6: Effectiveness Comparison for Different Methods

Method	Max(di/dt) [A/ns]		Method / Comment
	At +8 kV	At -8 kV	
Stainless steel reference	4.3~203.2	2.6~157.3	Reference based on IEC 61000-4-2
Cold plasma	2.8~3.7	2.0~3.8	Effective, creates NOx
Ionizer	2.7~4.0	1.6~3.1	Effective
UV laser	2.9~5.3	1.8~3.6	Effective, avoid eye and skin exposure
Moist air	3.2~4.5	0.7~3.3	Humidity very high, fog conditions
Gasket	2.0~7.8	1.3~8.3	Effective and easy to implement
Graphite	3.5~174.9	2.5~8.7	Only effective for negative voltages
Mylar tape	1.3~4.5	1.5~13.9	Depends on the DUT geometry, not practical as change to the IEC 61000-4-2 test standard

8.2.7 Summary

This section improves understanding of air discharge testing, its related problems, and may lead to an improved IEC 61000-4-2 standard. The research points at possible solutions to the problem of bad reproducibility in air discharge testing. Multiple methods to stabilize the spark initiation during air discharge testing have been investigated. Testing compared the maximal current derivative during the air spark to the reference event, which is the discharge between the ESD generator's air discharge tip and the ESD current target. All methods, such as UV exposure, cold plasma, using a graphite tip, high humidity, and modified surface materials, improved the reproducibility. However, only the usage of an ionizer and modified surface materials may be useful in ESD testing.

8.3 Testing Inside a Shielded Room

ESD testing is often performed inside a shielded room. The shielding of the room has no effect on the testing, reflections of the electromagnetic waves in the room are also not critical as the distance from the ESD generator to the DUT is much smaller than the path from the ESD generator to the walls and then to the DUT. However, the grounded metal walls change the capacitance of the HCP. A larger capacitance of the HCP to ground may influence the test results if the HCP to ground voltage is relevant. For example, if a keyboard is tested by discharges to the HCP (indirect ESD testing), then the ESD generator will charge the HCP to a voltage which is determined by the capacitance ratio of the ESD generator's internal capacitance and the HCP to ground capacitance. The keyboard, here assumed to be grounded, will see a different E-field. The capacitance of the HCP to ground is also affected by any DUTs or support equipment that are placed upon it. If this equipment is grounded, e.g., via a power cord, the capacitance of the HCP to ground will be increased by the test equipment that is placed upon the HCP. This change is part of the specific test setup, and not of concern. However, increased capacitance due to the proximity of a metal wall should be avoided.

In a display-down test situation, a phone for example, the capacitance from a small phone to the HCP is in the range of 50-100 pF, this is the same range of the HCP to ground capacitance. If there is a discharge to the phone's back side, then the voltage across the display (the voltage reached after ringing is over, so the voltage after a few nanoseconds) is determined by the phone to HCP capacitance, the ESD generator's internal capacitance and the HCP to ground capacitance. In this case the voltage across the display will vary with HCP to ground capacitance (and with the flatness of the dielectric insulator).

For most DUTs, the effect of the HCP to ground capacitance may not be critical as they do not react to discharge to the HCP. However, an adequate distance to metal walls should be maintained.

8.4 Humidity

Humidity only affects air discharge, contact mode is not affected. The charge decay on the surface may be affected, however, within the 35-65 % RH range this may not be such a strong effect. Secondary ESD may also be affected. For homogeneous field discharge structures, the effect may be large, but most practical secondary ESD structures will have sharp edges, or plastic surface guided spark paths. Thus, the effect of humidity on the secondary ESD will be low, as the time lag is dominated by the edges or plastic guided surfaces [Wan2014]. Still, the present temperature/humidity range allows rather large variations which may affect air discharge.

8.5   Charge Removal

The underlying idea of the IEC 61000-4-2 test standard is that every ESD is unaffected by a previous discharge. This relates to software; all error corrections should be completed before the next ESD is applied, and to electrostatics; any charges introduced from a previous ESD should be removed.

Often there is confusion about the best method of charge removal, for example:

- If an ionizer is used during the testing to remove charges after the ESD, does the ionizer change the air discharge spark? Does it change the secondary ESD occurrence or severity?

- How long will an ionizer take to remove the charges?

- Can one connect a high impedance ground strap to the DUT while testing? Or does one need to remove the ground strap for each discharge?

With respect to attaching a ground strap during testing to the DUT, one should consider two possible effects of the ground strap:

- The attached conducting structure may affect the current distribution during the ESD. This will be the case if the wire has a long section before its first high impedance resistor is included into the wire. A wire < 5 cm should not cause any concern as the capacitance of a < 5 cm wire to ground is probably < 2 pF and the propagation delay along the section is several hundred pico-seconds. Thus, a wire that contains a high voltage, high impedance resistor after < 2 cm should not affect the current distribution during testing in any significant way and it can be left attached during testing. An alternative is to use a conductive grounding using a high impedance wire, such as a carbon fiber. For example, if such a wire has a resistance of >10 kilohm/m it is most likely invisible from an RF point of view. Such wires are used in EM-field sensors to connect the local RF detection diode's DC voltage to an analyzing instrument.

- The attached wire will change the time constant of the discharge of the DUT. This is the goal of introducing the wire. However, if there is secondary ESD in a power supply then it may take time for the secondary ESD to develop. Thus, the time constant should not be too small. A time constant of 1 ms will not influence any secondary ESD. A value often selected is 1 megohm, however, if the DUT to ground capacitance is 100 pF, then this leads to 0.1 ms time constant, a value a little bit smaller than the suggested value. For that reason, a discharge resistance value of 100 megohm seems more appropriate. The 100 megohm can be created by placing 1 megohm close to the DUT and 100 megohm at the other end of the wire. This may lead to a mechanically more suitable structure as a smaller 1 megohm can be used; the voltage drop across this resistor will be rather small due to the series connected circuit. The secondary ESD case may need further explanation. Consider a DUT that is only "grounded" via a power-brick which is connected to a 2-wire power cord. In this case there may be no DC return path from the DUT to the ground. Now the ESD to the DUT will increase the voltage inside the power brick which may lead to a breakdown inside the power brick. This is a commonly observed secondary ESD situation. The power brick may be damaged and fail, or worse, overvoltage the DUT.

After considering the effect of charge removal methods that are applied during the injection for their possible effect on the ESD current it is suggested to simply use a static voltmeter to test if the charge removal method used (e.g., wire, brush, ionizer) is enough. If the static meter shows that the charges are removed to < 10 % of the initial charge value one can reasonably assume that the remaining charge has no effect on the next ESD pulse.

8.6   Multiple Pulse ESD

Although it is known that a real ESD often consists of a series of pulses, it is reasonable to base a test standard on single pulses independent of each other. Due to software correction and recovery it is plausible to argue that those series of pulses with millisecond time spacing will act differently than pulses which have seconds of time spacing. One needs to consider two aspects:

- Multiple pulses mainly occur if discharges from the skin are considered, discharges from a metal part usually have one dominating pulse and much smaller pulses which follow. Human-metal pulses are much more severe than HBM pulses, thus, it is reasonable to base the testing on the more severe case although it shows mainly one pulse

- The fact that an air discharge often consists of a series of pulses has been well documented. In general, the first discharge transfers most of the charge, it has the longest rise time and the largest peak value. Later discharges have faster rise times, but lower peak values. The reason for having multiple discharges is that the spark may extinguish if the current falls below a value on the first discharge. As the finger approaches the spark re-ignites at a lower voltage. The initial spark may have created a hole in the insulation of the skin, such that the rise time is faster upon re-ignition.

8.7   Documentation

8.7.1   Discharge Current

An advanced ESD system level test setup captures the discharge current during testing. Here a current clamp, such as an F-65/F-65A, is a good choice for capturing the current with sufficient bandwidth, and capturing the current using a flat frequency response such that the voltage shown by an oscilloscope can be expressed as current by simply multiplying it by the probe's transfer impedance. Any probes having a strong frequency dependent transfer impedance in the relevant frequency range from about 1 MHz to 1 GHz may require a more complex deconvolution to convert the captured voltage to the current waveform.

The current measurement serves the following purposes:

- It can be used to verify the ESD generator's performance prior to ESD testing. It is suggested to test at a low voltage, such as 250 volts, and at the highest voltage of 25 kilovolts. The two voltages are suggested as the typical failure mechanisms in the high voltage relays can cause either low or high voltage waveform problems. If the relay is worn, thus, its contacts have eroded over many ESD pulses, the low voltage waveform changes. However, the waveform at 4 kilovolts for example, may still be OK. If an insulation problem occurs in the ESD generator it is visible at the high voltage setting. Thus, these two voltage levels cover the most common failure types in ESD generator relays.

- If a DUT failure is observed during testing and the waveform is recorded, much better data is available for root cause analysis. For example, in an air discharge, it is quite common that the waveform will vary strongly. Thus, if the waveform is known which caused a failure it may help to understand the reason or the correlation to ESD test levels. This knowledge will also help in understanding spark-less ESD discharges as they occur on displays and provide information in disputes about passing / failing ESD tests between OEM manufactures and system integrating companies.

- Secondary ESD can be detected from the discharge waveform.

8.7.2 Video Recording

A problem often seen is a dispute between an OEM and a system integrator about passing or failing an ESD test. While there can be many reasons for test result variation, it is certainly not helpful if the test documentation is insufficient. It is known that the approach speed, and the angle of the ESD generator, routing of cables etc. can all affect test results. To improve the documentation, add a video recording to the ESD testing. Labs have placed video recording equipment on flexible arms such as the ones found in dentist office. This easily allows moving the camera to the region of interest. The camera will then record:

- Test setup

- ESD generator approach and location

- Parts of the DUT response.

These systems can be foot activated, such that the recording only captures the relevant seconds. If no error occurred, the recording may be disregarded. Such a system should be combined with a current measurement using a current clamp around the tip of the ESD generator, or at least on the ground strap. This captured current could then be correlated with the failure, such that the test report shows the video, the current waveform of the ESD that failed the DUT and possibly the DUT response.

8.8   Robotic ESD Testing

ESD testing can be performed by hand, robotically supported hand testing, and fully automated. In hand testing the operator will perform the testing and observe the DUT for failure indications. In robotically supported testing the robot will perform the testing. However, the operator will observe the DUT for failure indications. In fully automated ESD testing the robotic system will perform the testing, DUT supervision, DUT reset etc. See Figure 77.

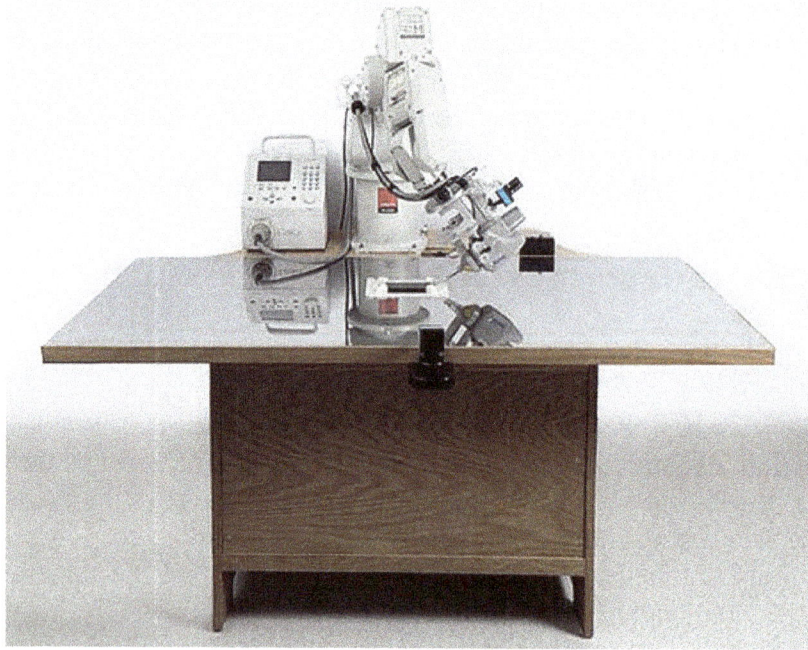
Figure 77: Robotic IEC 61000-4-2 test system (courtesy of API, www.amberpi.com)

8.8.1 Advantages of Robotic Testing

Repeatability

Air discharge testing depends strongly on the length of the arc. The length of the arc can vary significantly even for the same test point and at the same voltage. Those variations are partially statistical in nature, but also are influenced by the way the approach is performed. The rise time will be lower, and the peak values will be higher on average for faster approach speeds. Some careless people even drag the charged air discharge tip across the product (being in contact with the plastic surface) to see if a discharge occurs. This is a very risky practice, as this can lead to an unrealistically low air discharge rise time. It is important to control the approach speed (straight to the point of expected discharge), and to control the angle that the ESD generator is held.

The robotic system allows every test parameter, such as, approach speed, angle and discharge point, to be precisely controlled and accurate.

Test depth

The advantages in test depth mainly come from a large number of consistent discharges. The number of zaps per test point is set to 10 by IEC 61000-4-2. There are brief periods of time in which a DUT is much more sensitive to ESD, and ten discharges is a very low number to capture rare failures (or windows of opportunity) that happen only at certain condition of a DUT during normal operation. A larger number of discharges (in the range of 100) at each test point with good repeatability provides more statistically reliable test results.

A DUT is usually controlled by a DUT controller that sets the DUT in many different modes of operation. Automated robotic testers enable tests of such various modes of operation without interruption or skipping some modes, which happens frequently during manual tests.

The voltage is often set in large increments, like 2, 4, and 8 kilovolts and only pass or fail is reported. It is important to know pass or fail, but it is just as important to know the failure level. Classifying pass or fail based on a large test voltage separation can cause inconsistency in the passing (or failing) level. For example, if a DUT's real passing voltage is 8010 volts. It can pass at one test lab, but easily fail at a different test lab.

Documentation

All test actions can be recorded automatically and pulled out as necessary. This can help in understanding correlation analysis of failures and test parameters, such as discharge waveforms. Simulator waveform verification can be integrated into a test flow, and drifts can be flagged from long term trend monitoring. It is possible to integrate a current measurement into the testing, such that the ESD current of each pulse is recorded, and the current waveforms which caused an ESD failure are included in the test documentation.

Test speed

Test speed advantage can get larger for a complicated test flow that requires many different actions, or when the number of test points is high, or the number of same test samples is large.

8.8.2 Disadvantages of Robotic Testing

The main disadvantage of a robotic test is the difficulty of automatically monitoring various failures and the programming of the test points. Manual tests may be required to understand the types of failures for proper automatic failure detection by a robotic tester, or the robot is used for repeatable testing, while an operator observed the DUT for failures. In addition, few companies offer robotic test systems and the cost of such systems may be prohibitive for many companies.

Appendix A: Body-worn Equipment

Dr. David Pommerenke, Graz University of Technology
Marathe Shubhankar, MS&T EMC laboratory
Yingjie Gan, MS&T EMC laboratory

The core difference between the postures assumed for IEC 61000-4-2 human metal discharge and a discharge to a wearable device is the impedance between the charged body and the grounded structure discharged to. This leads to much higher currents for the same voltages [Zho2018, Ish2015, Ish2016, Ish2017 and Koh2018]. The difference in impedance is a result of the geometry. This is especially true for the waist worn device in which a larger portion of the body is close to the grounded structure, thus the geometry forms a much lower impedance (higher capacitance) which will lead to higher currents, see Figure A1.

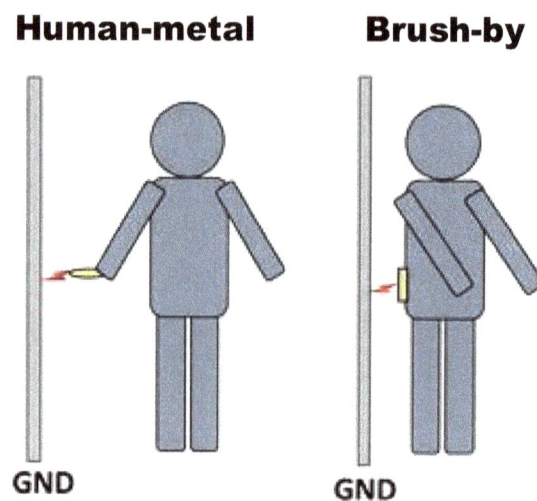

Figure A1: Different ESD scenarios for Brush-by and Human-metal Discharge

The discharge currents for different positions of the body-worn devices at 1 kilovolt are shown in Figure A2. The peak current for the body-worn devices are higher than the human-metal discharge case. The current for the waist-worn device can reach levels twice as high as in the human-metal discharge scenario as shown in Figure A3.

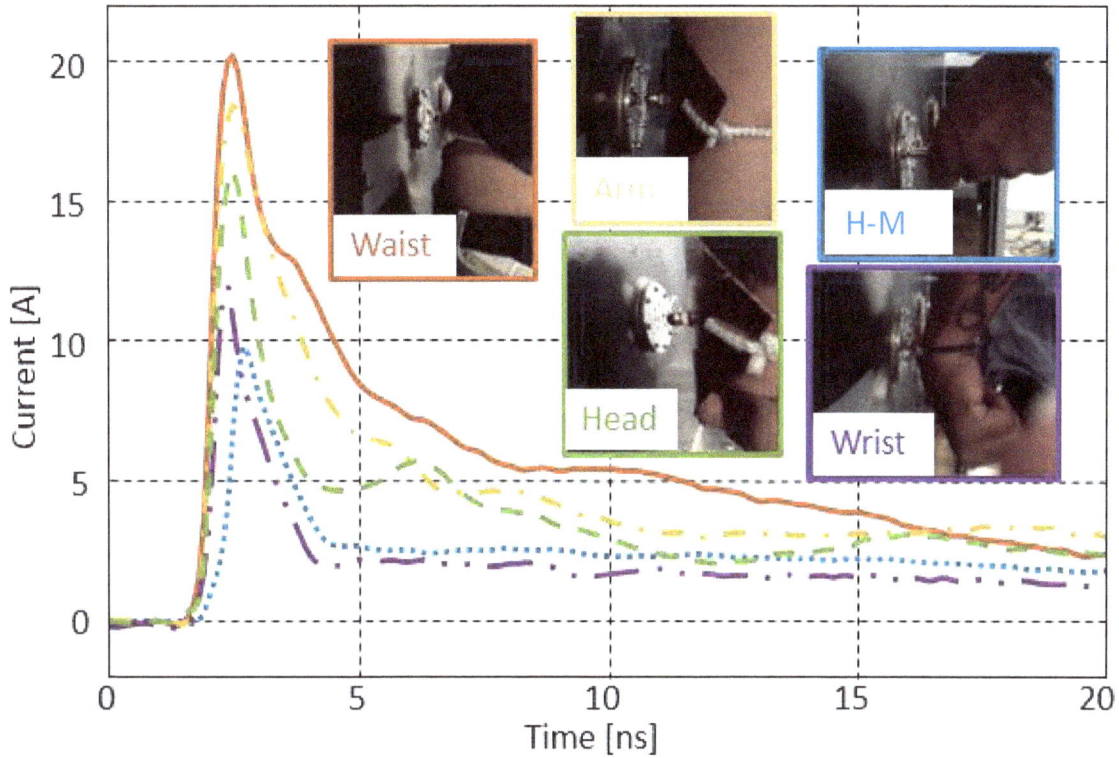

Figure A2: Discharge Current for Human-metal and body-worn Devices at 1 kV.

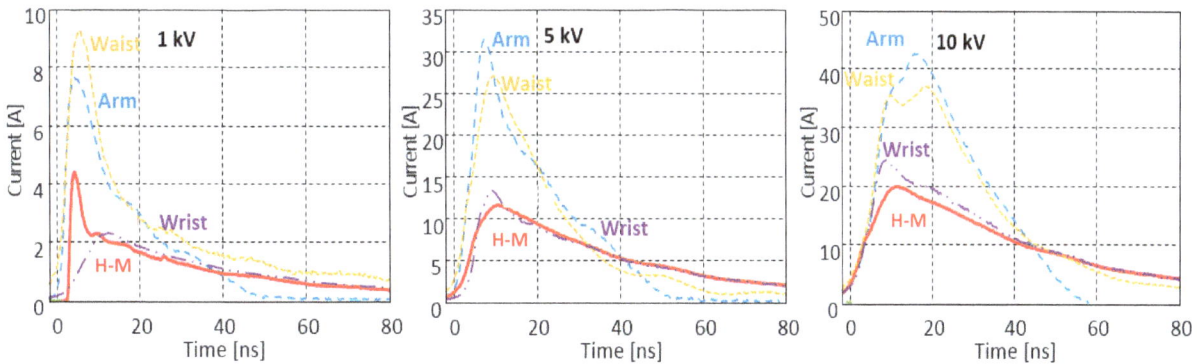

Figure A3: Discharge current for Human-metal and body-worn devices at different voltage levels.

When the voltage level increases, despite the variability for the air discharge, in most cases the current will be higher than 3.75 A/kV as specified for contact mode ESD. The current levels of the body-worn devices measured in the brush-by scenarios indicate the IEC 61000-4-2 standard setup is probably insufficient for ensuring the robustness of body-worn devices.

The capacitance variations need to be investigated for the cases when there are clothes or a plastic enclosure between the device and the human body. Data for a moist thin cloth or direct contact lead to similar waveforms. A dry insulating cloth and the data obtained using a 0.5 mm plastic insulator show similar waveforms at 1 kilovolt as shown in Figure A4. The insulation between the device and the body changes the total capacitance, but little effects were observed on the peak value or the rise time.

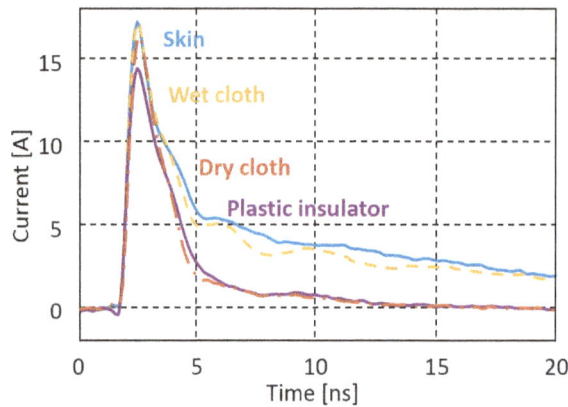

Figure A4: Discharge current for different connections at 1kV for waist worn device

The impedance formed by the body and the vertical ground plane can be used to determine the step response under the assumptions that the spark acts as an ideal switch and linearity. The impedance is measured using a vector network analyzer for different positions of the body-worn devices (see Figure A5) using a coax connection having the same dimensions as the ESD current target to ensure similar postures during the impedance measurements and the current measurements.

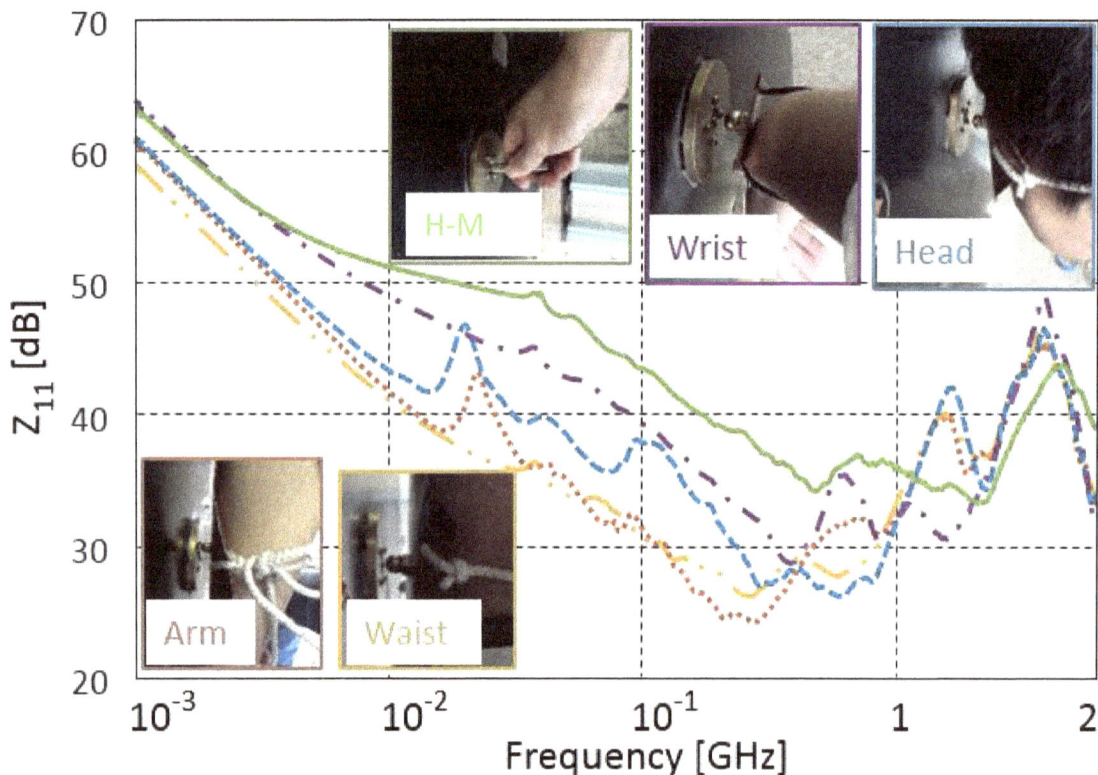

Figure A5: Measured Impedance Z_{11} for different positions of the wearable devices versus Frequency. Observe the capacitive behavior at lower frequencies, and the more complex behavior at higher frequencies caused by the local geometry around the discharge point.

By assuming that the spark is an ideal switch, the current can be reconstructed using the measured impedance. A 1 kilovolt step voltage source is used as excitation, the reconstructed currents for different positions are shown in Figure A6. The larger value of the current peak of the simulation data results from the assumption that the spark is an ideal switch. Simulations that model the actual time dependent spark resistance using Rompe and Weizel's spark resistance law can provide a better match to the measurements. For a system design, the main lesson is that the currents can be much larger than the currents during the IEC 61000-4-2 testing for the same voltage. The reason is the local impedance, which is much higher for a handheld metal part, relative to waist mounted metal.

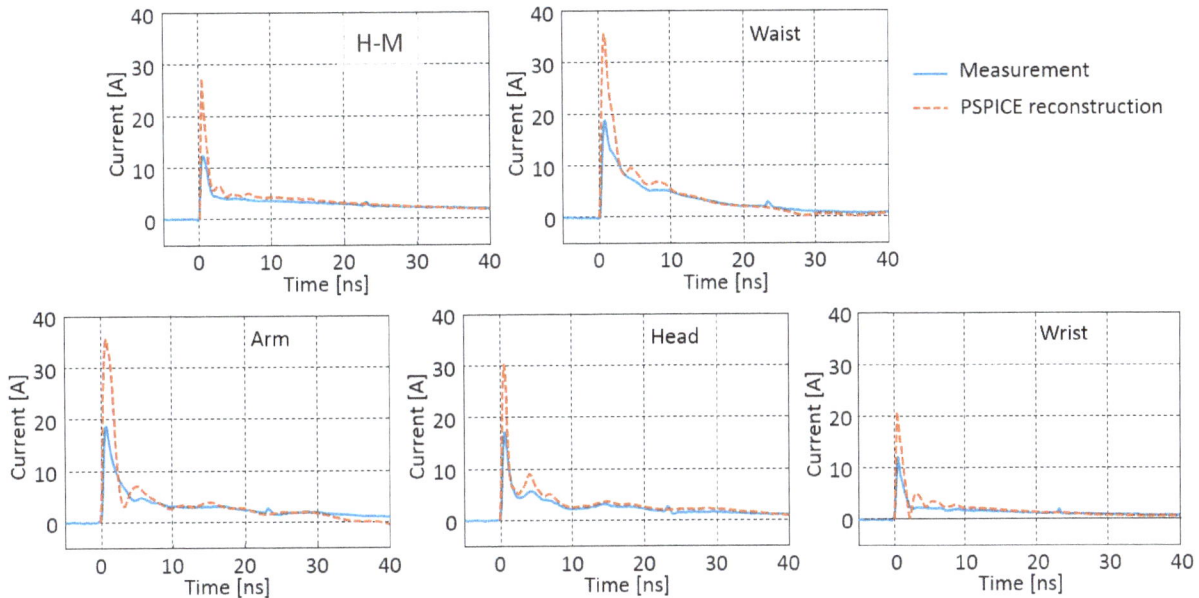

Figure A6: Comparison of currents obtained using PSPICE with measurements for different positions of the wearable devices and Human-metal discharge scenario at 1 kV.

The impedance can be approximated as an RLC circuit as shown in Figure A7. The parameters in the RLC models are tuned to achieve the best match with the measured impedance for each position, the values of the RLC parameters are shown in Table A1. C1 describes the capacitance of the body to ground. It dominates the impedance behavior in the frequency range from 1 MHz to 200 MHz.

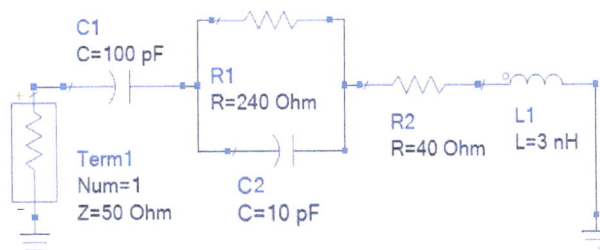

Figure A7: RLC model for Human-metal Discharge.

Table A1: RLC Parameters of the Wearable Devices Modeling for Different Positions

Position	R1 [Ω]	R2 [Ω]	C1 [pF]	C2 [pF]	L1 [nH]
Waist	240	24	220	150	4
Arm	380	20	220	125	9
Head	220	24	150	40	8
Wrist	350	29	120	20	8
H-M	240	40	100	10	3

After developing the linear parts of the models, the spark resistance is taken into consideration during the simulation. It has been shown that Rompe and Weizel's law can provide good predictions for the spark resistance having a spark length around or lower than the Paschen's value.

$$r(t) = \frac{l}{\sqrt{2K_R \int_0^t i\left(t'\right)^2 dt'}}$$

where $r(t)$ is the spark resistance at time t, l is the spark length, and K_R is a constant that is related to the gas pressure and type. Typically, $\mathrm{Kr} = (0.5{\sim}1) \times 10^{-4} m^2/(V^2 s)$.

An example is given for the waist-worn position at 10 kilovolts when the constant value is set to $0.5 \times 10^{-4} \, m^2/(V^2 s)$ for the Paschen case, see Figure A8.

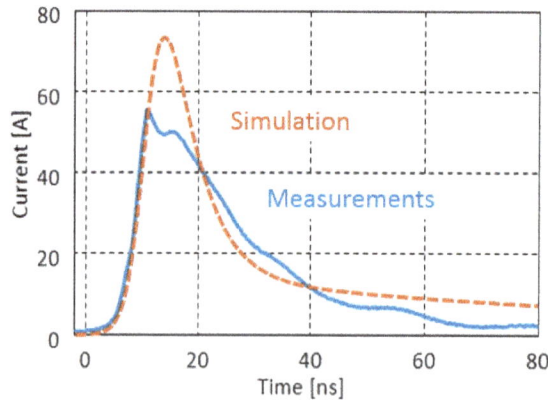

Figure A8: Simulated and measured current for waist-worn device at 10 kV, l = 2.7 mm, Kr = 0.5e-4 $m^2/(V^2 s)$.

The current levels for the body-worn devices can be twice as high compared to the discharges from a handheld metal or the contact mode IEC 61000-4-2 specifications. This indicates that testing of body-worn systems according to the IEC 61000-4-2 standard may underestimate the real risk. Two possible solutions may ensure product quality. The first solution would be to use a higher voltage level in testing body-worn equipment. This will correct for the current magnitude, but it may cause a breakdown through plastic gaps that would not happen at the actual voltage. The second solution is to develop an ESD generator that better reflects the currents seen in body-worn equipment.

Appendix B: Display Testing

Jianchi Zhou, MS&T EMC laboratory,
Dr. David Pommerenke, Graz University of Technology

B.0 Introduction

An electrostatic discharge can cause soft and hard failures of displays and touch screens [Koo2012, Kim2011, Li2106]. It is not possible for the spark to penetrate the glass, however, a variety of entry and coupling paths exists between the ESD and the display or its driving circuits [Shi2016].

The spark may reach the metal frame which is part of a flex cable connected display. If the edge encapsulation of the glass layers suffers an electrical breakdown, the spark may enter between the glass layers at the edge of the display into the touchscreen or display. If the display is not surrounded by a well-insulated structure, the spark may reach the flex cable, which connects the display to the main board. The spark may also reach other metallic structures around the display, such as the metal frame not connected to the display.

Especially in cases in which the spark can reach the flex cable or the encapsulation between glass layers, it will damage or upset the display. However, many designs surround the display with insulating structures, such that no direct discharge to a grounded structure is possible. For such a well-designed display, no sparking would be observed if a discharge to the display is attempted using an ESD generator in air discharge mode. However, not observing a spark often leads to the false conclusion of not having a current flow. The approaching ESD generator tip can cause surface charging that can reach currents of up to 10 amperes with <1 ns rise time.

During the IEC 61000-4-2 test the DUT is placed on a 0.5 mm insulating sheet, which is on the horizontal coupling plane (HCP). For discharge points on the back side of a phone this leads to a situation in which the display is facing the HCP. The capacitances formed between the phone and the HCP varies a lot depending on the on the size of the phone, its screen flatness, and the flatness of the insulator which may deteriorate over time. The regions with high local capacitance will receive more current, leading to reproducibility problems. Additionally, the possible corona at the edge of the phone can contribute several amperes to the current at higher voltage.

The discharges to the phone lead to a large displacement current flowing through the display. This current has multiple paths to the body of the phone: via the touch electronics, via the display electronics, and directly to the body of the phone. As these currents can reach 30 amperes (at 8 kV contact mode) they can lead to upset and damage of both the display and the touch layers.

The flatness of the insulating spacer is important. Currently, many laboratories use plastic that tends to wrinkle. This changes the distance to the DUT. If a polycarbonate plastic is selected, then a flat surface is created. It is suggested that for all tests in which the 0.5 mm thick insulator is used, a flat polycarbonate surface should be required.

B.1   Spark-less Surface Discharges

The corona current which spreads across the surface of the glass will cause a rapid change of the surface potential on the glass surface. This rapid surface potential change will cause a displacement current to flow via the capacitances to the inner structures of the display, such as, the touch layer or the display layer. These spark-less current paths are illustrated in Figure B1.

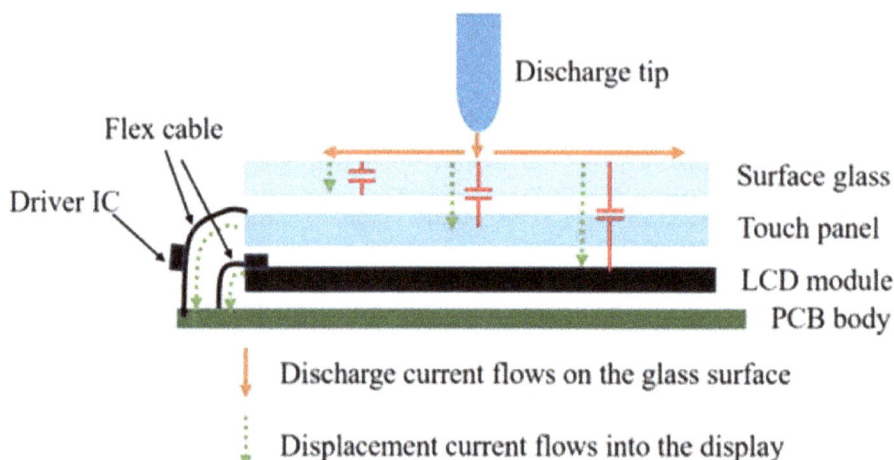

Figure B1: Spark-less surface corona current and displacement current paths into the display. The F-65 current probe is placed around the discharge pin.

The discharge current measured by the F-65 current probe at different voltage levels and polarities are shown in Figure B2. Because the displays' surfaces usually are made from insulating material such as glass or plastic, and discharges occur while a charged human is moving toward the display, air discharge mode was applied for this investigation.

The peak currents and total charges increase with discharge voltage level. The positive discharge current (about 11 amperes at +12 kilovolts) could be four times larger than the negative one (about –2.8 amperes at –12 kilovolts) while the rise time is similar (about 1 ns).

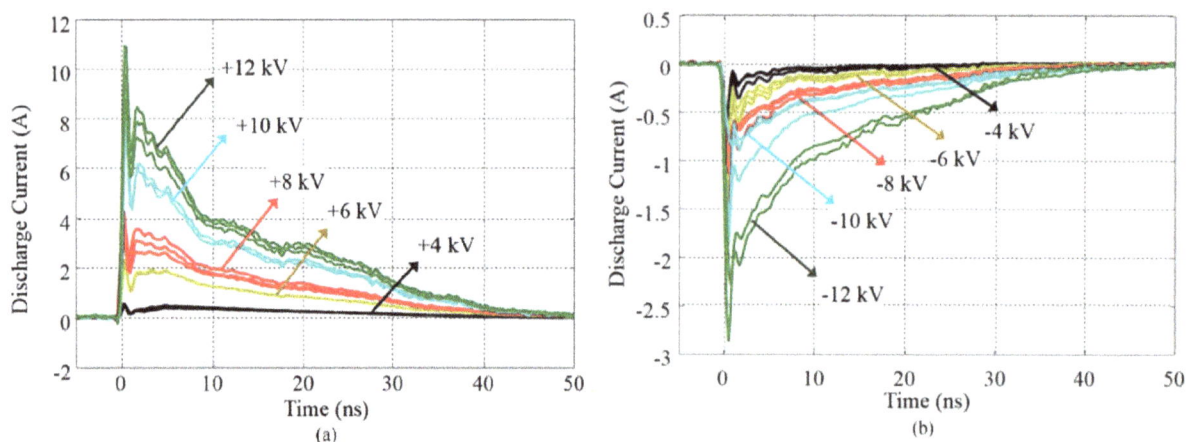

Figure B2: Discharge currents for different voltage levels and polarities, ESD generator discharge in air mode, 0.3 m/s approach speed: (a) positive and (b) negative discharge. Each voltage level was tested two to three times for repeatability resulting in multiple waveforms for each voltage level.

Lichtenberg's dust figure method can be used to help visualize the charge distribution and residual charges after surface discharging. An example is shown in Figure B3 for positive and negative discharges on a display. Area A is the corona discharge close to the electrode. In area B the streamers generate many branches, which are electron avalanche channels. Figure B3 (a) details a single branch. Typical negative dust figures do not show branching, see Figure B3 (b). The sizes of the negative dust figures are much smaller. This is due to the lower speed of the ions compared to the fast electrons dominating the positive discharge.

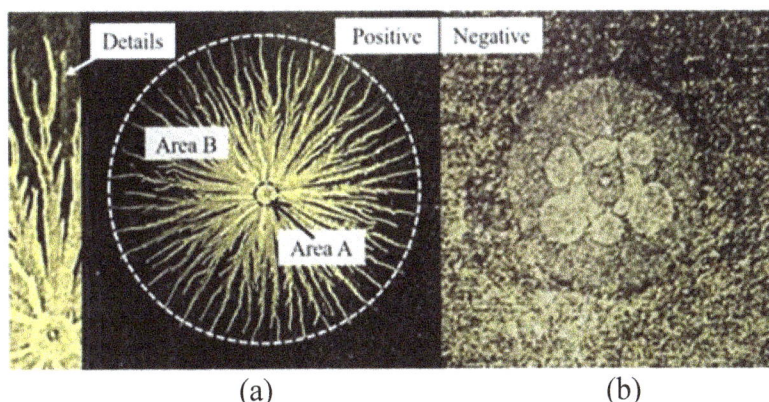

(a)   (b)

Figure B3: Typical dust figures for (a) positive (note white circle is a 35 mm diameter) and (b) negative (note round area in photo is 15 mm diameter) discharges on a display.

In summary, spark-less surface discharges on displays can inject up to 10 amperes of discharge current at hundreds of picoseconds rise times which may upset or damage the display. These discharge currents depend strongly on the polarity, voltage level, approaching speed, humidity, and the surface glass characteristics. The detailed discussion can be found in "Experimental Characterization and Modeling of Surface Discharging for an Electrostatic Discharge (ESD) to a Liquid Crystal Display (LCD)"[Gan2017].

B.2   IEC Test for Display Face Down

The general set-up situation is shown in Figure B4. The phone is placed on the HCP. The discharge to the body of the phone will force displacement current from the body of the phone to the HCP and via the resistive / inductive networks of the display and the touch layers to the HCP.

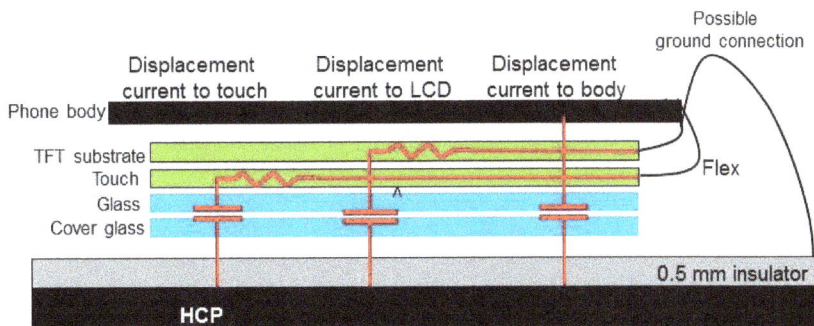

Figure B4: General setup of a phone or tablet in the display down configuration and indication of possible current paths. The F-65 current probe is placed around the discharge pin.

The currents depend strongly on the flatness of the insulating layer. The experimental results comparing the discharge currents when the cell phone is under different configurations, as shown in Figure B5, reveals the capacitance variations. The currents are measured using an F-65 current probe.

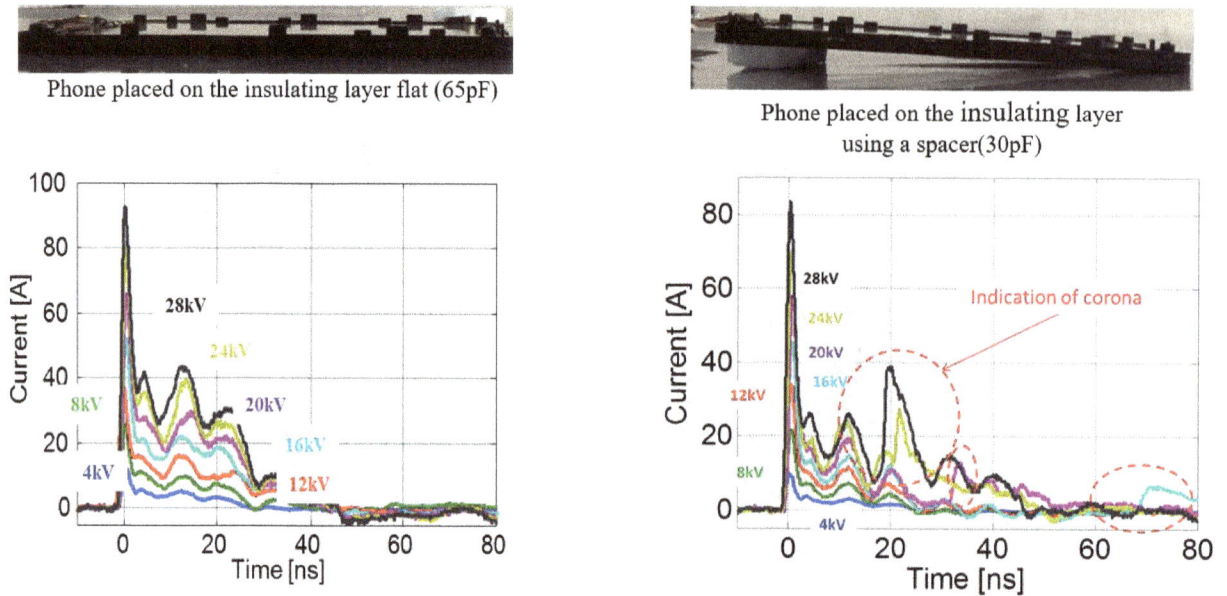

Phone placed on the insulating layer flat (65pF)

Phone placed on the insulating layer using a spacer(30pF)

Figure B5: Currents for different arrangements of the phone.

A nonlinear current rise around 20 ns after the peak, indicating corona discharge, is shown in Figure B5. Corona is a process that does not repeat well, this will also lead to the low repeatability of test results if the response of the display is affected by the corona. Thus, test setups which are prone to corona should be avoided. In general, corona caused by surface discharges on the thin plastic layer covering the HCP need to be considered for higher voltage ESD.

For higher test voltages, the voltage between the HCP and the phone is large enough to cause corona discharge on the insulating layer, shown in Figure B6. Surface discharges will cause additional currents at the edge of the phone and can spread multiple centimeters away from the phone. Prior to starting, the 0.5 mm insulator needs to be free of any surface charge.

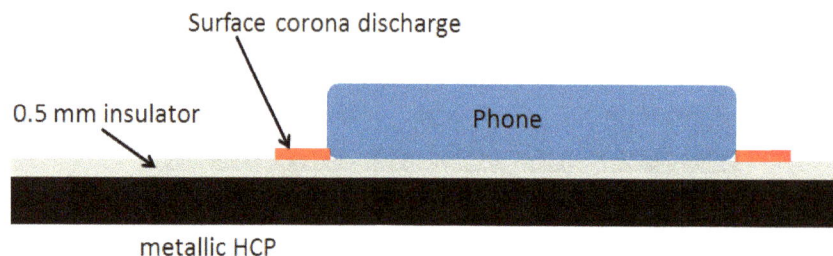

Figure B6: Illustration of the surface discharges along the top surface of the insulating layer used on top of the HCP.

The charges deposited on the surface can be visualized by using the Lichtenberg dust figure method. An example of such charge deposition measured at 25 kilovolts is shown in Figure B7 and Figure

B8. The phone is lifted from the HCP after discharge and the dust deposited. The surface discharge begins at around 20 ns after the initial pulse and, according to the measured data, adds tens of amperes which can increase the likelihood of fused bridges in the touch screen layer and other damage or upsets in the touch and the display.

The corona discharges form tree-like structures. Their propagation velocity is about 1 mm/ns and they can contribute to several amperes of additional current. A detailed photo is shown in Figure B8.

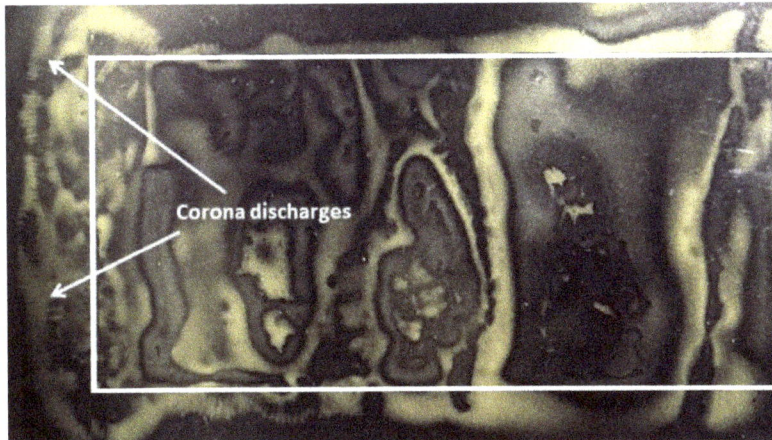

Figure B7: Distribution of charges on the insulating layer after discharging to a phone (display down). Charges deposited by corona discharge are visible outside the area the phone was placed in (white box).

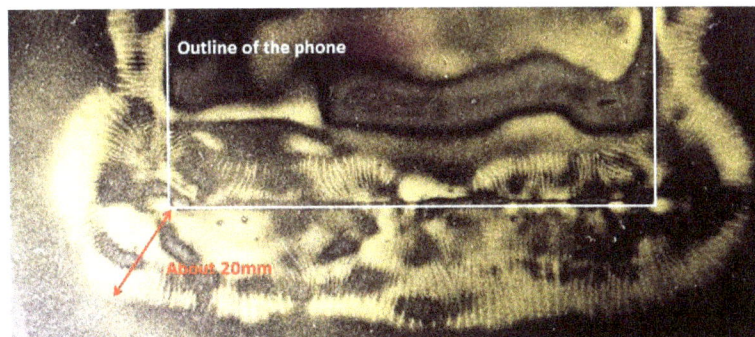

Figure B8: Detailed photo of the distribution of charges on the insulating layer after discharging to a phone (display down).

As discussed previously, the upside-down test situation is somewhat unrealistic, as few phones or tablets will be placed on a flat metal surface in a display down position. Besides being unrealistic, the test set-up may also lead to difficulty in reproducing results as the capacitance depends strongly on the thickness and flatness of the insulator used on the HCP. Increasing the insulator thickness to 5 mm is suggested for display down testing as it is more realistic to real usage. There would be less capacitance variations due to the flatness of the insulator. The corona will be less likely to occur due to lower field strengths. Fewer failures of the display would be expected, allowing introduction of advanced display and touch technologies faster. This only reduces the failure rate for the present un-realistic face down test. The ESD risk caused by discharges to the display (spark-less) is still included in the test sequence of the IEC 61000-4-2 standard as air discharge to the display must be performed.

Appendix C: Investigations into Triboelectric Charging of Cables

Dr. David Pommerenke, Graz University of Technology
Marathe Shubhankar, MS&T EMC laboratory
Yingjie Gan, MS&T EMC laboratory

C.1   Investigation on Triboelectric Charging of Cables

In this appendix, the details regarding the measurement sequence and the relationship of cable charging to CDE events is discussed.

C.1.1   Test Preparation and Discussion

Inside a climate chamber, a Faraday cup was used to measure the charge that cables accumulated after they had been rubbed against a sweater for about 5 seconds. The test sequence included:
1. Store all materials inside the chamber for at least 24 hours to allow the humidity to equalize within the materials.
2. Rub the cables against a sweater for a few seconds.
3. Measure the accumulated charge using the Faraday cup.
4. Repeat the measurement five times to check repeatability.
5. Adjust the environmental conditions in the chamber to test repeatability and dependency on humidity.
6. Identify the highest charge and voltage levels.

As shown in Table C1, four different sweaters were used, each having a different level of charge affinity (as based on the triboelectric series). Even though the main cable of interest in this work was USB 2.0 (29 cables tested), other cables were also tested (3 USB mouse cables, 3 digital visual interface (DVI) cables, 2 coax cables, 2 coax RG400 cables, 6 banana cables, and 2 power cables). In total, 47 cables were tested.

Table C1: Sweater and materials according to the labels (taken from https://www.alphalabinc.com/triboelectric-series)

Sweater	Ingredient	Charge affinity (nC/J)
Orange sweater	100% Polyester	-10
Brown sweater	Wool and Nylon mix (50/50)	+30
Black sweater A& D	Wool and Acrylic mix (50/50)	-10
Black sweater	100% Cotton	+5

As is shown in Figure C1, the measurement setup consists of a charge cup loaded by a 10 nF capacitor. The charge accumulated on the cable, and the corresponding voltage, are obtained from Eq. (1) and (2), respectively.

$$Q_{cable} = C_{Faraday-Cup} * V_{DVM} \qquad (1)$$

$$V_{cable-Plug-in-Scenario} = Q_{cable} / C_{cable} \qquad (2)$$

Where $C_{Faraday-Cup}$= 10 nF and C_{cable}=20 pF. C_{cable} is the capacitance of a cable to ground or the capacitance between a person holding a cable and the cable's inner structure itself.

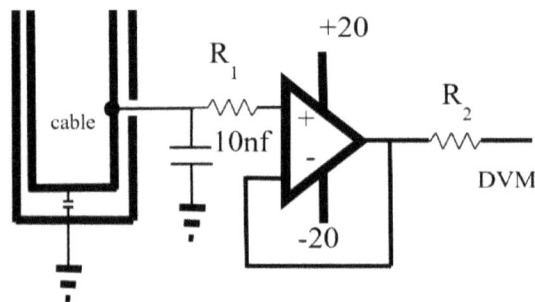

Figure C1: Measurement set-up, charge cup, and high impedance amplifier

To understand the uncertainty of the results, each test was repeated five times. Five different humidity conditions were studied as well. Although all tribocharging experiments on materials of normal daily use show large variations, it is of interest to identify general relationships between the charge and environmental conditions. The experimental results, effect of materials on charge, empirical relationship between charge and wet-bulb temperature, functional relationship between charge, relative humidity and temperature are explained in detail in [Rez2017].

C.2   Relationship to Cable Discharge Event (CDE)

To convert the charge levels into a voltage, a capacitance value must be assumed for the capacitance of the charged cable to ground. Based on measurements, 20 pF was selected. For any other capacitance value, the voltages would scale inversely to the capacitance. Using 20 pF, voltages were obtained for the plug-in cable discharge event. Figure C2 shows maximum and minimum voltages generated, versus wet bulb (WB) temperature, in USB cables for all sweaters. Figure C3 shows maximal and minimal generated voltages in all cables and four different sweaters. The largest voltage among all cables was -7 kilovolts, wet bulb (WB) = 6 °C, by the combination of coax_58 (RG400) cable and *the Brown sweater*. The largest generated voltage for USB 2.0 was 4 kilovolts under WB = 6 °C, using *the Orange sweater*. A secondary thought is needed to estimate the risk of upset or damage to an electronic system caused by inserting a cable charged to these voltage levels. As this study specifically focuses on USB 2.0 cables, one needs to take the connector specifications and the resulting pin sequencing into account. This is explained further in the USB cable plug-in measurements section of Chapter 2.

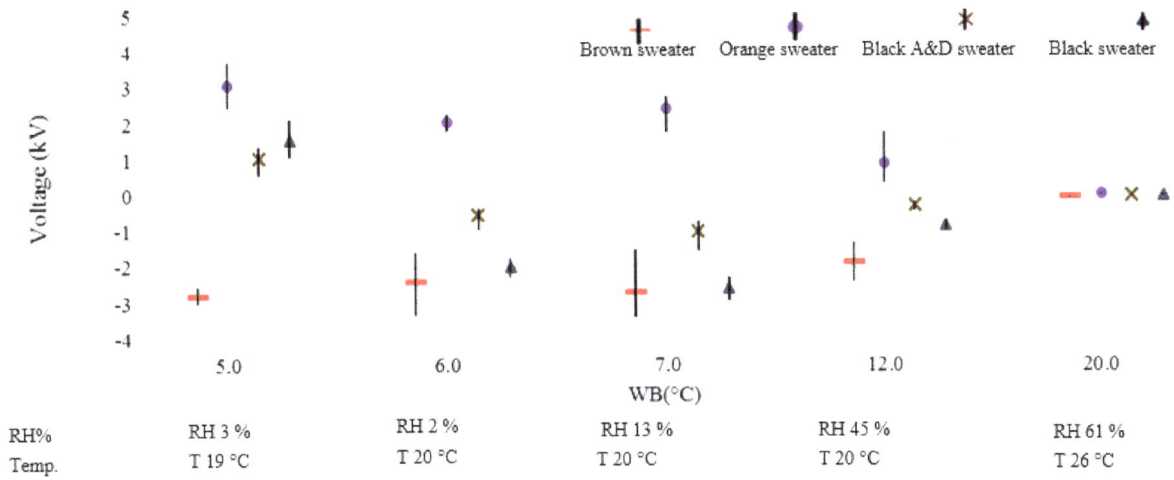

Figure C2: Maximum generated voltage with USB cables using 20 pF cable capacitance

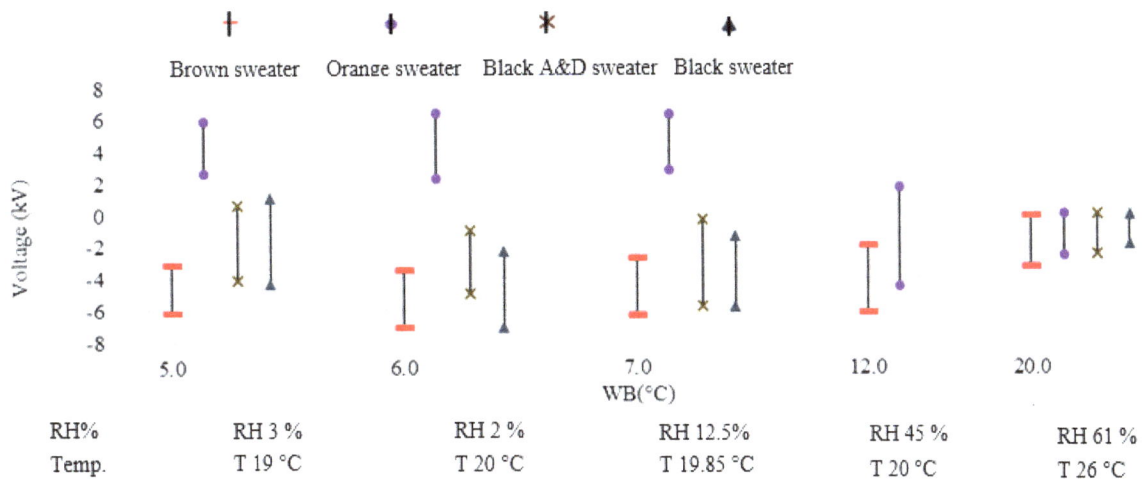

Figure C3: Maximum and minimum generated voltages with different sweaters versus WB temperature using 20 pF cable capacitance.

Appendix D: ESD Generator Full-wave Simulation Models

Ahmad Hosseinbeig, MS&T EMC laboratory
Dr. David Pommerenke, Graz University of Technology

D.1   Full-wave Based ESD Generator Models

Full-wave modeling of system level ESD is of interest for predicting ESD behavior in complex systems. A full-wave model solves an electromagnetic problem by implementing Maxwell's equations. Thus, it contains the coupling between the time changing magnetic field and the time changing electric field. In most cases the prediction is not absolute, but relative, asking if a design variant will improve the robustness, and if yes, by how much.

A complex system consists of multilayer PCBs for the electronic device and/or the system mechanical design of the device, including the connectors, LCD, battery, and chassis. In order to be able to simulate ESD phenomena in a complex system, a full-wave model of the ESD generator should be used. The full wave model may include the coupling of the transient fields, or it can be limited to just injecting the ESD discharge current. By using the full-wave model of an ESD generator, the field coupling to the system can also be predicted. If the simulation of ESD is done to predict a soft failure in the system, then the field coupling from the ESD generator should be considered in the simulation in most cases. Different full-wave ESD generator models have been presented [Cen2003, Wan2003 and Can2006a]. In the model presented in [Cen2003] which can be used with standard electromagnetic software [CST Microwave Studio], the structure was excited with a step function with a 1 ns rise time. In difference to a real ESD generator that does not model the relay and its associated low pass filters. Thus, radiation of strong ESD transient fields, with frequency components higher than 300 MHz caused by the fast voltage collapse (with 50~100 ps rise time) in the relay, was not considered in this model. A highly detailed model which was based on a finite-difference time-domain (FDTD) method has been presented in [Wan2003]. In this model (Figure D1), the currents and fields from an ESD simulator in contact mode were obtained by using the geometry and charge voltage. In the FDTD algorithm, the time-dependent material properties were controlled to model the physical stages of charging and discharging. This model shows the best accuracy, however, it required time dependent materials which are not a standard feature of most EM software.

A more detailed full-wave ESD generator model considering the details of the relay and the low pass filter has been presented in [Cai2008] (Figure D2). Although this model was verified by comparing some simulation results with measurements, the injected current into a large ground plane was not the same as measurement (Figure D3). In [She2014], a full-wave model of the ESD generator was developed and it was combined with the IC behavioral model to predict the failure of an 18 MHz D flip-flop IC.

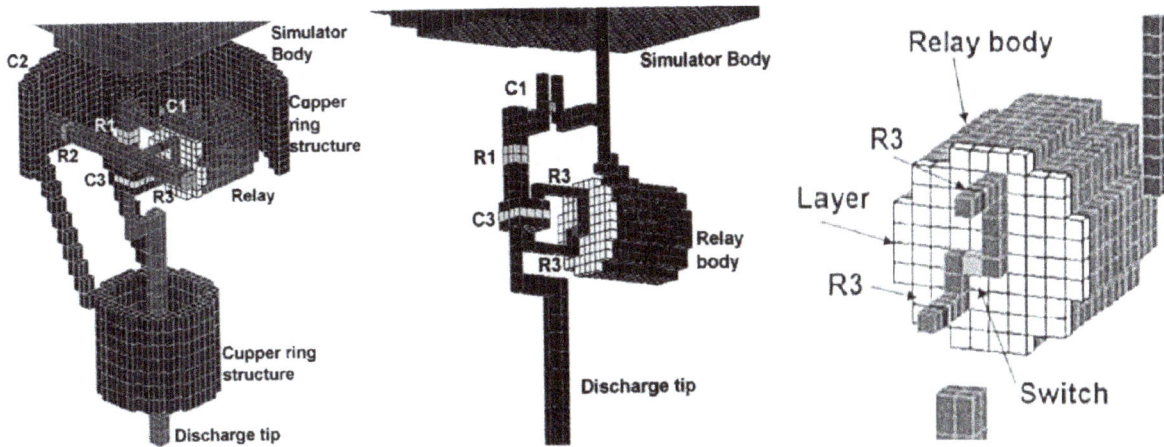

Figure D1: ESD generator model in [Wan2003].

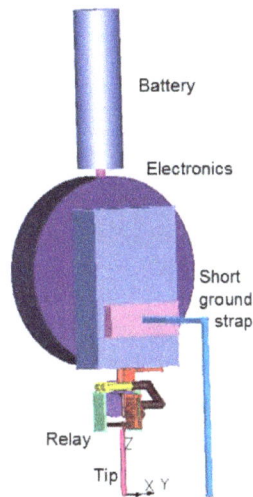

Figure D2: ESD generator full wave model overview [Cai2008].

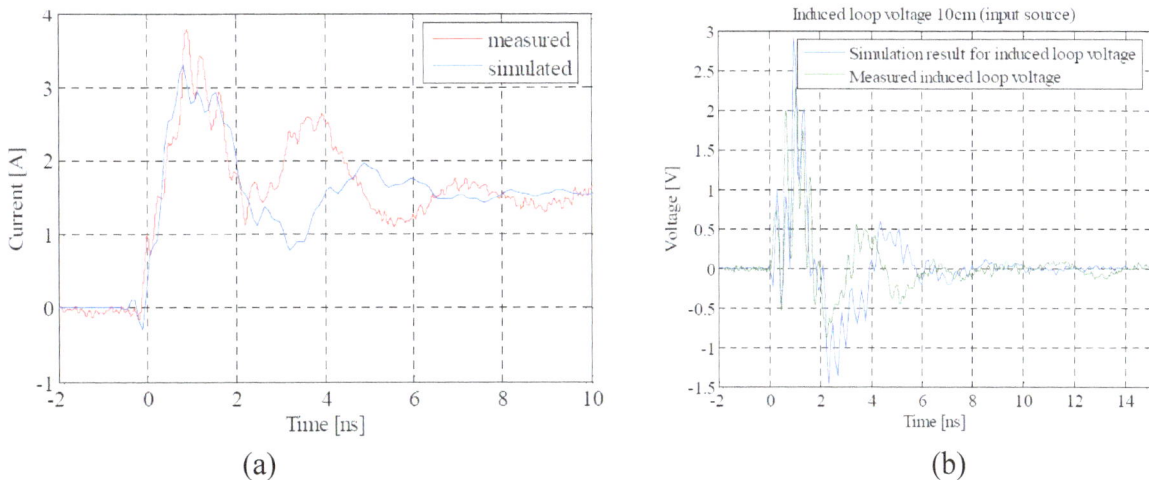

(a)   (b)

Figure D3: Simulated and measured (a) ESD discharge current, (b) induced loop voltage [Cai2008].

Although ESD generators from different manufacturers have almost the same injected current, they may excite significantly different electric and magnetic fields [Fre1998]. Therefore, for a full-wave model of an ESD generator to be able to simulate the transient ESD electromagnetic fields, the model should be created based on the real ESD generator of interest. In [Liu2009], a full-wave model for the Noiseken ESD generator has been presented (Figure D4). This model was developed in Computer Simulation Technology (CST) Microwave Studio and could simulate the discharge current and transient fields of the ESD generator (Figure D5). The generator's individual components such as the relay, capacitor unit, coil, ferrite rings and polyethylene disks were accurately modeled in this work. This model was verified by comparing the simulated and measured discharge current and induced loop voltage. A full-wave model of the Teseq ESD generator has been created in CST Microwave Studio by MS&T EMC Laboratory in 2011 (Figures D6-D7). This model can simulate the transient electromagnetic fields precisely since the main components of the generator such as the relay, RC block and coil are considered in the model.

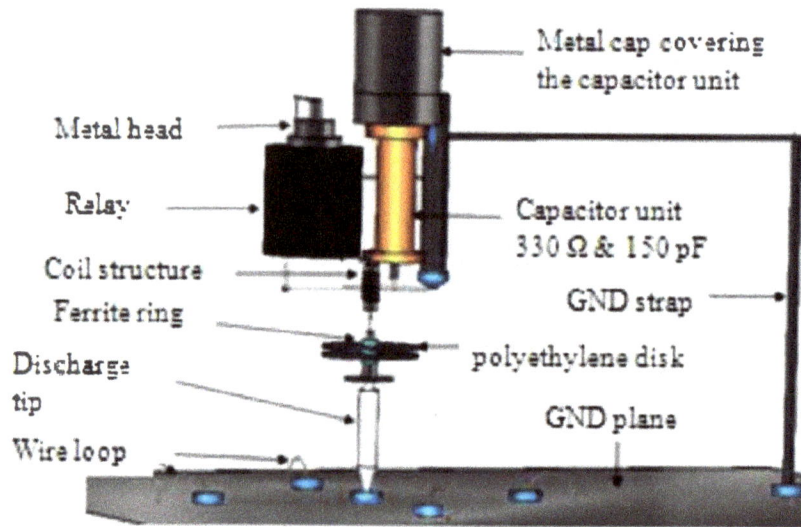

Figure D4: Full-wave model of Noiseken ESD generator [Liu2009].

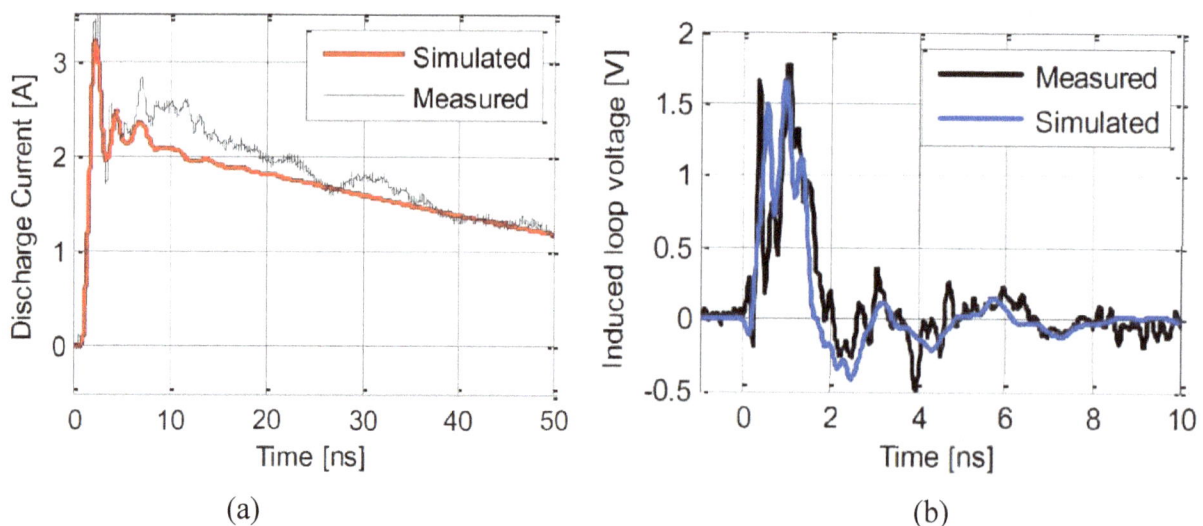

(a)   (b)

Figure D5: Comparison of the simulated and measured (a) discharge current, (b) induced loop voltage at 0 degree with 10 cm distance [Liu2009].

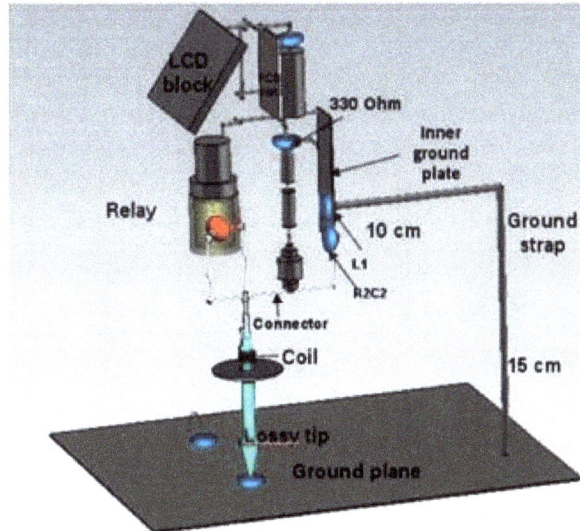

Figure D6: Full-wave model of Teseq ESD generator.

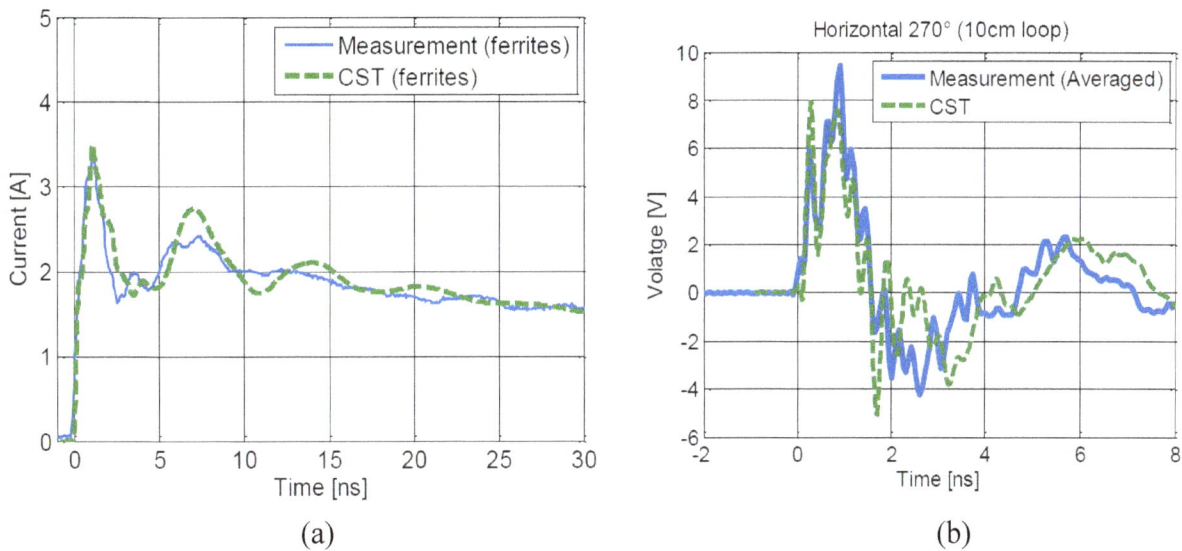

(a)

(b)

Figure D7: Comparison of the simulated and measured (a) discharge current, (b) induced loop voltage at 10 cm distance for Teseq ESD generator.

If the ESD injection into a complex system is investigated, the transient fields of an ESD generator are often not necessary. It is enough to consider only the ESD injected current in the full-wave model. For this case, another ESD generator model was developed by MS&T EMC Laboratory which has reduced dimension (63x) and volume (90000 x) compared to the Teseq ESD generator model. Figure D8 (a) shows the small ESD generator model. Due to its very small size, the less complex ESD generator model is suitable for modeling ESD injection into complex systems consisting of PCBs where there are many thin traces, vias, dielectric layers and passive components with affordable runtime. A comparison between the ESD current injected by the Teseq and the small ESD generators to a metal plate is presented in Figure D8 (b). A good agreement between the injected currents from both small and big models is observed.

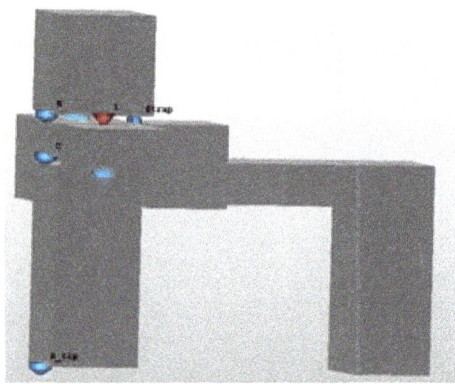

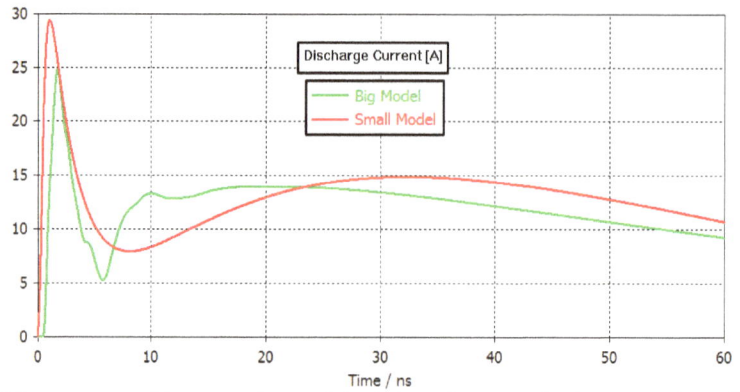

Figure D8: (a) Full-wave model of small ESD generator, (b) comparison of the injected current by Teseq and small ESD generator models.

Besides using the full-wave models of ESD generators in simulating ESD in contact discharge mode, these models can also be used in simulating air discharge ESD. In [Liu2011], the Noiseken ESD generator model presented in [Liu2009] has been used to simulate an air discharge ESD into an MP3 player (Figure D9). In this work, the linear ESD generator model was combined with the nonlinear arc resistance model and the currents and fields in air discharge mode were simulated (Figures D10 and D11). In the presented method, first the S-parameters of the linear section of the circuit were obtained from the full-wave simulation. Then, they were combined with the nonlinear part of the circuit in SPICE. In this method, the S-parameters were obtained once, and they could be reused when only the arc parameters were changed. In [Li2017], two methods of simulating the air discharge ESD current and fields have been presented by using the Teseq ESD generator model developed by the MS&T EMC Laboratory. The setup, which consisted of a discharging rod, is shown in Figure D12. The first simulation method was a two-step process in which the impedance between the tip of the rod and the ground was first simulated and then they were combined with the arc model in a circuit simulation. This method was the same as the work presented in [Liu2011] but the ESD generator model and the measurement setup are different. In the second method, the simulation was combined with the arc resistance of Rompe-Weizel (RW) directly by exchanging the voltage and current information in every time step. In this method, the transient electromagnetic co-co-simulation of CST Microwave Studio was used to simulate the currents in the discharging rod. The simulation simultaneously solved Maxwell's equation in the time-domain and the arc-resistance equation to estimate currents and fields for a given geometry, voltage and arc-length. It was shown that the simulated currents and current derivatives obtained from the presented methods matched well with the measurement results.

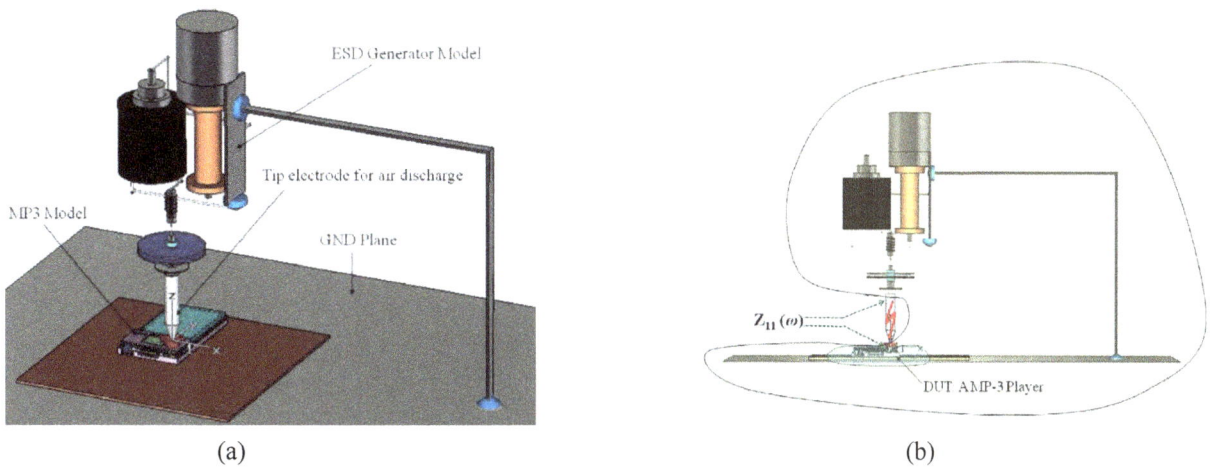

(a)   (b)

Figure D9: (a) Full-wave model of the Noiseken ESD generator and MP3 player (air discharge mode), (b) location of Z11 port [Liu2011].

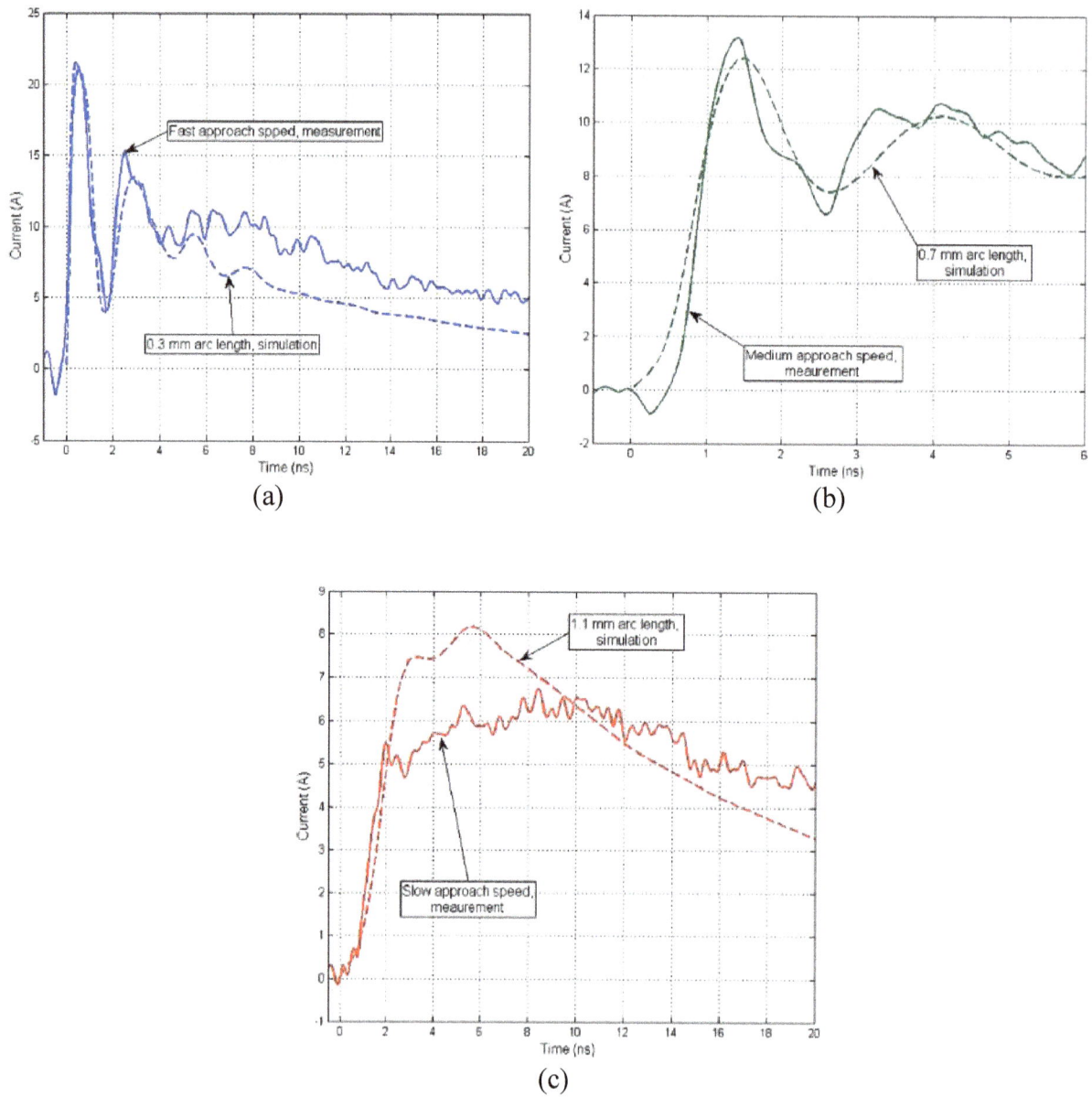

Figure D10: (a) Simulated discharge current for a 0.3-mm arc length and measured current for a fast approach speed, (b) Simulated discharge current for a 0.7-mm arc length and measured current for a medium approach speed, (c) Simulated discharge current for a 1.1-mm arc length and measured current for a slow approach speed [Liu2011].

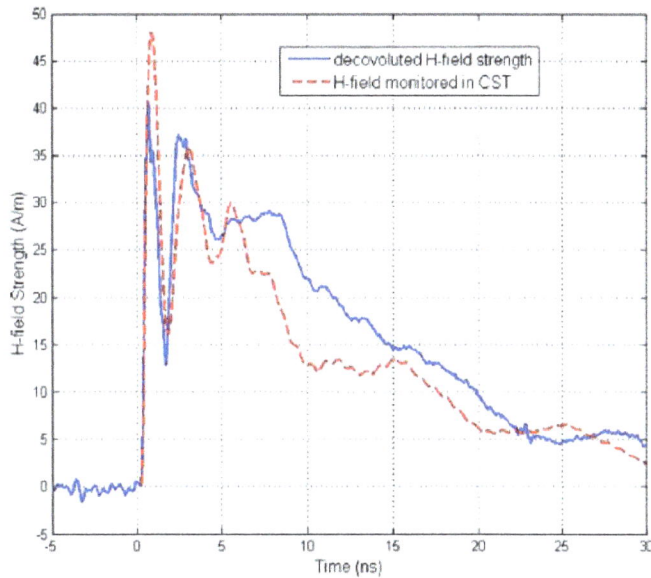

Figure D11: Measured and simulated magnetic field at 5 cm away from the discharge point [Liu2011].

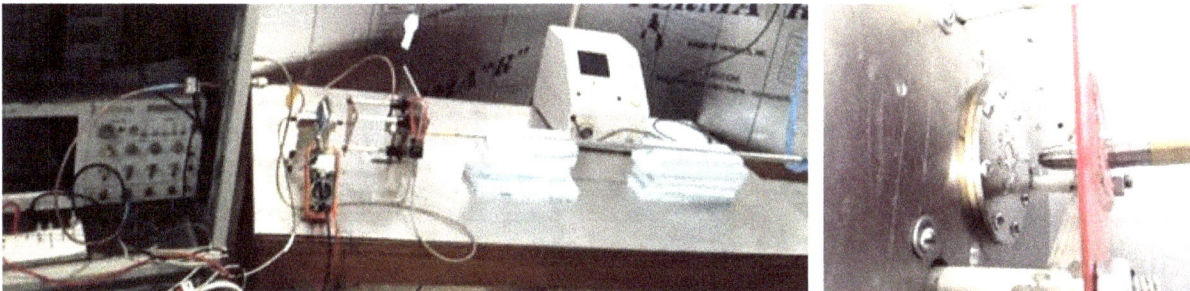

Figure D12: Test setup of the air discharge experiment in [Li2017].

When a full-wave ESD generator model is used to simulate ESD into a complex system, the proper solver of the full wave electromagnetic simulator should be used. Both frequency- and time domain solvers may be suitable. Each offers a set of advantages.

If the geometry can be gridded into hexahedral grids (meaning that grid lines are parallel to the X,Y and Y coordinates, but the distance between grid lines can change throughout the volume) then time domain solvers might be superior relative to frequency domain solvers as the time domain solvers can solve rather large volumes without facing memory issues. Additional advantages can be offered by different methods to reduce the number of cells after initial gridding. Here each vendor has its own methodology, such as sub-gridding, partially filled cells, and the 'lumper' in the transmission line method that combines cells and often reduces the number of cells by 99% or more.

If the structure has curvature or, e.g., two PCBs are arranged at a 30° angle to each other, the hexahedral grids will need very many cells. In these cases, frequency domain solvers will be attractive. These offer automatic iterative meshing and often lead to more accurate results. Both methods allow the inclusion of linear and non-linear circuit components. This can be achieved by calculating the linear part of the geometry first to create a linear S-parameter description that is then used in a circuit level time domain simulation in conjunction with the non-linear components, or the inclusion of non-linear components can be performed in lock-step during a time domain full wave simulation. This whitepaper cannot give a judgement as to which methods are superior, it can only encourage researching different methods and to identify the best method for the specific problem at hand.

CST Microwave Studio has two different time domain solvers: finite integration technique (FIT) and transmission line matrix (TLM). FIT is based on a consistent discretization scheme of an integral form of Maxwell's equations while TLM is based on the analogy between field propagation and transmission lines where the computational domain is considered as a mesh of transmission lines interconnected at nodes. The main difference between these two-time domain methods from the user point of view is the discretization of the structure. Unlike FIT, TLM utilizes a sub-gridding based on the geometry which avoids discretizing the regions which do not need fine grids. Therefore, for structures consisting of regions which are not needed to be meshed with fine grids, using the TLM solver can save computational space and time. For some structures, both FIT and TLM take almost the same time and space, but for some structures the only possible solver which can be used is TLM. As an example of the FIT and TLM usage, the Teseq ESD generator model of Figure D6 discharged on a ground plane is simulated. The simulation information is summarized in Table D1. The comparison between the ESD injected current obtained from TLM and FIT solvers are given in Figure D13.

Table D1: Simulation information for the Teseq ESD generator discharged on a ground plane solved by CST FIT and TLM solvers.

CST Solver	# of meshcells		# GPUs used	Simulation runtime
FIT	5.8×10^5		2	10 minutes
TLM	before lumping 104×10^6	After lumping 4×10^5	4	10 minutes

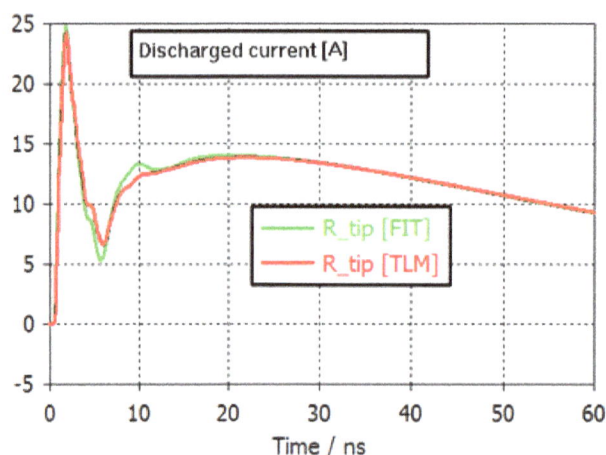

Figure D13: Comparison between discharged current on a ground plane obtained by CST (a) FIT and (b) TLM solvers.

Recently, MS&T EMC Laboratory worked on full-wave modeling of the secondary discharge in complex systems such as electronic devices [Mar2017a]. The developed model consists of the full-wave model of the Teseq ESD generator, a decorative metal in which ESD is discharged in contact mode, a metal screw mounted on the decorative metal and a current target at a distance from the tip of the screw. The distance between the current target and the screw is considered as the spark gap. One of the main challenges in developing this model was controlling the initiation of the arc resistance. In the circuit modeler of the FW simulator, a switch is used to control the on/off state of the Rompe-Weizel (RW) model. This is shown in Figure D14. This switch is necessary since the RW model does not have any explicit dependence on the Paschen value and requires decision making control to cause the initiation of the arc resistance. More details of the full-wave model can be found in [Mar2017a]. Table D2 summarizes the simulation results of secondary discharge parameters with the measurement results.

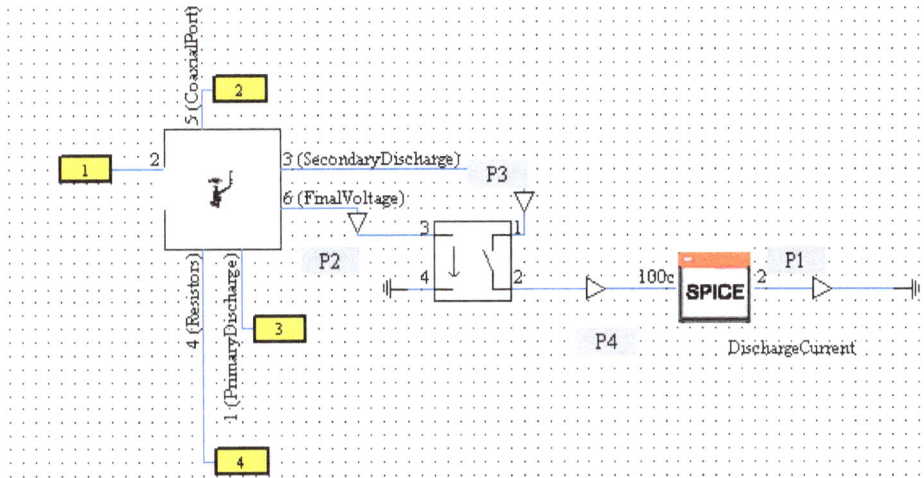

Figure D14: Switch-controlled RW model implementation in the circuit modeler of CST.

Table D2: Comparison of simulation and measurement results for the secondary discharge on the decorative metal [Mar2017a].

Parameters for 6 kV at 0.8 mm spark gap	Measurement	Simulation
Primary charging current peak	20.96 A	18.98 A
Primary charging current rise time (20% - 80%)	650 ps	550 ps
Secondary ESD peak current	68.73 A	82.18 A
Secondary ESD current rise time (20% - 80%)	550 ps	520 ps
Statistical time lag between the primary charging current and Secondary ESD	61.1 ns	5.05 ns
Paschen breakdown voltage	3.910 kV	3.910 kV
Peak metal plate voltage	5.58 kV	4.74 kV

List of References

B

[**Bar2008**] Baran, Janusz, and Jan Sroka. "Uncertainty of ESD pulse metrics due to dynamic properties of oscilloscope." IEEE Transactions on Electromagnetic Compatibility 50.4 (2008): 802-809.

[**Bar2010**] Baran, Janusz, and Jan Sroka. "Distortion of ESD generator pulse due to limited bandwidth of verification path." IEEE Transactions on Electromagnetic Compatibility 52.4 (2010): 797-803.

[**Bor2008**] M. Borsero, S. Caniggia, A. Sona, M. Stellini and A. Zuccato, "A new proposal for the uncertainty evaluation and reduction in air electrostatic discharge tests," 2008 International Symposium on Electromagnetic Compatibility - EMC Europe, Hamburg, 2008, pp. 1-6.

C

[**Cai2008**] Qing Cai, Jayong Koo, Argha Nandy, David Pommerenke, Jong Sung Lee, and Byong Su Seol. "Advanced full wave ESD generator model for system level coupling simulation." In Electromagnetic Compatibility, 2008. EMC 2008. IEEE International Symposium on, pp. 1-6. IEEE, 2008.

[**Can2006a**] S. Caniggia and F. Maradei, "Circuit and Numerical Modeling of Electrostatic Discharge Generators," in IEEE Transactions on Industry Applications, vol. 42, no. 6, pp. 1350-1357, Nov.-dec. 2006.

[**Can2006b**] S. Caniggia, F. Maradei, "Circuital and Numerical Modeling of Electrostatic Discharge Generators", Industry Applications Conference, 2005, Vol. 2, pp. 1119-1123, Oct. 2005.

[**Cen2003**] F. Centola, D. Pommerenke, K. Wang, T. V. Doren and S. Caniggia, "ESD excitation model for susceptibility study," IEEE Int. Symp. Electromagnetic Compatibility, vol.1, pp.58 – 63. 18- 22 Aug. 2003.

[**Chu2004**] Chundru, R.; Pommerenke, D.; Kai Wang; Van Doren, T.; Centola, F.P.; Jiu Sheng Huang; 'Characterization of human Metal ESD reference discharge event and correlation of generator parameters to failure levels-part I: reference event' Electromagnetic Compatibility, IEEE Transactions on Volume 46, Issue 4, Nov. 2004 Page(s):498 – 504

[**CST Microwave Studio**] https://www.3ds.com/products-services/simulia/products/cst-studio-suite/

D

[**Dwy1990**] V.M. Dwyer, A.J. Franklin and D.S. Campbell, "Thermal Failure in Semiconductor Devices, Solid--State Electronics", Vol. 33, No. 5, pp. 553-560, 1990

E

[**ESDATR26**] Technical Report of Working Group 26 of ESD Association; to be published in 2021.

F

[**Fot2006**] G. P. Fotis, I. F. Ganos, and I. A. Stathopulos, "Determination of discharge current equation parameters of ESD using genetic algorithms," Electronic Letters, vol. 42, no. 14, Jul. 2006.

[**Fre1998**] S. Frei and D. Pommerenke, "Fields on the horizontal coupling plane excited by direct ESD and discharges to the vertical coupling plane," J. Electrostat., pp. 177–190, 1998.

[**Fre2013**] Frei, S., D. Pommerenke, and B. Arndt. "High-voltage coupling in electronic devices and control units." FAT 249. 2013.

[**Fri1999**] S. Frei, M. Senghaas, R. Jobava, and W. Kalkner, "The influence of speed of approach and humidity on the intensity of ESD," in Proc. 13th Int. Zurich Symp. EMC, Zurich, Switzerland, Feb. 1999, pp. 105–110.

G

[Gan2016] Gan, Y., Xu, X., Maghlakelidze, G., Yang, S., Huang, W., Seol, B. and Pommerenke, D., "System-Level Modeling Methodology of ESD Cable Discharge to Ethernet Transceiver Through Magnetics," in IEEE Transactions on Electromagnetic Compatibility, vol. 58, no.5, pp.1407-1416, July 2016.

[Gan2017] Y. Gan; A. Talebzadeh; X. Xu; S. Shinde; Y. Zeng; K. H. Kim; D. Pommerenke, "Experimental Characterization and Modeling of Surface Discharging for an Electrostatic Discharge (ESD) to an LCD Display," in IEEE Transactions on Electromagnetic Compatibility , vol. PP, no.99, pp.1-11]

[Gos2010] Gossner, Harald, Werner Simbürger, and Matthias Stecher. "System ESD robustness by co-design of on-chip and on-board protection measures." Microelectronics Reliability 50, no. 9-11 (2010): 1359-1366.

[Gos1985] Y. Gosho and M. Saeki, "Role of water vapor in the breakdown of atmospheric air gap," in Proc. 8th Int. Conf. Oxford, 1985, pp. 161–162.

H

[Har2016] Harberts, Dick W. "Recommended sample size of ESD and surge immunity tests." Electromagnetic Compatibility-EMC EUROPE, 2016 International Symposium on. IEEE, 2016.

[Hya1993] Hyatt HM. The resistive phase of an air discharge and the formation of fast risetime ESD pulses. Journal of electrostatics. 1993 Dec 1;31(2-3):339-56.

I

[Ish2015] T. Ishida, S. Nitta, F. Xiao, Y. Kami, and O. Fujiwara, "An experimental study of electrostatic discharge immunity testing for wearable devices." in Proc. Int. Symp. IEEE Electromagnetic Compatibility, 2015, pp. 839-842.

[Ish2016] T. Ishida, F. Xiao, Y. Kami, O. Fujiwara and S. Nitta, "An alternative air discharge test in contact discharge of ESD generator through fixed gap," 2016 IEEE International Symposium on Electromagnetic Compatibility (EMC), Ottawa, ON, 2016, pp. 46-50.

[Ish2017] Ishida T, Okajima H, Endo K, Kinoshita M, Fujiwara O, Nitta S. Combined effects of relative humidity and temperature on air discharges of electrostatic discharge generator. In Electromagnetic Compatibility & Signal/Power Integrity (EMCSI), 2017 IEEE International Symposium on 2017 Aug 7 (pp. 135-139). IEEE.

[Iri2017] Oganezova, Irina, et al. "Human body impedance modelling for ESD simulations." 2017 IEEE International Symposium on Electromagnetic Compatibility & Signal/Power Integrity (EMCSI). IEEE, 2017.

[IEC2008] IEC, Std 61000-4-2, "Electromagnetic Compatibility (EMC), Part 4: Testing and measurement techniques – Section 2: Electrostatic discharge immunity test", Ed. 2.0, 2008.

[IEC2001] IEC, Std 61000-4-2, "Electromagnetic Compatibility (EMC), Part 4: Testing and measurement techniques – Section 2: Electrostatic discharge immunity test", Ed. 1.2, 2001.

[ISO2001] ISO, Std 10605, "Road vehicles – Test methods for electrical disturbances from electrostatic discharge", 2001

[InCWP3PI] Industry Council on ESD Target Levels, "White Paper 3 System Level ESD Part I: Common Misconceptions and Recommended Basic Approaches," December 2010, at www.esda.org or JEDEC publication JEP161, "System Level ESD Part I: Common Misconceptions and Recommended Basic Approaches", www.jedec.org

J

[John2011] D. Johnsson, H. Gossner, Study of system ESD co-design of a realistic mobile board, in EOS/ESD Symposium 2011, Sept 2011, pp. 1–10

K

[Kat2010] P.S. Katsivelis, G.N. Psarros, I.F. Gonos, I.A. Stathopulos, "Estimation of parameters for the electrostatic discharge current equation with real human discharge events reference using genetic algorithms," Inst. Phys. Publ. Meas. Sci. Technol., vol. 17, no. 10, pp.1-6, Oct. 2010.

[Kim2011] K. Kim and Y. Kim, "Systematic analysis methodology for mobile phone's electrostatic discharge soft failure," IEEE Trans. Electromagn. Compat., vol. 53, no. 3, pp. 611–618, Aug. 2011.

[Kin1981] King, W. Michael, and David Reynolds. "Personnel electrostatic discharge: Impulse waveforms resulting from ESD of humans directly and through small hand-held metallic objects intervening in the discharge path." *Electromagnetic Compatibility, 1981 IEEE International Symposium on*. IEEE, 1981.

[Koh2018] Kohani, M., Bhandare, A., Guan, L., Pommerenke, D., & Pecht, M. G. (2018). Evaluating Characteristics of Electrostatic Discharge (ESD) Events in Wearable Medical Devices: Comparison With the IEC 61000-4-2 Standard. *IEEE Transactions on Electromagnetic Compatibility*, *60*(5), 2018, 1304-1312.

[Koo2008] J. Koo et al., "Correlation Between EUT Failure Levels and ESD Generator Parameters," in IEEE Transactions on Electromagnetic Compatibility, vol. 50, no. 4, pp. 794-801, Nov. 2008

[Koo2012] J. Koo, K. Kim, and H. Kim, "System-level ESD immunity of mobile display driver IC to hard and soft failure," Elect. Electron. Eng., vol. 13,no. 1, pp. 329–335, 2012.

[Koh2017] Kohani M, Bhandare A, Guan L, Pommerenke D, Pecht MG. Evaluating Characteristics of Electrostatic Discharge (ESD) Events in Wearable Medical Devices: Comparison With the IEC 61000-4-2 Standard. IEEE Transactions on Electromagnetic Compatibility. 2017 Nov 28.

L

[Leu2001] Leuchtmann P, Sroka J. Enhanced field simulations and measurements of the ESD calibration setup. In Electromagnetic Compatibility, 2001. EMC. 2001 IEEE International Symposium on 2001 (Vol. 2, pp. 1273-1278). IEEE.

[Li2015] T. Li et al., "System-Level Modeling for Transient Electrostatic Discharge Simulation," in IEEE Transactions on Electromagnetic Compatibility, vol. 57, no. 6, pp. 1298-1308, Dec. 2015.

[Li2016] Z. Li, P. Maheshwari, and D. Pommerenke, "Measurement methodology for field-coupled soft errors induced by electrostatic discharge," IEEE Trans. Electromagn. Compat., vol. 58, no. 2, pp. 1–8, Mar. 2016.

[Li2017] Darwin Li, Jianchi Zhou, Ahmad Hosseinbeig, David Pommerenke, "Transient Electromagnetic Co-simulation of Electrostatic Air Discharge," EOS/ESD Symposium, 2017.

[Liu2009] Liu, Dazhao, Argha Nandy, David Pommerenke, Soon Jae Kwon, and Ki Hyuk Kim. "Full wave model for simulating a Noiseken ESD generator." In Electromagnetic Compatibility, 2009. EMC 2009. IEEE International Symposium on, pp. 334-339. IEEE, 2009.

[Liu2011], Liu D, Nandy A, Zhou F, Huang W, Xiao J, Seol B, Lee J, Fan J, Pommerenke D. Full-wave simulation of an electrostatic discharge generator discharging in air-discharge mode into a product. IEEE Transactions on electromagnetic compatibility. 2011 Feb;53(1):28-37.

[Lev1982] S. J. Levinson and E. E. Kunhardt, "Investigation of the statistical and formative time lags associated with the breakdown of a gas in a gap at high overvoltage," IEEE Trans. Plasma Sci., vol. 10, no. 4, pp. 266–270, Dec. 1982.

M

[**Mar2017a**] S. Marathe, D. Li, A. Hosseinbeig, H. Rezaei, P. Wei, J. Zhou, and D. Pommerenke, "On secondary ESD event monitoring and full-wave modeling methodology," 2017 39th Electrical Overstress/Electrostatic Discharge Symposium (EOS/ESD), 2017.

[**Mar2017b**] S. Marathe, H. Rezaei, D. Pommerenke, and M. Hertz, "Detection methods for secondary ESD discharge during IEC 61000-4-2 testing," 2017 IEEE International Symposium on Electromagnetic Compatibility & Signal/Power Integrity (EMCSI), 2017.

[**Mar2017c**] S. Marathe, P. Wei, S. Ze, L. Guan, and D. Pommerenke, "Scenarios of ESD discharges to USB connectors," 2017 39th Electrical Overstress/Electrostatic Discharge Symposium (EOS/ESD), 2017.

[**Mar2018**] S. Marathe, G. Maghlakelidze, H. Rezaei, D. Pommerenke, and M. Hertz, "Software-Assisted Detection Methods for Secondary ESD Discharge During IEC 61000-4-2 Testing," in IEEE Transactions on Electromagnetic Compatibility, vol.PP, no.99, pp.1-8, Jan. 2018.

[**Mar2011**] Mardiguian, Michel. Electro Static Discharge: Understand, Simulate, and Fix ESD Problems. John Wiley & Sons, 2011.

[**Mor2011**] Morando A, Borsero M, Sardi A, Vizio G. Critical aspects in calibration of ESD generators. Measurement Science Review. 2011 Jan 1;11(1):23-8.

[**Mor1953**] C. G. Morgan and D Harcombe, "Fundamental processes of the initiation of electrical discharges," in Proc. Phys. Soc., 1953, pp. 665–679.

[**Mee1978**] J. M. Meek and J. D. Craggs, Electrical Break Down of Gases. New York, NY, USA: Wiley, 1978.

[**Mer2012**] Robert Mertens, Hans Kunz, Akram Salman, Gianluca Boselli, Elyse Rosenbaum, "A Flexible Simulation Model for System Level ESD Stresses with Application to ESD Design and Troubleshooting," 2012 34th Electrical Overstress/Electrostatic Discharge Symposium (EOS/ESD), 2012, pp. 1-6.

N

[**Not2016**] G. Notermans, private communication, Sept. 2016.

[**Nie2010**] F. zur Nieden, B. Arndt, J. Edenhofer and S. Frei, "Impact of setup and pulse generator on automotive component ESD testing results," 2010 Asia-Pacific International Symposium on Electromagnetic Compatibility, Beijing, 2010, pp. 475-478.

[**Nie2009**] F. zur Nieden, B. Arndt, J. Edenhofer, S. Frei. Vergleich von ESD-SystemLevel Testmethoden für Packaging und Handling. ESD-Forum 2009.

P

[**Pat2017**] Patnaik A, Hua R, Pommerenke D. Characterizing ESD stress currents in human wearable devices. In Electrical Overstress/Electrostatic Discharge Symposium (EOS/ESD), 2017 39th 2017 Sep 10 (pp. 1-6). IEEE.

[**Pep1991**] Pepe RC. ESD multiple discharges. InElectromagnetic Compatibility, 1991. Symposium Record., IEEE 1991 International Symposium on 1991 Aug 12 (pp. 253-258). IEEE.

[**Pom1993**] D. Pommerenke, "On the influence of the speed of approach, humidity and arc length on ESD breakdown," in Proc. ESD Forum, Grainau, Germany, 1993, pp. 103–111.

[**Pom1995**] D. Pommerenke, "ESD: transient fields, spark simulation and rise time limit," J. Electrostat., vol. 36, pp. 31–54, Oct. 1995.

[**Pom1996**] D. Pommerenke and M. Aidam, 'ESD: waveform calculation, field and current of human and simulator ESD', Journal of Electrostatics, Vol. 38, Nov. 1996, pp. 33-51

[Pom1998] S. Frei, D.Pommerenke, 'Fields on the horizontal coupling plane excited by direct ESD and discharges to the vertical coupling plane', Journal of Electrostatics 44 (1998), pp.177-190

[Plasma] "Nonthermal plasma" [Online]. Available: https://en.wikipedia.org/ wiki/Nonthermal_plasma.

R

[Ren1993] Renninger, Robert G. "Improved statistical method for system-level ESD tests." Electromagnetic Compatibility, 1993. Symposium Record., 1993 IEEE International Symposium on. IEEE, 1993.

[Rez2017] H. Rezaei, L. Guan, A. Talebzadeh, S. Marathe, P. Wei, and D. Pommerenke, "Effect of relative humidity and materials on triboelectric charging of USB cables," 2017 IEEE International Symposium on Electromagnetic Compatibility & Signal/Power Integrity (EMCSI), 2017.

[Rit1992] Ritenour, T. J., and Franz Gisin. "Performing statistical ESD tests using the new ANSI C63. 1 6-1991 guide for ESD test methodologies." Electromagnetic Compatibility, 1992. Symposium Record., IEEE 1992 International Symposium on. IEEE, 1992.

[Rit2015] H. M. Ritter, L. Koch, M. Schneider and G. Notermans, "Air-discharge testing of single components," 2015 37th Electrical Overstress/Electrostatic Discharge Symposium (EOS/ESD), Reno, NV, 2015, pp. 1-7.

S

[Shi2016] S. Shinde et al., "ESD to the display inducing currents measured using a substition PC board," in Proc. IEEE Int. Symp. Electromagn. Compat., 2016, pp. 707–712.

[Sro2003] Sroka, Jan, and W. L. Klampfer. "Target influence of the calibration uncertainty of ESD simulators." Electromagnetic Interference and Compatibility, 2003. INCEMIC 2003. 8th International Conference on. IEEE, 2003.

[Son2003] Songlin S., B. Zengjun, T. Minghong, L. Shange, "A new analytical expression of current waveform in Standard 61000-4-2," High Power Laser and Particle Beams, vol.15, no.5, pp. 464-466, 2003.

[Sek2013] T. Sekine and H. Asai, "A framework for the simulation of electrostatic discharge immunity using the unified circuit modeling technique," 2013 IEEE International Symposium on Electromagnetic Compatibility, Denver, CO, 2013, pp. 869-874.

[She2014] Shen, Guangyao, Victor Khilkevich, Sen Yang, David Pommerenke, Hermann Aichele, Dirk Eichel, and Christoph Keller. "Simple D flip-flop behavioral model of ESD immunity for use in the ISO 10605 standard." In Electromagnetic Compatibility (EMC), 2014 IEEE International Symposium on, pp. 455-459. IEEE, 2014.

[She2015] G. Shen et al., "ESD Immunity Prediction of D Flip-Flop in the ISO 10605 Standard Using a Behavioral Modeling Methodology," in IEEE Transactions on Electromagnetic Compatibility, vol. 57, no. 4, pp. 651-659, Aug. 2015.

[Sve2002] Bönisch, Sven, David Pommerenke, and Wilfried Kalkner. "Broadband measurement of ESD risetimes to distinguish between different discharge mechanisms." Journal of Electrostatics 56.3 (2002): 363-383.

[Stad2017] Stadler, Wolfgang, Josef Niemesheim, Andreas Stadler, Sebastian Koch, and Harald Gossner. "Risk assessment of cable discharge events", 2017 39th Electrical Overstress/Electrostatic Discharge Symposium (EOS/ESD), pp. 1-9. IEEE, 2017.

[Sta1998] R. B. Standler, "Technology of fast spark gaps," Park Communications and Space Science Lab, Pennsylvania State Univ., Philadelphia, PA, USA, 1998

T

[**Tam2016**] P. Tamminen, private communication, Sept. 2016.

[**Tal2016**] A. Talebzadeh, M. Moradian, Y. Han, D. E. Swenson and D. Pommerenke, "Effect of Human Activities and Environmental Conditions on Electrostatic Charging," in IEEE Transactions on Electromagnetic Compatibility, vol. 58, no. 4, pp. 1266-1273, Aug. 2016.

W

[**Wan2003**] K. Wang, D. Pommerenke, R. Chundru, T. V. Doren, J. L. Drewniak, and A. Shashindranath, "Numerical modeling of electrostatic discharge generators," IEEE Trans. Electromagn. Compat., vol. 45, no. 2, pp. 258– 271, May 2003.

[**Wan2004**] Wang, Kai, et al. "Characterization of human metal ESD reference discharge event and correlation of generator parameters to failure levels-part II: Correlation of generator parameters to failure levels." *IEEE Transactions on Electromagnetic Compatibility* 46.4 (2004): 505-511.

[**Wan2013a**] Wan, Fayu, Michael Hillstrom, Carlton Stayer, David Swenson, and David Pommerenke. "The effect of humidity on static electricity induced reliability issues of ICT equipment in data centers-- Motivation and setup of the study." ASHRAE Transactions 119, no. 2 (2013): 341-358.

[**Wan2013b**] Wang, Jin Shan, Ke Wang, and Xiao Dong Wang. "Four order ESD generator circuit model and its simulation." Applied Mechanics and Materials. Vol. 239. Trans Tech Publications, 2013.

[**Wan2014**] F. Wan, V. Pilla, J. Li, D. Pommerenke, H. Shumiya, K. Araki, "Time Lag of Secondary ESD in Millimeter-Size Spark Gaps," IEEE Trans. Electromagn. Compat., vol. 56, no. 1, pp. 28–34, Feb. 2014.

[**Whi2012**] White Paper 3 System Level ESD Part II: Implementation of Effective ESD Robust Designs. 2012. [Online]. Available:
http://www.esdindustrycouncil.org/ic/docs/Industry%20Council%20White%20Paper%203%20PII%20Rev1%20Sep%202012.pdf

[**Wen1999**] Wendsche, Steffen, Ralf Vick, and Ernst Habiger. "Immunity testing of computerized equipment to fast electrical transients using adaptive test methods." Electromagnetic Compatibility, 1999 International Symposium on. IEEE, 1999.

[**Wol2015**] H. Wolf, H. Gieser, "Secondary discharge – A potential risk during system level ESD testing," in Electrical Overstress/Electrostatic Discharge Symposium (EOS/ESD), 2015 37th, vol., no., pp.1-7, 27 Sept.-2 Oct. 2015.

[**Wu2013**] Qimeng Wu, Ming Wei, "A mathematical expression for air ESD current waveform using BP neural network," Journal of Electrostatics, vol. 71, pp. 125-129, 2013

X

[**Xia2011**] J. Xiao, D. Pommerenke, J. L. Drewniak, H. Shumiya, T. Yamada, and K. Araki, "bmodel of secondary ESD for a portable product," Electromagnetic Compatibility (EMC), 2011 IEEE International Symposium on, 14-19 Aug. 2011, pp. 56-61.

[**Xia2012**] J. Xiao, D. Pommerenke, J. L. Drewniak, H. Shumiya, J. Maeshima, T. Yamada, and K. Araki, "Model of secondary ESD for a portable electronic product," IEEE Trans. Electromagn. Compat., vol. 54, no. 3, pp. 546–555, Jun. 2012.

Y

[**Yan2017**] S. Yang, A. Bhandare, H. Rezaei, W. Huang, and D. Pommerenke, "Step-Response-Based Calibration Method for ESD Generators in the Air-discharge Mode," in IEEE Transactions on Electromagnetic Compatibility, vol.PP, no.99, pp.1-4, Dec. 2017.

[**Yan2018**] S. Yang, A. Bhandare, H. Rezaei, W. Huang, and D. Pommerenke, "Step response- based calibration method for ESD generators in the air discharge mode," *IEEE Trans. Electromagn. Compat.*, 2018.

[**Yua2006**] Zhiyong Yuan, Tun Li, Jinliang He, Shuiming Chen and Rong Zeng, "New mathematical descriptions of ESD current waveform based on the polynomial of pulse function," in IEEE Transactions on Electromagnetic Compatibility, vol. 48, no. 3, pp. 589-591, Aug. 2006.

[**Yua2010**] Q. Yuan, S. Liu, X. Zhang, Z. Wu and M. Wei, "The Effect of Approach Speed and Charge Voltage on an Air Discharge," in IEEE Transactions on Electromagnetic Compatibility, vol. 52, no. 4, pp. 985-993, Nov. 2010.

[**You2017**] Yousef et al., "Efficient Circuit and EM Model of Electrostatic Discharge Generator," 2017 IEEE International Symposium on Electromagnetic Compatibility, Washington DC, 2017.

Z

[**Zhou2014**] Y. Zhou and J. J. Hajjar, "A circuit model of electrostatic discharge generators for ESD and EMC SPICE simulation," 2014 IEEE International Conference on Electron Devices and Solid-State Circuits, Chengdu, 2014, pp. 1-2.

[**Zho2016**] J. Zhou et al., "IEC 61000-4-2 ESD test in display down configuration for cell phones," in Proc. IEEE Int. Symp. Electromagn. Compat., 2016, pp. 713–718.

[**Zho2017**] Zhou J, Gan Y, Jin H, Pommerenke D. ESD Spark Behavior and Modeling for Geometries Having Spark Lengths Greater Than the Value Predicted by Paschen's Law. IEEE Transactions on Electromagnetic Compatibility. 2017 May 30.

[**Zho2018**] Zhou, J., Ghosh, K., Xiang, S., Yan, X., Hosseinbeig, A., Lee, J., & Pommerenke, D. Characterization of ESD Risk for Wearable Devices. *IEEE Transactions on Electromagnetic Compatibility*, *60*(5), 2018, 1313-1321.

[**Zhou2018**] J. Zhou, Y. Gan, H. Jin, D. Pommerenke, "ESD Spark Behavior and Modeling for Geometries Having Spark Lengths Greater Than the Value Predicted by Paschen's Law," in *IEEE Trans. Electromagn. Compat.*, vol.60, no.1, pp.115-121, Feb. 2018.

[**Zhou2019**] Zhou, Jianch, Keyu Zhou, Xin Yan, Li Shen, David Pommerenke, and Guang-xiao Luo. "Investigations into Methods to Stabilize the Spark in Air Discharge ESD." In 2019 IEEE International Symposium on Electromagnetic Compatibility, Signal & Power Integrity (EMC+ SIPI), pp. 147-151. IEEE, 2019.

Revision History

Revision	Changes	Date of Release
1.0	Initial Release	September 2020
1.1	Address typos in Sections 3.3, 5.3.2, 6.1.11 & D.1, updated JEP equivalent document to JEP163, logo page updated	May 2021
1.2	Update reference of JEDEC version to JEP164	October 2022

www.ingramcontent.com/pod-product-compliance
Lightning Source LLC
Chambersburg PA
CBHW051749200326
41597CB00025B/4492